AGRICULTURE AND ENVIRONMENT SERIES

BIOLOGICAL and BIOTECHNOLOGICAL CONTROL of INSECT PESTS

AGRICULTURE AND ENVIRONMENT SERIES

Jack E. Rechcigl
Editor-in-Chief

Agriculture is an essential part of our economy on which we all depend for food, feed, and fiber. With the increased agricultural productivity in this country as well as abroad, the general public has taken agriculture for granted while voicing its concern and dismay over possible adverse effects of agriculture on the environment. The public debate that has ensued on the subject has been brought about, in part, by the indiscriminate use of agricultural chemicals and, in part, by misinformation, based largely on anecdotal evidence.

At the national level, recommendations have been made for increased research in this area by such bodies as the Office of Technology Assessment, the National Academy of Sciences, and the Carnegie Commission on Science, Technology, and Government. Specific issues identified for attention include: contamination of surface and groundwater by natural and chemical fertilizers, pesticides, and sediment; the continued abuse of fragile and nutrient-poor soils; and suitable disposal of industrial and agricultural waste.

Although a number of publications have appeared recently on specific environmental effects of some agricultural practices, no attempt has been made to approach the subject systematically and in a comprehensive manner. The aim of this series is to fill the gap by providing the synthesis and critical analysis of the state of the art in different areas of agriculture bearing on environment and vice versa. Efforts will also be made to review research in progress and comment on perspectives for the future. From time to time methodological treatises as well as compendia of important data in handbook form will also be included. The emphasis throughout the series will be on comprehensiveness, comparative aspects, alternative approaches, innovation, and worldwide orientation.

Specific topics will be selected by the Editor-in-Chief with the council of an international advisory board. Imaginative and timely suggestions for the inclusion in the series from individual scientists will be given serious consideration.

BIOLOGICAL and BIOTECHNOLOGICAL CONTROL of INSECT PESTS

Jack E. Rechcigl
University of Florida
Soil and Water Science Department
Research and Education Center
Ona, Florida

Nancy A. Rechcigl
Yoder Brothers, Inc.
Parrish, Florida

CRC Press
Taylor & Francis Group
Boca Raton London New York

CRC Press is an imprint of the
Taylor & Francis Group, an informa business

CRC Press
Taylor & Francis Group
6000 Broken Sound Parkway NW, Suite 300
Boca Raton, FL 33487-2742

First issued in paperback 2019

ISBN-13: 978-1-56670-479-3 (hbk)
ISBN-13: 978-0-367-39944-3 (pbk)
Library of Congress catalog number: 99-31226

Library of Congress Cataloging-in-Publication Data

Catalog record is available from the Library of Congress

**Visit the Taylor & Francis Web site at
http://www.taylorandfrancis.com**

**and the CRC Press Web site at
http://www.crcpress.com**

Preface

Pest and disease management continues to be an important challenge to the agricultural community. Confronted with the shifts in pest pressure and the rise in new pest and crop problems, coupled with public concern over pesticide use and more stringent environmental regulations, today's crop producer must exhibit good stewardship and stay current with new technologies in order to produce high-quality crops in a profitable manner.

Concerns over environmental health and public safety, which were responsible for the removal of some highly effective broad-spectrum chemicals from the agricultural market, have led private companies and the research community to seek alternative approaches to improving crop protection. As a result, we have seen the development and registration of new reduced risk crop protection products. Products with this classification tend to have a more narrow spectrum of activity by targeting specific life stages or pest species. They are generally considered to be less toxic to the environment and can be integrated more easily into pest management systems that are based on biological control. Suppression of pest organisms by their natural enemies is recognized as one of the most suitable long-term pest management strategies for many production systems. Consequently, great effort has been exerted toward identification of natural enemies to effectively suppress various pests in different types of production systems. As more information is learned and these systems become more refined, we will see even more applications of this technology used in the future.

The purpose of this book is to present an overview of various alternative measures to traditional pest management practices, utilizing the biological control approaches as well as biotechnology. Other alternative measures using chemical insecticides, such as ecology control and integrated pest management, are the subject of a separate volume and consequently will not be discussed here.

The book is comprised of four sections. The first contains individual chapters concerning the use of various biological control agents. Specifically, there are chapters on insect parasitoids and predators, pathogenic microorganisms, semiochemicals, including pheromones, botanical insecticides, and insect growth regulators. The second deals with physiological and genetic approaches, namely the genetic control of insect pests and plant resistance to insects. The third section is devoted to various ways of making biological control of insect pests more effective, utilizing the latest advances in biotechnology. One chapter deals with the genetic engineering of insect resistance in plants and the second chapter with the genetic engineering of biocontrol agents of insects. A separate chapter is devoted to environmental impact of genetically engineered materials. The last section covers various aspects of governmental regulations when using biological control agents, as well as procedures governing the use of the recombinant DNA technology.

The individual chapters were written by experts in their fields of endeavor. The book should be of great interest not only to students, teachers, and researchers but also to agricultural practitioners, policy makers, and intelligent laymen concerned with food security and public safety. The book's subjects cover aspects of entomology, agricultural microbiology, plant physiology, plant biochemistry, economic botany, genetics and plant breeding, plant resistance, genetic engineering, environmental science, public policy, and law.

This publication should be a useful resource to students and professionals in the fields of entomology, agronomy, horticulture, and environmental sciences and those concerned with environmental issues in agriculture.

The editors wish to thank the individual contributors for the time and effort they put into the preparation of their chapters. In addition, special thanks are due to the Ann Arbor Press and CRC Press Staff and Editorial Board.

Jack E. Rechcigl
Nancy A. Rechcigl

The Editors

Jack E. Rechcigl is a Professor of Soil and Environmental Sciences at the University of Florida and is located at the Research and Education Center in Ona, FL. He received his B.S. degree (1982) in Agriculture from the University of Delaware, Newark, DE and his M.S. (1983) and Ph.D. (1986) degrees in Soil Science from Virginia Polytechnic Institute and State University, Blacksburg, VA. He joined the faculty of the University of Florida in 1986 as Assistant Professor, in 1991 was promoted to Associate Professor, and in 1996 attained Full Professorship. In 1999, he was named a University of Florida Research Foundation Professor.

Dr. Rechcigl has authored over 200 publications, including contributions to books, monographs, and articles in periodicals in the fields of soil fertility, environmental quality, and water pollution. His research has been supported by research grants totaling over $3 million from both private sources and government agencies. Dr. Rechcigl has been a frequent speaker at national and international workshops and conferences and has consulted in various countries, including Canada, Brazil, Nicaragua, Venezuela, Australia, New Zealand, Taiwan, Philippines, France, and the Czech Republic. He also serves on a number of national and international boards, including the University of Cukurova Mediterranean International Center for Soils and Environment Research in Turkey.

He is currently Editor-in-Chief of the *Agriculture and Environment Book Series*, Associate Editor of the Soil and Crop Science Society Proceedings, and until recently Associate Editor of the *Journal of Environmental Quality*. Most recently he has edited *Insect Pest Management: Techniques for Environmental Protection* (Lewis Publishers, 2000), *Environmentally Safe Approaches to Crop Disease Control* (Lewis Publishers and CRC Press, 1997), *Soil Amendments: Impacts on Biotic Systems* (Lewis Publishers and CRC Press, 1995), and *Use of By-Products and Wastes in Agriculture* (American Chemical Society, 1997). He is also serving as an invitational reviewer of manuscripts and grant proposals for scientific journals and granting agencies.

Dr. Rechcigl is a member of the American Chemical Society, Soil Science Society of America, American Society of Agronomy, International Soil Science Society, Czechoslovak Society of Arts and Sciences, various trade organizations, and the honorary societies of Sigma Xi, Gamma Sigma Delta, Phi Sigma, and Gamma Beta Phi.

Dr. Rechcigl has been the recipient of numerous awards, including the Sigma Xi Research Award, University of Philippines Research Award, University of Florida Research Honor Award, and University of Florida Research Achievement Award. Most recently he was elected a Fellow of the American Society of Agronomy, Fellow of the Soil Science Society of America, and the recipient of Honorary Professorship from the Czech Agricultural University in Prague.

Nancy A. Rechcigl holds the position of entomologist with Yoder Bros. Inc., Parrish, FL, specializing in plant disease and entomological problems of floricultural crops. Prior to joining Yoder Bros., Nancy worked for the University of Florida (1989–1994) as a County Horticultural Agent, providing diagnostic services and information on cultural practices and pest management to horticultural, landscape, and pest control industries. As an Extension Agent she was also responsible for supervising the County Master Gardener Program, providing instructional classes and operating a Plant Clinic that was popular with the urban community. From 1986 to 1989, she worked for Ball PanAm Inc., Parrish, FL as a Plant Pathologist responsible for the disease certification program of ornamental plants.

Over the past 12 years, Ms. Rechcigl has given numerous lectures on the identification and control of disease and pest problems of turf and ornamentals. In addition to writing a weekly gardening column "Suncoast Gardening" for the urban community, she frequently contributes articles to local trade and professional journals. Most recently she has co-edited the books *Environmentally Safe Approaches to Crop Disease Control* (Lewis Publishers and CRC Press, 1997), and *Insect Pest Management: Techniques for Environmental Protection* (Lewis Publishers, 2000).

Ms. Rechcigl received her B.S. degree (1983) in Plant Pathology from the University of Delaware, Newark, DE. She did her graduate work at Virginia Polytechnic Institute & State University, Blacksburg, VA, receiving her M.S. degree in 1986, specializing in Plant Virology.

Ms. Rechcigl is an active member of the American Phytopathological Society, Entomological Society of America, Florida Nurserymen and Growers Association, Czechoslovak Society of Arts and Sciences, and the Honorary Society of Phi Kappa Phi.

Contributors

Nancy E. Beckage
Department of Entomology
University of California
Riverside, California

Diane L. Belnavis
Longwood Gardens
Kennett Square, Pennsylvania

Bryony C. Bonning
Department of Entomology
Iowa State University
Ames, Iowa

J. Lindsey Flexner
Agricultural Products Department
Stine-Haskell Research Center
E. I. DuPont de Nemours Co., Inc.
Newark, Delaware

Angharad M. R. Gatehouse
Department of Biological Sciences
University of Durham
Durham, United Kingdom

John A. Gatehouse
Department of Biological Sciences
University of Durham
Durham, United Kingdom

Robert L. Harrison
Department of Entomology
Iowa State University
Ames, Iowa

Phillip O. Hutton
Office of Pesticides Program
U.S. Environmental Protection Agency
Washington, D.C.

G. Karg
Faculty of Biology
University of Kaiserslautern
Kaiserslautern, Germany

John L. Kough
Office of Pesticides Program
U.S. Environmental Protection Agency
Washington, D.C.

J. Thomas McClintock
Office of Pollution Prevention and
  Toxics
U.S. Environmental Protection Agency
Washington, D.C.

Michael L. Mendelsohn
Office of Pesticides Program
U.S. Environmental Protection Agency
Washington, D.C.

David B. Orr
Department of Entomology
North Carolina State University
Raleigh, North Carolina

Alan S. Robinson
Entomology Unit
FAO/IAEA Agriculture and
  Biotechnology Laboratory
International Atomic Energy Agency
Seibersdorf, Austria

Robert G. Shatters, Jr.
Horticultural Research Laboratory
USDA/ARS
Fort Pierce, Florida

C. Michael Smith
Department of Entomology
Kansas State University
Manhattan, Kansas

D. M. Suckling
The Horticulture and Food Research
  Institute of New Zealand, Ltd.
Lincoln, Canterbury
New Zealand

Charles P.-C. Suh
Department of Entomology
North Carolina State University
Raleigh, North Carolina

Nikolai A. M. van Beek
Agricultural Products Department
Stine-Haskell Research Center
E. I. DuPont de Nemours Co., Inc.
Newark, Delaware

Richard A. Weinzierl
Department of Crop Sciences
University of Illinois
Urbana, Illinois

Dedication

To our parents and our family for their love and support.

Table of Contents

Biological Control Agents

CHAPTER 1

Parasitoids and Predators

David B. Orr and Charles P.-C. Suh

CONTENTS

1.1   INTRODUCTION

Parasitoids and predators have been employed in the management of insect pests for centuries. The last century, however, has seen a dramatic increase in their use as well as an understanding of how they can be manipulated for effective, safe use in insect pest management systems. Despite this long history, it wasn't until 1919 that the term *biological control* was apparently used for the first time. The term was coined by the late Harry Smith of the University of California, who defined "biological control" as the suppression of insect populations by the actions of their native or introduced natural enemies (Smith, 1919). There has been recent debate regarding the scope and definition of biological control (see Nordlund, 1996), mainly as a result of technological advances in the tools available for pest management. In this chapter we will follow the definition presented by Van Driesche and Bellows (1996) as "... the use of parasitoid, predator, pathogen, antagonist, or competitor populations to suppress a pest population, making it less abundant and thus less damaging than it would otherwise be."

It is widely accepted that there are three general approaches to biological control: importation, augmentation, and conservation of natural enemies (DeBach, 1964; Van Driesche and Bellows, 1996). Importation biological control is often referred to as "classical biological control," reflecting the historical predominance of this approach. It generally involves importation and establishment of non-native natural enemy populations for suppression of non-native or native organisms. Augmentation includes activities in which natural enemy populations are increased through mass culture, periodic release (either inoculative or inundative) and colonization, for suppression of native or exotic pests. Inoculative releases are made with the intent of colonizing natural enemies early in a crop cycle so that they and their offspring will provide pest suppression for an extended period of time. Inundative releases are conducted to provide rapid pest suppression by the released individuals only, with no expectation of suppression by their offspring. These two approaches represent extremes on a continuum of activities, with most augmentative releases being a hybrid of the two. Conservation biological control can be defined as the study and modification of human influences that allow natural enemies to realize their potential to suppress pests. There are two general aspects of natural enemy conservation. The first is the identification and remediation of negative influences that suppress natural enemies. The second is the enhancement of systems (e.g., agricultural fields) as habitats for natural enemies. While augmentation deals with laboratory reared natural enemies, conservation deals with resident natural enemy populations.

Currently, the "classical" approach is probably the most recognized and heralded form of biological control among biological control practitioners. However, in the eyes of the general public, augmentation is more visible and recognized as a result of the wide availability of natural enemies in garden catalogs and nurseries. Partly as a result of this, conservation of beneficial organisms (especially in relation to home gardens) is also becoming more widely recognized by the general public.

This chapter is not intended to be an exhaustive review of research involving parasitoids and predators. Instead, we will try to focus on actual implementation of arthropod natural enemies in insect pest management. We begin with a brief history of the use of parasitoids and predators, and the development of the field of biological control. The next three sections deal with general concepts and challenges facing the use of parasitoids and predators within each of the three general approaches to biological control. Recent case histories are presented that illustrate points made within each of these sections. Finally, because of the controversial nature of the subject, we also summarize the current debate over the potential for nontarget impacts of parasitoids and predators used in pest management programs.

1.2 HISTORICAL USE OF PARASITOIDS AND PREDATORS

Predatory and parasitic relationships among insects existed long before the appearance of humans, and it is uncertain when these entomophagous habits were first recognized. The first accounts of predators being used as an insect management tool date back as early as 900 A.D. when Chinese citrus growers placed ants (*Oecophylla smaragdina* F.) on trees to protect them from other insects (McCook, 1882; Sweetman, 1958; Doutt, 1964; DeBach, 1974; Coulson et al., 1982). The Chinese also aided intertree movement of the ants by placing bamboo rods as runways or bridges between trees. The ants built large paper nests in the trees and were apparently quite effective at suppressing various lepidopterous pests of citrus. These ants were reportedly available for purchase up to at least the 1970s (DeBach, 1974).

Date growers in Yemen also employed ants as far back as the 1700s (Forskal, 1775; Botta, 1841). Colonies of predacious ants were moved from the mountains to lowland date palms each year for suppression of phytophagous ants. This example marks the first written record of predatory insects being moved from one location to another for pest control (Clausen, 1936; Fleschner, 1960; DeBach, 1974; van den Bosch et al., 1982). Another example is the 19th century practice of collecting and selling ladybugs for release in hops, a practice that may have been conducted for centuries (Doutt, 1964).

While the predatory behavior of some insects was recognized long ago and taken advantage of for pest management, the recognition and utilization of parasitic insects did not occur until relatively recently. Early observations of wasps emerging from butterfly larvae, such as those by Aldrovandi in 1602 and Redi in 1668, were misinterpreted as transformations of the butterfly larva into another larval stage through metamorphosis (Bodenheimer, 1931). These two workers also mistakenly identified the pupae of the wasps as eggs of the butterfly (Silvestri, 1909). Credit for the first correct interpretation of parasitism has changed through the years. Silvestri (1909) credited Vallisnieri, who in 1706 correctly identified the association between the parasitic wasp *Cotesia* (= *Apanteles*) *glomerata* (L.) and the cabbage butterfly, *Pieris rapae* (L.). DeBach (1974) indicated that Van Leuwenhoeck made mention of and illustrated a parasitoid of a sawfly that feeds on willow in 1701, while Van Driesche and Bellows (1996) note that Van Leuwenhoeck also correctly interpreted parasitism of aphids by a species of *Aphidius* wasp in 1700. Currently, the earliest reported correct interpretation of parasitism was by Martin Lister who in 1685 realized that adult ichneumon wasps emerging from caterpillars came from eggs laid in the caterpillars by adult female wasps (Van Driesche and Bellows, 1996).

The earliest reported successful introduction of a natural enemy from one country to another to control insect pests occurred in 1762 and involved the transportation of mynah birds *Gracula religiosa* L. from India to control red locust, *Nomadacris septemfasciata* Serville, in Mauritius (DeBach, 1974). In Europe, one of the first written proposals to use insect predators for pest control was given by Carl Linnaeus in 1772, who stated "every insect has its predator which follows it and destroys it. Such predatory insects should be caught and used for disinfecting crop plants" (Hörstadius, 1974). The first insect natural enemies purposely used in Europe to control an insect pest were predacious stinkbugs *Picromerus bidens* (L.), which were reportedly used with some success to control bedbugs as early as 1776 (DeBach, 1974). By the early 1800s, others such as Erasmus Darwin were advocating use of syrphid flies and coccinellid beetles to control aphids in greenhouses (Kirby and Spence, 1815; DeBach, 1974).

Reports on the value of entomophagous insects in suppressing agricultural and forest pests began appearing in Europe early in the 19th century (Kollar, 1837; Ratzeburg, 1844; Riley, 1931). The concept of using insect parasitoids for pest control in Europe also developed during this period (van den Bosch et al., 1982). Following the recognition of parasitic wasps emerging from caterpillars, Hartig in 1826 proposed that parasitized caterpillars be collected and stored in order to harvest adult wasps, which could then be later released to control cabbage butterflies (Sweetman, 1936). Actual efforts to experimentally manipulate populations of natural enemies (carabid and staphylinid predators) in agricultural settings began with Boisgiraud in 1840 and were continued by Villa in 1844 (Trotter, 1908).

An influx of European insect species that became serious agricultural pests in the United States during the 19th century prompted U.S. entomologists to consider reasons for the difference in pest status of these insects between the two continents. Asa Fitch (1856) first suggested that insect pests of European origin reached their pest status in the U.S. because of the lack of their indigenous natural enemies, and suggested that importing those enemies would provide the remedy for these pest outbreaks. Efforts at utilizing natural enemies in American agriculture began soon thereafter.

The first deliberate movement of parasitoids from one location to another was conducted by C.V. Riley, who distributed parasitoids of the weevil *Conotrachelus nenuphar* (Herbst) around the state of Missouri in 1870 (Doutt, 1964). In 1871, LeBaron shipped parasitized (by *Aphytis mytilaspidis* (LeBaron) oyster-shell scales) *Lepidósaphes úlmi* (L.) between two towns in Illinois (Doutt, 1964). The first predatory arthropod to be transported from one continent to another was the mite, *Tyroglyphus phylloxerae*, which was shipped from the U.S. and established in France in 1873 (Fleschner, 1960; Doutt, 1964). However, it did not suppress populations of the target pest, the grape phylloxera, *Daktulosphaira vitifoliae* (Fitch).

It was not until 1883 that the first parasitoid, *Cotesia (= Apanteles) glomeratus* (L.), was successfully moved and established from one continent to another (England to U.S.) for suppression of *P. rapae* by the U.S. Department of Agriculture (Riley, 1885; Riley, 1893). Other early international movements of natural enemies include shipment of several aphidophagous natural enemies to New Zealand (including *Coccinella undecimpunctata* L., which became established) (Doutt, 1964), and importation of *Trichogramma* spp. from the U.S. into Canada for control of the gooseberry sawfly, *Nématus ríbesii* (Scopoli) (Saunders, 1882). While a variety of international movements of insects for pest control occurred in the late 1800s, none of them achieved complete economic control (Fleschner, 1960).

It is generally accepted that the first successful case in terms of complete and sustained economic control of an insect pest by another insect is control of the cottony cushion scale, *Icerya purchasi* Maskell, in California during the late 1800s (Fleschner, 1960; Doutt, 1964; DeBach, 1974; van den Bosch et al., 1982). In 1869, *I. purchasi* was introduced into California, and by 1886, threatened to destroy the entire southern California citrus industry (DeBach, 1974). Efforts to find natural enemies of *I. purchasi* were made in Australia, native home of the pest, during 1887 and 1888. Two insects, the vedalia beetle, *Rodolia cardinalis* Mulsant (Coleoptera:

Coccinellidae), and a parasitic fly, *Cryptochetum iceryae* (Williston) (Diptera: Cryptochètidae), showed promise and were imported to California for field release in 1888. Within two years after release, *I. purchasi* was under complete control throughout the state. Although the vedalia beetle is mostly credited for controlling the cottony cushion scale, once established, the parasitic fly became the major control factor of the pest in the coastal areas of the state (Van Driesche and Bellows, 1996). This classic example is presented in many books dealing with insect biological control (e.g., DeBach, 1964, 1974; van den Bosch et al., 1982; Van Driesche and Bellows, 1996), and set the stage for future biological control programs. This example also has associated with it a behind-the-scenes review of the people involved in the project that includes an unfortunate love affair, political intrigue, and diamond jewelry (Doutt, 1958).

Although *C. iceryae* played a role in the suppression of *I. purchasi*, it has been somewhat overlooked, perhaps in part because it provides suppression only over a limited portion of the target pest's range. Greathead (1986) considered the importation of *Encarsia berlesi* (Howard) into Italy from the U.S. in 1906 for control of the mulberry scale, *Pseudaulacaspis pentagona* Targioni-Tozzetti, to be the first successful introduction of a parasitoid from one country to another for insect pest control.

Following the success of the cottony cushion scale project, numerous biological control efforts ensued worldwide (Clausen, 1978; Luck, 1981; van den Bosch et al., 1982; Greathead, 1986; Greathead and Greathead, 1992) some of which were just as successful. Although the primary focus of early efforts in biological control was importation of natural enemies, other methods of manipulating parasitoids and predators were also considered. While the concept of mass rearing insects for future releases was proposed as early as 1826 by Hartig, the first practical attempt toward augmentation of natural enemies in western Europe was probably made in 1899 by Decaux, who devised a complete management program for apple orchards, including releases of field-collected inchneumonid wasps (Decaux, 1899; Biliotti, 1977). Mass culture and periodic release of natural enemies in North America began with the 1916 discovery that mealybugs and black scale could be reared successfully on potato sprouts (Smith and Armitage, 1931). Another early augmentation effort involved *Cryptolaemus montrouzieri* Mulsant, which was mass-reared in an insectary and distributed to citrus groves for control of mealybugs (Armitage, 1919, 1929). It is interesting to note that this ladybeetle, also known as the mealybug destroyer, is commonly sold by commercial insectaries nowadays for suppression of mealybugs (Hunter, 1997).

Concerted efforts in augmentation of insect pests in North America, Europe, and Asia did not begin until the mid-20th century when the procedure was evaluated for pest suppression on a variety of fruit, vegetable, field, and forest crops (Beglyarov and Smetnik, 1977; Biliotti, 1977; Ridgway et al., 1977). Conservation and enhancement of natural enemy populations was considered by a variety of workers throughout the history of biological control. However, just as with augmentation, concerted efforts toward conservation of natural enemies did not begin in earnest until the mid-20th century.

1.3 IMPORTATION BIOLOGICAL CONTROL

From 1890 through 1960, approximately 2300 species of parasitoids and predators were introduced in approximately 600 different situations worldwide for suppression of arthropod pests (Hall et al., 1980). The overall level of establishment of these natural enemies was calculated to be 34%, with complete suppression of target pests occurring in 16% of situations, and some level of pest suppression achieved in an additional 42% of situations (Hall and Ehler, 1979; Hall et al., 1980). These rates have apparently not increased over the last 100 years (Hall and Ehler 1979, Hall et al. 1980), although the percentage of successful projects that are complete successes has reportedly risen since the 1930s (Hokkanen, 1985).

While the statistical validity of analyses of historical data for success or failure rates of biological control has been examined (Stiling, 1990) and questioned (Van Driesche and Bellows, 1996), the results of these analyses have prompted some to call for a more in-depth understanding of the reasons for failure in order to improve the success level of importation biological control (e.g., Hopper, 1996). A variety of reasons have been proposed for the level of failures in classical biological control programs, including inadequate procedures (Beirne, 1985), climate, predation or parasitism by native fauna, lack of alternative hosts or food (Stiling, 1993), feeding niche of target pest (Gross, 1991), inadequate knowledge of natural enemy and target pest taxonomy (Hanson, 1993), and an insufficient amount of time or effort expended on these projects (Greathead, 1986; Waterhouse and Norris, 1987). However, few experiments have been conducted to test hypotheses regarding failures in importation biological control, although attractive opportunities exist for conducting these experiments (Hopper, 1996).

Procedures followed in natural enemy introduction programs are fairly standard and generally include exploration for agents in areas of pest origin; preintroduction studies (such as species identification, biological and ecological characterization, and rearing procedures); quarantine of agents for introduction; release; and evaluation. Van Driesche and Bellows (1996) detail these steps and specific methodology involved in implementing each. A variety of systematic examinations of empirical data have been conducted through the years to search for features associated with high rates of success in biological control programs (Stiling, 1990). Some of these are briefly discussed below.

The proper selection of natural enemies for use in any biological control program may be critical for success. From a theoretical standpoint, selection of natural enemies for introduction biological control programs has been described as following two general approaches. The "reductionist" approach involves selecting natural enemies based on a particular set of biological or life history characteristics, such as fecundity or searching efficiency, while the "holistic" approach focuses on the enemies' interaction with other mortality factors in the pest's life cycle (Waage, 1990). These parameters can be examined in population models either independently or in an integrated manner as a way to make comparisons between species, and theoretically derive criteria to select biological control agents (Waage, 1990). From a practitioner's standpoint, introduction strategies can be described by a continuum of activities that range from empirical to predictive approaches (Ehler, 1990). While

it is appropriate to move away from a strictly empirical approach to species introductions and strike a balance between the two, the predictive approach is currently
somewhat limited because the theoretical framework for biological control is relatively underdeveloped (Ehler, 1990). In practice, the selection of natural enemies is
often constrained by the availability of time and funding, or by the need to quickly
solve a serious pest problem (Waage, 1990; Ehler, 1990). As a result a small, often
arbitrary, selection of enemies is made, and studies of their effect on hosts in the
area of origin may not be possible (Waage, 1990). In fact, many studies may be
terminated before the best agent is found (Waage, 1990).

Proper identification of both pest and natural enemy as well as at least a basic
understanding of their biologies are critical to many biological control projects.
Improved tools for use in systematics, such as molecular techniques (e.g., Hoy, 1994)
have been applied to identification of difficult groups of natural enemies (e.g., van
Kan et al., 1996). Advances in understanding of host selection by parasitoids (see
papers introduced by Vinson et al., 1998), and parasitoid and predator reproduction
(Werren, 1997) have direct applications to biological control projects. Improved
understanding of how to manage the genetics of species being introduced (Hopper
et al., 1993) can lead to improved rates of establishment, especially if experiments
are conducted to continue this understanding (Hopper, 1996).

The use of parasitoids in classical biological control programs far outweighs the
use of predators, and there has been some debate regarding the comparative value
and successful use of each of these natural enemy groups (Hall and Ehler, 1979;
van den Bosch et al., 1982; Greathead, 1986). Hall and Ehler (1979) and Hall et al.
(1980) report that the rates of establishment (34%) and complete success (14%)
were identical for both parasitoids and predators. Greathead (1986) reported that
570 parasitoid species have been released on 2110 occasions, resulting in establishment in 860 cases involving 393 species, against 274 pest species in 99 countries.
On 216 of these occasions, parasitoids either alone or in combination with predators
provided complete or satisfactory pest suppression and another 52 cases resulted in
a "useful" reduction in pest numbers. In contrast, Greathead (1986) also reported
that of the 302 occasions in which insect predators became established, 89 resulted
in complete or partial pest suppression either as a result of predators by themselves
or in combination with parasitoids, and 12 provided "useful" reductions.

Many adventive insect species become pests because they are unaccompanied
by natural enemies from their native home. Traditionally, importation biological
control has sought to reestablish "old" associations of adventive organisms in new
environments with natural enemies from their area of origin (Nechols and Kauffman,
1992). However, Hokkanen and Pimentel (1984) reported a 75% greater probability
of successful biological control when "new associations" are established between
natural enemies and pests. Despite criticism of this suggested approach (Nechols
et al., 1992; Greathead, 1986), in their reanalysis of data, Hokkanen and Pimentel
(1989) again concluded that the new association approach for selecting biological
control agents was not only environmentally and statistically sound, but suggested
this approach would be especially successful for classical biocontrol of native pests.
However, this "neoclassical" approach to biological control, especially when native
pest species are targeted, has been sharply criticized for its potential to detrimentally

impact nontarget organisms (Lockwood, 1993a; Simberloff and Stiling, 1996). Aldrich (1995) suggests that the new associations approach be taken a step further with the idea of "teaching" physiologically competent endemic natural enemies to recognize adventive pests as hosts. This may provide a more environmentally and sociologically sound method of biological control that avoids the controversies surrounding potential nontarget impacts of imported natural enemies (Aldrich, 1995).

There is no clear consensus for whether release of single or multiple species is best in a classical biocontrol program (Hassell, 1978; Myers et al., 1989). DeBach (1974) argued that there is usually one best natural enemy for a particular pest in a given habitat, which alone can often sufficiently suppress pest populations. However, greater pest suppression may be obtained with multiple agents if they attack target pests in different locations, seasons, life stages, or host densities (DeBach, 1974; Murdoch et al., 1984; Tagaki and Hirose, 1994). DeBach (1974) concluded that although competition between multiple species could affect the efficiency of each species, overall host population suppression is greater when multiple species are combined. Others contend that pest suppression may be compromised if multiple natural enemies interfere with one another's foraging activities (Force, 1974; Ehler and Hall, 1982; Briggs, 1993) or if they attack one another (Polis and Holt, 1992; Rosenheim et al., 1993). Field data exist that support both scenarios (Messenger et al., 1976). Ehler (1990) suggested that species selection should not be approached from the traditional dichotomy of single versus multiple species introductions. Rather, he urges determination of the appropriate species or species group for a given situation.

Beirne (1975) reported trends indicating a positive relationship between rate of natural enemy establishment and total numbers of natural enemies released. Ehler and Hall (1982) reported similar trends looking at success of programs and total numbers of natural enemies released. Hoy and Herzog (1985) suggested that many unsuccessful attempts failed because not enough exotic natural enemies were released, and that these programs should be reevaluated if no other reason for failure can be determined. Thus, it appears that the more insects released, the greater the chance of establishment and success. However, Greathead (1986) pointed out that these trends did not take into account the various combination of releases in time and space nor the amount of effort involved in making the releases, both of which can affect establishment rates. Additionally, Greathead (1986) cites several examples in which natural enemies have been established with only a single release of less than 50 individuals.

The selection of target pests and the criteria used to do so are also critical aspects of biological control programs. Often the target species selected for biological control efforts have not been the ones most susceptible to biological control, or have not been the most urgent pest problems (Harris, 1984). A variety of factors have been described as important to determining the suitability of a target pest, including biological, economic, administrative and institutional, and social issues (Waterhouse and Norris, 1987; Barbosa et al., 1997). There have been few in-depth evaluations of multiple target pests based on these factors (e.g., Van Driesche and Carey, 1987; Barbosa et al., 1994). Barbosa et al. (1997) proposed a questionnaire that defines these issues and assigns a numerical rating to the response to provide a quantitative measure of the suitability of pest species as targets for biological control. This evaluation index for setting priorities among potential candidate targets for biological control includes a variety of assessments under three general categories: biological control feasibility, economic assessment, and institutional and administrative assessment.

As with natural enemy selection, a variety of systematic examinations of empirical data have been conducted of target pest and habitat selection considerations that are associated with high success rates for biological control. These include consideration of whether pests are direct or indirect, native or exotic, sedentary or mobile, and the geographic location of the target pests' habitat, the degree of habitat stability, and the climate (Hall and Ehler, 1979; Hall et al., 1980; Hokkanen, 1985; Greathead, 1986; Stiling, 1990, 1993).

Ultimately, the success or failure of a program will be determined by its economic success. Economic assessments of classical biological control programs are multidimensional, with a range of economic effects, and can therefore be a complex undertaking (Tisdell, 1990). Tisdell (1990) cites several components that should be considered in economic assessments such as cost savings for producers, profit increases, cost of program, value of land, and lowered cost of product. Barbosa et al. (1997) emphasize crop importance, pest importance, and project cost in their economic assessments. Reichelderfer (1981) suggests that the economic benefits of a biological control project are a function of the type and degree of damage by a pest, efficiency of the biological control agent, market price for the crop, and risk aversion of producers. The most common method of determining the economic benefits of biological control programs is through cost-benefit analyses. This approach offers a systematic way to determine if the use of biological control results in a net gain (Headley, 1985; Tisdell, 1990). Habeck et al. (1993) presented an economic model that determines how large average expected economic benefits of importation biological control projects need to be for benefits to exceed costs.

Economic assessments of the use of introduced natural enemies have been made for several arthropod pests (Ervin et al., 1983; Norgaard, 1988; Voegele, 1989; Tisdell, 1990). Undoubtedly, classical biological control programs have produced some of the highest benefit-to-cost ratios of any pest management approach, exceeding billions of dollars in terms of total savings (Tisdell, 1990). For example, a recent introduction program initiated against the ash whitefly in California resulted in a benefit: cost ratio ranging between $270:1 and $344:1 (Jetter et al., 1997). Marsden et al. (1980) reported an average benefit: cost ratio (for the period 1960–2000) of 9.4:1 for three importation biological control programs conducted by CSIRO Division of Entomology against insect pests in Australia, compared to a 2.5:1 benefit-cost ratio for non-biological control projects conducted by the agency during the same time period. The economic benefits of classical biological control are enhanced by the fact that programs are self-sustaining and permanent, so that benefits continue to accrue annually without additional cost.

1.3.1   Case History: Cassava Mealybug in Africa

The cassava mealybug, *Phenacoccus manihoti* Matile-Ferrero, was accidentally introduced from South America to Africa in the early 1970s (Hahn and Williams, 1973; Sylvestre, 1973; Matile-Ferrero, 1978; Leuschner, 1982). Unaccompanied by its native natural enemies, the cassava mealybug spread throughout Africa (Lema and Herren, 1982; Herren et al., 1987), becoming a serious pest of cassava, a major food staple in Africa (Leuschner, 1982; Sylvestre and Arraudeau, 1983). It attacks the roots and leaves of the plant, causing tuber yield losses up to 84% (Herren, 1981; Nwanze, 1982) and nearly 100% loss of foliage (Lema and Herren, 1985). Despite the effectiveness of chemical insecticides, the low crop value of cassava along with socioeconomic constraints and farmer's inexperience in handling and applying insecticides dictated that other control measures be sought in order to provide a safe and economical long-term solution (Singh, 1982; Lema and Herren, 1982). Thus, a biological control program was initiated in Africa (Lema and Herren, 1982, 1985; Herren and Lema, 1982) to complement existing research and developments in host plant resistance.

In the mid-1970s, scientists began exploring South America for natural enemies of the cassava mealybug (Norgaard, 1988). Although a complex of natural enemies was discovered (Lohr et al., 1990), the parasitoid *Apoanagyrus (Epidinocarsis) lopezi* Desantis (Hymenoptera: Encyrtidae), found in Paraguay by M. Yaseen, was selected for use in a biological release program for cassava mealybug control. With funding from the International Fund for Agriculture Development, the Africa-wide Biological Control Project (ABCP) was initiated at the International Institute of Tropical Agriculture (IITA) headed by Hans Rudolf Herren in 1980 (Herren, 1987; Herren et al., 1987), and in 1981, parasitoids were imported to Nigeria for propagation and field release (Herren and Lema, 1982; Lema and Herren, 1985). Within 3 years after initial releases, parasitoids had spread over 200,000 km² in southwestern Nigeria, and by the end of 1985, over 50 releases had been made in 12 African countries (Herren et al., 1987). By 1990, *A. lopezi* was established in 24 countries covering an area of more than 12.7 million km² (Neuenschwander et al., 1990).

Based on exclusion experiments (Neuenschwander et al., 1986; Cudjoe et al., 1992), follow up field studies (Hammond et al., 1987; Neuenschwander and Hammond, 1988; Neuenschwander et al., 1990), and a computer simulation model (Gutierrez et al., 1987), it was concluded that *A. lopezi* was responsible for declines in cassava mealybug populations and damage to plants. Currently, *P. manihoti* has been virtually eliminated in 30 African countries and no longer poses a serious threat to most cassava-growing regions.

The cost and benefit of the release program in Africa, accumulated over 40 years (1974–2013), was estimated to be $49 million and $9.4 billion, respectively, with a cost:benefit ratio ranging between 1:170 to 1:431 depending on the scenario (Schaab et al., 1996). In a worse case scenario, Norgaard (1988) conservatively estimated a cost-benefit ratio of 1:149. The release program, under the direction of Hans Rudolf Herren, not only saved one of Africa's major staple crops and farmers billions of dollars, but demonstrated again that the classical approach to biological control can be a very successful method for controlling serious insect pests in agriculture.

Because of the program's immense success, Hans Herren was awarded the prestigious World Food Prize in 1995.

1.4 AUGMENTATION BIOLOGICAL CONTROL

Microorganisms have been more amenable to laboratory culture and manipulation than arthropod natural enemies. As a result, most larger-scale commercial ventures into utilizing natural enemies have employed microorganisms, their genes, or gene products for development of bio-pesticides or transgenic crops. However, a small but growing industry has developed around the use of arthropods for augmentation biological control, and implementation of augmentation has increased significantly in recent years.

The augmentation of natural enemies is a practice that is widely recognized by the general public in the U.S. mainly as a result of widespread availability of arthropod natural enemies such as lady beetles (especially *Hippodamia convergens* Guerin-Meneville) and mantids through garden catalogs and nurseries (Cranshaw et al., 1996). The industry providing these organisms has grown tremendously in the past 20 years. Ridgway and Vinson (1977) reported 50 North American suppliers of natural enemies, while Hunter (1992) reported 95 suppliers and 102 different organisms sold for biological control. More recently, Hunter (1997) reported 142 commercial suppliers and over 130 different species of beneficial organisms, of which 53 are arthropod predators and 46 are parasitoids. The categories of organisms listed by Hunter (1997) included 17 predatory mites, 4 stored product pest parasites and predators, 17 aphid parasites and predators, 9 whitefly parasites and predators, 23 parasites and predators for greenhouse pests, 7 scale and mealybug parasites and predators, 12 insect egg parasites, 6 moth and butterfly larval parasites, 8 filth fly parasites, 4 "other" insect parasites, and 21 general predators. Anonymous (1998) lists biological control products and companies that provide them worldwide. The suppliers listed by Hunter (1997) and Anonymous (1998) do not reflect the "countertop" sales of organisms such as ladybeetles and mantids by many local nurseries and some large discount home-improvement centers. Annual sales of natural enemies in the U.S. amount to approximately $9–10 million (U.S. Congress OTA, 1995), and approximately $60 million worldwide (Leppla and King, 1996).

Augmentative releases of parasitoids and predators are currently included in a variety of pest management programs around the world (Leppla and King, 1996). Although natural enemies are sold for suppression of pests in several different systems (e.g., manure management, urban environments, stored product protection, pastures and forests), detailed estimates of implementation are readily available only for food cropping systems. In Europe, approximately 40–60,000 ha of orchard, vineyard, and vegetable crops are treated with natural enemies annually (Bigler, 1991). Approximately 5000 ha of greenhouses worldwide utilize some form of augmentative biological control (van Lenteren et al., 1997). In the U.S., augmentative releases of natural enemies take place on approximately 10% of greenhouses, 8% of nurseries, 19% of cultivated fruit and nut acreage, and 3% of the cultivated vegetable acreage (U.S. Congress OTA, 1995). On one high-value crop, strawberries, beneficial mites are used for spider mite suppression on approximately 50–70% of acreage in California alone (U.S. Congress OTA, 1995). Egg parasitoids in the genus *Trichogramma* are the most widely produced and released arthropods in augmentative biological control. These parasitoids have several advantages, including relative ease of rearing and the fact that they kill their host in the egg stage before it causes feeding injury (Wajnberg and Hassan, 1994). Worldwide, approximately 20 species of *Trichogramma* are regularly used in augmentative biological control programs to control primarily lepidopterous pests in at least 22 crops and trees on an estimated 32 million ha (Li, 1994). Most of these species are released in large numbers, i.e., inundative releases, in order to rapidly suppress the target pest.

Although the practice of augmenting parasitoids and predators for insect management has seen modest but notable implementation throughout the world during the past two decades, impediments to increased implementation remain. These include the continued need for development of economically viable large-scale rearing technology, a lack of experimental data for release strategies to provide more predictable results, and the need for widespread adoption of effective mechanisms to ensure consumers receive quality products that perform as advertised.

Efficient mass production systems are necessary for augmentation of arthropods to become more commonly accepted as a pest management tool (Nordlund and Greenberg, 1994). A variety of impediments to these systems include a lack of artificial rearing media and systems for efficient delivery of artificial media, limited automation, as well as the need for properly designed facilities, effective quality controls, and improved management systems (Nordlund and Greenberg, 1994). As technology develops, however, especially in regard to rearing, this form of biological control may become more widespread in the future (Hoy et al., 1991; Moffat, 1991; Parrella et al., 1992; Nordlund and Greenberg, 1994; Leppla and King, 1996; Nordlund et al., 1998).

Although augmentation has been shown to be effective against a variety of pests and cropping situations, there is a lack of clear experimental data that supports the use of many of the arthropod natural enemies currently on the market. An example of conflicting data involves a 'product' that has been on the market for decades, the lady beetle *H. convergens*. Use of this beetle has long been considered ineffectual mainly as a result of concerns over dispersal of beetles following release (DeBach and Hagen, 1964). Recently, however, Flint et al. (1995) demonstrated that although *H. convergens* dispersal does occur, significant reductions of aphid numbers could be obtained in potted roses. The lack of clear or sufficient efficacy data has led to commercial suppliers making recommendations for the use of some products based on limited or anecdotal evidence. This situation has in turn prompted some growers to experiment with products on their own in order to find strategies that work for their specific situation.

Because of the lack of supporting data for many augmentation approaches, recommendations still cannot be made regarding rates and application methodologies that provide predictable results (Parrella et al., 1992). Several authors have called for development of predictive models to assist in implementation of augmentation biological control (Huffaker et al., 1977; Stinner, 1977; King et al., 1985; van

Lenteren and Woets, 1988; Ehler, 1990), but this has only rarely been done (see for example Parrella et al., 1992). There may be several explanations for the lack of experimental work supporting augmentation. One is certainly the tremendous logistical difficulties involved in conducting the large-scale, statistically valid, detailed studies that are required to effectively evaluate natural enemy augmentation (Luck et al., 1988). Another may be a perceived similarity between augmentative releases and the insecticide paradigm that has discouraged research interest in this area (Parrella et al., 1992).

Poor quality of released natural enemies or incorrect release rates can lead to unsatisfactory pest suppression and contribute to the unpredictability of augmentation biological control (Hoy et al., 1991). A variety of general approaches have been presented in the literature for quantifying the quality of commercially produced natural enemies (e.g., Boller and Chambers, 1977; King and Leppla, 1984; King et al., 1985; Bigler, 1991, 1994; Nicoli et al., 1994). Three components of insect mass production quality controls were identified in Bigler (1991) as (1) production control, monitoring general procedures of rearing processes (e.g., equipment maintenance and environmental conditions); (2) process control, monitoring quality of the unfinished product (e.g., % hatch, larval, and pupal weights, % pupation); and (3) monitoring quality of the final product (e.g., quantity, size, sex ratio, fecundity, and other biological and behavioral characteristics).

These tests, though necessary, can be very labor intensive and add to production costs. As a result, several authors have suggested relatively quick, inexpensive approaches to quality control testing that include flight-testing (Couillien and Gregoire, 1994; Dutton and Bigler, 1995; Doodeman et al., 1996; van Lenteren et al., 1996), DNA "fingerprinting" (Santiago-Blay et al., 1994), evaluation of host size (Purcell et al., 1994), "walking" ability (Vereijssen et al., 1997), and behavioral analysis (Lux and Bigler, 1991). A review of the current status of standardized quality control test for natural enemies sold for greenhouse pest management is presented by van Lenteren (1996a).

Although natural enemy producers and researchers can routinely conduct quality control tests, it may not be possible or desirable for consumers to conduct these tests themselves. Various efforts have been made to assist consumers in ensuring that they release quality products. Recommendations for using augmentative biological controls have begun to include suggestions for quick and easy quality control checks, such as in the use of *Neoseiulus fallacis* (Garman) on strawberries (e.g., Coop et al. 1997). In addition, several extension resources have been made available to assist the public in making informed consumer decisions regarding suppliers, and product quality and efficacy when purchasing natural enemies (e.g., Orr and Baker, 1997a, b; Knutson, 1998).

Very few independent studies have actually sampled the quality of commercially produced natural enemy "products" (Losey and Calvin, 1995; Fernandez and Nentwig, 1997; O'Neil et al., 1998; Schmidt, 1998). In general, what these studies highlight is that quality of commercial arthropod natural enemies received by consumers is frequently not satisfactory. It has been suggested that a solution to quality control concerns may lie in the development of regulatory legislation that includes provisions to govern quality of commercial natural enemies (van Lenteren, 1996b). There are currently very few countries that regulate the arthropods used in augmentative biological control (e.g., Martin and Wearing, 1990; Nedstam, 1994; Bigler, 1997; Blumel and Womastek, 1997).

Where government regulations do not exist, industry self regulation has in some cases been adopted that "includes quality control manuals with tests for production, process and product control, total quality management with shared authority and responsibility for quality, and the International Organization for Standards (ISO 9000) programme of full management" (Leppla and King, 1996). Institutional support for self-regulation comes from several public and private organizations (Leppla and King, 1996). Where government regulation of these organisms does exist, there are both advantages and disadvantages to the regulation process. Registration of macroorganisms used in augmentative biological control in Switzerland, for example, requires data on the organism's efficacy, and "bioecology," an assessment of risks to humans and the environment, and information on evaluation and registration in neighboring countries (Bigler, 1997). The advantages are that quality control protocols are followed, ineffective products are not marketed, and environmental and human hazards are assessed. However, higher costs associated with registration may delay or prevent implementation of products.

1.4.1 Case History: *Trichogramma* Wasps

Because *Trichogramma* wasps are the most widely augmented arthropod natural enemies, they make a good example to study mechanisms by which companies produce and consumers receive biological control products from commercial suppliers. The development and sale of *Trichogramma* products in western Europe and the U.S. differ dramatically, and provide an interesting comparison, with the European situation serving as a good model for future commercial development of augmentation biological control.

Most *Trichogramma* used in western Europe are for suppression of the European corn borer, *Ostrinia nubilalis* (Hübner). Currently, approximately 58,000 ha of corn are treated annually with *Trichogramma brassicae* Bezdenko, 84% of which is supplied by BIOTOP in Valbonne, France (F. Kabiri, personal communication). In 1998, between 400,000 and 500,000 ha of corn in France were treated with some type of pesticide product for *O. nubilalis* suppression, approximately 10% (45,000 ha) of which was *T. brassicae* (F. Kabiri, personal communication). Sales of *T. brassicae* by BIOTOP have consistently increased each year from 1988, *when T. brassicae* first appeared on the market in France as a crop protection product.

There are several reasons for this success. High product quality is maintained each year through careful selection and maintenance of founder populations of parasitoids, and sophisticated quality controls are employed throughout the host and parasitoid production process (Frandon et al., 1991; Bigler, 1994). The encapsulated formulation of *T. brassicae* was registered as a crop protection product with 10 years of efficacy and quality control data available for the registration process (Anonymous, 1988; F. Kabiri, personal communication). A single pest species, *O. nubilalis*, was targeted initially and efficacy data collected prior to commercial sales demonstrated suppression equal to pesticides (Kabiri et al., 1990; Frandon and Kabiri, 1998). Efficacy data is collected annually by BIOTOP and its research partners and release methods have been constantly improved so that consistent suppression of an *O. nubilalis* generation can now be achieved with a single release rather than the original three releases at a cost that is comparable to pesticides (Kabiri et al., 1990; Frandon and Kabiri, 1998). This efficacy data is advertised publicly, in a manner similar to pesticide products. Distribution of product to customers is tightly controlled by BIOTOP. All orders for *T. brassicae* within a given season are processed well before the beginning of the season. Prior to product shipment, new customers are provided with extensive technical information on *T. brassicae* biology, rearing, handling, and release, in order to maximize their understanding of the product, and their chances for success. The company monitors temperature data and *O. nubilalis* population development in all areas where releases will take place to ensure synchrony between *O. nubilalis* oviposition and *T. brassicae* releases. Customers are informed in advance of their delivery and release dates. All material is shipped to customers by overnight delivery in double-boxed Styrofoam containers ice packs, or in refrigerated trucks.

In contrast to the western European situation, an estimated 140,000 to 350,000 ha of various crops are treated with *Trichogramma* annually in the U.S. (Li, 1994). *Trichogramma* in the U.S. have traditionally been sold as general parasitoids, often for home garden use, with the target pest commonly listed in supplier catalogs as "caterpillars." Often there is little efficacy data to support the use of *Trichogramma*, and no efficacy data appears in advertising. For example, one of the target pests *Trichogramma* sold in the U.S. is the cotton bollworm (usually a mixture of two species, *Heliothis virescens* (F.) and *Helicoverpa zea* (Boddie)) (Knutson, 1998). In contrast to most pest/crop combinations for which *Trichogramma* are sold in the U.S., a considerable amount of field research has been done in cotton to determine appropriate release methodologies and assess efficacy. Results from large-scale *Trichogramma* releases in cotton have been quite variable, with some suggesting good pest suppression and others indicating poor suppression (e.g., Ables et al., 1979; Jones et al., 1977; Stinner et al., 1974; King et al., 1985). A recent reevaluation of *Trichogramma* releases in cotton (Suh et al., 1998) demonstrated that even though parasitism significantly increased egg mortality of bollworms, density-dependent mortality in the larval stages compensated for this mortality resulting in no reduction in cotton damage. These authors concluded that the egg stage of bollworms is an inappropriate target for biological control efforts in cotton.

Despite the availability of considerable technical knowledge and institutional support (Leppla and King, 1996), there appears to be little implementation of quality control procedures in U.S. production of *Trichogramma*. Delivery to customers also is not carefully controlled. Schmidt (1998) reported quality and shipment data for *Trichogramma* ordered covertly from 12 U.S. suppliers, for *H. zea* suppression in tomatoes. The average number of days material spent in transit was 2.06 ± 0.64 days. Packaging ranged from padded envelopes to cardboard boxes with newspaper packing to Styrofoam boxes with cold packs. Approximately 41% of shipments contained no product information or instructions, 52% of shipments contained basic information on biology and release, and 7% of shipments contained detailed information on biology, release, host, and general information. Of the 24 shipments sampled for species composition, 46% had a mix of species or had species other than that which was claimed. There was no difference in sex ratio between endemic and insectary-produced *Trichogramma*. However, there was an approximately 14-fold increase in brachyptery of female *Trichogramma* in insectary-reared *Trichogramma* when compared to endemic (field-collected) specimens. Although actual prices ranged from $19.50 to $126.99 per 100,000 *Trichogramma* (both sexes), when quality parameters were included, the cost to obtain 100,000 releasable macropterous females ranged from $81.37 to $3,694.16.

This example highlights the need for more widespread implementation of well-documented quality control procedures throughout production and delivery of *Trichogramma* products in the U.S., and the need for strong data to support the use of products prior to their reaching the market. Because consumer awareness of quality and efficacy issues in the U.S. may not be high, it also suggests the need for much more stringent regulation of the U.S. biological control industry, either through a major modification of the current voluntary approach, or legislated through public agencies. Long-term success of the industry may depend on this change.

1.5   CONSERVATION BIOLOGICAL CONTROL

Conservation biological control includes identification and remediation of negative human influences that suppress natural enemies, as well as enhancement of systems as habitats for natural enemies. The most common human influence that negatively impacts natural enemies in many systems is pesticide application. As a result, modifications of pesticide use practices are the most commonly implemented form of conservation biological control, and have long been considered an important component of integrated pest management programs (Newsom and Brazzel, 1968). Pesticide use in the U.S. is modified to favor natural enemies on 37% of cultivated vegetable acreage, 22% of cultivated fall potato acreage, and 57% of cultivated cotton acreage (U.S. Congress OTA, 1995).

Pesticide use can be modified to favor natural enemies in a variety of ways, including treating only when economic thresholds dictate, use of active ingredients and formulations that are selectively less toxic to natural enemies, use of the lowest effective rates of pesticides, and temporal and spatial separation of natural enemies and pesticides (Hull and Beers, 1985; Poehling, 1989; Ruberson et al., 1998). Pesticide use decisions in IPM programs for insect pests are typically based on sampling pest populations to determine if they have reached economic threshold levels (Pedigo, 1989), although some work has been done to incorporate natural enemy sampling into these pesticide use decisions.

Sampling for natural enemy populations or their effect on pests can be used to revise economic thresholds to more accurately determine the need or timing for pesticide applications within a pest generation (Ostlie and Pedigo, 1987), or to predict the need for treatment of a future pest generation (Van Driesche et al., 1994).

An example is a sequential sampling plan that takes into account parasitized *H. zea* eggs when estimating this pest's population levels in tomatoes (Hoffman et al., 1991).

However, the use of economic thresholds alone in IPM does not necessarily imply natural enemy conservation, if for example a broad-spectrum pesticide is used for treating the pest population that exceeded the threshold (Ruberson et al., 1998). The use of selective pesticides is perhaps the most powerful tool by which pesticide use decisions can be modified to favor natural enemies (Hull and Beers, 1985), and the one most readily available to growers (Ruberson et al., 1998). While the concepts behind modifying pesticide use are relatively straightforward, implementing these modifications is not necessarily straightforward. One obstacle is that the primary source of information regarding IPM is probably extension services, yet at least in the U.S., there are a variety of competing sources for information regarding pesticide use (Rajotte et al. 1987)

A variety of other approaches to conservation of parasitoids and predators have been studied, and are comparatively complex. These include management of soil, water, and crop residue; modification of cropping patterns; manipulation of noncrop vegetation; and direct provision of resources to natural enemies (Van Driesche and Bellows, 1996). In general, these approaches are aimed at enhancing the density of resident natural enemy populations or communities to increase their effectiveness in pest suppression.

Approaches to conservation biological control other than pesticide use modification are the focus of much of the current research in this area (see papers in Barbosa, 1998). This research is continuing toward developing an understanding of the ecological processes affecting natural enemies at spatial scales ranging from individual fields to entire landscapes (Ferro and McNeil, 1998; Landis and Menalled, 1998; Letourneau, 1998). Many of these alternatives have been shown to be effective (Barbosa, 1998), but are rarely implemented (U.S. Congress OTA, 1995), highlighting an extreme gap between research and implementation for this type of biological control (Ehler, 1998). Often cited examples of conservation biological control are the practices of strip harvesting hay alfalfa or intercropping alfalfa and cotton in California (Stern et al., 1964, 1976; van den Bosch and Stern, 1969). Both practices act to conserve natural enemy populations in cotton by preventing migration of lygus bugs to cotton and subsequent pesticide applications for this pest. While both practices can be highly effective, both pose operational problems, are more expensive, and as a result are not widely implemented by growers (Ehler, 1998).

1.5.1 Case History: The Cotton Aphid

The cotton aphid, *Aphis gossypii* Glover, was not considered a major cotton pest in the U.S. until population outbreaks, attributed to the elimination of natural enemies, followed insecticide applications for boll weevil control in the 1940s (Slosser et al., 1989). Outbreaks became less frequent and more easily controlled following the development and use of organophosphate insecticides (USDA, 1960). However, during the past decade the cotton aphid has reemerged as one of the most important pests of cotton in much of the cotton growing areas in the U.S. (Hardee and O'Brien,

1990) due to the development of insecticide-resistant populations (Grafton-Cardwell, 1991) and the elimination of natural enemies with broad-spectrum insecticides targeted toward other insect pests (Edelson, 1989; Kerns and Gaylor, 1991, 1993). From 1995 to 1997 an average of approximately 89 million pounds of cotton were lost to aphids nationwide, with approximately 29% of U.S. cotton being treated for aphids (Williams, 1998).

Since insecticide resistance was detected (Lee, 1992) and surveys begun eight years ago in North Carolina, less than 1% of cotton acreage in North Carolina has been treated annually for aphids (the exception was 1997 with 3.5%) (J. S. Bacheler, personal communication). Three factors are predominantly responsible for the difference in the North Carolina average and the national average during the past few years (J. S. Bacheler, personal communication).

First, the majority of aphid populations in North Carolina are resistant to insecticides, with the exception of imidacloprid (Bacheler, 1997). Second, the cotton aphid is attacked by a variety of natural enemies, the most important of which appears to be the parasitoid wasp, *Lysiphlebus testaceipes* (Cresson), and the entomopathogenic fungus, *Neozygites fresenii* (Nowakowski) (Frazer, 1988, Steinkraus et al., 1991, 1995; Kidd et al., 1994; Knutson and Ruberson, 1996). This natural enemy complex is very effective in keeping aphid populations below treatment thresholds, and scouting recommendations emphasize inclusion of natural enemy observations, and treatment recommendations call for pesticide applications only in the event of very high aphid populations without either *L. testaceipes* or *N. fresenii* present (Bacheler, 1997; Toth, 1998).

The third reason is that there are very few early to mid-season pest populations in cotton that require treatment with insecticides in North Carolina. The boll weevil has been eliminated as an economic pest for 15 years, over 90% of cotton acreage is treated with an at-planting systemic insecticide, and mid-season pest populations are not usually abundant enough to warrant treatment (Jack Bacheler, personal communication). This allows natural enemy populations to increase and maintain aphid populations below treatment thresholds. Generally, it is not until mid to late July, when foliar insecticides are first applied for heliothine control, that arthropod natural enemies are disrupted. However, by this time the entomopathogen *N. fresenii* is present in most fields, producing epizootics and suppressing late season aphid populations.

This example illustrates several points that may reflect the actual approach to implementing many conservation biological control efforts within IPM programs. Even though conservation of cotton aphid natural enemies saves growers approximately $5,000,000 annually on the approximately 670,000 acres of cotton grown in North Carolina (J. S. Bacheler, personal communication), recommendations which serve to conserve beneficial arthropods may exist more because of a set of fortuitous circumstances than purposeful planning. There are essentially no detailed ecological data to provide a foundation for this program. The recommendations evolved as the result of careful, though anecdotal, field observations of the high efficacy of natural control factors. Aphid population values provided for scouting and treatment decisions are qualitative rather than quantitative, ranging from "low" to "very high" (Bacheler, 1997). Also, as pointed out above, North Carolina is in a unique situation in terms of pest pressure when compared with the rest of the cotton belt. Finally, pesticides for cotton insect management in North Carolina are chosen primarily on the basis of cost and effectiveness, rather than selectivity for beneficials. It is fortuitous that these pesticides are spatially and temporally selective toward natural enemies (i.e., at-planting systemic insecticides early and bollworm treatments applied late in the season).

1.6 NONTARGET IMPACTS OF PARASITOIDS AND PREDATORS

The past two decades have brought increasing concerns and discussion over the impact of invasive, exotic organisms on native flora and fauna throughout the world (U.S. Congress OTA, 1993). Parasitoids and predators used for biological control have been included in these concerns and discussions. Therefore, we present a brief summary of the current status of studies and debate regarding potential nontarget impact of parasitoids and predators.

The use of genetically altered organisms in pest management programs has always drawn intense debate regarding safety and unintended nontarget impacts (e.g., de Selincourt, 1994; Gould, 1994; Barrett et al., 1996; Bengtsson and Ort, 1997; Hoy et al., 1998). In contrast, biological control employing naturally occurring organisms has historically been considered an environmentally safe and effective means of managing insect pests (Doutt, 1972; DeBach, 1974; Caltagirone, 1981). Recently, however, the potential impact of naturally occurring parasitoids and predators on nontarget organisms has also come under scrutiny from a variety of sources (Howarth, 1983, 1991; Simberloff, 1992; Lockwood, 1993 a,b; Simberloff and Stiling, 1996, 1998; Lockwood, 1997).

These concerns have prompted a much-needed discussion regarding the potential for nontarget effects and approaches to addressing the issue (Carruthers and Onsager, 1993; Simberloff and Stiling, 1996, 1998; Duan and Messing, 1997; Van Driesche and Hoddle, 1997; Follett, 1999; Frank, 1998). Most of the discussion has focused on the risks posed by the importation and release of exotic natural enemies against exotic pests, i.e., classical biological control (see papers in Follett, 1999). Far less attention has been paid to potential for nontarget effects from augmentative releases of natural enemies (Orr et al., 1999), and conservation biological control has apparently not raised any issues related to nontarget impacts.

Simberloff and Stiling (1996) summarize the controversy associated with importation biological control and highlight potential risks such as predation or parasitism of nontarget species, competition with native species, community and ecosystem effects, and unexpected effects such as loss of species dependent on the target species of biological control efforts. Simberloff and Stiling (1996) argue that the few documented cases of nontarget impacts, compared with the number of natural enemy introductions, may be more the result of a lack of monitoring and documentation than a lack of actual impacts. These authors also suggest that current regulations and protocols are inadequate, and should do more to assess potential effects on noneconomic species and ecosystems, as well as compare the effects of target pests with the potential nontarget impacts of natural enemies prior to release.

Despite these concerns, the management of exotic pests with importation biological control is being advocated by some as a tool to assist conservation of natural areas (Frank and Thomas, 1994; U.S. Congress OTA, 1995; Van Driesche, 1994). The invasion of alien species appears likely to continue, and perhaps worsen, as a result of international trade and travel (Sailer, 1978; Frank and McCoy, 1992). For many of these alien species that develop into pest problems, importation biological control practiced in a scientifically sound manner may be the only economically viable, long-term solution.

Orr et al. (1999) discuss the potential for nontarget impacts of augmentative releases of arthropod natural enemies, especially *Trichogramma* species. In general, it appears that the measured potential for nontarget impacts from augmentative releases is strongly a function of how measurements are taken. As studies progress from simple no-choice tests in laboratories to more biologically realistic cage and field trials, the observed potential impact of parasitoids on nontarget organisms typically declines (e.g., Duan and Messing, 1997; Orr et al., 1999). This suggests that, at least in the cases studied, the actual nontarget impacts of augmentative releases in field conditions are probably negligible. Risks from augmented arthropods would primarily be of concern if released organisms were non-native. However, with native organisms there might also be potential for subtler impacts from "genetic pollution," i.e., one race, strain, biotype, ecotype, etc. being introduced into an area it previously did not occupy. The biological soundness of mixing and resulting hybridization of different populations as a result of movement and release has been questioned (e.g., Pinto et al., 1992).

The controversy over potential nontarget impacts of biological control is far from resolved. It has, however, prompted some biological control researchers to collect more data relevant to potential nontarget impacts as part of their work plans. Sufficient data collection of this type should allow resolution of these conflicts, at least on a case-by-case basis, and ensure that parasitoids and predators can continue to contribute to insect pest management in as safe a manner as possible.

1.7 CONCLUDING REMARKS

The use of parasitoids and predators in pest management systems has had a long, rich history. While there are a variety of impediments, there also exist many opportunities for a continuing and expanding role for parasitoids and predators in insect pest management. The continual influx of arthropod species from increased international trade results each year in new pests of agriculture and forestry (Sailer, 1978; Frank and McCoy, 1992) as well as major threats to nature conservation (U.S. Congress OTA, 1993). Changes in pest management tactics are resulting from environmental and human safety concerns, development of insecticide resistance, and increases in pesticide cost and availability. Public concerns over pesticide use have resulted in government action such as a mandated 50% cut in European countries' pesticide use (Matteson, 1995), the EPA, USDA and FDA initiative to implement IPM in the U.S. (U.S. Congress OTA, 1995), and FIFRA and FQPA requirements and tolerances for pesticides in the U.S. (EPA, 1997; Klassen, 1998).

Pesticides will remain a major component of IPM programs into the foreseeable future. However, the concerns outlined above dictate movement from pesticide-based pest management systems to more truly integrated insect pest management approaches, creating opportunities for increased inclusion of biologically based pest management tools such as parasitoids and predators.

REFERENCES

Ables, J. R., S. L. Jones, R. K. Morrison, V. S. House, D. L. Bull, L. F. Bouse, and J. B. Carlton. 1979. New developments in the use of *Trichogramma* to control lepidopteran pests of cotton, pp. 125–127. *In*: Proceedings, Beltwide Cotton Production Research Conference. National Cotton Council of America. Memphis, TN.

Aldrich, J. R. 1995. Testing the "new associations" biological control concept with a tachinid parasitoid (*Euclytia flava*). J. Chem. Ecol. 21: 1031–1042.

Anonymous. 1988. The maize pyralid. Phytoma 403: 54.

Anonymous. 1998. 1998 Directory of Least-Toxic Pest Control Products. The IPM Practitioner 19 (11/12). 52 pp.

Armitage, H. 1919. Controlling mealybugs by the use of their natural enemies. Calif. Stat. Hort. Comm. Monthly Bulls. 8: 257–260.

Armitage, H. M. 1929. Timing field liberations of *Cryptolaemus* in the control of the citrophilus mealybug in the infested citrus orchards of southern California. J. Econ. Entomol. 22: 910–915.

Bacheler, J. S. 1997. Insect management on cotton, pp. 137–157. *In*: Cotton information. Publ. No. AG-417. N.C. Coop. Ext. Serv., North Carolina State Univ., Raleigh.

Barbosa, P. [ed.]. 1998. Conservation biological control. Academic Press, San Diego. 396 pp.

Barbosa, P., S. M. Braxton, and A. Segarra-Carmona. 1994. A history of biological control in Maryland. Biological Control 4: 185–243.

Barbosa, P., S. M. Braxton, and A. Segarra-Carmona. 1997. Guidelines for evaluating the status and prospects of biological control on a statewide basis. Agricultural Experiment Station, Univ. of Maryland, College Park. 139 pp.

Barrett, J. A., R. Dardozzi, and C. Ramel. 1996. Biotechnology and the production of resistant crops. Science and the Total Environment 188: Supplement 1: 106–111.

Beglyarov, G. A. and A. I. Smetnik. 1977. Seasonal colonization of entomophages in the USSR., pp. 283–328. *In*: R. L. Ridgway and S. B. Vinson [eds.], Biological control by augmentation of natural enemies: insect and mite control with parasites and predators. Plenum Press, New York.

Beirne, B. P. 1975. Biological control attempts by introductions against pest insects in the field in Canada. Can. Entomol. 107: 225–236.

Beirne, B. P. 1985. Avoidable obstacles to colonization in classical biological control of insects. Can J. Zool. 63: 743–747.

Bengtsson, B. O. and D. R. Ort. 1997. Pros and cons of foreign genes in crops. Nature, London 385: 6614, 290.

Bigler, F. 1991. [ed.]. Quality control of mass reared arthropods. Proceedings 5th workshop of the IOBC working group "quality control of mass reared arthropods," March 25–28, 1991, Wageningen, The Netherlands.

Bigler, F. 1994. Quality control in *Trichogramma* production, pp. 93–111. *In*: E. Wajnberg and S. A. Hassan [eds.], Biological control with egg parasitoids, CAB International, Wallingford, Oxon, U.K.

Bigler, F. 1997. Use and registration of macroorganisms for biological control protection. Bull. OEPP 27: 95–102.

Biliotti, E. 1977. Augmentation of natural enemies in western Europe, pp. 341–348. *In*: R. L. Ridgway and S. B. Vinson [eds.], Biological control by augmentation of natural enemies. Plenum Press, New York.

Blumel, S. and R. Womastek. 1997. Authorization requirements for organisms as plant protection products in Austria. Bull. OEPP 27: 127–131.

Bodenheimer, F. A. 1931. Zur fruhgeschichte der enforschung des insektenparasitismus (to the early history of the study of insect parasitism). Arch. Gesch. Math. Naturwis. Technol. 13: 402–416.

Boller, E. F. and D. L. Chambers. 1977. Quality aspects of mass-reared insects, pp. 219–235. *In*: R. L. Ridgway and S. B. Vinson [eds.], Biological control by augmentation of natural enemies. Insect and mite control by parasites and predators. Plenum Press, New York.

Botta, P. E. 1841. Relation d'un voyage dans l'Yemen. Duprat, Paris.

Briggs, C. J. 1993. Competition among parasitoid species on a stage-structured host and its effect on host suppression. Am. Nat. 141: 372–397.

Caltagirone, L. E. 1981. Landmark examples in classical biological control. Annu. Rev. Entomol. 26: 213–232.

Carruthers, R. I. and J. A. Onsager. 1993. Perspective on the use of exotic natural enemies for biological control of pest grasshoppers (Orthoptera: Acrididae). Environ. Entomol. 22: 885–903.

Clausen, C. P. 1936. Insect parasitism and biological control. Ann. Entomol. Soc. Am. 29: 201–223.

Clausen, C. P. [ed.]. 1978. Introduced parasites and predators of arthropod pests and weeds: a world review. Agric. Handbook No. 480. USDA, Washington, D.C.

Coop, L., R. Rosetta, and B. Croft. 1997. Release calculator and guidelines for using *Neoseiulus fallacis* to control two-spotted spider mites in strawberry. *http://www.orst.edu/Dept/entomology/ipm/mcalc.html*. Oregon State University, Corvallis.

Couillien, D. and J. C. Gregoire. 1994. Take-off capacity as a criterion for quality control in mass-produced predators, *Rhizophagus grandis* (Col.: Rhizophagidae) for the biocontrol of bark beetles, *Dendroctonus micans* (Col.: Scolytidae). Entomophaga 39: 385–394.

Coulson, J. R., W. Klaasen, R. J. Cook, E. G. King, H. C. Chiang, K. S. Hagen, and W. G. Yendol. 1982. Notes on biological control of pests in China, 1979. *In*: Biological control of pests in China. U.S. Dept. Agric., Washington, D.C.

Cranshaw, W., D. C. Sclar, and D. Cooper. 1996. A review of 1994 pricing and marketing by suppliers of organisms for biological control of arthropods in the United States. Biological Control 6: 291–296.

Cudjoe, A. R., P. Neuenschwander, and M. J. W. Copland. 1992. Experimental determination of the efficiency of indigenous and exotic natural enemies of the cassava mealybug, *Phenacoccus manihoti* Mat.-Ferr. (Hom., Pseudococcidae), in Ghana. J. Appl. Ent. 114: 77–82.

DeBach, P. [ed.]. 1964. Biological control of insect pests and weeds. Chapman and Hall, London. 844 pp.

DeBach, P. 1974. Biological control by natural enemies. Cambridge Univ. Press, London. 323 pp.

DeBach, P. and K. S. Hagen. 1964. Manipulation of entomophagous species, pp. 429–458. *In*: P. DeBach [ed.], Biological control of insect pests and weeds. Chapman and Hall, London.

Decaux, F. 1899. Destruction rationnelle des ins ectes qui attaquent les arbres fruitiers par l'emploi simultane des insecticides, des insectes auxiliares, et par la protection et l'elevage de leurs ennemis naturel les parasites. J. Soc. Hort. Fr. 22: 158–184.

de Selincourt, K. 1994. Genetic engineers target: Third World crops. Genetic Engin. Biotechnol. 14: 155–176.

Doodeman, C. J. A. M., J. C. van Lenteren, I. Sebestyen, and Z. Ilovai. 1996. Short-range flight test for quality control of *Encarsia formosa*. Proceedings of the Section Experimental and Applied Entomology of the Netherlands Entomol. Soc. 7: 153–158.

Doutt, R. L. 1958. Vice, virtue and the Vidalia. Bull. Entomol. Soc. Am. 4: 119–123.

Doutt, R. L. 1964. The historical development of biological control, pp. 21–42. *In*: P. DeBach [ed.], Biological control of insect pests and weeds. Chapman and Hall, London.

Doutt, R. L. 1972. Biological control: parasites and predators. *In*: Pest Control Strategies for the Future (National Academy of Sciences). National Academy of Sciences Printing and Publishing Office, Washington, D.C.

Duan, J. J. and R. H. Messing. 1997. Biological control of fruit flies in Hawaii: factors affecting nontarget risk analysis. Agric. and Human Values 14: 227–236.

Dutton, A. and F. Bigler. 1995. Flight activity assessment of the egg parasitoid *Trichogramma brassicae* (Hym.: Trichogrammatidae) in laboratory and field conditions. Entomophaga 40: 223–233.

Edelson, J. V. 1989. Control of secondary pests, 1988. Insecticide and Acaricide Tests 14: 230.

Ehler, L. E. 1990. Some contemporary issues in biological control of insects and their relevance to the use of entomophagic nematodes, pp. 1–19. *In*: R. Gaugler and H. K. Kaya [eds.], Entomopathogenic nemetodes in biological control. CRC Press, Boca Raton, FL.

Ehler, L. E. 1998. Conservation biological control: past, present, and future, pp. 1–8. *In*: P. Barbosa [ed.], Conservation biological control. Academic Press, New York.

Ehler, L. E. and R. W. Hall. 1982. Evidence for competitive exclusion of introduced natural enemies in biological control. Environ. Entomol. 11: 1–4.

EPA (U.S. Environmental Protection Agency). 1997. Raw and processed food schedule for pesticide tolerance Reassessment. Federal Register (1997) 62 (149, 4 Aug.) 42020–42030. Office of the Federal Register, Washington, D.C.

Ervin, R. T., L. J. Moffitt, and D. E. Meyerdirk. 1983. Comstock mealybug (Homoptera: Pseudococcidae): cost analysis of a biological control program in California. J. Econ. Entomol. 76: 605–609.

Fernandez, C. and W. Nentwig. 1997. Quality control of the parasitoid *Aphidius colemani* (Hym., Aphidiidae) used for biological control in greenhouses. J. Appl. Entomol. 121: 447–456.

Ferro, D. N. and J. N. McNeil. 1998. Habitat enhancement and conservation of natural enemies of insects, pp. 123–132. *In*: P. Barbosa [ed.], Conservation biological control. Academic Press, San Diego.

Fitch, A. 1856. Sixth, seventh, eighth, and ninth reports on the noxious, beneficial and other insects of the state of New York. Albany, NY. 259 pp.

Fleschner, C. A. 1960. Parasites and predators for pest control, pp. 183–199. *In*: Biological and chemical control of plant and insect pests. Am. Assoc. Advance. Science, Washington, D.C.

Flint, M. L., S. H. Dreistadt, J. Rentner, and M. P. Parrella. 1995. Lady beetle release controls aphids on potted plants. Calif. Agric. 49: 5–8.

Follett, P. [ed.]. 1999. Non-Target Effects.

Force, D. C. 1974. Ecology of insect host-parasitoid communities. Science 184: 624–632.

Forskal, P. 1775. Descriptiones animalium, avium, amphibiorum, piscium, insectorum, vermium; quae in itinere orientali observavit P. Forskal, post mortem auctoris edidit, Carsten Niebuhr. Hauniae, Moeller (pt. 3).

Frandon, J. and F. Kabiri. 1998. Biological control against the European corn borer with *Trichogramma* — technical improvements for large scale application, pp. 291–298. *In*: Proceedings, 1er Colloque Transnational sur les luttes biologique, integree et raisonnee, Jan. 21–23, Lille, France. SRPV NORD, Pas de Calais, France.

Frandon, J., F. Kabiri, J. Pizzol, and J. Daumal. 1991. Mass-rearing of *Trichogramma brassicae* used against the European corn borer *Ostrinia nubilalis*, pp. 146–151. *In*: F. Bigler [ed.], Proceedings 5th workshop of the IOBC working group "quality control of mass reared arthropods," March 25–28, 1991, Wageningen, The Netherlands.

Frank, J. H. 1998. How risky is biological control? Comment. Ecology 79: 1829–1834.

Frank, J. H. and E. D. McCoy. 1992. The immigration of insects to Florida, with a tabulation of records published since 1970. Florida Entomol. 75: 1–28.

Frank, J. H. and M. C. Thomas. 1994. *Metamasius callizona* (Chevrolat) (Coleoptera: Curculionidae), an immigrant pest, destroys bromeliads in Florida. Florida Entomol. 78: 619–623.

Frazer, B. D. 1988. Predators, pp. 217–230. *In*: A. K. Minks and P. Harrweijn [eds.], World crop pests. Aphids: their biology, natural enemies and control. Vol. 2B. Elsevier, New York.

Gould, F. 1994. Potential and problems with high-dose strategies for pesticidal engineered crops. Biocont. Sci. Technol. 4: 451–461.

Grafton-Cardwell, E. E. 1991. Geographical and temporal variation in response to insecticides in various life stages of *Aphis gossypii* (Homoptera: Aphididae) infesting cotton in California. J. Econ. Entomol. 84: 741–749.

Greathead, D. J. 1986. Parasitoids in classical biological control, pp. 289–318. *In*: J. Waage and D. Greathead [eds.], Insect parasitoids. Academic Press, New York.

Greathead, D. J. and A. H. Greathead. 1992. Biological control of insect pests by parasitoids and predators: the BIOCAT database. Biocontrol News and Information 13: 61–68.

Gross, P. 1991. Influence of pest feeding niche on success rates in classical biological control. Environ. Entomol. 20: 1217–1227.

Gutierrez, A. P., B. Wermelinger, F. Schulthess, J. U. Baumgärtner, J. S. Yaninek, H. R. Herren, P. Neuenschwander, B. Lohr, W. N. O. Hammond, and C. K. Ellis. 1987. An overview of a system model of cassava and cassava pests in Africa. Insect Sci. Applic. 8: 919–924.

Habeck, D. H., S. B. Lovejoy, and J. G. Lee. 1993. When does investing in biological control research make economic sense? Florida Entomol. 76: 96–101.

Hahn, S. K. and R. J. Williams. 1973. Enquete sur le manioc en Republique du Zaire, 12–20 March 1973. Report to the Minister of Agriculture of the Republique of Zaire, 12 pp.

Hall, R. W. and L. E. Ehler. 1979. Rate of establishment of natural enemies in classical biological control. Bull. Entomol. Soc. Am. 25: 280–282.

Hall, R. W., L. E. Ehler, and B. Bisabri-Ershadi. 1980. Rate of success in classical biological control of arthropods. Bull. Entomol. Soc. Am. 26: 111–114.

Hammond, W. N. O., P. Neuenschwander, and H. R. Herren. 1987. Impact of the exotic parasitoid *Epidinocarsis lopezi* on Cassava mealybug (*Phenacoccus manihoti*) populations. Insect Sci. Appl. 8: 887–891.

Hanson, P. 1993. The importance of taxonomy in biological control. Manejo-Integrado-de-Plagas 29: 48–50.

Hardee, D. D. and P. J. O'Brien. 1990. Cotton aphids: current status and future trends in management, pp. 169–171. *In*: Proceedings, Beltwide Cotton Production Research Conference. National Cotton Council of America. Memphis, TN.

Harris, P. 1984. Current approaches to biological control of weeds, pp. 95–103. *In*: J. S. Kelleher and M. A. Hulme [eds.], Biological control programmes against insects and weeds in Canada. CAB. Slough, U.K.

Hassell, M. P. 1978. The dynamics of arthropod predator-prey systems. Princeton University Press, Princeton, NJ.

Headley, J. C. 1985. Cost benefit analysis: defining research needs, pp. 53–63. In: M. A. Hoy and D. C. Herzog [eds.], Biological control in agricultural IPM systems. Academic Press, New York.

Herren, H. R. 1981. Biological control of the cassava mealybug, pp. 79–80. In: E. R. Terry, K. A. Oduro, and F. Caveness [eds.], Tropical root crops, research strategies for the 1980s. Proceedings, First Triennial Root Crop Symposium, 8–12 September, 1980, Ibadan, Nigeria. Intl. Dev. Res. Ctr., Ottawa, Canada.

Herren, H. R. 1982. Recent advances in the biological control of the cassava mealybug: Homopt. Pseudococcidae. IITA Annual Report. Ibadan, Nigeria.

Herren, H. R. 1987. Africa-wide biological control project of cassava pests. A review of objectives and achievements. Insect Sci. Applic. 8: 837–840.

Herren, H. R., P. Neuenschwander, R. D. Hennessey, and W. N. O. Hammond. 1987. Introduction and dispersal of Epidinocarsis lopezi (Hym., Encyrtidae), an exotic parasitoid of the cassava mealybug, Phenacoccus manihoti (Hom., Pseudococcidae), in Africa. Agric. Ecosystems Environ. 19: 131–144.

Hoffman, M. P., L. T. Wilson, F. G. Zalom, and R. J. Hilton. 1991. Dynamic sequential sampling plan for Helicoverpa zea (Lepidoptera: Noctuidae) eggs in processing tomatoes: parasitism and temporal patterns. Environ. Entomol. 20: 1005–1012.

Hokkanen, H. M. T. 1985. Success in classical biological control. CRC Crit. Rev. Plant Sci. 3: 35–72.

Hokkanen, H. M. T. and D. Pimentel. 1984. New approach for selecting biological control agents. Can. Entomol. 116: 1109–1121.

Hokkanen, H. M. T. and D. Pimentel. 1989. New associations in biological control: theory and practice. Can. Entomol. 121: 829–840.

Hopper, K. R. 1996. Making biological control introductions more effective, pp. 59–76. In: Proceedings, Biological control introductions–opportunities for improved crop production. British Crop Protection Council, Farnham, U.K.

Hopper, K. R., R. T. Roush, and W. Powell. 1993. Management of genetics of biological control introductions. Annu. Rev. Entomol. 38: 27–51.

Hörstadius, S. 1974. Linnaeus, animals and man. Biol. J. Linn. Soc. 6: 269–275.

Howarth, F. G. 1983. Classical biocontrol: panacea or Pandora's Box? Proc. Hawaii Entomol. Soc. 24: 239–244.

Howarth, F. G. 1991. Environmental impacts of classical biological control. Annu. Rev. Entomol. 36: 485–509.

Hoy, C. W., J. Feldman, F. Gould, G. G. Kennedy, G. Reed, and J. A. Wyman. 1998. Naturally occurring biological controls in genetically engineered crops, pp. 185–205. In: P. Barbosa [ed.], Conservation biological control. Academic Press, New York.

Hoy, M. A. 1994. Insect molecular genetics: an introduction to principles and applications. Academic Press, San Diego. 546 pp.

Hoy, M. A. and D. C. Herzog [eds.]. 1985. Biological control in agricultural IPM systems. Academic Press, New York. 589 pp.

Hoy, M. A., R. M. Nowierski, M. W. Johnson, and J. L. Flexner. 1991. Issues and ethics in commercial releases of arthropod natural enemies. Am. Entomol. 37: 74–75.

Huffaker, C. B., R. L. Rabb, and J. A. Logan. 1977. Some aspects of population dynamics relative to augmentation of natural enemy action, pp. 3–38. In: R. L. Ridgway and S. B. Vinson [eds.], Biological control by augmentation of natural enemies. Plenum Press, New York.

Hull, L. A. and E. H. Beers. 1985. Ecological selectivity: modifying chemical control practices to preserve natural enemies, pp. 103–122. *In*: M. A. Hoy and D. C. Herzog [eds.], Biological control in agricultural IPM systems. Academic Press, Orlando, FL.

Hunter, C.D. 1992. Suppliers of beneficial organisms in North America. Publ. PM 92-1. California Environ. Protec. Agency, Dept. Pesticide Regul., Sacramento, CA.

Hunter, C.D. 1997. Suppliers of Beneficial Organisms in North America. Publ. PM 97-01. California Environ. Protec. Agency, Dept. Pesticide Regul., Sacramento, CA.

Jetter, K., K. Klonsky, and C. H. Pickett. 1997. A cost/benefit analysis of the ash whitefly biological control program in California. J. Arboriculture 23: 65–72.

Jones, S. L., R. K. Morrison, J. R. Ables, and D. L. Bull. 1977. A new and improved technique for the field release of *Trichogramma pretiosum*. Southwest. Entomol. 2: 210–215.

Kabiri, F., J. Frandon, J. Voegele, N. Hawlitzky, and M. Stengel. 1990 [Evolution of a strategy for inundative releases of *Trichogramma* brassicae Bezd. (Hym *Trichogrammatidae*) against the European corn borer, *Ostrinia nubilalis* Hbn. (Lep. Pyralidae)]. In proceedings, ANPP — Second International Conference on Agricultural Pests, Versailles, 4–6 Dec. 1990. (French, translated by P. Matteson).

Kerns, D. L. and M. J. Gaylor. 1991. Induction of cotton aphid outbreaks by the insecticide sulprofos, pp. 699–701. *In*: Proceedings Beltwide Cotton Conference. National Cotton Council of America, Memphis, TN.

Kerns, D. L. and M. J. Gaylor. 1993. Induction of cotton aphid outbreaks by insecticides in cotton. Crop Prot. 12: 387–393.

Kidd, K.A., C.A. Nalepa, and K.R. Ahlstrom. 1994. Parasitoids of cotton aphids, p. 49. *In*: C.A. Nalepa and K.A. Kidd [eds.], 1994 report of activities, biological control laboratory, North Carolina Department of Agriculture, Raleigh, NC.

King, E. G. and N. C. Leppla. [eds.]. 1984. Advances and challenges in insect rearing. USDA-ARS, New Orleans, LA.

King, E. G., K. R. Hopper, and J. E. Powell. 1985. Analysis of systems for biological control of crop pests in the U.S. by augmentation of predators and parasites, pp. 201–227. *In*: M. A. Hoy and D. C. Herzog [eds.], Biological control in agricultural IPM systems. Academic Press, New York.

Kirby, W. and W. Spence. 1815. An introduction to entomology. Longman, Brown, Green and Longmans, London. 607 pp.

Klassen, P. 1998. An "alternative" view. Farm Chemicals, Sept., 18–23.

Knutson, A. 1998. The *Trichogramma* manual: a guide to the use of *Trichogramma* for biological control with species reference to augmentative releases for control of bollworm and budworm in cotton. Publ. No. B–6071. Texas Agric. Ext. Serv., Texas A&M Univ. Res. and Ext. Center, Dallas, TX.

Knutson, A. and J. Ruberson. 1996. Field guide to predators, parasites and pathogens attacking insect and mite pests of cotton. Publ. No. B-6046. Texas Agric. Ext. Serv., Texas A&M Univ., College Station, TX.

Kollar, V. 1837. *In*: London's Gardener's Magazine, 1840.

Landis, D. A. and F. D. Menalled. 1998. Ecological considerations in the conservation of effective parasitoid communities in agricultural systems, pp. 101–121. *In*: P. Barbosa, [ed.], Conservation biological control. Academic Press, San Diego, CA.

Lee, S. Y. 1992. Physiological, biochemical and behavioral aspects of insecticide resistance in North Carolina populations of the cotton aphid, *Aphis gossypii* Glover. MSc. Thesis, North Carolina State Univ., Raleigh. 88 pp.

Lema K. M. and H. R. Herren. 1982. Biological control of cassava mealybug and cassava green mite: front-line release strategy, pp. 68–69. *In*: Root crops in eastern Africa: Proceedings of a workshop held in Kigali, Rwanda, 23–27 Nov., 1980. Intl. Dev. Res. Ctr., Ottawa, Canada.

Lema, K. M. and H. R. Herren. 1985. Release and establishment in Nigeria of *Epidinocarsis lopezi* a parasitoid of the cassava mealybug, *Phenacoccus manihoti*. Entomol. Exp. Appl. 38: 171–175.

Leppla, N. C. and E. G. King. 1996. The role of parasitoid and predator production in technology transfer of field crop biological control. Entomophaga: 41: 343–360.

Letourneau, D. K. 1998. Conservation biology: lessons for conserving natural enemies, pp. 9–38. *In*: P. Barbosa [ed.], Conservation biological control. Academic Press, San Diego.

Leuschner, K. 1982. Pest control for cassava and sweet potato, pp. 60–64. *In*: Root crops in eastern Africa: Proceedings of a workshop held in Kigali, Rwanda, 23–27 Nov. 1980. Intl. Dev. Res. Ctr., Ottawa, Canada.

Li, Li-Yang. 1994. Worldwide use of *Trichogramma* for biological control on different crops: a survey, pp. 37–54. *In*: E. Wajnberg, and S. A. Hassan [eds.], Biological control with egg parasitoids. CAB International, Wallingford, U.K.

Lockwood, J. A. 1993a. Environmental issues involved in biological control of rangeland grasshoppers (Orthoptera: Acrididae) with exotic agents. Environ. Entomol. 22: 503–518.

Lockwood, J. A. 1993b. Benefits and costs of controlling rangeland grasshoppers (Orthoptera: Acrididae) with exotic organisms: search for a null hypothesis and regulatory compromise. Environ. Entomol. 22: 904–914.

Lockwood, J. A. 1997. Competing values and moral imperatives: an overview of ethical issues in biological control. Agric. and Human Values 14: 205–210.

Lohr, B., A. M. Varela, and B. Santos. 1990. Exploration for natural enemies of the cassava mealybug, *Phenacoccus manihoti* (Homoptera: Pseudococcidae), in South America for the biological control of this introduced pest in Africa. Bull. Entomol. Res. 80: 417–426.

Losey, J. E. and D. D. Calvin. 1995. Quality assessment of four commercially available species of *Trichogramma* (Hymenoptera: Trichogrammatidae). J. Econ. Entomol. 88: 1243–1250.

Luck, R. F. B. 1981. Parasitic insects introduced as biological control agents for arthropod pests, pp. 125–284. *In*: D. Pimentel [ed.], CRC Handbook of Pest Management in Agriculture. CRC Press, Boca Raton, FL.

Luck, R. F. B., M. Shepard, and P. E. Penmore. 1988. Experimental methods for evaluating arthropod natural enemies. Annu. Rev. Entomol. 33: 367–391.

Lux, S. and F. Bigler. 1991. Diagnosis of behaviour as a tool for quality control of mass reared arthropods, pp. 66–92. *In*: F. Bigler [ed.], Quality control of mass reared arthropods. Proceedings 5th Workshop of the IOBC Global Working, March 1991. Wageningen, Netherlands.

Marsden, J. S., G. E. Martin, D. J. Parham, T. J. Risdell Smith, and B. G. Johnston. 1980. Returns on Australian Agricultural Research. CSIRO, Canberra, Australia.

Martin, N. A. and C. H. Wearing. 1990. Natural enemies for inundative and seasonal inoculative release: policy issues, pp. 209–211. *In*: Proceedings, 43rd New Zealand weed and pest control conference. New Zealand Weed and Pest Control Society. Palmerston North. New Zealand.

Matile-Ferrero, D. 1978. Cassava mealybug in the People's Republic of Congo. *In*: K. F. Nwanze and K. Leuschner [eds.], pp. 29–46, Proceedings Inter. Workshop on the Cassava Mealybug *Phenacoccus manihoti* Mat.-Ferr. (Pseudococcidae), 26–29 June 1977. M'vuazi, Zaire. Intl. Instit. Trop. Agric., Ibadan, Nigeria.

Matteson, P. C. 1995. The "50% pesticide cuts" in Europe: a glimpse of our future? Am. Entomol. 41: 210–220.

McCook, H. 1882. Ants as beneficial insecticides. Proc. Acad. Natl. Sci. Philadelphia, pp. 263–271.

Messenger, P. S., E. Biliotti, and R. van den Bosch. 1976. The importance of natural enemies in integrated control, pp. 543–563. In: C. B. Huffaker and P. S. Messenger [eds.], Theory and practice of biological control. Academic Press, San Fransisco.

Moffat, A. S. 1991. Research on biological pest control moves ahead. Science 252: 211–212.

Murdoch, W. W., J. D. Reeve, C. B. Huffaker, and C. E. Kennett. 1984. Biological control of olive scale and its relevance to ecological theory. Am. Nat. 123: 371–392.

Myers, J. H., C. Higgins, and E. Kovacs. 1989. How many insect species are necessary for the biological control of insects? Environ. Entomol. 18: 541–547.

Nechols, J. R. and W. C. Kauffman. 1992. Introduction and overview, pp. 1–5. In: W. C. Kauffman and J. E. Nechols [eds.], Selection criteria and ecological consequences of importing natural enemies. Entomol. Soc. Am., Lanham, MD.

Nechols, J. R., W. C. Kauffman, and P. W. Schaefer. 1992. Significance of host specificity in classical biological control, pp. 41–52. In: W. C. Kauffman and J. E. Nechols [eds.], Selection criteria and ecological consequences of importing natural enemies. Entomol. Soc. Am., Lanham, MD.

Nedstram, B. 1994. Registration of beneficials for biological control in Sweden. SP–Rapport 7: 321–323.

Neuenschwander, P. and W. N. O. Hammond. 1988. Natural enemy activity following the introduction of Epidinocarsis lopezi (Hymenoptera: Encyrtidae) against the cassava mealybug Phenacoccus manihoti (Homoptera: Pseudococcidae), in southwestern Nigeria. Environ. Entomol. 17: 894–902.

Neuenschwander, P., W. N. O. Hammond, O. Ajuonu, A. Gado, N. Echendu, A. H. Bokonon-Ganta, R. Allomasso, and I. Okon. 1990. Biological control of the cassava mealybug, Phenacoccus manihoti (Hom., Pseudococcidae) by Epidinocarsis lopezi (Hym., Encyrtidae) in West Africa, as influenced by climate and soil. Agriculture, Ecosystems and Environment 32: 39–55.

Neuenschwander, P., F. Schulthess, and E. Madojemu. 1986. Experimental evaluation of the efficiency of Epidinocarsis lopezi, a parasitoid introduced into Africa against the cassava mealybug, Phenacoccus manihoti. Entomol. Exp. Appl. 42: 133–138.

Newsom, L. D. and J. R. Brazzel. 1968. Pests and their control, pp. 367–405. In: F. C. Elliot, M. Hoover, and W. K. Porter [eds.], Advances in production and utilization of quality cotton: principles and practices. Iowa State Univ., Ames.

Nicoli, G., M. Benuzzi, and N. C. Leppla [eds.]. 1994. Quality control of mass reared arthropods. Proceedings 7th Workshop of the IOBC Global Working Group. September 13–16, 1993. Rimini, Italy. IOBC/OILB. 238 pp.

Nordlund, D. A. 1996. Biological control, integrated pest management and conceptual models. Biocontrol News and Information 17: 35–44.

Nordlund, D. A. and S. M. Greenberg. 1994. Facilities and automation for the mass production of arthropod predators and parasitoids. Biocont. News and Inform. 4: 45–50.

Nordlund, D. A., Z. X. Wu, A. C. Cohen, and S. M. Greenberg. 1998. Recent advances in the in vitro rearing of Trichogramma spp. Proceedings 5th Intl. Symposium, Trichogramma and other egg parasitoids.

Norgaard, R. B. 1988. The biological control of cassava mealybug in Africa. Am. Agric. Econ. Assoc. (May): 366–371.

Nwanze, K. F. 1982. Relationships between cassava root yields and infestations by the mealybug, Phenacoccus manihoti. Trop. Pest Management 28: 27–32.

O'Neil, R. J., K. L. Giles, J. J. Obrycki, D. L. Mahr, J. C. Legaspi, and K. Katovich. 1998. Evaluation of the quality of four commerically available natural enemies. Biological Control 11: 1–8.

Orr, D. and J. Baker. 1997a. Biological Control: Purchasing Natural Enemies. Pub. No. AG-570-1. NC Coop. Ext. Serv., North Carolina State Univ., Raleigh.

Orr, D. and J. Baker. 1997b. Biological Control: Application of Natural Enemies. Pub. No. AG-570-2. NC Coop. Ext. Serv., North Carolina State Univ., Raleigh.

Orr, D. B., C. Garcia-Salazar, and D. A. Landis. 1999. *Trichogramma* nontarget impacts: a method for biological control risk assessment. *In*: Follett, P. [ed.], Non-target effects.

Ostlie, K. R. and L. P. Pedigo. 1987. Incorporating pest survivorship into economic thresholds. Bull. Entomol. Soc. Am. 33: 98–102.

Parrella, M. P., K. M. Heinz, and L. Nunney. 1992. Biological control through augmentative release of natural enemies: a strategy whose time has come. Am. Entomol. 38: 172–179.

Pedigo, L. P. 1989. Entomology and pest management. Macmillan, New York. 646 pp.

Pinto, J. D., D. J. Kazmer, G. R. Platner, and C. A. Sassaman. 1992. Taxonomy of the *Trichogramma minutum* complex (Hymenoptera: Trichogrammatidae): allozymic variation and its relationship to reproductive and geographic data. Ann. Entomol. Soc. Am. 85: 413–422.

Poehling, H. M. 1989. Selective application strategies for insecticides in agricultural crops, pp. 151–175. *In*: P. C. Jepson [ed.], Pesticides and nontarget invertebrates. Intercept, Wimborne, U.K.

Polis, G. A. and R. D. Holt. 1992. Intraguild predation: the dynamics of complex trophic interactions. Trends Ecol. Evol. 7: 151–154.

Purcell, M. F., K. M. Daniels, L. C. Whitehand, and R. H. Messing. 1994. Improvement of quality control methods for augmentative releases of the fruit fly parasitoids, *Diachasmimorpha longicaudata* and *Psyttalia fletcheri* (Hymenoptera: Braconidae). Biocontrol, Science and Technology 4: 155–166.

Rajotte, E. G., R. F. Kazmierczak, Jr., G. W. Norton, M. T. Lambur, and W. A. Allen. 1987. The national evaluation of extension's integrated pest management (IPM) programs. VCES Publication 491–010. Virginia Cooperative Extension Service, Virginia State University, Petersburg.

Ratzeburg, J. T. C. 1844. Die ichneumonen der forstinsekten in forslicher and entomologischer beziehung; ein anhang zur abbildung und beschreibung der forstinsekten. Nicolaischen Burchhandlung, Berlin. 3 vols.

Reichelderfer, K. H. 1981. Economic feasibility of biological control of crop pests, pp. 403–417. *In*: G. C. Papavisas [ed.], Biological control in crop protection. Allanheld, Osman, Totowa, NJ.

Ridgway, R. L. and S. B. Vinson. 1977. Commercial sources of natural enemies in the U.S. and Canada (Appendix), pp. 451–453. *In*: R. L. Ridgway and S. B. Vinson [eds.], Biological control by augmentation of natural enemies. Plenum Press.

Ridgway, R. L., E. G. King, and J. L. Carrillo. 1977. Augmentation of natural enemies for control of plant pests in the Western Hemisphere, pp. 379–416. *In*: R. L. Ridgway and S. B. Vinson [eds.], Biological control by augmentation of natural enemies. Plenum Press, New York.

Riley, C. V. 1885. Fourth report of the U.S. Entomological Commission, p. 323. *In*: S. H. Scudder (1889). Butterflies of eastern United States and Canada. Cambridge Univ. Press, U.K.

Riley, C. V. 1893. Parasitic and predaceous insects in applied entomology. Insect Life 6: 130–141.

Riley, W. A. 1931. Erasmus Darwin and the biological control of insects. Science 73: 475–476.

Rosenheim, J. A., L. R. Wilhoit, and C. A. Armer. 1993. Influence of intraguild predation among generalist predators on the suppression of an herbivore population. Oecologia 96: 439–449.

Ruberson, J. R., H. Nemoto, and Y. Hirose. 1998. Pesticides and conservation of natural enemies in pest management, pp. 207–220. In: P. Barbosa [ed.], Conservation biological control. Academic Press, San Diego.

Sailer, R. I. 1978. Our immigrant insect fauna. Bull. Entomol. Soc. Am. 24: 3–11.

Santiago-Blay, J. A., F. Agudelo-Silva, and C. Orrego. 1994. DNA "fingerprints" for Trichogramma. IPM Practitioner 16: 7, 13.

Saunders, W. 1882. Address of the President of the Entomological Society of Ontario. Can. Ent. 14: 147–150.

Schaab, R. P., P. Neuenschwander, J. Zeddies, and H. R. Herren. 1996. Economics and ecology of biological control of the cassava mealybug Phenacoccus manihoti (Mat.-Ferr.) (Hom., Pseudococcidae) in Africa. Paper presented at the International Symposium on Food Security and Innovations: Successes and Lessons Learned, Univ. of Hohenheim, 11–13 Mar. 1996. Stuttgart, Germany.

Schmidt, V. B. 1998. Assessing the impact of beneficial insect populations on organic farms. M.Sc. Thesis, North Carolina State Univ., Raleigh. 264 pp.

Silvestri, F. 1909. Consideration of the existing condition of agricultural entomology in the United States of North America, and suggestions which can be gained from it for the benefit of Italian agriculture. Bull. Soc. Ital. Agric. 14: 305–367. (Reprinted in Hawaiian Forester and Agric. 6: 287–336.)

Simberloff, D. 1992. Conservation of pristine habitats and unintended effects of biological control, pp. 103–117. In: W. C. Kauffman and J. E. Nechols [eds.], Thomas Say Proceedings: Selection criteria and ecological consequences of importing natural enemies. Entomol. Soc. Am., Lanham, MD.

Simberloff, D. and P. Stiling. 1996. How risky is biological control? Comment. Ecology 77: 1965–1974.

Simberloff, D. and P. Stiling. 1998. How risky is biological control? Reply. Ecology 79: 1834–1836.

Singh, T. P. 1982. The mealybug problem and its control, pp. 70–72. In: Root crops in eastern Africa: Proceedings of a workshop held in Kigali, Rwanda, 23–27 Nov. 1980. Intl. Dev. Res. Ctr., Ottawa, Canada.

Slosser, J. E., W. E. Pinchak, and D. R. Rummel. 1989. A review of known and potential factors affecting the population dynamics of the cotton aphid. Southwest. Entomol. 14: 302–313.

Smith, H. S. 1919. On some phases of insect control by the biological method. J. Econ. Entomol. 12: 288–292.

Smith, H. S. and H. M. Armitage. 1931. The biological control of mealybugs attacking citrus. Calif. Agr. Exp. Stn. Bull. 509.

Steinkraus, D. C., R. G. Hollingsworth, and P. H. Slaymaker. 1995. Prevalence of Neozygites fresenii (Entomophthoralis: Neozygitaceae) on cotton aphids (Homoptera: Aphididae) in Arkansas cotton. Environ. Entomol. 24: 465–474.

Steinkraus, D. C., T. J. Kring, and N. P. Tugwell. 1991. Neozygites fresenii in Aphis gossypii on cotton. Southwest Entomol. 16: 118–123

Stern, V. M., P. L. Adkisson, G. O. Beingolea, and G. A. Viktorov. 1976. Cultural controls, pp. 593–613. In: C. B. Huffaker and P. S. Messenger [eds.], Theory and practice of biological control. Academic Press, New York.

Stern, V. M., R. van den Bosch, and T. F. Leigh. 1964. Strip cutting alfalfa for lygus bug control. Calif. Agric. 18: 4–6.

Stiling, P. 1990. Calculating the establishment rates of parasitoids in classical biological control. Am. Entomol. 36: 225–230.

Stiling, P. 1993. Why do natural enemies fail in classical biological control programs? Am. Entomol. 39: 31–37.

Stinner, R. E. 1977. Efficacy of inundative releases. Annu. Rev. Entomol. 22: 515–531.

Stinner, R. E., R. L. Ridgway, J. R. Coppedge, R. K. Morrison, and W. A. Dickerson, Jr. 1974. Parasitism of *Heliothis* eggs after field releases of *Trichogramma pretiosum* in cotton. Environ. Entomol. 3: 497–500.

Suh, C. P., D. B. Orr, J. W. Van Duyn. 1998. Reevaluation of *Trichogramma* releases for suppression of heliothine pests in cotton, pp. 1098–1101. *In*: Proceedings, Beltwide Cotton Production Research Conference. National Cotton Council of America. Memphis, TN.

Sweetman, H. L. 1936. The biological control of insects. Comstock Publ. Assoc., Ithaca, NY. 461 pp.

Sweetman, H. L. 1958. The principles of biological control. Wm. C. Brown, Dubuque, IA. 560 pp.

Sylvestre, P. 1973. Aspects agronomiques de la production du manioc a' la ferme d'etat de Mantsumba (Rep. Pop. Congo), Mission Report. IRAT, Paris. 35 pp.

Sylvestre, P. and M. Arraudeau. 1983. Le manioc, p. 32. *In*: R. Coste [ed.], Techniques agricoles et production tropicales. Maisonneuve and Larose, Paris.

Takagi, M. and Y. Hirose. 1994. Building parasitoid communities: the complementary role of two introduced parasitoid species in a case of successful biological control, pp. 437–448. *In*: B. A. Hawkins and W. Sheehan [eds.], Parasitoid community ecology. Oxford Univ. Press, New York.

Tisdell, C. A. 1990. Economic impact of biological control of weeds and insects, pp. 301–316. *In*: M. Mackauer, L. E. Ehler and J. Roland [eds.], Critical issues in biological control. Intercept, Andover, U.K.

Toth, S. J., Jr. [ed.]. 1998. 1998 North Carolina agricultural chemicals manual. College of Agriculture and Life Sciences, North Carolina State Univ., Raleigh. 454 pp.

Trotter, A. 1908. Due precursori nell' applicazione degli insetti carnivori a difesa delle plante coltivate. Redia 5: 126–132.

U.S. Congress OTA (Office of Technology Assessment). 1993. Harmful non-indigenous species in the United States. OTA-F-565. U.S. Congress, Office of Technology Assessment, Washington, D.C.

U.S. Congress OTA (Office of Technology Assessment). 1995. Biologically based technologies for pest control. OTA-ENV-636. U.S. Congress, Office of Technology Assessment, Washington, D.C.

USDA. 1960. The cotton aphid: how to control it. U.S. Dept. Agric. Leaflet 467.

van den Bosch, R. and V. M. Stern. 1969. The effect of harvesting practices on insect populations in alfalfa, pp. 47–54. *In*: R. Komarek [ed.], Proceedings Tall Timbers Conf. On Ecol. Anim. Contr. By Habitat Manag. Tall Timbers Research Station, Tallahassee, FL.

van den Bosch, R., P. S. Messenger, and A. P. Gutierrez. 1982. An introduction to biological control. Plenum Press, New York. 247 pp.

Van Driesche, R. G. 1994. Classical biological control of environmental pests. Florida Entomol. 77: 20–33.

Van Driesche, R. G. and T. S. Bellows, Jr. 1996. Biological control. Chapman and Hall, New York. 539 pp.

Van Driesche, R. G. and E. Carey. 1987. Opportunities for increased use of biological control in Massachusets. Res. Bull. 718. Massachusets Experiment Station, Univ. of Massachusets, Amherst. 141 pp.

Van Driesche, R. G. and M. Hoddle. 1997. Should arthropod parasitoids and predators be subject to host range testing when used as biological control agents? Agric. and Human Values 14: 211–226.

Van Driesche, R. G., J. S. Elkinton, and T. S. Bellows, Jr. 1994. Potential use of life tables to evaluate the impact of parasitism on population growth of the apple blotch leafminer (Lepidoptera: Gracillaridae). *In*: C. Maier [ed.], Integrated management of tentiform leafminers, Phyllonorycter (Lepidoptera: Gracillaridae) spp., in North American apple orchards. Thomas Say Publications in Entomology, Entomol. Soc. Am., Lanham, MD.

Van Kan, F. J. P. M., I. M. M. S. Silva, M. Schilthuizen, J. D. Pinto, and R. Stouthamer. 1996. Use of DNA-based methods for the identification of minute wasps of the genus *Trichogramma*. Proceedings of the section Experimental and Applied Entomology of the Netherlands Entomological Society 7: 233–237.

van Lenteren, J. C. 1996a. Quality control tests for natural enemies used in greenhouse biological control. Bull. OILB-SROP 19: 83–86.

van Lenteren, J. C. 1996b. Regulatory issues related to biological control in Europe. Bull. OILB-SROP 19: 79–82.

van Lenteren, J. C. and J. Woets. 1988. Biological and integrated control in greenhouses. Annu. Rev. Entomol. 33: 239–269.

van Lenteren, J. C., Y. C. Drost, H. J. W. van Roermund, C. J. A. M. Posthuma-Doodeman. 1997. Aphelinid parasitoids as sustainable biological control agents in greenhouses. J. Appl. Econ. 121: 473–485.

van Lenteren, J. C., C. J. A. M. Posthuma-Doodeman, M. Roskam, and G. Wessels. 1996. Quality control of *Encarsia formosa:* flight tests. Bull. OILB-SROP 19: 87–90.

Vereijssen, J., I. Silva, J. Honda, and R. Stouthamer. 1997. Development of a method to predict the biological control quality of *Trichogramma* strains. Proceedings of the Section Experimental and Applied Entomology of the Netherlands Entomol. Soc. 8: 145–149.

Vinson, S. B., F. Bin, and L. E. M. Vet. 1998. Critical issues in host selection by parasitoids. Biological Control 11: 77–78.

Voegele, J. M. 1989. Biological control of *Brontispa longissima* in western Samoa: an ecological and economic evaluation. Agriculture, Ecosystems and Environment 27: 315–329.

Waage, J. K. 1990. Ecological theory and the selection of biological control agents, pp. 135–157. *In*: M. Mackauer, L. E. Ehler, and J. Roland, Critical issues in biological control. Intercept, Andover, U.K.

Wajnberg, E. and S. A. Hassan. [eds.]. 1994. Biological control with egg parasitoids. CAB Intl., Oxon, U.K. 286 pp.

Waterhouse, D. F. and K. R. Norris. 1987. Biological control, Pacific prospects. Inkata Press, Melbourne, Australia.

Werren, J. H. 1997. Biology of *Wolbachia*. Annu. Rev. Entomol. 42: 587–609.

Williams, M. R. 1998. Beltwide cotton insect losses: 1997, pp. 904–925. *In*: Proceedings, Beltwide Cotton Production Research Conference. National Cotton Council of America, Memphis, TN.

CHAPTER 2

Microbial Insecticides

J. Lindsey Flexner and Diane L. Belnavis

CONTENTS

2.1 INTRODUCTION

The first documentation of insect diseases is usually attributed to the descriptions of honeybee maladies recorded by Aristotle somewhere between 330 and 323 B.C. It should be noted, however, that observations of diseased silkworms were recorded in China as far back as 2700 B.C. (Steinhaus 1956 and 1975). The majority of early descriptive insect pathology concentrated on these two domesticated insects — the honeybee (*Apis mellifera*) and the silkworm (*Bombyx mori*). This is understandable when you consider that the silkmoth was the foundation of the Far Eastern textile industry and the honeybee was the major source of sweetener and the base ingredient for alcoholic beverages during early civilization. The husbandry of these species dates back to the advent of written language and probably beyond. Aristotle also recorded the diseases of other invertebrate life forms, including ants, oysters, scallops, and lobsters. He may be considered to be the first invertebrate pathologist.

The purpose of this chapter, however, is not the history and current state of invertebrate diseases (i.e., invertebrate pathology). This is far too large a topic to be covered adequately in a single book chapter. The history of invertebrate pathology has been documented by one of the founders of modern invertebrate pathology, Edward A. Steinhaus, in his book *Disease in a Minor Cord* (1975). For a recent text providing an overview of invertebrate pathology we direct the readers to Tanada and Kaya (1993) and Lacey (1997). The focus of this chapter is to document how microbial pathogens may be used to impact the population dynamics of invertebrate pests of economic importance, part of a discipline known as microbial control.

Microbial control has been the subject of several books (Burges and Hussey 1971; Burges 1981a; Kurstak 1982) and recently reviewed by Lacey and Goettel (1995). This chapter concentrates on those microorganisms that have been made commercially available to control arthropod pests — defined here as microbial insecticides. These products must compete with their synthetic organic counterparts (i.e., chemical insecticides) in the pest management marketplace. Their success depends on how broadly they are used when consumers are given the choice of many different alternatives for pest control.

The majority of commercially available pathogens target insects. However, research and development of pathogens to control weeds (Te Beest et al. 1992) and plant diseases (Sivan and Chet 1992) have increased dramatically in the past decade; there are currently over 20 pathogens commercially available for weed and plant disease control (Copping 1998).

To find the roots of microbial insecticides we can jump forward in history almost 2000 years after Aristotle's descriptions to one of the founding fathers of the germ theory of disease, Agostino Maria Bassi. Working with the white muscardine fungus originally isolated from silkworms (later named *Beauveria bassiana*), Bassi showed that this disease could be artificially passed by needle to many different species of insects (Steinhaus 1975). According to Steinhaus, Bassi declared that "such transmissions could be accomplished whenever desired! Herein lies the germ of the thought that man can communicate agents of disease to susceptible (insect) pests at will."

Before the advent of modern light optics most of the early descriptions of invertebrate pathology were fungi since the fruiting bodies were easily visible with the naked eye. The "Chinese plant worm" described and illustrated in 1726 by Rene-Antoine Ferchault de Reaumur appears to be the first published record of a diseased insect. He believed that the "stemlike vegetable growth" emerging from a noctuid larva was the root of a plant. In 1749 a Franciscan friar in Cuba described dead wasps with little trees growing out of their bellies. This was a fungus that we know today as a member of the genus *Cordyceps*. In the quarter of a millennium that has elapsed since the discovery of these "little trees" we are still trying to consistently use microorganisms to effectively control economically important insect pests.

In 1971 and 1981 two books laid the foundation for microbial control as it exists today (Burges and Hussey 1971; Burges 1981a). The books contained contributions from a total of 99 researchers working in applied invertebrate pathology. The chapters focused on the major organisms believed to be potential control agents. The first book was dedicated entirely to the control of insects and mites (Burges and Hussey 1971). The second focused on microbial control of arthropods; however, it contained one chapter on the use of microorganisms to control plant diseases (Corke and Rishbeth 1981). These books were organized by the major taxonomic groups of pathogens infecting insects: bacteria, viruses, fungi, nematodes, and microsporidia. In addition they contained chapters on identification, safety, production, formulation, and economics. They were written at a time when *Bacillus thuringiensis* was being used to control over 50% of the cabbage looper on California cole crops, several baculoviruses were being registered as commercial insecticides, and a plant was to be built in Russia to produce enough *B. bassiana* to treat 5 million acres (Burges and Hussey 1971). It would appear that we were about to develop an entire new set of tools to be used in agricultural pest management. However, in the concluding remarks in both books the editor(s) sounded cautiously optimistic about the future of microbial pesticides.

The time between 1980 to 1990 may be viewed as the decade of disappointment for microbial insecticides. Many of the microbial insecticides registered during the 1970s were discontinued. Market share of *B.t.* declined in many agricultural areas and several registered viral insecticides were discontinued. By the mid-1980s, 70% of the $10 million *B.t.* market was being used against forest pests (Navon 1993). The introduction of the synthetic pyrethroid insecticides in the mid-1970s had provided a very cheap, relatively safe and effective means for broad-spectrum insect pest management in many crops around the world (Hirano 1989).

During the 1990s microbial pesticides experienced somewhat of a renaissance due in part to the advent of pathogens for plant disease control. In 1998 there were 59 species of microbial pathogens registered as pesticides worldwide. Thirty-six species were registered to control animal pests (i.e., insects, mites, slugs and nematodes). These included 11 viruses, 9 bacteria, 9 nematodes, 6 fungi, and 1 microsporidia. Twenty species were registered to control plant diseases; 15 fungi and 5 bacteria. Three fungal species were registered as mycoherbicides (Copping 1998). This chapter will discuss these microbial insecticides by their major pathogen groups.

2.2 BACTERIA

Most of the pathogenic entobacteria are found in the families Bacillaceae, Pseudomonadaceae, Enterobacteriaceae, Streptococcaceae, and Micrococaceae (Tanada and Kaya 1993). Although there are many different types of bacteria that are known to acutely or chronically infect insects, only members of two genera of the order Eubacteriales, *Bacillus* (Bacillaceae) and *Serratia* (Enterobacteriaceae), have ever been registered to control insects. *Bacillus* is by far the most important microbial pesticide genus. In 1948, the first microbial insecticides registered in the U.S. were bacteria from the genus *Bacillus*. Two different species of *Bacillus* were registered to control the Japanese beetle, *Popillia japonica* (*B. popilliae popilliae* and *B. popilliae lentimorbus*). Another species of the Bacillaceae, *Bacillus thuringiensis* (*B.t.*), has been the most widely used and successful microbial pesticide ever registered. The sales of *B.t.* since its registration in 1961 account for the majority of all microbial pesticide sales worldwide. In 1997 the total sales of sprayable *B.t.* were estimated to be $145 million (Wood Mackenzie 1998). *Bacillus thuringiensis* subsp. *kurstaki* (*B.t.k.*) is registered for control of many important lepidopterous pests with over 30 different trade names by more than a dozen major companies (Copping 1998). A list of these products and companies have been compiled in Table 2.1 (CDMS 1998; CEPA/DPR 1998; Copping 1998; CPCR 1998). This table does not include a list of all *B.t.* products sold worldwide. There are currently over 100 *B.t.k.* products that have active registration labels in the U.S. alone. This does not include over 100 different *B.t.k.* strains or subspecies, but does denote over 100 different registered products. There is a high degree of redundancy between many of these registrations. For example, there are currently at least one dozen companies that sell Abbott's product Dipel®. Abbott also has 16 products based on different concentrations and formulations of Dipel®. Table 2.1 attempts to show the unique or larger volume products. A complete list of registered *B.t.k.* products can be found by searching the O.P.P. code 006402 at the following web site — http://www.cdpr.ca.gov/docs/epa/m2.htm.

With the advent of genetic engineering, truncated forms of the *B.t.* endotoxins have been expressed in the tissues of several economically important plants. This new technology, transgenic plants, has the potential to revolutionize crop protection through the next millennium (Estruch et al. 1997). Though it may utilize proteins, peptides, or small molecules produced by microbial pathogens, this technology is not an example of microbial insecticides and will not be covered further in this chapter. Transgenic plants and crop protection will be discussed in detail later in this volume.

There are several important subspecies of *B.t.* that have been registered and marketed as separate products on different insect pests. These include *Bacillus thuringiensis* subsp. *aizawai* (*B.t.a*) for control of other Lepidoptera less susceptible to *B.t.k.*, *Bacillus thuringiensis* subsp. *israelensis* for control of mosquitoes and blackflies, and *Bacillus thuringiensis* subsp. *tenebrionis* for control of leaf beetles in the family Chrysomelidae. Currently *B.t.* is being manufactured and/or sold in every country where Lepidoptera are important economic pests. However, sprayable *B.t.*s still make up less than 2% of the total global insecticide market (Wood Mackenzie 1998).

Another species of *Bacillus*, *Bacillus sphaericus*, has been registered for mosquito control (Copping 1998). The only non-Bacillus bacterial microbial insecticide is *Serratia entomophila*, currently registered for pastureland grub control in New Zealand.

2.2.1 *Bacillus thuringiensis*

Members of the family Bacillaceae are gram positive motile or non-motile rods that produce an endospore. There are two major genera in the family: *Bacillus* and

Table 2.1 Commercially Avaliable *B.t.* for Lepidoptera Control

Commercial Name	Current Producer	*B.t.* Subsp./strain
Bactospeine	Abbott	*kurstaki/* HD-1
Biobit	Abbott	*kurstaki/* HD-1
Dipel	Abbott	*kurstaki/* HD-1
Florbac	Abbott	aizawai
Foray	Abbott	*kurstaki/* HD-1
XenTari	Abbott	aizawai
Cordalene	Agrichem	*kurstaki/* HD-1
BMP 123	Becker	*kurstaki*/HD263
Biobest-Bt	Biobest	*kurstaki/* HD-1
Bacticide	Cequisa	*kurstaki/* HD-1
Worm Wipper	Cape Fear Chemicals	*kurstaki/* HD-1
Collapse	Calliope	*kurstaki/* HD-1
Baturad	Cequisa	*kurstaki/* HD-1
Condor	Ecogen	*kurstaki/* EG2348
Crymax	Ecogen	*kurstaki/* EG7841
Cutlass	Ecogen	*kurstaki/* EG2371
Lepinox	Ecogen	*kurstaki/* EG7826
Raven	Ecogen	*kurstaki/* EG2424
Ecotech Bio	Ecogen/ AgrEvo	*kurstaki/* EG2371
Ecotech Pro	Ecogen/ AgrEvo	*kurstaki/* EG2348
Jackpot	Ecogen/ Intrachem	*kurstaki/* EG2424
Rapax	Ecogen/ Intrachem	*kurstaki/* EG2348
Forwarbit	Forward International	*kurstaki/* HD-1
Bio-Worm Killer	Green Light Co	*kurstaki/* HD-1
Bactospeine Koppert	Koppert	*kurstaki/* HD-1
Guardjet	Mycogen/ Kubota	*kurstaki/* Cry1 Ac
Maatch	Mycogen	*kurstaki*/Cry1 Ac & *aizawai*/Cry1 C
M/C	Mycogen	*aizawai*/Cry1 C
M-Peril	Mycogen	*kurstaki/* Cry1 Ac
MVP	Mycogen	*kurstaki/* Cry1 Ac
Bactec BT 16	Plato Industries	*kurstaki/* EG2348
Bactec BT 32	Plato Industries	*kurstaki/* HD-1
Insectobiol	Samabiol	*kurstaki/* HD-1
Bactosid K	Sanex	*kurstaki/* HD-1
Soilserv BT	Soil Serv Inc	*kurstaki/* HD-1
Agrobac	Tecomag	*kurstaki/* HD-1
Able	Thermo Trilogy	*kurstaki/* M-200
Agree	Thermo Trilogy	*aizawai/* GC-91
Costar	Thermo Trilogy	*kurstaki/* SA-12
Delfin	Thermo Trilogy	*kurstaki/* SA-11
Design	Thermo Trilogy	*aizawai/* GC-91
Javelin	Thermo Trilogy	*kurstaki/* SA-11
Thuricide	Thermo Trilogy	*kurstaki/* HD-1
Turex	Thermo Trilogy	*aizawai/* GC-91
Vault	Thermo Trilogy	*kurstaki/* SA-11
Larvo-Bt	Troy Biosciences	*kurstaki/* HD-1
Troy-Bt	Troy Biosciences	*kurstaki/* HD-1
Ringer BT	Verdant Inc	*kurstaki/* HD-1
Safor BT	Verdant Inc	*kurstaki/* HD-1
BT 320	Wilbur Ellis Inc	*kurstaki/* HD-1

Based on Copping (1998), CPCR (1998), CDMS (1998), CEPA/DPR (1998)

Clostridium. Members are separated mainly by their oxygen requirements with most members of *Bacillus* being aerobic and *Clostridium* being anaerobic. Both genera form rod-shaped cells that may occur in chains (Tanada and Kaya 1993). Some members of the family, such as *B.t.* and *B. sphaericus*, form parasporal inclusions (i.e., crystal) during sporulation. This crystal has been shown to contain a number of protein toxins that infect a variety of important insect pests. The toxin(s) when ingested act as a stomach poison. The osmotic balance of the midgut epithelial cells is disrupted when the toxins bind to the surface of the microvillar membranes. The insects usually stop feeding shortly after ingestion of the bacteria and death occurs one to five days later. In some cases the bacterial spore may also be necessary for the full larvicidal effect (Lacey and Goettel 1995).

The pathogen *B. thuringiensis* subsp. *kurstaki* was probably first described from diseased silkworms by Pasteur sometime in the mid to late 1860s (Steinhaus 1960). It was first isolated from silkworms by Ishiwata around the turn of the century. It was reisolated by Berliner from the Mediterranean flour moth, *Anagasta kuehniella*, and given its current name in 1915 (Tanada and Kaya 1993). It was not until the mid 1920s that *B.t.* was field-tested for efficacy against lepidopterous hosts. The first trials were against the European corn borer, *Ostrinia nubilalis*. In 1942, three years before the introduction of DDT, Steinhaus attracted the attention of applied entomologists worldwide by demonstrating the potential of this bacterium in insect pest management (Tanada and Kaya 1993). Since that time there have been over 30 subspecies of *B.t.* classified, containing over 140 described crystalline toxins (Yamamoto and Powell 1993; Crickmore et al. 1998a,b). These toxins have not only shown activity against Lepidoptera, Diptera, and Coleoptera but recent isolates have been discovered that show activity against nematodes, mites, lice, aphids, and ants (Lacey and Goettel 1995).

B. thuringiensis subspecies have been differentiated by various methods, including biochemical tests, H serotypes, parasporal bodies, antigens, esterase production, antibiotic production, enzymes, phages, and lectin grouping (Tanada and Kaya 1993; Yamamoto and Powell 1993). The taxonomy of *B.t.* will not be discussed in this chapter. The authors will refer to the oversimplified pathotypes first described by Krieg et al. (1983) to classify the bacterial microbial insecticides. Although these pathotypes may not be appropriate for *B.t.* taxonomy (i.e., many subspecies have toxins that overlap the various pathotypes), we believe they are a valid way to describe the various *B.t.* commercial offerings. Pathotype A includes subspecies pathogenic for Lepidoptera (i.e., *B.t.a.* and *B.t.k.*); Pathotype B are pathogenic for Diptera (i.e., *B.t.i.*); and Pathotype C infect Coleoptera (i.e., *B.t.t.*). For in-depth reviews of *B. thuringiensis* toxins and development of genetically improved strains, see Ellar et al. (1986), Carlton (1988), Hofte and Whiteley (1989), Adang (1991), Yamamoto and Powell (1993), Pietrantonio et al. (1993), Van-Nguyen (1995) and Schnepf et al. (1998).

Production of *B.t.* is done in large capacity industrial scale bioreactors. The scale of production and media cost make it the most inexpensive of all the microbial insecticides to produce. However, it is still expensive compared to many synthetic insecticides. In addition, media must be optimized for each new strain put into commercial production. Fermentation optimization has the potential to greatly influence cost of production and ultimately commercial competitiveness. Recombinant technology may be a powerful tool for improving the commercial production of *B.t.* For an overview of *B.t.* production see Bernhard and Utz (1993).

2.2.2 Other Bacteria

There are two other species of bacteria currently registered as microbial insecticides: *Bacillus sphaericus* and *Serratia entomophila*. *B. sphaericus* is a strict aerobic spore former that is registered for mosquito control (VectoLex®). It is particularly effective against members of the genus *Culex* (Copping 1998). The species is genetically heterogeneous with many saprophytic strains; however, several are vertebrate pathogens and more than 50 isolates are mosquito pathogens (Tanada and Kaya, 1993). The bacterium is ubiquitous in soil and aquatic habitats. The effect of the toxin was described in the mid 1970s (Singer 1973; Davidson et al. 1975). Because of its persistence in the field compared to *B.t.i.*, *B. sphaericus* has become a promising candidate for mosquito abatement.

S. entomophila is a non-spore forming, gram-negative, anaerobic, rod-shaped bacteria in the family Enterobacteriaceae. Members of the genus *Serratia* are commonly found in nature as saprophytes in soil and water (Grimont and Grimont 1978). Several species have been isolated from foodstuffs; still others are vertebrate pathogens. *S. entomophila* is the causal agent of amber disease in the New Zealand grass grub, *Costelytra zealandica*. It is registered for control of *C. zealandica* in New Zealand pasture (Invade).

2.2.3 Pathotype A

Most of the current products registered for Lepidoptera control are based on *B.t.k.* strain HD-1 (Table 2.1), first introduced by Abbott Laboratories in 1971 (Navon 1993). This strain has been shown to have activity on over 100 lepidopteran species. Another strain, HRD12 (Javelin®), was introduced by Sandoz (now licensed to Thermo Trilogy) to improve the control of Lepidoptera not particularly sensitive to HD-1. Javelin® was later changed to strain SA-11. Strains of *B.t.a.* were commercialized later as they are more effective on *Spodoptera* sp. and certain key pests that had developed resistance to *B.t.k.* (Tabashnik et al. 1998). In addition, several companies have genetically engineered *B.t.* strains to broaden the host range or increase the field persistence.

Ecogen and Novartis developed products that were based on the conjugation of two distinct strains (i.e., *kurstaki, aizawai,* or *tenebrionis*). These strains usually had increased host range and were assigned new strains: EG2348, EG2349, EG2371, EG2424, CG-91 and CGA 237218 (Copping 1998; Navon 1993). This type of technology yielded products such as Foil® (conjugate of *B.t.k.* and *B.t.t.*), which was active on lepidopteran and coleopteran pests of potatoes (Foil is no longer available). Mycogen has encapsulated the *B.t.* toxin(s) from *B.t.k., B.t.a.,* and *B.t.t.* in the bacteria *Pseudomonas fluorescens* so that it is killed in such a way that it forms a rigid microcapsule around the toxin(s) (Copping 1998). This technology can also be used to design commercial product with expanded host range (e.g., Maatch®).

The biological activity of *B.t.* is standardized for commercial control of Lepidoptera and regulatory purposes using the potency bioassay. There are at least two assays that have been established to quantify lepidopteran activity based on a reference standard. A French bioassay was established with the Mediterranean flour moth, *A. kuehniella*, and the *B.t.* strain E61. An American bioassay was based on the cabbage looper, *Trichoplusia ni*, and a reference strain of HD-1 (Navon 1993). However, as new toxins are discovered and new commercial targets identified the official test insects have been increased or replaced by other species. These assays were not quantified by amount of active protein but on a unit of activity known as an international unit (IU). See Navon (1993) and Dulmage et al. (1971) for bioassay details.

Registration of *B.t.*s has been a relatively inexpensive process, because they are naturally occurring insect pathogens with no known effects on nontarget arthropods or vertebrates (Burges 1981b; Meadows 1993). Thus, they have been registered on a wide variety of crops, including many niche markets not available to synthetic insecticides. Currently in the U.S., *B.t.*s are registered on over 50 vegetable crops, 10 species of small fruits and berries, 20 species of nuts, citrus, pome and stone fruit, 9 legumes, 30 field crops, 15 herbs and spices, bedding plants, flowers, ornamentals, turf, forestry, landscape trees and shrubs, and various tropical crops. These registrations cover most of the economically important Lepidoptera: many loopers, armyworms, cutworms, leafrollers and borers, diamondback moth, gypsy moth, spruce budworm, and tobacco budworm.

2.2.4　Pathotype B and Dipteran Active Bacteria

B.t. israelensis was discovered by Margalit in 1976 (Becker and Margalit 1993). It was the first *B.t.* subspecies discovered that had activity only on Diptera. Of the over 100 strains of *B.t.i.* that have been discovered since 1976, none have shown significantly better mosquitocidal activity than the original isolate (Becker and Margalit 1993). It has been used worldwide as an alternative to synthetic organic insecticides for control of mosquitoes and blackflies. It was first registered in the U.S. in 1981 and has recently been used extensively in the Upper Rhine Valley in Germany where *Aedes* spp. are a major nuisance pest. It is also used extensively by the World Health Organization (WHO) and United Nations (UN) in their vector control programs. Toxicity of *B.t.i.* is similar to *B.t.k.* in that activity is due to four major protein toxins produced in the parasporal body. High toxicity is based on synergism between the 27 kDa protein and one or more of the higher molecular weight proteins. Potency of formulated *B.t.i.* is based on standardized bioassays and international units similar to *B.t.k.* For details see Dulmage et al. (1990).

Bacillus sphaericus was first described by Kellen et al. in 1965, who isolated it from a dead mosquito larva from the U.S. (Berry et al. 1991). Since that time more highly toxic strains have been isolated from around the world including strain 1593 from Indonesia, strain 2297 from Sri Lanka, and strain 2362 from Nigeria (Berry et al. 1991). Because it provides very good control of *Culex* spp. mosquitoes, the use of *Bacillus sphaericus* is increasing. Under certain environmental conditions it provides increased residual activity by maintaining itself in the environment. It also provides control in highly polluted waters. Table 2.2 lists currently registered products.

Table 2.2 Commercially Avaliable *Bacillus* for Diptera Control

Commercial Name	Current Producer	*Bacillus* Sp
Bactimos	Abbott	*B.t. israelensis*
Gnatrol	Abbott	*B.t. israelensis*
Skeetal	Abbott	*B.t. israelensis*
VectoBac	Abbott	*B.t. israelensis*
VectoLex GC	Abbott	*B. sphaericus*
Acrobe	American Cyanamide	*B.t. israelensis*
Aquabac	Becker Microbial	*B.t. israelensis*
BMP	Becker Microbial	*B.t. israelensis*
Bactis	Caffaro	*B.t. israelensis*
BTI Granules	Clarke Mos. Cont.	*B.t. israelensis*
Prehatch SG	Meridian	*B.t. israelensis*
Vectocide	Sanex	*B.t. israelensis*
Summit Bactimos	Summit Chemicals	*B.t. israelensis*
Summit Mosquito Bits	Summit Chemicals	*B.t. israelensis*
Tekar	Thermo Trilogy	*B.t. israelensis*

Based on Copping (1998), CPCR (1998), CDMS (1998), CEPA/DPR (1998)

Formulation technology has been particularly important for the Dipteran active bacteria. Long residual activity is very important for successful vector control. In addition, because these products target the larvae of biting flies rather than the adults, they must disperse in water and maintain themselves at the air–water interface to be highly effective. Thus, these products have been registered with a number of different formulation types, including powders, oil-based liquids, and floating briquettes and granules. For more detail on bacterial control of mosquitoes and blackflies see de Barjac and Sutherland (1990).

2.2.5 Pathotype C and Coleopteran Active Bacteria

Although *B. popilliae* and *B. lentimorbus* were the first registered microbial insecticides in the U.S., there are no longer any active registrations for *B. popilliae* in the U.S. as of December 1998. It is possible that *B. popilliae* is still being sold elsewhere in the world; however, it was not listed as a registered pesticide in the recently released biopesticide manual (Copping 1998). Currently, only subspecies or conjugates of *B.t.* are registered for beetle control in the U.S. *B.t. tenebrionis* was first isolated and identified by Huger and Krieg in 1982 (Keller and Langenbruch 1993). It was the first *B.t.* strain isolated with a δ-endotoxin that demonstrated coleopteran activity. The coleopteran activity cited in other strains/subspecies of *B.t.* was probably due to the α and/or β-exotoxins present in these strains (Keller and Langenbruch 1993). Since 1982 at least two other novel strains/subspecies have been found with unique coleopteran active δ-endotoxins. *B.t.t.* is most active against leaf beetles from the family Chrysomelidae. In the U.S. and Baltic States the most economically important species in this family is the Colorado potato beetle (CPB), *Leptinotarsa decemlineata*. However, several other leaf beetles have been targeted successfully with these products, including alder leaf beetle (*Agelastica alni*), cottonwood leaf beetle (*Chrysomela*

Table 2.3   Commercially Avaliable Bacteria for Coleoptera Control

Commercial Name	Current Producer	Bacteria Sp
Ditera	Abbott	*B.t. tenebrionis*
Novodor	Abbott	*B.t. tenebrionis*
Raven	Ecogen	*B.t. kurstaki* (EG2424)
Invade	Industrial Research Ltd	*Serratia entomophila*
M-Trak	Mycogen	*B.t. tenebrionis/Cry 3A*
Trident	Thermo Trilogy	*B.t. tenebrionis*

Based on Copping (1998), CPCR (1998), CDMS (1998), CEPA/DPR (1998)

scripta), cranberry tree leaf beetle (*Galerucella viburni*), eucalyptus tortoise beetle (*Paropsis charybdis*), and the elm leaf beetle (*Xanthogaleruca luteola*). Ferro and Gelernter (1989) describe a bioassay similar to other *B.t.*-based assays using a standard freeze-dried technical powder and potency units (IU) based on the CPB. Similar assays have been developed for other important chrysomelid species (see Keller and Langenbruch (1993) for details). Commercially available bacteria for Coleoptera control are listed in Table 2.3.

There is only one bacterial product that is registered for beetle control on species other than Chrysomelidae. *Serratia entomophila* is the causal agent of amber disease in the New Zealand grass grub, *C. zealandica* (Coleoptera: Scarabaeidae). The bacterium was isolated from the grass grub in New Zealand and documented in 1982 by Trought et al. (1982). It is currently registered for use in New Zealand pastureland.

2.3 VIRUSES

The first accounts of baculovirus infections come from diseased silkworms in ancient China. The first description in western civilization is found in a poem written by a 16th century Italian bishop (Miller 1997b). By the 19th century microscopy was able to detect highly reflective crystals floating in the "melted" remains of dead caterpillars and other insects. Research during the first half of the 20th century found that these crystals were composed of polyhedra that encapsulated or occluded the actual infectious particles (i.e., virions). It was also during this time that the large nuclear polyhedrosis viruses (NPVs) were distinguished from the much smaller granular looking viruses (i.e., granulosis viruses (GVs)). Bergold was the first to characterize the rod-shaped virions found within the occlusion bodies in the 1930s and 1940s. It was also during this time that the utility of baculoviruses as biocontrol agents was described (Miller 1997b). During the 1950s through the 1970s the use of baculoviruses for biological control of insects was championed by Steinhaus and carried on by his students. In the last two decades molecular baculovirology has flourished and is currently a dynamic and diverse field that encompasses expression of heterologous genes for pharmaceutical use as well as improvement of the insecticidal function by increase in the speed of kill. Molecular baculovirology has been reviewed in several chapters in recent books edited by Miller 1997a; Clem 1997;

Friesen 1997; Jarvis 1997; Lu et al. 1997; Lu and Miller 1997; Miller and Lu 1997; O'Reilly 1997; Possee and Rohrmann 1997. For a recent review of baculoviruses as expression vectors see Possee (1997).

Viruses have been isolated from more than a thousand species of insects from at least 13 different insect orders (Martignoni and Iwai 1986; Onstad 1996, 1997; Tanada and Kaya 1993). Entomopathogenic viruses from almost a dozen viral families have been isolated: Ascoviridae, Baculoviridae, Birnaviridae, Iridoviridae, Nodaviridae, Parvoviridae, Picornaviridae, Poxviridae, Reoviridae, Rhabdoviridae, and Tetraviridae (Adams and Bonami 1991; Evans and Shapiro 1997; Kurstak 1991; Murphy et al. 1995). However, the number of currently registered viral insecticides is made up of a relatively small number of entomopathogenic viruses exclusively from the family Baculoviridae. For a comprehensive review of the baculoviruses we direct the readers to the two-volume set edited by Granados and Federici (1986a,b).

All of these commercially available baculoviruses are targeted for lepidopteran hosts with the exception of one forestry product for sawfly control (Hymenoptera: Diprionidae) (Table 2.4). Most are very specific, infecting one or several insects often within the same genus. Most wild-type baculoviruses kill their hosts very slowly, sometimes taking 7 to 10 days to kill later instars. All baculoviruses are extremely susceptible to degradation under UV light, making them ephemeral in direct sunlight when applied as insecticides without protective formulations.

The first viral insecticide was the single-nucleocapsid nucleopolyhedrovirus of *Helicoverpa (Heliothis) zea* (HzSNPV) registered in 1971 (Ignoffo 1973). This virus was developed by the USDA in the mid-1960s for control of Heliothines (e.g., Tobacco Budworm (*Heliothis virescens*), corn earworm (*H. zea*), and *Heliothis armigera*) on cotton, row crops, fruits, and vegetables. It was commercialized by International Minerals and Chemical Corporation (Viron-H®), who sold it to Sandoz in the mid-1970s. Sandoz sold HzSNPV under the trade name Elcar® beginning in 1975 and discontinued it in the early 1980s. HzSNPV is currently produced and marketed as Gemstar® by Thermo Trilogy. Another virus registered in the mid-1970s was the type 1 Cypovirus of *Dendrolimus spectailis* (DsCPV-1) in the family Reoviridae. This Cypovirus (formally Cytoplasmic Polyhedrosis Virus) was the only non-baculovirus ever registered and the only viral insecticide ever registered in Japan (Katagiri 1981). To our knowledge it is no longer actively registered in Japan, although this and other CPVs may still be used to control forest pests in China. Two additional viruses were registered to control forest pests in the 1970s. The U.S. Forest Service (USFS) registered the multi-nucleocapsid nucleopolyhedrovirus of *Orygia pseuedotsugata* (OpMNPV) in 1976 (TM Biocontrol-1) and the multi-nucleocapsid nucleopolyhedrovirus of *Lymantria dispar* (LdMNPV) in 1978 (GYPCHEK). In 1983 the USFS registered another virus for control of sawflies using a combination of two viruses isolated from the European pine sawfly, *Neodiprion sertifer*, and the redheaded pine sawlfy, *Neodiprion lecontei* (CEPA/DPR 1998; Copping 1998). However, this virus could only be produced in wild field populations of sawflies and registration was cancelled in the U.S. in 1991. According to Copping (1998) this virus combination is still registered in Finland, Canada, and the U.K. (Table 2.4).

Table 2.4 Commercially Avaliable Viruses for Lepidopteran Control

Commercial Name	Current Producer	Viral Sp	Target Pests
Granupom	AgrEvo	*Cydia pomonella* GV	Codling moth
Carposin	Agrichem	*Cydia pomonella* GV	Codling moth
VPN-80	Agricola El Sol	*Autographa californica* MNPV	Lepidopteran larvae
Polygen	Agroggen	*Anticarsia gemmatalis* MNPV	Velvetbean caterpillar, sugar cane borer
Capex 2	Andermatt Biocontrol	*Adoxophyes orana* GV	Summer fruit tortrix
Madex 3	Andermatt Biocontrol	*Cydia pomonella* GV	Codling moth
Ness-A	Applied Chemical	*Spodotera exigua* MNPV	Beet armyworm
Ness-E	Applied Chemical	*Spodotera exigua* MNPV	Beet armyworm
Leconteivirus	Canadian Forestry Service	*Neodiprion sertifer/N. lecontei* MNPV	Sawfly larvae
Multigen	EMBRAPA	*Anticarsia gemmatalis* MNPV	Velvetbean caterpillar, sugar cane borer
Monisarmiovirus	Kemira	*Neodiprion sertifer/N. lecontei* MNPV	Sawfly larvae
Virin-EKS	NPO Vector	*Mamestra brassicae* MNPV	Lepidopteran larvae
Virin-GYAP	NPO Vector	*Cydia pomonella* GV	Codling moth
Carpovirusine	NPP/Calliope	*Cydia pomonella* GV	Codling moth
Mamestrin	NPP/Calliope	*Mamestra brassicae* MNPV	Lepidopteran larvae
Virox	Oxford Virology	*Neodiprion sertifer/N. lecontei* MNPV	Sawfly larvae
Elcar	Novartis	*Helicoverpa zea* SNPV	Heliothines
Cyd-x	Thermo Trilogy	*Cydia pomonella* GV	Codling moth
Gemstar	Thermo Trilogy	*Helicoverpa zea* SNPV	Heliothines
not determined	Thermo Trilogy	*Anagrapha falcifera* MNPV	Lepidopteran larvae
Spod-X	Thermo Trilogy	*Spodotera exigua* MNPV	Beet armyworm
Gypcheck	U.S. Forest Service	*Lymantria dispar* MNPV	Gypsy moth
TM Biocontrol-1	U.S. Forest Service	*Orygia pseudotsugata* MNPV	Douglas-fir tussock moth

Based on Copping (1998), CPCR (1998), CDMS (1998), CEPA/DPR (1998)

By far the most successful viral insecticide used to date is the multi-nucleocapsid nucleopolyhedrovirus of *Anticarsia gemmatalis* (AgMNPV) applied to over 1 million ha annually for control of the velvetbean caterpillar, *A. gemmatalis,* in Brazil (Moscardi 1989). AgMNPV is typical of the viral insecticides in that it has a very narrow host range and has been produced and marketed to control only two lepidopterous pests: *A. gemmatalis* and the sugar cane borer, *Diatreae saccaralis*. The majority of the viral insecticides have very narrow host ranges and thus require a fairly significant pest to make their commercialization a successful venture. This is the case for two other viral insecticides that have been recently registered, the multi-nucleocapsid nucleopolyhedrovirus of *Spodoptera exigua* (SeMNPV) and the granulovirus of *Cydia pomonella* (CdGV). There are currently six different companies that are marketing CdGV just for codling moth control in pome fruit (Table 2.4).

There are several viral insecticides that have been registered that have a broader host range then the baculoviruses described above. These include the multi-nucleocapsid nucleopolyhedrovirus of *Mamestra brassicae* (MbMNPV), *Autographa californica* (AcMNPV), and *Anagrapha falcifera* (AnfaNPV). MbMNPV is registered on a diverse group of insects, which includes *M. brassicae, H. armigera, Phthorimaea operculla,* and *Plutella xylostella.* AcMNPV and AnfaNPV have been shown to infect as many as 30 different lepidoptera species from as many as 10 separate families (Adams and McClintock 1991). However, their commercial potential is far less than this number and probably only represents two to five economically important species at the most.

Recently, the nucleopolyhedroviruses such as AcMNPV have been genetically modified for an increased killing speed. The most promising of these genetically modified viruses encode and express insect-selective toxins and have resulted in an approximate 20 to 40% reduction in the killing time of their insect hosts (Hammock et al. 1993; Maeda et al. 1991; McCutchen et al. 1991; Stewart et al. 1991; Tomalski 1992; Tomalski and Miller 1991). These genetically modified baculoviruses have the potential to greatly improve the utility of baculoviruses as insecticides and will be discussed in more detail in later chapters.

2.3.1 Baculoviruses

The taxonomy of double-stranded DNA baculoviruses has been recently revised (Volkman et al. 1995). The family currently consists of two genera, the Nucleopolyhedroviruses (NPVs) and Granuloviruses (GVs). These new genera replace the Nuclear Polyhedrosis Viruses (Subgenus A) and Granulosis Viruses (Subgenus B), respectively. The Nonoccluded Baculoviruses (Subgenus C) are no longer part of the family. For a description of the previous taxonomy see Bilimoria (1986).

The Nucleopolyhedroviruses consist of packaged nucleocapsids, either as one (SNPV) or many (MNPV), within a single viral envelope (virion). These are randomly distributed within the electron dense occlusion body (OB). OBs may vary in size from 0.15 to 15 µm. The type species of the genus is AcMNPV. There are 15 recognized species and 482 tentative species (Volkman et al. 1995). The Granuloviruses are on average smaller than the NPVs and consist of a single nucleocapsid within a virion each individually occluded. They range in size from 0.13 to 0.5 µm and contain a single molecule of dsDNA. The type species is *Plodia interpunctella* granulovirus (PiGV). There are four recognized species and 131 tentative species (Volkman et al. 1995). For details on baculovirus structure see Funk et al. (1997).

A typical nucleopolyhedrovirus infection cycle in a lepidopteran host is shown in Figure 2.1. OBs dissolve in the highly alkaline host midgut and the virus infects the host epithelial cells. The virus replicates within the nuclei of susceptible tissue cells. Tissue susceptibility varies greatly between viruses with some NPVs being capable of infecting almost all tissue types and most GVs being tissue-specific replications (e.g., fat body cell only). Early in the infection cycle a budded form of the virus migrates between cells, spreading infection from cell to cell while other virions remain in the host nucleus and are eventually occluded. For a recent review of pathogenesis see Federici (1997).

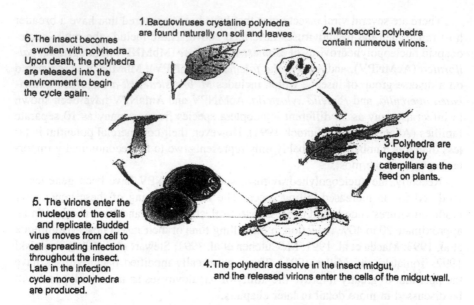

1. Baculoviruses crystaline polyhedra are found naturally on soil and leaves.

6. The insect becomes swollen with polyhedra. Upon death, the polyhedra are released into the environment to begin the cycle again.

2. Microscopic polyhedra contain numerous virions.

3. Polyhedra are ingested by caterpillars as the feed on plants.

5. The virions enter the nucleous of the cells and replicate. Budded virus moves from cell to cell spreading infection throughout the insect. Late in the infection cycle more polyhedra are produced.

4. The polyhedra dissolve in the insect midgut, and the released virions enter the cells of the midgut wall.

Figure 2.1   Illustration by Cheryl Giandalia. Reprinted by permission of E. I. DuPont.

The ecology, epizootiology, and population dynamics of baculoviruses are an interesting area of research that has received relatively little attention compared to other areas of baculovirology. Possible adaptation strategies, host range, ecological and physiological associations, and evolution of insect viruses were discussed and reviewed by Evans and Entwistle (1987). More recently, Cory et al. (1997) have reviewed the ecology of baculoviruses and have included discussions on genetic variation, viral transmission, temporal patterns, and risk assessment of genetically modified baculoviruses. For additional reviews on the ecology of baculoviruses see Dwyer (1991), Dwyer (1992), Evans (1986), Fuxa (1989 and 1990), Kaupp and Sohi (1985), and Hostetter and Bell (1985).

Efficacy of baculoviruses is measured in the laboratory by an insect bioassay. OBs are quantified usually under a light microscope using a haemocytometer. Other standardized techniques include dry counting, impression film, and electron microscope estimates (Evans and Shapiro 1997). Bioassays are usually based on OBs/ml for liquid solutions and OBs/gr of formulated or dry powders. There are several different types of bioassays using synthetic diet or leaf material. These protocols have been summarized by Evans and Shapiro (1997) and Hughes and Wood (1986).

Production of baculoviruses has been done using *in vivo* and *in vitro* techniques. Currently all commercially produced baculoviruses are produced *in vivo*. Techniques vary from producing AgMNPV in wild populations of *A. gemmatalis* in Brazilian soybean fields to the automated production of HzSNPV in *H. zea* on artificial diet reared in trays produced by a form fill and seal machine. Baculovirus production is reviewed by Black et al. (1997), Shapiro (1986), and Weiss and Vaughn (1986).

2.4 FUNGI

As with the bacteria, the diversity of fungi known to infect insects is great. Entomopathogenic fungi are found in the division Eumycota in the subdivisions: Mastigomycotina, Zygomycotina, Ascomycotina, and Deuteromycotina (see tables in Tanada and Kaya (1993) and McCoy et al. (1988) for details). The two most important orders are the Entomophthorales (Zygomycotina: Zygomycetes) and the Moniliales (Deuteromycotina: Hyphomycetes syn. Deuteromycetes). Recently there has been a move to reclassify the imperfect fungi (i.e., Deuteromycotina) as mitosporitic fungi because their sexual stages are unknown or no longer exist and they cannot be effectively classified. Deuteromycotina is an artificial classification and as sexual stages of these fungi are found they are moved to the correct subdivision (usually Ascomycotina).

All of the entomogenous fungi that have been registered as insecticides have been mitosporitic fungi with the exception of the water fungus (Mastigomycotina: Oomycetes) *Lagenidium giganteum* currently used for mosquito control. However, it should be noted that certain members of the Entomophthorales are extremely pathogenic to many insects and mites and have been shown to cause dramatic epizootics in the field (Steinkraus et al. 1995). They have never been commercialized because there are currently no economically viable production systems available for this important group. Mass production of entomogenous fungi is reviewed by Bartlett and Jaronski (1988).

There are numerous biotic and abiotic constraints on the ability of fungi to infect their hosts. These include desiccation, UV light, host behavior, temperature, pathogen vigor, and age, etc. (Lacey and Goettel 1995). Many of these constraints are currently being addressed by advances in formulation technology (Lacey and Goettel 1995). Unlike the bacteria and viruses, which must be consumed, toxicity from entomopathogenic fungi most often occurs from contact of the fungal conidia with the host cuticle. This necessitates thorough coverage of the pests and foliage. Brownbridge et al. (1998) found that high volume sprays using a hydraulic sprayer provided the best levels of control. Although there are no standardized bioassays, those that have been designed are often based on a direct conidial shower (Papierok and Hajek 1997). Assay techniques for most of the mitosporitic fungi are very similar and follow the general guidelines described by Papierok and Hajek (1997) for the Entomophthorales. However, assays for the water molds are somewhat unique. Bioassay techniques for the Oomycetes are described by Kerwin and Petersen (1997).

The first registered mycoinsecticide was *Hirsutella thompsonii* produced by Abbott Labs under the tradename Mycar®. *H. thompsonii* had been known to cause dramatic epizootics in spider mites as early as the 1920s (McCoy 1981). It received full registration in the U.S. in 1981. Registration of this product was cancelled in 1988. The next fungal insecticide, *Verticillium lecanii*, was registered and produced by the British firm Tate and Lyle. They registered two strains of *V. lecanii*. Vertalec® was effective on aphids and Mycotal® was targeted for greenhouse whitefly control.

Table 2.5   Commercially Avaliable Fungi for Insect Control

Commercial Name	Current Producer	Fungi/ Strain	Target Pests
Ago Bio. Bassiana	Ago Biocontrol	*Beauveria bassiana*	Soft-bodied insects
Ago Bio. Metarhizium 50	Ago Biocontrol	*Metarhizium anisopliae*	Coleoptera & Lepidoptera
Ago Bio. Verticillium 50	Ago Biocontrol	*Verticillium lecanii*	Greenhouse pests
Laginex AS	Agraquest Inc	*Lagenidium giganteum*	Mosquitoes
Engerlingspilz	Andermatt	*Beauveria brongniartii*	*Hoplochelis marginalis* & *Melolontha melolontha*
Lagenidium giganteum	CA Dept of Health	*Lagenidium giganteum*	Mosquitoes
Bioblast	EcoScience	*Metarhizium anisopliae*/ ESF 1	Termites
Biopath Roach Chamber	EcoScience/ Terminex	*Metarhizium anisopliae*/ ESF 1	Cockroaches
Green Muscle	Internat. Instit. Of Bio. Con.	*Metarhizium flavoviride*	Grasshoppers & locusts
Mycotal	Koppert	*Verticillium lecanii*/ whitefly strain	Whitefly
Vertalec	Koppert	*Verticillium lecanii*/ aphid strain	Aphids
Botanigard	Mycotech	*Beauveria bassiana*/ GHA	Whitefly, thrips, aphids, & mealybugs
Mycotrol	Mycotech	*Beauveria bassiana*/ GHA	Whitefly, thrips, aphids, & mealybugs
Betel	NPP/Calliope	*Beauveria brongniartii*	*Hoplochelis marginalis* & *Melolontha melolontha*
Ostrinil	NPP/Calliope	*Beauveria bassiana*/ Bb-147	*O. nubialis* & *O. funacalis*
PFR-97	Thermo Trilogy	*Pacilomyces fumosoroseus*/ Apopka 97	Whitefly, thrips, aphids, & spider mites
Naturalis	Troy Bioscience	*Beauveria bassiana*/ ATCC 74040	Homoptera, Coleoptera, & Heteroptera

Based on Copping (1998), CPCR (1998), CDMS (1998), CEPA/DPR (1998)

Both of these products are still marketed and manufactured by Koppert in Europe *V. lecanii* is also registered in South America by the Colombian company Ago Biocontrol. Efficacy of *V. lecanii* is reviewed by Hall (1981).

Currently, the most widely used fungal insecticide is *Beauveria bassiana*. There are at least three major strains of *B. bassiana* being marketed in the U.S., Europe, and South America (Table 2.5). *B. bassiana* is also probably manufactured and sold in China and Russia, although the authors have no specific information on available products. *B. bassiana* strain Bb-147 is registered on maize (Ostrinil®) in Europe for control of the European corn borer, *O. nubilalis*, and the Asiatic corn borer, *Ostrinia furnacalis*. In contrast, *B. bassiana* strain GHA is registered in the U.S. for control of whitefly, thrips, aphids, and mealybugs (Mycotrol® and Botanigard®) (Copping 1998). *B. bassiana* strain ATCC 74040 is registered against many soft-bodied insects in the orders Homoptera and Coleoptera (Naturalis®). *B. bassiana* is also registered by Ago Biocontrol. Another strain of *Beauveria*, *B. brongiartii*, is registered on sugar cane and barley for control of white grubs and cockchafers (Betel® and Engerlingspilz®). In barley it is delivered to cockchafer grubs through inoculated seed (Copping 1998).

There are currently two registered fungal insecticides in the genus *Metarhizium*. *M. anisopliae* was one of the first fungi used in biological control experiments. Metchnikoff produced large quantiies of *M. anisopliae* spores to control the wheat cockchafer, *Anisoplia austriaca*, in Russia in 1879 (McCoy et al.1988). It is currently registered in the U.S. for use in roach traps (Biopath®) and for control of termites (Bioblast®). It is also produced by Ago Biocontrol for control of coleopteran and lepidopteran greenhouse pests. *M. flavoviride* has been registered by the International Institute of Biological Control for grasshoppers and locusts in Africa and the tropics.

Paecilomyces fumosoroseus has been recently registered in the U.S. for control of whitefly, thrips, aphids, and spider mites (PFR-97®). This strain (Apopka 97) was originally isolated from the mealybug, *Phenacocus solani*, in Apopka FL. It was commercialized by W. R. Grace and recently licensed to Thermo Trilogy. It is being developed in Europe by Biobest.

The water mold *Lagenidium giganteum* was first registered for mosquito control by the California Department of Health Services in 1991. It is currently produced under the tradename Laginex® by Agraquest. *L. giganteum* is a facultative mosquito parasite with a large host range and worldwide distribution (Federici 1981). The Mastigomycotina are a primitive fungal group that are almost entirely aquatic and show the closest phylogenetic resemblance to the protozoa because they produce a zoospore. It is the zoospore of *L. giganteum* that kills its host when it invades the hemocoel and forms an extensive mycelium throughout the body (McCoy et al. 1988). Commercial production of *L. giganteum* is probably the most difficult of any of the current fungal insecticides. *L. giganteum* has a very narrow host range with distinct sterol and nutritional requirements (Bartlett and Jaronski 1988).

Production of the Deuteromycetes is much less expensive. They have a much broader host range and are able to grow and sporulate on many generalized media. They can adapt to a wide variety of growing conditions. This has made them the fungi most amenable to mass production. For descriptions of mass production of *B. bassiana, B. brongiartii, M. anisopliae*, and *P. fumosoroseus* see Bartlett and Jaronski (1988).

2.5 NEMATODES

Nematodes are not actually microbial organisms but are always included in discussions, reviews, or symposia on microbial control of insects. According to Gaugler and Kaya (1990) they occupy a middle ground "between predators, parasitoids and microbial pathogens; a position that endows them with a unique combination of biological control attributes." They are probably second only to bacteria (i.e., *B.t.*) in terms of commercial importance of microbial insecticides. Table 2.6 lists the currently available nematodes for insect control. It represents eight species of nematodes from three genera being sold and/or manufactured by over a dozen companies offering 36 different products. This list probably greatly underestimates

Table 2.6   Commercially Available Nematodes for Insect Control

Commercial Name	Current Producer	Nematode	Target Pests
Dickmaulrussler-nematoden	Andermatt	*Heterorhabditis megidis*	Soil insects primarily *Otiorhynchus sulcatus*
Traunem	Andermatt	*Steinernema feltiae*	Sciarid flies, and other soil insects
EcoMask	BioLogic	*Steinernema carpocapsae*	Root weevils, cutworms, fleas, borers, fungus gnats
Heteromask	BioLogic	*Heterorhabditis bacteriophora*	*P. japonica* larvae and other insects
Hortscan	BioLogic	*Steinernema carpocapsae*	Root weevils, cutworms, fleas, borers, fungus gnats
Termask	BioLogic	*Steinernema carpocapsae*	Termites
Scanmask	BioLogic/ IPM Laboratories	*Steinernema feltiae*	Sciarid flies, and other soil insects
Nema-BIT	BIT	*Heterorhabditis bacteriophora*	*P. japonica* larvae and other insects
Cruiser	Ecogen	*Heterorhabditis bacteriophora*	*P. japonica* larvae and other insects
Otinem S	Ecogen	*Steinernema scapterisci*	Mole crickets
Nema-top	e-nema	*Heterorhabditis bacteriophora*	*P. japonica* larvae and other insects
Nema-green	e-nema	*Heterorhabditis bacteriophora*	*P. japonica* larvae and other insects
Nema-plus	e-nema	*Steinernema feltiae*	Sciarid flies, and other soil insects
Steinernema glaseri	Greenfire	*Steinernema glaseri*	White grubs (Scarabaeidae)
Guardian	Hydro-Gardens	*Steinernema carpocapsae*	Root weevils, cutworms, fleas, borers, fungus gnats
Lawn Patrol	Hydro-Gardens	*Heterorhabditis bacteriophora*	*P. japonica* larvae and other insects
Steinernema glaseri	Integrated Pest Management	*Steinernema glaseri*	White grubs (Scarabaeidae)
Larvanem	Koppert	*Heterorhabditis megidis*	Soil insects primarily *Otiorhynchus sulcatus*
Entonem	Koppert	*Steinernema feltiae*	Sciarid flies, and other soil insects
Sciarid	Koppert	*Steinernema feltiae*	Sciarid flies, and other soil insects
Nemasys	MicroBio/ Biobest	*Steinernema feltiae*	Sciarid flies, and other soil insects
Nemasys-H	MicroBio/ Biobest	*Heterorhabditis megidis*	Soil insects primarily *Otiorhynchus sulcatus*
Nemasys-M	MicroBio/ Biobest	*Steinernema feltiae*	Sciarid flies, and other soil insects
Nemaslug	MicroBio	*Phasmarhabditis hermaphrodita*	Slugs
Heterorhabditis megidis	Neudorff	*Heterorhabditis megidis*	Soil insects primarily *Otiorhynchus sulcatus*
Steinernema carpocapsae	Neudorff	*Steinernema carpocapsae*	Root weevils, cutworms, fleas, borers, fungus gnats

Table 2.6 (continued)   Commercially Available Nematodes for Insect Control

Commercial Name	Current Producer	Nematode	Target Pests
Steinernema feltiae	Neudorff	*Steinernema feltiae*	Sciarid flies, and other soil insects
Exhibit SC-WDG	Novartis BCM	*Steinernema carpocapsae*	Root weevils, cutworms, fleas, borers, fungus gnats
Exhibit SF-WDG	Novartis BCM	*Steinernema feltiae*	Sciarid flies, and other soil insects
Steinernema glaseri	Praxis	*Steinernema glaseri*	White grubs (Scarabaeidae)
Bio Safe WG	Thermo Trilogy	*Steinernema carpocapsae*	Root weevils, cutworms, fleas, borers, fungus gnats
Millenium	Thermo Trilogy	*Steinernema carpocapsae*	Root weevils, cutworms, fleas, borers, fungus gnats
Savior WG	Thermo Trilogy	*Steinernema carpocapsae*	Root weevils, cutworms, fleas, borers, fungus gnats
Magnet	Thermo Trilogy	*Steinernema feltiae*	Sciarid flies, and other soil insects
X-Gnat	Thermo Trilogy	*Steinernema feltiae*	Sciarid flies, and other soil insects
Bio Vector 355	Thermo Trilogy	*Steinernema riobrave*	Root weevils, mole crickets, and plant nematodes
Devour	Thermo Trilogy	*Steinernema riobrave*	Root weevils, mole crickets, and plant nematodes

Based on Copping (1998), CPCR (1998), CDMS (1998)

the nematode products worldwide as many countries, including the U.S., do not require EPA registration for any species or strains of *Steinernema* or *Heterorhabditis* (Hominick and Reid 1990). Thus, many companies are easily overlooked when attempting to compile such a table.

The first reports of insect parasitic nematodes come from the mid-1700s (Tanada and Kaya 1993). This is the same time period as "little trees" were being described growing out of dead wasps. These "little worms" remained not much more than a curiosity for the next 200 years. Then in 1929 R. W. Glaser isolated *Steinernema glaseri* from a Japanese beetle (*P. Japonica*) and was able to culture it on artificial media and use it in field tests against the beetle. The nematodes significantly reduced populations in several trials (Tanada and Kaya 1993). In the mid-1950s *Steinernema carpocapsae* was discovered and by the mid-1970s *Heterorhabditis* had been described. The first commercially available nematode was *Romanomermis culicivorax*, registered for mosquito control in 1976. The commercial venture failed because of supply chain problems (i.e., production, storage and transport) and because *B.t.i.* provided more effective biological control (Tanada and Kaya 1993).

Entomopathogenic nematodes are found within two different classes of Nemata (i.e., Adenophorea and Secernentea) spanning five orders (i.e., Adenophorea: Stichosomida, Secernentea: Rabditida, Secernentea: Diplogasterida, Secernentea: Tylenchida, and Secernentea: Aphelenchida). As is the case with all other microbial insecticides, the commercially available nematodes comprise a much smaller taxonomic subset. Currently there are eight commercially available species from three families within the Rabditida (i.e., Rhabditidae, Steinernematidae, and Heterorhabditidae) (Table 2.6).

Nematodes parasitize their hosts by direct penetration either through the cuticle or natural opening in the host integument (i.e., spiracles, mouth, or anus) (Kaya and Gaugler 1993). Insect death is not due to the nematode itself but a symbiotic bacteria that is released by the dauer juvenile upon entry into the host (Akhurst and Boemare 1990). The symbionts are specific with members of the genus *Xenorhabdus* associated with the steinernematids and *Photorhabdus* associated with the heterorhabditids (Lacey and Goettel, 1995). There is currently a great deal of interest in the toxins of these bacteria as researchers continue to look for new proteins capable of being expressed in transgenic plants.

A standardized bioassay for entomopathogenic nematodes has not been developed. Assay techniques for *Romanomermis culicivorax*, steinernematids, and heterorhabditids are reviewed by Kaya and Stock (1997). Probably one of the most successful nematodes in terms of production and marketing is *Steinernema carpocapsae*. It is sold under at least nine different names targeting a diverse group of soil-inhabiting insects including termites, fleas, cutworms, and root weevils. Probably some of the most consistent efficacy trails have been against root weevils such as the black vine weevil, *Otiorhynchus sulcatus* (Klein 1990).

In general, both steinernematids and heterorhabditids tend to do best against soil-inhabiting insects and borers. There has been limited success when applying nematodes to foliage, aquatic habitats, or on pests inhabiting manure (Begley 1990). New formulations and strain selection may be able to address these limitations. *Steinernema feltiae* is sold for control of sciarid flies particularly in mushroom production. This nematode has been shown to persist in compost and give control of the sciarid *Lycoriella auripila* equivalent to diflubenzuron (Begley 1990). *Steinernema glaseri* is still marketed in the U.S. for grub control by Praxis. *Steinernema riobravis* has been recently registered by Thermo Trilogy for control of mole crickets, *Scapteriscus spp*, the sugarcane rootstalk borer *Diaprepes abbreviatus*, citrus root weevil, *Pachnaeus litus*, and the blue-green weevil, *P. opalus*, in citrus (Copping 1998). *Steinernema scapterisci* was isolated from mole crickets in South America. It was originally produced by BioSys (now Thermo Trilogy) who licensed it to Ecogen. It is listed in *The Biopesticide Manual* (Copping 1998) but is not listed by Ecogen on their worldwide web site as a currently produced product.

There are two species of heterorhabditids that are currently commercially available. *Heterorhabditis megidis* is not available in the U.S. and is sold mostly in Europe for weevil and soil insect control. *Heterorhabditis bacteriophora* is effective on a wide range of soil insect pests but is sold primarily for Japanese beetle control (Copping 1998).

Two major hurdles that had to be overcome for commercialization of entomopathogenic nematodes was the development of large scale *in vitro* rearing systems and formulations that would allow for adequate shelf life and infectivity in the field. Glaser began rearing nematodes on artificial diet almost 70 years ago. He discovered early on that nematode yields could be drastically affected by microbial contaminants that periodically appeared in his media. He also discovered that if he allowed the natural flora from the nematode to establish prior to inoculation, contaminants were greatly reduced and consistency of production was dramatically improved (Friedman 1990). We now know that the natural flora that colonized Glaser's plates was the symbiotic bacteria *Xenorhabdus*. Glaser developed the first large-scale production process using yeast, bovine ovarian substance, and additional antimicrobials. His costs were high and his yields inconsistent, but he had laid the foundation for mass production of nematodes. Now some 60 years later, nematodes are grown in large-scale bioreactors similar to those used for the production of *B.t.* or antibiotics. Formulation of nematodes has been another challenge that needed to be overcome to create a commercial venture with entomopathogenic nematodes. Nematodes must be held in a chilled state prior to formulation and then mixed with materials that will enhance their handling, application, persistence, and storage. These materials include alginate, clays, activated charcoals, and polyacrylamide gels. For a review of nematode formulations see Georgis (1990).

2.6 PROTOZOA

Entomogenous protozoa are an extremely diverse group with relationships ranging from commensal to pathogenic. They are generally slow acting and debilitating rather than quick and acute. Although they are undoubtedly important in natural biological regulation of insect populations, they do not possess the attributes necessary for a successful microbial insecticide. They can be extremely effective at reducing the fitness and fecundity of insects reared in culture so that they may have a very real negative effect on microbial insecticides (such as viruses) produced *in vivo*. Most protozoan infections cause sluggishness, irregular or slowed growth, resulting in reduced feeding, vigor, fecundity, and longevity (Lacey and Goettel 1995; Tanada and Kaya 1993).

Currently in the U.S. there is one registered protozoan insecticide targeted against grasshoppers and the Mormon cricket. The microsporidian *Nosema locustae* is known to infect at least 60 different species of grasshoppers and crickets. It is sold in the western U.S. as Nolo Bait® by M & R Durango and Grasshopper Control Semaspore Bait® by Beneficial Insectary. Both products are formulated on bran that is consumed by the grasshoppers. Death follows shortly after lethal infection and egg production is reduced by 60 to 80% in surviving adults (Copping 1998). Its utility as a grasshopper control agent remains questionable and after a decade of use its effectiveness in the U.S., Canada, and Africa continues to be assessed with mixed results (Lacey and Goettel 1995). However, assessment of sublethal effects in highly nomadic insects (such as grasshoppers) is problematic at best. Bioassay techniques to determine lethal and sublethal effects are described by Undeen and Vavra (1997).

2.7 THE FUTURE OF MICROBIAL PESTICIDES

Entomopathogens often regulate insect populations under natural conditions. Most often, however, pest insect epizootics do not reduce populations below the necessary economic threshold. Inoculation of a pest population with an insect pathogen has provided classical biological control in some situations (Tanada and Kaya 1993), but the inundative use of microbial pesticides has proven to be a more effective method for reducing insect pest populations. Alternative methods of pest control have become an integral part of IPM programs as the concern over the effects of conventional chemical pesticides has increased in the past two decades. As these compounds are removed from the marketplace, there is an ongoing search for safe, effective products that have minimal impact on beneficial organisms. Microbial pesticides are presently experiencing a renaissance as regulatory constraints are resolved and mass production technology is improved. Their success, however, will ultimately depend on the willingness of consumers to employ new methods of pest control and integrate microbials into their IPM programs.

REFERENCES

Adams, J. R., and Bonami, J. R. (eds.) (1991a). *Atlas of Invertebrate Viruses*, CRC Press, Boca Raton, FL.

Adams, J. R., and McClintock, J. T. (1991b). "Baculoviridae, Nuclear Polyhedrosis Viruses Part 1. Nuclear Polyhedrosis Viruses of Insects." *Atlas of Invertebrate Viruses*, J. R. Adams and J. R. Bonami, eds., CRC Press, Boca Raton, FL, 87–204.

Adang, M. J. (1991). "*Bacillus thuringiensis* insecticidal crystal proteins: gene structure, action, and utilization." *Biotechnology for Biological Control of Pests and Vectors*, K. Maramorosch, ed., CRC Press, Boca Raton, FL, 4–24.

Akhurst, R. J., and Boemare, N. E. (1990). "Biology and taxonomy of *Xenorhabdus*." *Entomopathogenic Nematodes in Biological Control*, R. Gaugler and H. K. Kaya, eds., CRC Press, Boca Raton, FL, 75–90.

Bartlett, M. C., and Jaronski, S. T. (1988). "Mass production of entomogenous fungi." *Fungi in Biological Control*, M. N. Burge, ed., Manchester University Press, Manchester, U.K., 61–85.

Becker, N., and Margalit, J. (1993). "Use of *Bacillus thuringiensis israelensis* against mosquitoes and blackflies." Bacillus thuringiensis, *an Environmental Biopesticide: Theory and Practice*, P. F. Entwistle, J. S. Cory, M. J. Bailey, and S. Higgs, eds., John Wiley and Sons, Chichester, U.K., 147–170.

Begley, J. W. (1990). "Efficacy against insects in habitats other than soil." *Entomopathogenic Nematodes in Biological Control*, R. Gaugler and H. K. Kaya, eds., CRC Press, Boca Raton, FL, 215–231.

Bernhard, K., and Utz, R. (1993). "Production of *Bacillus thuringiensis* insecticides for experimental and commercial uses." Bacillus thuringiensis, *an Environmental Biopesticide: Theory and Practice*, P. F. Entwistle, J. S. Cory, M. J. Bailey, and S. Higgs, eds., John Wiley and Sons, Chichester, U.K., 256–267.

Berry, C., Hindley, J., and Oei, C. (1991). "The *Bacillus sphaericus* toxins and their potential for biotechnology development." Bacillus thuringiensis *Insecticidal Crystal Proteins: Gene Structure, Action, and Utilization*, K. Maramorosch, ed., CRC Press, Boca Raton, FL, 35–51.

Bilimoria, S. L. (1986). "Taxonomy and identification of baculoviruses." *The Biology of Baculoviruses: Volume I, Biological Properties and Molecular Biology*, R. R. Granados and B. A. Federici, eds., CRC Press, Boca Raton, FL.

Black, B. C., Brennan, L. A., Dierks, P. M., and Gard, I. E. (1997). "Commercialization of baculoviral insecticides." *Baculoviruses,* L. K. Miller, ed., Plenum Press, New York, 341–387.

Brownbridge, M., Skinner, M., and Parker, B. L. (1998). Factors affecting the efficacy of fungal preparations in ornamental pest management. Ohio Florist's Association Bulletin no. 824: 14–16.

Burges, H. D. (ed.) (1981a). *Microbial Control of Pest and Plant Diseases 1970–1980,* Academic Press, London.

Burges, H. D. (1981b). "Safety, safety testing and quality control of microbial pesticides." *Microbial Control of Pests and Plant Diseases 1970–1981,* H. D. Burges, ed., Academic Press, London, 738–763.

Burges, H. D., and Hussey, N. W. (eds.) (1971). *Microbial Control of Insects and Mites,* Academic Press, London.

Carlton, B. C. (1988). "Development of genetically improved strains of *Bacillus thuringiensis*: A biological insecticide." *Biotechnology for Crop Protection,* P. A. Hedin, J. J. Menn, and R. M. Hollingworth, eds., American Chemical Society, Washington, D.C., 260–279.

CDMS. (1998). "Crop Data Management Systems." *http://www.cdms.net/manuf/manuf.asp* (12/10/98).

CEPA/DPR. (1998). "USEPA/OPP Pesticide Products Database." *http://www.cdpr.ca.gov/docs/epa/m2.htm* (12/10/1998).

Clem, R. J. (1997). "Regulation of programmed cell death by baculoviruses." *Baculoviruses,* L. K. Miller, ed., Plenum Press, New York, 237–266.

Copping, L. G. (ed.) (1998). *The Biopesticide Manual,* British Crop Protection Council, Franham, Surrey, U.K.

Corke, A. T. K., and Rishbeth, J. (1981). "Use of Microorganisms to control plant diseases." *Microbial Control of Pests and Plant Diseases 1970–1981,* H. D. Burges, ed., Academic Press, London.

Cory, J. S., Hails, R. S., and Sait, S. M. (1997). "Baculovirus ecology." *Baculoviruses,* L. K. Miller, ed., Plenum Press, New York, 301–339.

CPCR. (1998). *"Crop Protection Chemicals Reference."* *http://www.greenbook.net/welcome.htm* (12/10/98).

Crickmore, N., Zeigler, D. R., Feitelson, J., Schnepf, E., Van Rie, J., Lereclus, D., Baum, J., and Dean, D. H. (1998a). *"Bacillus thuringiensis* toxin nomenclature." *http://www.biols.susx.ac.uk/Home/Neil_Crickmore/Bt/index.html* (12/10/1998).

Crickmore, N., Zeigler, D. R., Feitelson, J., Schnepf, E., Van Rie, J., Lereclus, D., Baum, J., and Dean, D. H. (1998b). "Revision of the nomenclature for the *Bacillus Thuringiensis* pesticidal crystal proteins." *Micro. Mol. Biol. Rev.,* 62, 807–813.

Davidson, E. W., Singer, S., and Briggs, J. D. (1975). "Pathogenesis of *Bacillus sphaericus* strain SSII-1 infectious in *Culex pipiens quinquefasciatus* (=*C. pipiens fatigans*) larvae." *J. Invertebr. Pathol.,* 25, 179–184.

de Barjac, H., and Sutherland, D. J. (eds.) (1990). *Bacterial Control of Mosquitoes and Blackflies,* Rutgers University Press, New Brunswick, NJ.

Dulmage, H. T., Boening, O. P., Rehnborg, C. S., and Hansen, G. D. (1971). "A proposed standardized bioassay for formulations of *Bacillus thuringiensis* based on the international unit." *J. Invert. Pathol.,* 18, 240–245.

Dulmage, H., Correa, J. A., and Gallegos-Morales, G. (1990). "Potential for improved formulations of *Bacillus thuringiensis israelensis* through standardization and fermentation development." *Bacterial Control of Mosquitoes and Blackflies,* H. de Barjac and D. J. Sutherland, eds., Rutgers University Press, New Brunswick, NJ, 110–133.

Dwyer, G. (1991). "The roles of density, stage, and patchiness in the transmission of an insect virus." *Ecology*, 72, 559–574.

Dwyer, G. (1992). "On the Spatial Spread of Insect Pathogens — Theory and Experiment." *Ecology*, 73(2), 479–494.

Ellar, D. J., Knowles, B. H., Haider, M. Z., and Drobniewski, F. A. (1986). "Investigation of the specificity, cytotoxic mechanisms and relatedness of *Bacillus thuringiensis* insecticidal δ-endotoxins from different pathotypes." *Bacterial Protein Toxins*, P. Falmagne, J. E. Alouf, F. J. Fehrenbach, J. Jeljaszewicz, and M. Thelestam, eds., Gustav Fischer Verlag, Stuttgart.

Estruch, J. J., Carozzi, N. B., Desai, N., Duck, N. B., Warren, G. W., and Koziel, M. G. (1997). "Transgenic plants: An emerging approach to pest control." *Nature Biotechnol.*, 15, 137–141.

Evans, H. F. (1986). "Ecology and Epizootiology of Baculoviruses." *The Biology of Baculoviruses:* Volume II, *Practical Application for Insect Control*, R. R. Granados and B. A. Federici, eds., CRC Press, Boca Raton, FL, 89–132.

Evans, H. F., and Entwistle, P. F. (1987). "Viral Diseases." *Epizootiology of Insect Diseases*, J. R. Fuxa and Y. Tanada, eds., John Wiley and Sons, New York.

Evans, H., and Shapiro, M. (1997). "Viruses." *Manual of Techniques in Insect Pathology*, L. Lacey, ed., Academic Press, San Diego, 17–53.

Federici, B. A. (1981). "Mosquito control by the fungi *Culicinomyces, Lagenidium*, and *Coelomomyces.*" *Microbial Control of Pest and Plant Diseases 1970–1980*, H. D. Burges, ed., Academic Press, London, 555–572.

Federici, B. A. (1997). "Baculovirus pathogenesis." *Baculoviruses*, L. K. Miller, ed., Plenum Press, New York, 33–59.

Ferro, D., and Gelernter, W. D. (1989). "Toxicity of a new strain of *Bacillus thuringiensis* to Colorado potato beetle." *J. Econ. Entomology*, 82, 750–755.

Friedman, M. J. (1990). "Commercial production and development." *Entomopathogenic Nematodes in Biological Control*, R. Gaugler and H. K. Kaya, eds., CRC Press, Boca Raton, FL, 153–172.

Friesen, P. D. (1997). "Regulation of baculovirus early gene expression." *Baculoviruses*, L. K. Miller, ed., Plenum Press, New York, 141–170.

Funk, C. J., Braunagel, S. C., and Rohrmann, G. F. (1997). "Baculovirus structure." *Baculoviruses*, L. K. Miller, ed., Plenum Press, New York, 7–32.

Fuxa, J.R. (1989). "Importance of epizootiology to biological control of insects with viruses." *Mem. Inst. Oswaldo Cruz, Rio de Janeiro*, 84 (suppl. III), 81–88.

Fuxa, J.R. (1990). "Fate of released entomopathogens with reference to risk assessment of genetically engineered microorganisms." *Bull. Ent. Soc. Amer.*, 35, 12–24.

Gaugler, R., and Kaya, H. K. (eds.) (1990). *Entomopathogenic Nematodes in Biological Control*, CRC Press, Boca Raton, FL.

Georgis, R. (1990). "Formulation and application technology." *Entomopathogenic Nematodes in Biological Control*, R. Gaugler and H. K. Kaya, eds., CRC Press, Boca Raton, FL, 173–191.

Granados, R., and Federici, B. A. (eds.) (1986a). *The Biology of Baculoviruses: Volume I, Biological Properties and Molecular Biology*, CRC Press, Boca Raton, FL.

Granados, R., and Federici, B. A. (eds.) (1986b). *The Biology of Baculoviruses:* Volume II, *Practical Applications for Insect Control*, CRC Press, Boca Raton, FL.

Grimont, P. A. D., and Grimont, F. (1978). "The genus *Serratia.*" *Annu. Rev. Microbiol.*, 32, 221–248.

Hall, R. A. (1981). "The fungus *Verticillium lecanii* as a microbial control against aphids and scales." *Microbial Control of Pests and Plant Diseases 1970–1980*, H. D. Burges, ed., Academic Press, London, 483–498.

Hammock, B. D., McCutchen, B. F., Beetham, J., Choudary, P. V., Fowler, E., Ichinose, R., Ward, V. K., Vickers, J. M., Bonning, B. C., Harshman, L. G., Grant, D., Uematsu, T., and Maeda, S. (1993). "Development of Recombinant Viral Insecticides by Expression of an Insect-Specific Toxin and Insect-Specific Enzyme in Nuclear Polyhedrosis Viruses." *Arch. Insect Biochem. Physiol.*, 22(3–4), 315–344.

Hirano, M. (1989). "Characteristics of Pyrethroids for insect control in Agriculture." *Pestic. Sci.*, 27, 353–360.

Hofte, H., and Whiteley, H. R. (1989). "Insecticidal crystal proteins of *Bacillus thuringiensis*." *Microbiol. Rev.*, 53, 242–255.

Hominick, W. M., and Reid, A. P. (1990). "Perspectives on entomopathogenic nematology." *Entomopathogenic Nematodes in Biological Control*, R. Gaugler and H. K. Kaya, eds., CRC Press, Boca Raton, FL, 327–348.

Hostetter, D.L., and Bell, M.R. (1985). "Natural dispersal of Baculoviruses in the environment." *Viral Insecticides for Biological Control*, K. Maramorosch and K.E. Sherman, eds., Academic Press, Orlando, 249–284.

Hughes, P. R., and Wood, H. A. (1986). "In vivo and in vitro bioassay methods for Baculoviruses." *The Biology of Baculoviruses:* Volume II, *Practical Application for Insect Control*, R. R. Granados and B. A. Federici, eds., CRC Press, Boca Raton, FL, 1–30.

Ignoffo, C. M. (1973). "Development of a viral insecticide: concept to commercialization." *Exp. Parasitol.*, 33, 380–406.

Jarvis, D. L. (1997). "Baculovirus expression vectors." *Baculoviruses*, L. K. Miller, ed., Plenum Press, New York, 389–431.

Katagiri, K. (1981). "Pest control by Cytoplasmic Polyhedrosis Viruses." *Microbial Control of Pest and Plant Diseases 1970–1980*, H. D. Burges, ed., Academic Press, London, 433–440.

Kaupp, W.J., and Sohi, S.S. (1985). "The role of viruses in the ecosystem." *Viral Insecticides for Biological Control*, K. Maramorosch and K.E. Sherman, eds., Academic Press, Orlando, 249–284.

Kaya, H. K., and Gaugler, R. (1993). "Entomopathogenic nematodes." *Annu. Rev. Entomol.*, 38, 181–206.

Kaya, H. K., and Stock, P. (1997). "Techniques in insect nematology." *Manual of Techniques in Insect Pathology*, L. L. Lacey, ed., Academic Press, San Diego, 281–324.

Keller, B., and Langenbruch, G. (1993). "Control of Coleopteran pests by *Bacillus thuringiensis*." Bacillus thuringiensis, *an Environmental Biopesticide: Theory and Practice*, P. F. Entwistle, J. S. Cory, M. J. Bailey, and S. Higgs, eds., John Wiley and Sons, Chichester, U.K., 171–191.

Kerwin, J. L., and Petersen, E. E. (1997). "Fungi: Oomycetes and Chytridiomycetes." *Manual of Techniques in Insect Pathology*, L. L. Lacey, ed., Academic Press, San Diego, 251–268.

Klein, M. G. (1990). "Efficacy against soil-inhabiting pests." *Entomopathogenic Nematodes in Biological Control*, R. Gaugler and H. K. Kaya, eds., CRC Press, Boca Raton, FL, 195–214.

Krieg, A., Huger, A. M., Langenbruch, G. A., and Schnetter, W. (1983). "*Bacillus thuringiensis* var. *tenebrionis*: Ein neur gegenuber Larven von Coleoptera wirksamer Pathotyp." *Z. Angew. Entomol.*, 96, 500–508.

Kurstak, E. (ed.) (1982). *Microbial and Viral Pesticides*, Marcel Dekker, New York.

Kurstak, E. (ed.) (1991). *Viruses of Invertebrates*, Marcel Dekker, New York.

Lacey, L. A., and Goettel, M. S. (1995). "Current developments in microbial control of insect pests and prospects for the early 21st century." *Entomophage*, 40, 3–27.

Lacey, L. L. (ed.) (1997). *Manual of Techniques in Insect Pathology*, Academic Press, San Diego.

Lu, A., Krell, P. J., Vlak, J. M., and Rohrmann, G. F. (1997). "Baculovirus DNA replication." *Baculoviruses*, L. K. Miller, ed., Plenum Press, New York, 171–191.

Lu, A., and Miller, L. K. (1997). "Regulation of baculovirus late and very late gene expression." *Baculoviruses*, L. K. Miller, ed., Plenum Press, New York, 193–216.

Maeda, S., Volrath, S. L., Hanzlik, T. N., Harper, S. A., Majima, K., Maddox, D. W., Hammock, B. D., and Fowler, E. (1991). "Insecticidal effects of an insect-specific neurotoxin expressed by a recombinant baculovirus." *Virology*, 184, 777–780.

Martignoni, M. E., and Iwai, P. J. (1986). "A catalogue of viral diseases of insects, mites, and ticks." *USFS Pac. Northwest Res. STN. Gen. Tech. Rep. PNW-195.*

McCoy, C. W. (1981). "Pest control by the fungus *Hirsutella thompsonii.*" *Microbial Control of Pests and Plant Diseases 1970–1980*, H. D. Burges, ed., Academic Press, London, 499–512.

McCoy, C. W., Samson, R. A., and Boucias, D. G. (1988). "Entomogenous Fungi." *Handbook of Natural Pesticides:* Volume V. *Microbial Insecticides*, Part A. *Entomogenous Protozoa and Fungi*, C. M. Ignoffo and N. Bhushan Mandava, eds., CRC Press, Boca Raton, FL, 151–236.

McCutchen, B. F., Choudary, P. V., Crenshaw, R., Maddox, D., Kamita, S. G., Palekar, N., Volrath, S., Fowler, E., Hammock, B. D., and Maeda, S. (1991). "Development of a recombinant baculovirus expressing an insect-selective neurotoxin: potential for pest control." *BioTechnology*, 9, 848–852.

Meadows, M. P. (1993). "*Bacillus thuringiensis* in the environment: Ecology and risk assessment." Bacillus thuringiensis, *an Environmental Biopesticide: Theory and Practice*, P. F. Entwistle, J. S. Cory, M. J. Bailey, and S. Higgs, eds., John Wiley and Sons, Chichester, U.K., 193–220.

Miller, L. K. (ed.) (1997a). *The Baculoviruses*, Plenum Press, New York.

Miller, L. K. (1997b). "Introduction to the baculoviruses." *The Baculoviruses*, L. K. Miller, ed., Plenum Press, New York, 1–6.

Miller, L. K., and Lu, A. (1997). "The molecular basis of baculovirus host range." *Baculoviruses*, L. K. Miller, ed., Plenum Press, New York, 217–235.

Moscardi, F. (1989). "Use of viruses for pest control in Brasil: the case of the nuclear polyhedrosis virus of the soybean caterpillar, Anticarsia gemmatalis." *Mem. Inst. Oswaldo Cruz, Rio de Janeiro*, 84(Suppl. III), 51–56.

Murphy, F. A., Fauquet, C. M., Bishop, D. H. L., Ghabrial, S. A., Jarvis, A. W., Martelli, G. P., Mayo, M. A., and Summers, M. D. (1995). *Virus Taxonomy: Classification and Nomenclature of Viruses*, Springer-Verlag, Wien, New York.

Navon, A. (1993). "Control of Lepidopterous pests with *Bacillus thuringiensis.*" Bacillus thuringiensis, *an Environmental Biopesticide: Theory and Practice*, P. F. Entwistle, J. S. Cory, M. J. Bailey, and S. Higgs, eds., John Wiley and Sons, Chichester, U.K., 125–146.

O'Reilly, D. R. (1997). "Auxiliary genes of baculoviruses." *Baculoviruses*, L. K. Miller, ed., Plenum Press, New York, 267–300.

Onstad, D.W. (1996). "Viral Diseases of Insects in the Literature (VIDIL)." *http://insectweb.inhs.uiuc.edu/Pathogens/VIDIL/index.html.*

Onstad, D.W. (1997). "Ecological Database of the World's Insect Pathogens (EDWIP)." *http://insectweb.inhs.uiuc.edu/Pathogens/EDWIP/index.html.*

Papierok, B., and Hajek, A. E. (1997). "Fungi: Entomophthorales." *Manual of Techniques in Insect Pathology*, L. L. Lacey, ed., Academic Press, San Diego, 187–212.

Pietrantonio, P. V., Federici, B. A., and Gill, S. S. (1993). "Interaction of *Bacillus thuringiensis* endotoxin with the insect midgut epithelium." *Parasites and Pathogens of Insects*, N. E. Beckage, S. N. Thompson, and B. A. Federici, eds., Academic Press, San Diego, 55–74.

Possee, R. D. (1997). "Baculoviruses as expression vectors." *Current Opinion in Biotechnology*, 8(5), 569–572.

Possee, R. D., and Rohrmann, G. F. (1997). "Baculovirus genome organization and evolution." *Baculoviruses*, L. K. Miller, ed., Plenum Press, New York, 109–140.

Schnepf, E., Crickmore, N., Van Rie, J., Lereclus, D., and Dean, D. H. (1998). "*Bacillus thuringiensis* and its pesticidal proteins." *Micro. Mol. Biol. Rev.*, 62, 775–806.

Shapiro, M. (1986). "In vivo production of Baculoviruses." *The Biology of Baculoviruses: Volume II, Practical Applications for Insect Control*, R. Granados and B. A. Federici, eds., CRC Press, Boca Raton, FL, 31–62.

Singer, S. (1973). "Insecticidal activity of recent bacterial isolates and their toxins against mosquito larvae." *Nature*, 244, 110–111.

Sivan, A., and Chet, I. (1992). "Microbial Control of Plant Diseases." *Environmental Microbiology*, R. Mitchell, ed., John Wiley and Sons, New York.

Steinhaus, E. A. (1956). "Microbial control. The emergence of an idea." *Hilgardia*, 26, 107–160.

Steinhaus, E. A. (1960). "Insect pathology: Challenge, achievement, and promise." *Bull. Entomol. Soc. Am.*, 6, 9–16.

Steinhaus, E. A. (1975). *Disease in a Minor Chord*, Ohio State University Press, Columbus.

Steinkraus, D. C., Hollingsworth, R. G., and Slaymaker, P. H. (1995). "Prevalence of *Neozygites fresenii* (Entomophthorales: Neozygitaceae) on cotton aphids (Homoptera: Aphidae) in Arkansas cotton." *Environ. Entomol.*, 24, 465–474.

Stewart, L. M. D., Hirst, M., Lopez Ferber, M., Merryweather, A. T., and Cayley, P. J. (1991). "Construction of an improved baculovirus insecticide containing an insect-specific toxin gene." *Nature*, 352, 85–88.

Tabashnik, B. E., Liu, Y. B., Malvar, T., Heckel, D. G., Masson, L., and Ferre, J. (1998). "Insect resistance to *Bacillus thuringiensis*: Uniform or diverse." *Philo. Trans. Roy. Soc. London Series B Biol. Sci.*, 353 (1376), 1751–1755.

Tanada, Y., and Kaya, H. K. (1993). *Insect Pathology*, Academic Press, San Diego.

Te Beest, D. O., Yang, X. B., and Cisar, C. R. (1992). "The status of biological control of weeds with fungal pathogens." *Annu. Rev. Phytopathol.*, 30, 637–657.

Tomalski, M. D., and Miller, L. K. (1991). "Insect paralysis by baculovirus-mediated expression of a mite neurotoxin gene." *Nature*, 352, 82–85.

Tomalski, M. D., and Miller, L.K. (1992). "Expression of a Paralytic Neurotoxin Gene to Improve Insect Baculoviruses as Biopesticides." 10, 545–549.

Trought, T. E. T., Jackson, T. A., and French, R. A. (1982). "Incidence and transmission of a disease of a grass grub (*Costelytra zealandica*) in Canterbury." *N. Z. J. Exp. Agric.*, 10, 79–82.

Undeen, A. H., and Vavra, J. (1997). "Research methods for entomopathogenic Protozoa." *Manual of Techniques in Insect Pathology*, L. L. Lacey, ed., Academic Press, San Diego, 117–151.

Van-Nguyen, A. (1995). "Biochemistry and molecular genetics of a novel entomocidal toxin from *Bacillus thuringiensis*," University of Cambridge, Cambridge, U.K.

Volkman, L. E., Blissard, G. W., Friesen, P., Keddie, B. A., Possee, R., and Theilmann, D. A. (1995). "Family Baculoviridae." *Virus Taxonomy: Classification and Nomenclature of Viruses*, F. A. Murphy, C. M. Fauquet, D. H. L. Bishop, S. A. Ghabrial, A. W. Jarvis, G. P. Martelli, M. A. Mayo, and M. D. Summers, eds., Springer-Verlag, Wien, New York, 104–113.

Weiss, S. A., and Vaughn, J. L. (1986). "Cell culture methods for large-scale propogation of Baculoviruses." *The Biology of Baculoviruses: Volume II, Practical Applications for Insect Control*, R. Granados and B. A. Federici, eds., CRC Press, Boca Raton, FL, 63–88.

Wood Mackenzie. (1998). "Wood Mackenzie Agricultural Products Database." Wood Mackenzie.
Yamamoto, T., and Powell, G. K. (1993). "*Bacillus thuringiensis* crystal proteins: Recent
    advances in understanding its insecticidal activity." *Advanced Engineered Pesticides*,
    L. Kim, ed., Marcel Dekker, New York, 3–43.

CHAPTER 3

Pheromones and
Other Semiochemicals

D.M. Suckling and G. Karg

CONTENTS

1-56670-479-0/00/$0.00+$.50
© 2000 by CRC Press LLC

3.1 INTRODUCTION

Insects rely on several sensory modalities to survive and reproduce, but olfactory information is one of the most important sources of information for many groups. Volatile and non-volatile cues often contain important information on the location of hosts or mates, and insects are well adapted to receiving and processing such information. The odors that trigger specific behavioral responses in the organism are called semiochemicals. This term includes pheromones, kairomones, and a wide range of other classes of behaviorally active compounds (below, and Nordlund, 1981; Howse et al., 1998). Semiochemicals can be used to mediate the behavior of the target organism in a wide range of ways. Their potential for use in pest management was recognized early, and pheromones and plant volatiles have been used for trapping insects for decades. Semiochemicals have also been widely tested in many other pest management applications. Their successful use requires a good understanding of the behavior of the target organism, including the underlying mechanisms that influence the behavior.

The application of semiochemicals for strategic pest control has made considerable progress since their first introduction. The increasing success rate of applications based on pheromones and other semiochemicals has occurred because of the development and application of new techniques of identification, chemical synthesis, new release techniques, and especially more detailed knowledge about the insect's behavior and the parameters required for their successful application. A survey on the role of pheromones in pest management reported by Shani (1993) indicated the optimism felt by researchers in this area, with a reported 1.3 million ha of crops (1% of the cultivated area) being treated in some way with pheromones in 1990. Although the share of the pest control market held by pheromones is still very small, it is increasing as new products and processes become commercially available.

There are several approaches in which pheromones and other semiochemicals can be used in pest management. The attraction of the insect to pheromone or other attractive lures is utilized in the majority of pest management systems involving pheromones or other semiochemicals. Monitoring the number of insects caught is the most widespread use of pheromones, and there are many different ways in which this information can be used. Flight activity can be recorded as the basis for timing of insecticide applications or other control tactics. Trapping can be used for efficiently monitoring the frequency or dispersion of insects or even their population traits such as insecticide resistance, and for the detection of low pest densities, for example in biosecurity or quarantine programs. Pheromone- or kairomone-based lures can also provide the basis for various direct control options. The group of direct control approaches using attractants includes mass trapping, "lure-and-kill," and "lure-and-infect" tactics.

Another highly developed direct control tactic is called "mating disruption." Here, the insect is not necessarily attracted to lures, but rather a large number of pheromone dispensers is deployed to interfere with orientation toward conspecifics and interrupt the life cycle of the insect by preventing mating.

This chapter reviews how insects detect odors, how they respond to different classes of semiochemicals, and the application of a wide range of such chemicals in different pest management tactics.

3.2 INSECT ORIENTATION TO SEMIOCHEMICALS

3.2.1 Chemoreception

Insect antennae carry a number of different types of receptors, including mechanoreceptors and chemoreceptors. Chemoreception is achieved by means of specialized hairlike organs called sensilla. Sensilla vary in shape, size, and the spectrum of volatiles they can detect, and sexual dimorphism is common in most insect groups. They can function as very efficient molecular sieves (e.g., antennae of male moths). Molecules caught by the sensilla on the antenna are transported into the interior through pores by diffusion. There they bind with pheromone-binding or general odorant binding proteins to form ligands, which are then transported across the lymph to receptor sites on the dendrite of the receptor cells, which extend through the lumen of each sensillum trichodeum (Figure 3.1). This process (called transduction) elicits a nervous potential, modifying the electrical conductance of the receptor cell membrane. This depolarization of the receptor potential spreads passively in the dendrites toward the spike generating zone, where a spike (action potential) is generated and transmitted to the antennal bulb in the insect brain. This information is processed along with other cues from the insect's internal and external environment, and is expressed in orientation and other behavior. A similar process applies to other chemosensory organs, such as larval mouthparts, fly tarsi, ovipositors, and so forth, which may use contact cues from non-volatile semiochemicals.

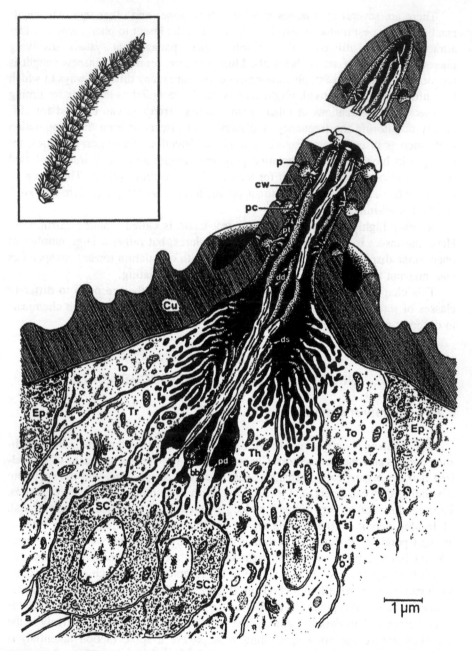

Figure 3.1 Schematic representation of a moth antenna (*left*), and details of a typical antennal olfactory sensillum trichodeum in Lepidoptera. Reconstructed from electron micrographs of *Yponeumeuta* spp. taken by P.L. Cuperus. Original drawing courtesy of Jan van der Pers. Cu, antennal cuticle; To, tormogen cell; Tr, trichogen cell; Th, thecogen cell; SC, sensory cell; sj, septate junction; bb, basal bodies; r, rootlet; pd, proximal part of dendrite; ds, dendritic sheet; dd, distal parts of dendrites; p, pore; pc, pore cavity; cw, cuticular wall of sensillum; pt, pore tubuli.

3.2.1.1   Orientation Toward Odor Sources

Insects use a range of other cues in locomotion, including visual, mechanical, and acoustic stimuli. Different types of odor-induced maneuvers toward odor sources have been identified using both free-flying and tethered insects, but many mechanisms involved remain to be determined (David, 1986). There are basic differences in orientation mechanisms used by walking and flying insects. An understanding of these mechanisms is useful in pest management, because they are important in the responses of insects to semiochemicals. Knowledge of these mechanisms can improve our ability to manipulate insect behavior in a desirable fashion. In most insects, the adult is the dispersive life stage and has an important role in host finding. Adult insects, such as female moths, are often responsible for host plant choice. In contrast, larvae usually have a lesser role, aiming to optimize foraging over a short distance by walking. Such differences in locomotory capability and orientation behavior have obvious implications for pest management and need to be considered in the design of control tactics.

3.2.2   Flying Insects

The structure of odor plumes is important for the orientation of flying insects and for an understanding of how pest management applications based on attraction may operate. Odor plumes are not continuous, time-averaged phenomena, but are better considered as filamentous structures that vary immensely in concentration with peaks and troughs. Surprisingly, the peak concentration has been found to be maintained over large distances downwind from point sources. This was originally shown using ions (Murlis and Jones, 1981), but is likely to be the case for odors. This type of filamentous plume structure and the cues provided by the rapidly changing concentrations in the plume are important for insect orientation (Baker ct al., 1985).

The orientation mechanism used in upwind flight of male moths is chemically triggered, optomotor-controlled anemotaxis. They use wind-borne cues, along with visual information (ground speed) and odor. It is widely believed that male moths in an odor plume use a "template," which characteristically produces the zigzag flight in the following way. After activation (odor detection by the antennae), male moths take off or turn upwind and begin casting sideways to detect the plume. Inside the filamentous plume, they have been shown to exhibit an upwind surge upon encountering pheromone, followed by a return to the lateral casting movement (with increasing amplitude) when the meandering plume filaments are temporarily lost (Figure 3.2). A sequence of lateral casting and forward surging movements in this way is thought to explain the orderly upwind progress observed toward the source (Mafra-Neto and Cardé, 1996; Vickers and Baker, 1996). The basic process is shown in Figure 3.2. The mechanisms by which orientation maneuvers are built into the full sequence of behavior leading to host location is less understood for other flying insects, including flies (e.g., Schofield and Brady, 1997) and wasps (Kerguelen and Cardé, 1997), where casting behavior is absent or not obvious. For tsetse flies and

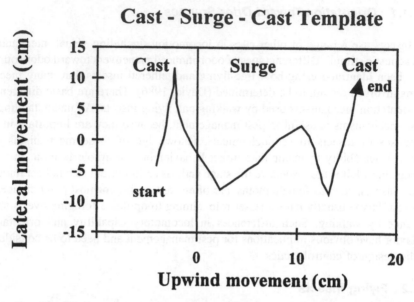

Figure 3.2   Flight template of a moth in a pheromone plume, with lateral casting followed by an upwind surge after encountering a pheromone filament. Redrawn after Vickers and Baker (1996).

other insects groups (e.g., other Diptera), mechanoreceptive anemotaxis is being discussed as a possible mechanism of host location.

3.2.3   Walking Insects

Walking insects do not require the same visual information as flying insects, because they are in touch with solid surfaces. In particular, they do not need to take visually derived assessment of ground speed into account, because mechanical information is sufficient to provide the basis for progress. Walking insects still require chemical cues and wind direction (as well as visual cues) in order to locate an odor source. The same process is used by adult and larval walking insects, which need to integrate additional physical information, such as edges or barriers. Short-distance orientation is based on local environmental features, which are often detected by the difference in input between a bilateral pair of chemoreceptors (tropotaxis).

3.3   PHEROMONES AND OTHER SEMIOCHEMICALS

Odorants serve many different functions for insects. Pheromones, which operate intraspecifically, are the best understood and most widely used class of semiochemicals in pest management. They are "substances which are secreted to the outside by an individual and received by a second individual of the same species, in which they release a specific reaction, for example, a definite behavior or a developmental process" (Karlson and Lüscher, 1959). Pheromones are usually classified by function

(e.g., sex pheromones, aggregation pheromones, trail pheromones, alarm pheromones, etc.). Kairomones, allomones, and synomones are semiochemicals that play
a role in interspecific communication. Kairomones are substances that are "adap-
tively favorable to the receiver, but not to the emitter" (Nordlund, 1981). This group
includes insect-insect and insect-plant interactions. Allomones are substances that
are favorable to the sender alone, such as defensive compounds. Synomones are
beneficial to both species, and include species isolating mechanisms, such as pheromone components, which act as behavioral inhibitors for related species, and plant
volatiles used to attract pollinators.

3.3.1 Pheromones

More than a thousand moth sex pheromones (Arn et al., 1992; 1998), and hundreds of other pheromones have been identified, including sex and aggregation
pheromones from beetles and other groups of insects (Mayer and McLaughlin,
1991). Pheromones have an important and well-established role in insect control,
especially within the framework of Integrated Pest Management (IPM). This section
offers a brief review of the main types of insect pheromones and their main properties
in relation to pest management opportunities.

3.3.1.1 Sex Pheromones

Long-range sex pheromones are released by either one (mainly the females) or
both genders for the purpose of mate attraction. The sex pheromone of an insect
usually consists of a blend of different components, although there are exceptions
to this. These components are volatile, specific to one species or a small number of
related species, and are very potent over considerable distances. This specificity
allows a targeted application to manage one specific insect, with minimal influence
on the rest of the ecosystem. Moth sex pheromones are usually simple molecules
(e.g., long-chained aliphatic, lipophilic, acetates, aldehydes, or alcohols), often with
one or two double bonds. In Diptera, Coleoptera, and other groups, sex pheromones
usually have more complex chemical structures (see below), which are comparatively
unstable and therefore much more difficult to synthesize and formulate, as well as
being expensive (Inscoe et al., 1998). There are therefore more pest management
applications using moth pheromones than pheromones of other insect orders. The
applications will be discussed later in this chapter.

3.3.1.2 Aggregation Pheromones

Aggregation pheromones are attractive to both sexes, and are best understood in
Coleoptera. They also tend to operate over a long range and can attract thousands
of individuals of either sex, offering good potential for mediating pest attack. Like
beetle sex pheromones, these aggregation pheromones generally have more complex
chemical structures (e.g., cyclic and/or chiral compounds) (see Inscoe et al., 1990;
Howse et al., 1998) and elicit a much more complex behavior that is less open to
manipulation. In a number of cases, aggregation pheromones are not very stable or

amenable to synthesis and deployment, and therefore have been less frequently used in pest management.

3.3.1.3 Alarm Pheromones

Alarm pheromones have been identified most frequently from social insects (Hymenoptera and termites) and aphids, which usually occur in aggregations. In many cases, they consist of several components. The function of this type of pheromone is to raise alert in conspecifics, to raise a defense response, and/or to initiate avoidance. Their existence has been known for centuries, with descriptions of bee stings attracting other bees to attack (Butler, 1609; cited in Free, 1987). More recently, Weston et al. (1997) showed a dose response of attractancy and repellency for several pure volatiles from the venom of the common and German wasps *Vespula vulgaris* and *V. germanica.* The compounds are usually highly volatile (low-molecular-weight) compounds such as hexanal, 1-hexanol, sesquiterpenes (e.g., (E)-β-farnesene for aphids), spiroacetals, or ketones (Franke et al., 1979). Some applications of alarm pheromones of aphids in combination with other agents are considered below.

3.3.1.4 Trail Pheromones

Trail pheromones are mainly known from Hymenoptera and larvae of some Lepidoptera. They have been identified from a range of sources in Hymenoptera, including abdominal, sting, and tarsal glands. They are essentially used for orientation to and from the nest, on foraging trails (e.g., in ants or termites). Trail pheromones are characteristically less volatile than alarm pheromones. The trails are replenished through continuous traffic, otherwise they dissipate. While trail pheromones are frequently associated with walking insects such as ants, they also exist for other insects. Bees use trail pheromones during foraging, both for marking attractive foraging sites and for scent marking of unproductive food sources (Free, 1987). Identification and synthesis of the trail pheromone for bumblebees could lead to increased efficiency in their use for pollination. It is also possible to manipulate trail following and recruitment of tent caterpillars (e.g., *Malocosoma americanum*) (Fitzgerald, 1993), that can be serious defoliators in North American forests. It remains to be seen whether the use of the trail pheromone compounds could lead to novel pest management solutions, and they will not be considered further here.

3.3.1.5 Host Marking Pheromones

Spacing or host marking (epidietic) pheromones are used to reduce competition between individuals, and are known from a number of insect orders (Papaj, 1994). One of the best studied is from the apple maggot *Rhagoletis pomonella* (Tephritidae). Females ovipositing in fruit mark the surface to deter other females (Prokopy, 1972). This behavior has also been studied in the related cherry fruit fly (*Rhagoletis cerasi*), and a commercial product using it is under development in Switzerland. The product is a non-volatile sprayable formulation of aqueous host marking pheromone applied weekly for control. It is used in combination with unsprayed trap trees containing yellow sticky traps deployed to prevent pest build up in the block. It is most likely to be appropriate for niche markets, such as eco-labeled fruit.

Egg laying is a key stage determining subsequent population density, so it is perhaps not surprising that there is considerable evidence of such pheromones affecting gravid females of herbivores (e.g., Schoonhoven, 1990). There is also exploitation of prey host marking and sex pheromones by parasitoids, which use the signal persistence of these intraspecific cues to find their hosts (Hoffmeister and Roitberg, 1997). Mating deterrent pheromones are also known from a number of insects, including tsetse flies, houseflies, and other Diptera (Fletcher and Bellas, 1988). These pheromones are released by unreceptive females to deter males from continuing mating attempts. Exploitation of these cues remains largely unexplored.

3.3.2 Other Semiochemicals

There are a number of different types of semiochemicals that operate between species, as defined above (allomones, synomones, kairomones). These types of compounds include compounds involved in floral attraction of pollinators, as well as compounds that function as species isolating mechanisms, such as sex pheromones of related species, where an inhibitor often functions to prevent mating among sympatric species. These types of compounds are only just beginning to be applied, but there are excellent prospects for their use in pest management if certain difficulties (e.g., formulation, below) can be overcome. Novel applications of kairomones have also been suggested in recent years. These include the application of the stimulo-deterrent diversionary or "push-pull" strategy (Miller and Cowles, 1990), and the use of attractants and repellents in various ways, considered below.

3.4 MONITORING WITH SEMIOCHEMICALS

Insects can be readily attracted using pheromones or other attractants. Combination of this attraction with a system of retaining the insects is necessary as the basis for trapping systems. While passive traps or other sampling systems can be successful at collecting actively mobile insects, trap efficiency can be increased many times by the use of a specific attractant. This occurs because the active space, or area of influence of the trap, can be greatly increased by the attraction of insects to semiochemicals. Regular inspection of the number of insects caught in such traps provides the basis for a monitoring system. While sex pheromones are most widely used as attractants in monitoring systems, other semiochemicals, such as host plant odors, have been used against certain insect groups for many years (e.g., fruit flies).

3.4.1 Aspects of Attraction and Trap Design

The ideal monitoring system must meet certain criteria. In principle, the trap efficiency (number of insects caught per visiting insect) should remain constant. If this is not the case, then the number of insects caught may not reflect the population

Table 3.1 Considerations for the Design of Semiochemically Based Insect Traps

Parameter	Ideal Features	Problems
Lure	Constant attraction	Changing release rate, blend, isomers, and active space
Physical shape	Noninhibitory, omnidirectional	Visual, physical, and plume structure interference
Color	Attractive or neutral	Nontarget catch (e.g., bees)
Durability	Long lived	UV; rain
Trapping surface	Constant retention rate	Saturation; glue aging (dust/insect parts); glue viscosity (temperature); evaporation (liquid trapping well)
Service frequency	Relatively infrequent	Labor cost
Cost	Low cost	Manufacturing volume; complex designs; short durability

density in a useful way. In practice, many factors can influence trap efficiency (Table 3.1). Traps using sticky surfaces to retain the insects can saturate, with reduced efficiency at higher catches. The release rate and stability of the attractive components are very important for the efficacy of the lures. Many insects only respond to semiochemicals over a certain concentration range or require exposure to a defined blend, and the efficacy can be hampered by the presence of isomers that may appear in the lure over time due to isomerization, oxidation, and polymerization. Trap efficiency can also be affected by the insect phenology. For example, in tortricid moths, earlier emergence of males leads to a changing rate of competition between traps and virgin females (Croft et al., 1985). Hence the proportion of the male moth population caught after females emerge is reduced.

Many different types of traps have been developed for monitoring insects (e.g., Jones, 1998). Some traps have a sticky trapping surface (e.g., pane traps, delta traps, wing traps, or tent traps). These designs are often used for small insects such as smaller moths and scale insects. Alternatively, other traps use some kind of flight barrier (e.g., funnel traps, drain pipe or slit traps often used for bark beetles), or a liquid trapping medium (e.g., McPhail traps used for wasps and fruit flies, and fermenting molasses traps for moths).

Attractive semiochemicals for use in traps have commonly been formulated on rubber septa or other simple types of passive carriers. These carriers simply function as a practical and cost-effective reservoir for the semiochemical. In practice, temperature and age are the most important factors affecting the release rate of lures. The release rate from many substrates cannot be readily controlled. The release rate changes significantly over time, often following a zero-order profile. Controlled release devices following a first-order profile have been proposed for some time (Weatherston, 1990; Leonhardt et al., 1990), and new developments are still emerging. The new dispensers are mostly based on polymers or laminated materials. These new developments also have the ability to protect the components from UV, which can otherwise lead to degradation and/or isomerization (Jones, 1998).

Kairomones have been very important for monitoring and control of fruit flies (Tephritidae), Japanese beetle (*Popillia japonica*, Scarabidae) and Diabroticite rootworm beetles (Chrysomelidae) (Metcalf and Metcalf, 1992). Kairomone-baited traps can be effective for monitoring, but like pheromone traps require similar levels of intensive development of trap design, deployment strategy, and lure formulation to overcome problems of low catches or intensive maintenance (e.g., Hoffmann et al., 1996). Further improvements are under investigation for monitoring and control of such pests using kairomones. For example, Vargas et al., (1997) reported equivalent captures of *Bactrocera dorsalis* (Tephritidae) to coffee juice compared to the standard kairomone lure, which may lead to new attractants. Teulon et al. (1993) compared the use of various kairomones for monitoring of thrips, and postulated on the use of mass trapping using *p*-anisaldehyde.

Monitoring has been applied extensively for timber and bark beetles (Borden, 1995). The attraction of bark beetles (Scolytidae) to host trees under natural conditions is very complex. It requires several steps, all involving the release of kairomones and other semiochemicals. These compounds often occur as different enantiomers, and stereoisomers, of which only some may be behaviorally active. For example, *Dentroctonus ponderosae* females are initially attracted to a suitable tree releasing a kairomone (myrcene). Those females will start releasing (+)-exo-brevicomin, a sex pheromone that attracts males. Males entering the tree release (–)-frontalin, an aggregation pheromone that attracts both males and females to the attacked tree in a mass attack, in order to rapidly overcome the host defense system (see Howse et al., 1998). In a new example involving the Scolytidae, traps or trap trees baited with the gas ethylene have been proposed for use in IPM of the olive beetle *Phloetribus scarabaeoides* (González and Campos, 1995; Peña et al., 1998).

Various trapping applications have been developed against Diptera. Tsetse flies (Glossinidae) are attracted to carbon dioxide, acetone, 1-octen-3-ol, 4-methylphenone, 3-propylpenole, cattle urine extracts (Carlson et al., 1978). These compounds are used to monitor tsetse flies and can increase trap catch of *Glossina pallidipes* (Hall, 1990). (Z)-9 tricosene has been identified as the sex pheromone of the house fly *Musca domestica* (Muscidae) (Carlson et al., 1971) and has been used for monitoring purposes (Browne, 1990). Houseflies can also be monitored using multicomponent lures releasing ethanol, skatole, ammonia, fermentation products, and other compounds (Jones, 1998). Cossé and Baker (1996) have identified several attractive constituents of pig manure that elicit upwind flight to the source in houseflies, and it may be feasible to formulate them into effective house fly baits in future.

There are other examples involving trapping of Diptera. Traps baited with isothiocyanates catch the brassica pod midge (*Dasyneura brassicae,* Cecidomyiidae), and may be used in future as part of an IPM program (Murchie et al., 1997). Olfactory attractants are the basis of all present fruit fly (Tephritidae) detection, monitoring and control strategies (Jang and Light, 1996). In the case of olive fruit fly, the method combines a food attractant, a phagostimulant, a male sex pheromone, a female aggregation pheromone with additional arrestment and other properties, and an insecticide-treated wood board (Haniotakis et al., 1991). A mosquito oviposition pheromone (*erythro*-6-acetoxy-5-hexadecanolide) has demonstrated uses in mosquito (Culicidae) control (Otieno et al., 1988), and further work is under way to develop "ovitraps" to capture gravid females. Behavioral and electrophysiological studies have shown the potential of oviposition pheromones of *Culex quinquesfasciatus* (Blair et al., 1994; Mordue et al., 1992) and habitat-related cues and field trials are under way. In field trials, the oviposition pheromone attracted *C. quinque-fasciatus* females from up to 10m.

3.4.2 Applications of Monitoring

The most widespread and simple application of semiochemicals involves monitoring the presence, seasonal phenology, distribution, density, or dispersion of pests. Monitoring is being used for a wide range of species and crops all over the world, especially in agricultural and horticultural ecosystems (e.g., Wall, 1989; Howse, 1998). Attractants thus offer the advantage of bringing the insect to the person, saving both sampling time and expense.

3.4.3 Survey

At the very simplest level, traps can be a very efficient way of determining the presence of insects, even at a very low density at a country, region, or farm level. This approach is the basis of biosecurity or quarantine surveys, where the aim is to determine the presence of a species, and prevent its establishment. This is especially important around airports and harbors, where alien pests can be easily and accidentally introduced to foreign ecosystems. Exotic pests can have disastrous environmental and economic consequences, and increasingly many countries are attempting to reduce the risk of such introductions. There is potential for establishing low-cost monitoring of exotic pests, integrated with existing schemes. For example, Schwalbe and Mastro (1988) discussed the addition of pheromones of exotic species to lures already in use for other purposes. They cited a number of examples of combinations where the additional lure had no adverse impact on the catch of either species.

Pheromone traps can provide growers and consultants with information on the distribution of specific pests in a region or on a farm. For example, Shaw et al. (1994) showed that the Nelson region of New Zealand was largely partitioned between two morphologically similar sibling species of leafrollers affecting apple orchards (Figure 3.3). The determination of the specific pest fauna present at a farm can lead to opportunities for tailoring of specific solutions, as we move toward more precise targeting of pest problems.

Insect biological control agents are now being trapped with sex pheromones (e.g., Brodeur and McNeil, 1994). As in other cases, this new tool will permit monitoring of the presence, phenology, and relative abundance of the biocontrol agent, and in future could give an indication to growers whether the population might be high enough for successful control of the codling moth. Pheromone traps are also being used for monitoring the establishment of a biological control agent for weeds, *Cydia succedana* (Tortricidae), introduced for control of gorse (*Ulex europeaus*) in New Zealand (Suckling et al., 1999). Kairomones have been used for monitoring scale biological control agents, because of the response of the parasitoids to scale insect pheromones (Gieselmann and Rice, 1990).

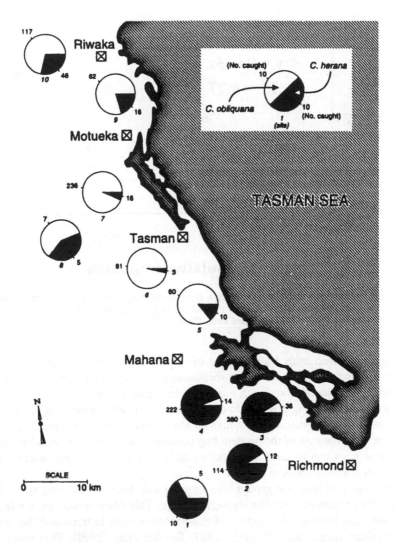

Figure 3.3 Geographical distribution of two sibling species of leafrollers in Nelson, New Zealand, based on a survey using pheromone trap catch (From Shaw et al., *N.Z. J. Zool.* 21:1–6, 1994).

3.4.4 Decision Support

Monitoring populations can give an early indication of outbreak in an established population. Thresholds of catch have been developed for a large number of insects and used as the basis for conventional pest management interventions (mainly insecticides) (Jutsum and Gordon, 1989). The aim has generally been to achieve control with reduced numbers of insecticide applications. The system works on the principle that an intervention, such as insecticide spray application, is only required if a certain defined sampling threshold is exceeded. In one recent example, Bradley et al. (1998) determined a threshold of leafroller pheromone trap catch from a

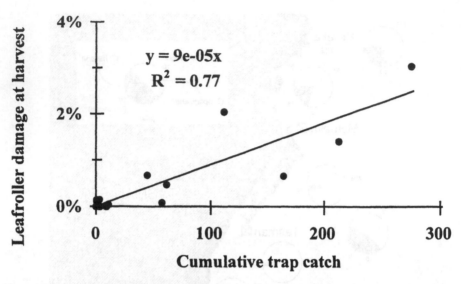

$$y = 9e\text{-}05x$$
$$R^2 = 0.77$$

Figure 3.4   Correlation between cumulative pheromone trap catch of *Epiphyas postvittana* (Lepidoptera: Tortricidae) and larval damage on apples at harvest. (From Bradley et al., *Proc. 51st N.Z. Plant Prot. Conf.*, 1998).

correlation of catch with fruit damage of apples at harvest (Figure3.4). Catches greater than the threshold led to the recommendation for application of a selective insecticide. Considerable research is needed to define a threshold for intervention, and a low-cost sampling protocol is usually necessary. Pheromone- or semiochemical-based trapping sometimes provides an attractive option as the sampling system of choice, as in the case of the mullein bug (*Campylomma verbasci*, Miridae), where pheromone traps can be used to predict nymphal densities or as the basis of a risk rating system (McBrien et al., 1994).

There are problems for groups like Lepidoptera, because trapping of males is several steps removed from the damaging stage. This often results in a relatively low correlation between the number of male moths caught in traps and the number of caterpillars found (e.g., Trumble, 1997; Bradley et al., 1998). This problem is particularly acute with polyphagous insects, which may arise from noncrop host plants (e.g., Izquierdo, 1996).

Successful examples of monitoring also come from food processing plants and warehouses, which are regularly infested by stored products pests (mainly moths and beetles). Monitoring of these pests using pheromone or food traps is used as a supplement for conventional inspection methods (Pinninger et al., 1984; Trematerra, 1989; Burkholder, 1990). Sex pheromones are used for moths, and aggregation pheromones and/or food baits are usually used for beetles.

Trapping has been used to monitor cyclical pests in forestry, such as spruce budworm (*Choristoneura fumiferana*, Tortricidae) to warn of impending outbreaks, and trigger action (Sanders and Lyons, 1993). Bark and timber beetles have also been controlled using this approach (Borden, 1995).

Blight et al. (1984) monitored pea and bean weevils (*Sitona lineatus*, Cucurlionidae) with the aggregation pheromone at overwintering sites. Catches were used to support decisions about the need for treatment and its timing. However, there was no direct relationship between numbers trapped at overwintering sites and leaf notching, although monitoring was still considered valuable (Biddle et al., 1996). Smart et al. (1996) used a mixture of isothiocyanates for monitoring the phenology of cabbage seed weevil (*Ceutorhynchus assimilis*) in oilseed rape, another example of traps based on plant volatiles, rather than pheromones.

Catch in traps baited with pheromones or other semiochemicals can be used with meteorological data as inputs for phenology models to predict the timing of flight activity or other life stages (Knight and Croft, 1991). This approach is likely to be particularly useful for biorationals and more selective insecticides, where the activity is specific to certain life stages.

3.4.5 Monitoring Resistance

Repeated application of insecticides can select for an increase in insecticide resistance frequency and dispersion in the population. Traditional methods to monitor the presence and distribution of insecticide resistant insects are often laborious, mainly because of the difficulties involving population sampling. A more elegant method for extracting independent samples of genotypes from the population is based on the use of pheromones. This approach is based either on attracting large numbers of moths for collection by sweep netting (Suckling et al., 1985), or to traps (Riedl et al., 1985). The adult males can then be tested for expression of resistance by topical application of insecticides, or in more refined versions through incorporation of insecticide into the sticky glue on the trap (Haynes et al., 1987).

3.5 DIRECT CONTROL OF PESTS USING SEMIOCHEMICALS

Direct control methods using semiochemicals against insects are selective and more environmentally benign tactics, compared with more broad-spectrum control tactics. Importantly, their success is density-dependent, they generally suffer from the risk of immigration of mated females, and they are less effective in polyphagous species with multiple matings. The methods outlined below require a high degree of success to provide pest management to below the economic threshold. Long-range attractants (especially sex and aggregation pheromones) are increasingly being applied in the direct control of insects, in several ways.

Different methods of direct control using sex pheromones against Lepidoptera are compared in Figure 3.5. One disadvantage of the use of semiochemicals in control of many insects is that they generally mediate the behavior of adults and therefore are not directly linked to the damaging larval stages. Some of these disadvantages do not apply to bark and timber beetles, and methods such as mass trapping have consequently enjoyed greater success against this group (below).

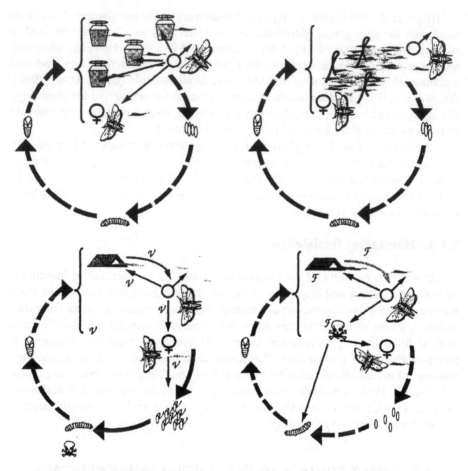

Figure 3.5   Comparison of the process used in three different methods for direct control of Lepidoptera, highlighting the life stage affected (clockwise from top left: mass trapping, mating disruption, and lure-and-infect (fungus and virus).

3.5.1 Mass Trapping

In mass trapping, a very high proportion of the pest must be caught before mating or oviposition to reduce the pest population. This reduction must then be translated into a reduction in damage to an economically acceptable level. A higher number of traps should theoretically lead to a greater reduction in the population. Success with this method requires that the lure is very attractive, eventually out-competing the naturally occurring attractant. For Lepidoptera, it is essential that males are trapped before mating, and it is most likely to succeed with insects that mate only once. In the case of Coleoptera, trapping based on aggregation pheromones aims to reduce the number of both sexes before eggs are laid or damage is done by feeding adults. Mass trapping of fruit flies (both sexes) is similar, except that it is based on kairomone attractants. In these cases, it is most important that there is minimal influx of the pest from outside the protected areas, unless luring the pests into a specific area is part of the control strategy.

Mass trapping suffers from a number of theoretical and practical deficiencies. In the case of moths, very high levels of male annihilation (e.g., >95%) are required for success (Knipling, 1979). Roelofs et al. (1970) showed that a 95% reduction of fecundity was only achievable with five traps per calling red-banded leafroller moth (*Argyrotaenia velutinana*, Tortricidae). A high pest population density, often with an aggregated distribution, means that a high number of competing attractive plumes are released by insects, as well as an increased possibility of accidental encounter of the other sex. Hence it can be seen that mass trapping would be more successful at lower pest densities. In addition, mass trapping is rather cost- and labor-intensive because of trap maintenance. As with other traps, there can also be problems with the blend, change of release rate, or trap efficiency over time (Table 3.1).

Furthermore, many pests are not restricted to the crop area, which may be surrounded by other host plants. Hence the required degree of isolation is hard to achieve under field conditions, unless the insect is limited to a defined and treatable area. In some cases it has been practical to treat the entire crop or habitat. For example, Ngamine et al. (1988) controlled a sugar cane pest (Elateridae) on an island using mass trapping. They used lures that were 50 to 200 times more attractive than virgin females, and reduced the population by 30 to 40%. Mass trapping is likely to work best for the eradication of small or confined populations. Isolation is more easily achievable in relatively confined situations like food processing plants and warehouses. The prospects for successful application of mass trapping may be better for pests of stored products, compared to many field situations (e.g., Trematerra, 1989).

In forestry, mass trapping has been successfully used against populations of *Gnathotrichus sulcatus* (Scolytidae) (e.g., Borden, 1990) and the mountain pine beetle *Dendroctonus ponderosae* (e.g., Borden and Lindgren, 1988) in western Canada and northern U.S. In Europe, mass trapping has also worked in Scandinavia against the spruce bark beetle *Ips typographus* (e.g., Bakke et al., 1983; Bakke and Lie, 1989). Mass trapping of bark and timber beetles is usually applied with other tactics, including use of trap trees, post-logging mop-up, anti-aggregation pheromones, and other variations (Borden, 1995).

Despite the problems that have occurred in the practical application of mass trapping, there are a number of examples of large-scale mass trapping efforts with sex pheromones or other lures (Table 3.2). Most cases have not been economically successful (Bakke and Lie, 1989). Nevertheless, despite the lack of much previous success, mass trapping is still being attempted for control of agricultural insects, including Lepidoptera. For example, Park and Goh (1992) reported less damage of onions with mass trapping of *Spodoptera exigua* in Korea. In Australian stonefruit orchards, James et al. (1996) achieved mass trapping of *Carpophilus* beetles (Nitidulidae) using water-based funnel traps.

As in forestry, mass trapping in agriculture may be most effective when combined with several other tactics. For eradication of fruit flies as biosecurity or quarantine pests, mass trapping is often combined with restricted movement of plant material and spot treatments of insecticide. Traps for mass trapping of palm weevils using

Table 3.2 Examples of Mass Trapping for Insect Pest Management

Common name	Species	Crop/commodity	References
Olive fruit fly	Dacus oleae	Olives	Jones, 1998
Olive moth	Prays oleae	Olives	Jones, 1998
Indian meal moth	Plodia interpunctella	Stored products	Trematerra, 1989
Tropical warehouse moth	Ephestia caudata	Stored products	Trematerra, 1989
Ambrosia beetle	Gnathotrichus sulcatus	Forests	Borden, 1990
Mountain pine beetle	Dendroctonus ponderosae	Forests	Borden and Lindgren, 1988
Beet armyworm	Spodoptera exigua	Onions	Park and Goh, 1992
Spruce bark beetle	Ips typographus	Forests	Bakke et al., 1983; 1989
Sugar cane wireworm	Melanotus okinawensis	Sugar cane	Ngamine et al., 1988

aggregation pheromone also contain insecticide-treated-food to retain and poison the insects (Hallett et al., 1993).

3.5.2 Lure and Kill

Attracticidal tactics combine lures with insecticides. Despite considerable research, there are few successfully commercialized attracticides. They share many problems in common with mass trapping (Figure 3.5). Haynes et al. (1986) showed that the effectiveness of an attracticide depended on males freely contacting the treated sources, rapid sublethal effects on the behavior response after contact, and the level of insecticide-induced mortality. McVeigh and Bettany (1986) reported a lure and kill technique against the Egyptian cotton leafworm (*Spodoptera littoralis*), that used treated filter papers as the substrate. More recently, lure and kill has been reported to work against codling moth (Charmillot and Hofer, 1997), and a commercial product ("Sirene," Novartis) is now registered in Switzerland.

Kairomonal attractants can also be used in this pest management tactic, as shown by the attracticide developed for control of *Amyelois transitella* (Pyralidae) in almonds (Phelan and Baker, 1987). It is also not necessary to use insecticides for success. Initial control of tsetse flies in Africa used odors released by oxen or buffalo urine to attract flies to cloth doped with insecticides (Vale et al., 1988). Later, flies were attracted to electrified nets with these and other odorants. An alternative to conventional insecticides could make use of insect pathogens as biopesticides if they can kill the attracted insect before mating occurs.

3.5.3 Lure and Infect

A more elegant development of this general approach is called "autodissemination," and combines insect pathogens with pheromone or other lures (Figure 3.5). The aim of this tactic is not to kill the insects right away, but rather to use them as vectors of the disease into the wider population. Different pathogens could be used, with slightly different pathways from virus (baculovirus or granulosis virus), fungus (e.g., *Zoopthora radicans*, Pell et al., 1993), or a bacterium (e.g., *Serratia ento-*

mophila, O'Callaghan and Jackson, 1993), or even entomopathogenic nematodes. There is a major advantage in this approach, if it can generate disease outbreaks that can multiply in the area and pest population affected. It is possible that fewer insects may need to be directly attracted to the pathogen stations, which could reduce the costs and labor required. Insect pathogens have a number of advantages over insecticides (previous chapter), and this combination approach appears to use the best properties of specificity from both the lure and the pathogen to provide an environmentally benign approach that can be integrated with other methods, including natural enemies.

This approach has been explored with nucleopolyhydrosis virus against tobacco budworm (Noctuidae) (e.g., Jackson et al., 1992), and a granulosis virus against codling moth (Tortricidae) (Hrdy et al., 1996). A fungus is also being developed for use against diamondback moth *Plutella xylostella* (Yponomeutidae) (Pell et al., 1993; Furlong et al., 1995). There are also examples of autodissemination of fungus being developed against Coleoptera (Japanese beetle, *Popillia japonica,* Scarabidae) (Klein and Lacey, 1998) and termites (Delate et al., 1995).

The critical requirements for success with pathogens may be difficult to achieve, and include operational factors such as formulation and delivery systems, as well as biological factors. For example, both bacterial toxins and viruses only become pathogenic upon consumption. In the case of Lepidoptera, virus-infected males must locate a mate and transfer the pathogens to females during copulation. Ovipositing females must then transfer virus to the surface of the eggs (Figure 3.6), and eventually to larvae during ingestion at eclosion. In the case of a grass grub beetle *Costelytra zealandica* (Scarabidae), it is possible that an aggregation pheromone could be used to increase the pathogenic bacterial count in the larval habitat (O'Callaghan and Jackson, 1993).

Fungi can be transferred between both adults and larvae. Infected adults (with gender depending on use of a sex or aggregation pheromone) can be attracted to a delivery station of some type. They must remain in the station long enough to pick up an adequate dose of pathogen before exiting. Upon their return to the field, they can spread the pathogen, die, and then sporulate to cause an epizootic within several days. Unfortunately, few pathogens are stable in the environment, due to UV radiation and dessication (fungi and nematodes). This, along with the need for economic pathogen production, and maintenance of delivery stations are likely to be the main constraints against this method.

3.5.4 Mating Disruption

Another direct control tactic using pheromones for the control of insects is called "mating disruption." Here, the aim is to prevent mating and hence reduce the incidence of larvae in the next generation. This is normally done by releasing a large amount of sex pheromone in the treated area. The behavior of male insects is disturbed by their exposure to synthetic pheromone released by dispensers (Cardé and Minks, 1995; Sanders, 1997). There has been considerable debate about the mechanisms underlying mating disruption (Bartell, 1982), although there is general

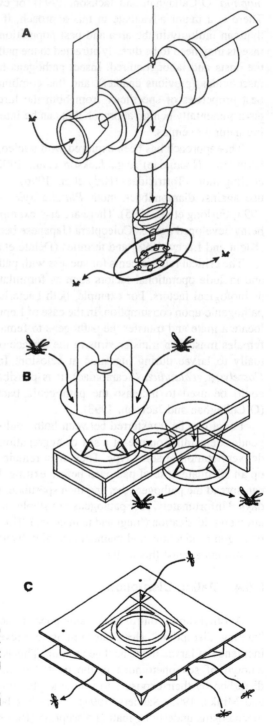

Figure 3.6 Comparison of three lure and infect delivery stations being developed for autodissemination of insect pathogens against (a) sap beetles and other insects (Vega et al.,1995), (b) Japanese beetle (Klein and Lacey, 1998), and (c) diamondback moth (Pell et al., 1993).

agreement now that more than one mechanism may be operational at the same time, and they may vary between species (Sanders, 1997).

The mechanisms involved in mating disruption fall into two main categories:

- "False trails": Male moths actively search, but fail to locate females because of the large number of competing pheromone trails released by dispensers. This mechanism relies on the use of an attractive pheromone blend.
- Sensory overload: Normal searching behavior is absent or terminated due to sensory fatigue, camouflage of natural plumes, repellency, or sensory imbalance (blend masking). This type of mechanism does not rely on the use of an attractive pheromone blend, and can include behavioral inhibitors (synomones).

3.5.4.1 Strength and Weaknesses

Mating disruption has a number of advantages as a pest management tactic. It is species specific, has a low environmental impact, and is more sustainable than broad-spectrum tactics, with no evidence (so far) of "resistance," which may occur with insecticides. It has proven to be one of the preferred control methods against insecticide-resistant populations. Control failure with insecticides can greatly increase the adoption of mating disruption (as happened in northwestern U.S. with codling moth in the 1990s).

Mating disruption also has several important disadvantages. The technique is operational against adults, while the damage is usually done by larvae in the following generation (i.e., Lepidoptera). Hence control is dependent on a low level of immigration. Even when the system has been proven technically, the most serious problem is often the cost, compared to broad-spectrum insecticides. This is particularly a problem when more complex and/or unstable components are used. Secondary pests, formerly controlled by insecticides, often emerge as important problems (Cardé and Minks, 1995). Monitoring the success is difficult, because pest control is not measurable until much later (below). Although "resistance" in the strict sense of genetic adaptation has not yet been observed in the field, there is a risk that insects will adapt to the application of pheromones if a sufficient proportion of the population is under selection.

3.5.4.2 Biological and Operational Factors

Several biological and operational factors are important for success. Insects that are most suitable for this approach are host specific, mate once, have limited fecundity, single generation, short life span and limited mobility, and have females that release small quantities of pheromone.

The concentration of the synthetic pheromone as well as the structure of the pheromone plume are also important for success, and several operational factors can contribute to these parameters.

In general, it is assumed that higher airborne pheromone concentrations in the treated areas lead to improved management of the pest. Unfortunately, the amount of atmospheric pheromone actually required for. disruption in the field (parts per

Table 3.3 Operational Parameters Affecting the Success of Mating Disruption

Parameter	Key feature	Problems
Dispenser release rate	Constant, long lived	Temperature dependency, instability, limited loading, ease of monitoring
Blend	Stable components	Differential volatility, instability
Dispenser type	Physical properties	Cost effectiveness, biodegradability
Application method	Rapid	Delivering optimum height in crop
Number of points per ha	Minimum number	Often unknown
Atmospheric concentration	Minimum effective level	Usually unknown, especially affected by wind
Seasonal timing	Preceding female emergence	Often unknown initially
Application height	Optimum for disrupting insect	Often unknown, difficult to achieve in tree fruits, mating height may be species specific
Price	Competitive with alternative controls	Relatively high monetary cost

billion) is virtually unknown in almost every case. There are a few examples of empirical studies where pheromone concentrations have been measured in the laboratory (Sanders, 1997) or on relatively few occasions in the field (Bengtsson et al., 1994), but these data are of limited value except as a broad guideline, because the considerable instantaneous fluctuations in the atmospheric concentration (Suckling and Angerilli, 1996; Karg and Suckling, 1997). Lack of knowledge of the required concentration of pheromone to consistently prevent mate location has led to an empirical approach in pheromone deployment ("rules of thumb"), and is a contributing factor to the challenge of achieving success with this tactic.

Many different pheromone formulations, including hollow-fibers, microencapsulated sprayables, laminate dispensers, polyethylene tubing, and aerosols have been developed for mating disruption (see Cardé and Minks, 1995; Howse et al., 1998). Most of the dispensers available at the moment do not accomplish at least one of the requirements for an ideal dispenser (Table 3.3). More recently, electronically activated aerosol formulations have been developed (Shorey and Gerber, 1996; Mafra-Neto and Baker, 1996). Here the constant and passive release of pheromone (which is potentially very wasteful) is replaced by an active application, which can be timed (for insect activity) by addition of sensors for light, wind and/or temperature. This approach seems very promising, especially indoors. At this stage, this novel formulation requires further evaluation. There are many examples of mating disruption being tested against different groups of insects from both agricultural and forest ecosystems. The number of successful applications is growing (Cardé and Minks, 1995). Details of commercially available pheromones are probably best obtained directly from suppliers, many of whom are on the internet.

Table 3.4 Some Established Cases of Mating Disruption against Lepidoptera

Common name	Species	Crop
Pink bollworm	Pectinophora gossypiella	Cotton
Oriental fruit moth	Grapholitha molesta	Stonefruit
Tomato pinworm	Keiferia lycopersicella	Tomato
Codling moth	Cydia pomonella	Apples
Rice stem borer	Chilo supressalis	Rice
Grape berry moth	Eupoecelia ambiguella	Grapes
Smaller tea tortrix	Adoxophyes spp.	Tea
Oriental tea tortrix	Homona magnanima	Tea
Lightbrown apple moth	Epiphyas postvittana	Apples
Gypsy moth	Lymantria dispar	Broadleaved/oak forests
Diamondback moth	Plutella xylostella	Brassicas

3.5.4.3 Targets of Mating Disruption

Mating disruption has been most frequently used against moths, and there are a number of successful cases (Cardé and Minks, 1995), of which there is a sample in Table 3.4Mating disruption has also been applied in a few cases to other insect orders, although such examples have not yet progressed to commercial use. For example, McBrien et al. (1997) reported population suppression of a Heteropteran. They were successful at disrupting Mullein bug, although further improvements to the deployment system were seen as necessary. The pheromone of California red scale (*Aonidiella aurantii*), and the citrus mealybug (*Plannococcus citri*) was found to be too expensive and unstable, even though reductions in mating were achieved (Hefetz et al., 1990). There are also examples from Coleoptera, such as the sweetpotato weevil ((*Cylas formicarius*, Cucurlionidae) Mason and Jansson, 1991; Miyatake et al., 1997). Leal et al. (1997) reported the development of mating disruption of a sugar cane pest (*Migdolus fryanus*, Cerambycidae), with over 18,000 ha of the crop treated in Brazil. Similar requirements for these insects are likely to apply, and success will be dependent on the development of stable, cost-effective formulations and other factors.

3.5.4.4 Assessment Methods

The success or failure of mating disruption is ultimately measured at harvest of the crop, although many other methods have been used to determine the level of success achieved before this stage (Table 3.5). Trap catch to pheromone lures can easily be disrupted and is therefore insufficient by itself to measure success, especially if females can still mate in the absence of catch (e.g., Suckling and Shaw, 1992). Use of calling females as lures or sentinel tethered females is labor intensive. Detailed harvest assessment to precisely quantify damage is also labor intensive and expensive, and if damage is present, it is too late to implement control tactics.

Table 3.5   Comparison of Methods for Assessing the Efficacy of Mating Disruption

Method	Advantages	Disadvantages	References
Trapping of males	Rapid and low cost	Wrong life stage, may not indicate female immigration or mating success	Charmillot and Vickers, 1991; Cardé and Minks, 1995
Tethered females	Indicates mating success	Costly, labor intensive, location may not reflect natural dispersion/density	Charmillot and Vickers, 1991; Cardé and Minks, 1995
Crop assessment	"Ultimate test"	Too late, in the event of failure	Charmillot and Vickers, 1991; Cardé and Minks, 1995
Field bioassays, behavioral observations	Insight into mechanisms	Labor intensive	Witzgall et al., 1996
Air sampling with chemical analysis	Measurement of field concentrations	High cost, long integration interval	Caro et al., 1980; Flint et al. 1990; Witzgall et al., 1996
Field EAG for mean concentrations	Corroborates chemistry, short time interval, spatial distribution of pheromone	Technically difficult, interpretation problems	Sauer et al., 1992; Karg and Sauer 1995; Suckling et al., 1994
Field EAG for plume structure	High temporal resolution, insight into insects' sensory environment	Technically difficult, interpretation problems	Suckling and Angerilli 1996; Karg and Suckling 1997; Sauer and Karg, 1998
Field single cell recording	Pheromone specific, high temporal resolution	Technically difficult, interpretation problems	van der Pers and Minks, 1993, 1997
Modeling	Portability as a general tool	Requires validation	Uchijima 1988; Suckling et al., 2000

3.5.5   Characterizing Atmospheric Pheromone Conditions

A more detailed understanding of the factors affecting the concentration and distribution of pheromones under field conditions would help to understand the underlying mechanism and may open up possibilities for improvement of the method. Three different techniques have been commonly used to measure pheromones in the field. These are chemical analysis, field electroantennogram recordings, and single sensillum recordings. Unfortunately, these methods do not indicate the success of the method directly. Rather, they are useful to characterize the conditions required for success, assessed using other methods.

3.5.5.1 Chemical Analysis

Pheromone concentration can be measured most accurately by using air-sampling methods, in conjunction with chemical analysis (Caro et al., 1980; Witzgall et al., 1996). Unfortunately, the temporal resolution of this method is very low, because several hours of sampling is required to obtain enough pheromone to be detectable. It is also relatively costly per sample, and often has detection problems, because pheromone concentrations are typically very low (parts per billion). Chemical analysis does not allow insight into the fine structure of the odor plumes, which is essential for the understanding and interpretation of insect behavior. This is currently best achieved with electrophysiological methods, using insect antennae.

3.5.5.2 Field Electroantennogram Recordings

Baker and Haynes (1989) used an electroantennogram (EAG) to record phero-mone plume structures using oriental fruit moth antennae. Their recordings showed that the pheromone concentrations were strongly fluctuating, confirming the results of Murlis and Jones (1981), who recorded using negative ions. Later, Sauer et al. (1992) described a portable device that was used to determine mean pheromone concentrations in vineyards (Karg and Sauer, 1995; Sauer and Karg, 1998), apple orchards (Karg et al., 1994; Suckling et al., 1996), pea fields (Bengtsson et al., 1994) and cotton fields (Färbert et al., 1996). Field EAGs indicated some differences between crops in the distribution of pheromone, with a more even distribution of pheromone inside the borders of treated vineyards compared to apple orchards.

Field EAGs have proven useful for describing instantaneous fluctuations in pheromone concentrations. More pheromone filaments were detected with a higher number of point sources in orchards (Suckling and Angerilli, 1996), and in the presence of higher wind speeds (Karg and Suckling, 1997). Removal of the dispens-ers caused the disappearance of the detectable filaments. These large-scale fluctua-tions do not seem to be required for mating disruption of *E. postvittana* (Karg and Suckling, 1997), because some disruption of trap catch occurred without filaments being detectable in the orchard air (Figure 3.6).

These studies highlighted the influence of plant canopy on the spatial and tem-poral distribution of pheromone in treated areas. Foliage acts to reduce wind speed, which reduces losses from the system. Pheromone is also taken up and released by foliage, and atmospheric concentrations are higher and more uniform when the plant canopy is fully developed. Therefore mating disruption is most successful when there is the maximum plant canopy present.

Field EAGs offer relatively rapid descriptions of the sensory environment expe-rienced by the insect over much shorter time intervals than other methods. The system's portability and the high sensitivity of the antenna to pheromones enables us to gain information concerning the three-dimensional distribution of pheromone in mating disruption plots. This method has some disadvantages that hinder the interpretation, including interaction of pheromone and host-plant detection, variance between individual antennae, and the nonlinearity of the detector outside a certain range (Rumbo et al., 1995). The importance of these problems seems to vary from species to species.

3.5.5.3 Single Sensillum Recording in the Field

Van der Pers and Minks (1993) developed a tool designed to carry out single-cell recordings in the field. The electrical responses are recorded from an individual pheromone-specific sensillum, which contains pheromone-sensitive receptor cells (sensillae trichodea, Figure 3.1). The individual sensillae respond to only one component or to a very narrow range of stimuli, which means that plant volatile reception is less likely to interfere with the response to pheromone components.

The major disadvantage of this technique is the use of a considerably more difficult and fragile preparation, which hampers its use in field measurements. In addition, the information acquired from recordings made from single olfactory hairs might be insufficient for the interpretation of sensory input, especially when the insect uses a pheromone blend. Basic studies are needed on adaptation, calibration, the influence of varying outdoor conditions, and background firing of cells to underpin this approach (van der Pers and Minks, 1997).

3.5.5.4 Modeling

The technical difficulties involved in measuring pheromone concentrations make the option of modeling attractive. There have been various attempts to model pheromone concentrations in the field, usually focusing on the concentrations downwind from a point source (Elkington and Cardè, 1988). A meteorological-based model was developed to describe the influence of weather (solar radiation, wind, etc.) on atmospheric concentration above tea fields (Uchijima, 1988). A statistical framework has recently been proposed, which takes into account plant canopy, meteorological conditions, and operational factors (Suckling et al.,1999). This approach should help to provide a sound basis for optimizing operational factors, such as dispenser application height, density, and so forth. This approach may be the best way of reducing the gap between current "rules of thumb" and optimal recommendations, provided that it is used in conjunction with behavioral assays and other studies to determine the threshold concentrations required for success.

3.6 OTHER APPLICATIONS OF SEMIOCHEMICALS

There are a number of relatively new applications of semiochemicals in pest management. Many are still in their early stages of development, but it is of interest to consider their potential to play a role in future.

3.6.1 Deterrents and Repellents

Floral fragrances are often attractant or repellent to insects (Mookherjee et al., 1993), but most commercial insect repellents are currently based on DEET or citronella. DEET has various adverse side effects on human health (Mumcuoglu et al., 1996), but some essential oils could have prospects for replacing it. For example, pine oil has been found to have repellent or oviposition deterrent properties against some insects (e.g., Ntiamoah et al., 1996). Constituents of pine oil, such as linalool, can reduce feeding and oviposition of houseflies (Maganga et al., 1996). In addition, α-pinene is being developed as an oviposition repellent where cherry bark tortrix (*Enarmonia formosana*) (Tortricidae) is a problem in urban Vancouver (McNair et al., 1997).

Inexpensive and stable chemicals such as palmitic and oleic acids could have potential for reducing oviposition of the European grape berry moth (*Lobesia botrana*) in the field (Gabel and Thiery, 1996). Oviposition-deterrency was also recorded to occur between species (Thiery and Gabel, 1993), and this approach could have potential against pest complexes of similar species.

3.6.2 Exploiting Natural Enemies

In the process of host location, natural enemies use a wide range of semiochemical cues from host-related products including scales and pheromones from female moths, oral secretions from larvae, aggregation and host-marking pheromones, and so forth (Lewis and Martin, 1990). Insect damage to plants also attracts many natural enemies by changing the quality or quantity of volatiles released after herbivory. Research is under way to understand and eventually exploit this and other aspects of semiochemically mediated interactions between plants and beneficial organisms, to improve the efficacy of biological control and enhance plant resistance (Dicke, 1998). It is also possible to attract generalist predators into the crop. Sant'Ana et al. (1997) reported the potential for this approach using spined soldier bug (*Podisus maculiventris*, Hemiptera).

Unfortunately, there are likely to be complications to this general approach due to the complexity of the natural systems, and success is proving somewhat elusive in most cases. For example, in the case of *Aphytis melinus* and California red scale, it appears that training of parasitoids to host kairomones through associative learning may be needed to make innundative releases most effective (Hare et al., 1997). The economics of such applications could be difficult.

Pure plant volatiles have also been examined as a way of increasing parasitoid activity. It was possible for Roland et al. (1995) to attract tachnid parasitoids of winter moth to apple trees using borneol, but no increase in oviposition was observed. This failure to obtain control illustrates the challenge in achieving the full range of behaviors required for a numerical response from natural enemies. It is essential to demonstrate that bringing natural enemies to an area will reduce pest density. Companion planting has long been viewed as a pest management tactic in organic gardening, but there is little scientific evidence for a mechanistic basis for it. Attracting natural enemies of pests by planting associated noncrop plants is being tested in a range of cropping systems (e.g., Wratten and van Emden, 1995). It is more likely to work if the companion plants offer a needed resource to the natural enemies, for example, if alternative nectar sources are a limiting factor to parasitoids.

3.6.3 Integration of Semiochemicals into Pest Management

This chapter has attempted to highlight examples of the trend toward integration of tactics, finally approaching the ideal of "integrated pest management." In principle, integrated combinations of pest management tactics are likely to produce better control than individual components alone. In an ecological sense, there is an increasing recognition that sustainable pest management will come from provision of the widest possible range of mortality factors against pests. However, it is also clear that the increasing complexity of these systems will place new demands on farmers.

The stimulo-deterrent diversionary strategy or "push-pull" strategy (Millar and Cowles, 1990), is an interesting concept being considered by several research groups. This strategy provides opportunities for enhanced control in sites where the pest becomes concentrated, by manipulating pest behavior using attractants and repellents. This could involve treatment with inhibitors or antifeedants on the crop, and attractants for predators or parasitoids in nearby trap crops. The trap crops may comprise attractive plants or cultivars, which could be treated with attractants and fungal pathogens (Pickett et al., 1995). For example, synthetic aggregation pheromone may be used to divert spined citrus bugs (*Biprorulus bibax*) to areas such as trap trees, for targeting with chemical control (James et al., 1996).

Management of insect resistance to transgenic plants has been recognized to be a concern, given the field evolution of resistance to Bt toxins in some species. Tactics based on semiochemicals could offer a useful way of managing this risk in some systems, reducing selection for resistance to the transgenic plants. For example, integration of mating disruption with deployment of transgenic apples would appear to offer prospects for delaying the evolution of resistance to Bt toxins in lightbrown apple moth (*Epiphyas postvittana,* Tortricidae) (Caprio and Suckling, 1995).

Colonization of hosts by bark beetles involves both pheromones and kairomones. Recognition of this complexity has led to the development of elegant multicomponent systems. For example, the inhibitor-based tactic to suppress southern pine beetle (*Dendroctonus frontalis*) infestations, uses the antiaggregation pheromone verbenone. This component does not affect the natural enemies, which also use host volatiles as part of their olfactory cue for locating host habitat (Salom et al., 1995). Furthermore, green leaf alcohols may also be useful to support it (Wilson et al., 1996).

Alarm pheromones have not been widely exploited to date. They are inherently volatile, with rapid onset and conclusion of the behavior, which is a challenge for practical use. Alarm pheromones of aphids have been examined for pest management. Combinations of alarm pheromone ((E)-β-farnesene) with contact pesticides (permethrin) or fungal pathogens have been tested against cotton aphids. They appear to work by increasing the activity of the pest, leading to increased exposure (Pickett et al., 1995). (E)-β-farnesene is now commercially available, and work is under way to develop it further.

Within the context of mating disruption, it is possible to consider multiple species formulations that are the true pheromones for one member of a pest complex, and synomones for other species. For example, mating disruption of several closely related species of New Zealand leafrollers can be achieved using a single blend of components that is attractive for only one species (potential false trails), but works using other mechanisms (e.g., habituation) against related species (Suckling and Burnip, 1996; 1997).

Thus we can see from the examples in this chapter that a very wide range of variations is possible, incorporating pheromones and other semiochemicals into pest management applications by themselves, or more often, in combination with other components.

3.7 THE FUTURE OF SEMIOCHEMICALS

Pheromones and other semiochemicals are powerful tools and are showing an increasingly important role in pest management. This is happening for several reasons. There is an increasing recognition of the cost-effectiveness of pheromones and other semiochemicals in reducing broad-spectrum pesticide use, in food and fiber production systems. This recognition is especially being driven by the increasing influence of consumers in developed countries, who are expressing greater interest than ever in safe food and more benign production systems. This trend toward market acceptance of alternatives is coupled with the increasing failures of broad-spectrum pest-management tactics due to insecticide resistance or other undesirable side effects of their use. At the same time, there is a rapidly increasing number of semiochemicals available, because of the development of more sophisticated equipment and progress with their identification and synthesis. An improved understanding of the operational, biological, and behavioral aspects of applied chemical ecology is also raising the prospects for success. Semiochemicals are increasingly being integrated with a range of other methods to produce new tactics or enhance more established tactics. In summary, it is clear that applications of semiochemicals will expand in the now classical fields of pest monitoring, mass trapping, and mating disruption, and we can also expect the development of more applications using kairomones, such as pest management by the attraction of predators and parasitoids.

REFERENCES

Aldrich, J.R. Chemical communication in the true bugs and parasitoid exploitation, in *Chemical Ecology of Insects*, Vol. 2, Cardé, R.T. and Bell, W.J., Eds., Chapman and Hall, New York, pp. 318–362, 1995.

Aldrich, J.R. and P. Puapoomchareon. Management of predaceous hemipterans with semiochemicals: Practice and potential, XX Int. Congress of Entomology, Firenze, Italy, Aug. 25–31, 1996. p. 625, 1996.

Arn, H., M. Tóth, and E. Priesner. http://www.nysaes.cornell.edu/pheronet/, 1998.

Arn, H., M. Tóth, and E. Priesner. List of Sex Pheromones of Lepidoptera and Related Attractants, *Organisation Internationale de Lutte Biologique, Section Régionale Ouest Paléarctique*, Montfavet, 1992.

Baker, T.C. and H.F. Haynes. Field and laboratory electroantennographic measurements of pheromone plume structure correlated with oriental fruit moth behaviour, *Physiol. Entomol.* 14, pp. 1–12, 1989.

Baker, T.C., M.A. Willis, K.F. Haynes, and P.L. Phelan. A pulsed cloud of sex pheromone elicits upwind flight in male moths, *Physiol. Entomol.* 10, pp. 257–265, 1985.

Bakke, A. and R. Lie. Mass trapping, in *Insect Pheromones in Plant Protection*, Jutsum, A.R., Gordon, R.F.S., Eds., John Wiley & Sons, New York, pp. 67–87, 1989.

Bakke, A., T. Seather, and T.M. Kvamme. Mass-trapping of the sprucebark beetle *Ips typographus*, Pheromone and trap technology. *Medd. Nor. Inst. Skogforsk.* 38, pp. 1–35, 1983.

Bartell, R.J. Mechanisms of communication disruption by pheromone in control of Lepidoptera, A review, *Physiol. Entomol.* 7, pp. 353–364, 1982.

Bengtsson, M., G. Karg, P.A. Kirsch, J. Löfqvist, A. E. Sauer, and P. Witzgall. Mating disruption of the pea moth *Cydia nigricana* F. (Lepidoptera, Tortricidae) with a repellent blend of sex pheromone and attraction inhibitors, *J. Chem. Ecol.* 20, pp. 871–887, 1994.

Biddle, A.J., L.E. Smart, M.M. Blight, and A. Lane. A monitoring system for the pea and bean weevil (*Sitona lineatus*), *Proc. Brighton Crop Prot. Conf. Pests and Dis.*, pp. 173–178, 1996.

Blair, J.A., A.J. Mordue, B.S. Hansson, L.J. Waldhams, and J.A. Pickett. A behavioural and electrophysiological study of oviposition attractants and host-odour cues for *Culex quinquefasciatus*, *Physiol. Entomol.* 18, pp. 343–348, 1994.

Blight, M.M., J.A. Pickett, M.C. Smith, and L.J. Wadhams. An aggregation pheromone of *Sitona lineatus*, *Naturwissenschaften* 71, pp. 480–481, 1984.

Borden, J.H. Use of semiochemicals to manage coniferous tree pests in western Canada, in *Behavior-Modifying Chemicals For Insect Management*, Ridgway, R.L., R.M. Silverstein, and M.N. Inscoe, Eds., Marcel Dekker, New York, pp. 281–315, 1990.

Borden, J.H. Development and use of semio-chemicals against bark and timber beetles, in *Forest Insect Pests in Canada*, Armstrong, J.A. and Ives, W.G.H., Eds., Natural Resources Canada, Ottawa, pp. 431–449, 1995.

Borden, J.H. and B.S. Lindgren. The role of semiochemicals in IPM of the mountain pine beetle, in *Integrated Control of Scolytid Bark Beetles*, Payne T.L. and H., Saarenmaa, Eds., Virginia Polytechnic Institute and State University, Blacksburg, Virginia, pp. 247–255, 1988.

Bradley, S.I., J.T.S. Walker, C.H. Wearing, P.W. Shaw, and A.J. Hodson. The use of pherome traps for leafroller action thresholds in pipfruit. *Proc. 51st N.Z. Plant Prot. Conf.*, pp. 173–178, 1998. http://www.hortnet.co.nz/publications/nzpps/proceeds.htm/

Brodeur, J. and J.N. McNeil. Seasonal ecology of *Phidius nigripes* (Hymenoptera: Aphididae), a parasitoid of *Macrosiphum euphorbiae* (Homoptera: Aphidiae), *Environ. Entomol.* 23, pp. 292–298, 1994.

Browne, L.E. The use of pheromones and other attractants in house fly control, in *Behavior-Modifying Chemicals for Insect Management*, Ridgway, R.L., R.M. Silverstein, and M.N. Inscoe, Eds., Marcel Dekker, New York, pp. 531–537, 1990.

Burkholder, W.E. Practical use of pheromones and other attractants for stored products insects, in *Behavior-Modifying Chemicals for Insect Management*, Ridgway, R.L., R.M. Silverstein, and M.N. Inscoe, Eds., Marcel Dekker, New York, pp. 5–18, 1990.

Butler, C. *The Feminine Monarchie, Or A Treatise Concerning Bees,* Joseph Barnes, Oxford, 1609, cited in Free, J.B. *Pheromones of Social Bees,* Combstock Publishing, Ithaca, New York, 1987.

Caprio, M.A. and D.M. Suckling. Mating disruption reduces the risk of resistance development to transgenic apple orchards: Simulations of the lightbrown apple moth, *Proc. 48th N.Z. Plant Prot. Conf.*, pp. 52–58, 1995. http://www.hortnet.co.nz/publications/nzpps/procceds.htm.

Cardé, R.T. Principles of mating disruption, in *Behavior-Modifying Chemicals for Insect Management*, Ridgway, R.L., R.M. Silverstein, and M.N. Inscoe, Eds., Marcel Dekker, New York, pp. 42–71, 1990.

Cardé, R.T. and A.K. Minks. Control of moth pests by mating disruption: Successes and constraints, *Annu. Rev. Entomol.* 40, pp. 559–585, 1995.

Carlson, D.A., P.A. Langley, and P.M. Huyton. Sex pheromone attractant of tsetse fly: isolation, identification and synthesis of the contact aphrodisiacs, *Science* 120, pp. 750–753, 1978.

Carlson, D.A., M.S. Mayer, and D.L. Silhacek. Sex pheromone attractant of the house fly: isolation, identification and synthesis, *Science* 174, pp. 76–78, 1971.

Caro, J.H., D.E. Glodfelty, and H.P. Freeman. (Z)-9-Tetradecen-1-ol formate. Distribution and persistence in the air within a corn crop after emission from a controlled-release formulations, *J. Chem. Ecol.* 6, pp. 229–239, 1980.

Charmillot, P.J., and D. Hofer. Control of codling moth, *Cydia pomonella* L., by an attract and kill formulation, in *Technology transfer in mating disruption, IOBC WPRS Bull. 20,* pp. 139–140, 1997.

Charmillot, P.J. and R. Vickers. Use of sex pheromones for control of tortricid pests in pome and stone fruits. in *Tortricid Pests: Their Biology, Natural Enemies and Control. World Crop Pests, Volume 5,* van der Geest, L.P.S and H.H. Evenhuis, Eds. Elsevier, Amsterdam, pp. 487–496, 1991.

Cornelius, M.L. and J.K. Grace. Semiochemicals extracted from a Dolichoderine ant affects the feeding and tunneling behavior of the Formosan subterranean termite (Isoptera: Rhinotermitidae), *J. Econ. Entomol.* 87, pp. 705–708, 1994.

Cossé, A.A. and T.C. Baker. House flies and pig manure volatiles: wind tunnel and behavioral studies and electrophysiological evaluation. *J. Agric. Entomol.* 113, pp. 301–317, 1996.

Croft, B.A., A.L. Knight, L. Flexner, and R. Miller. Competitive effect of caged virgin females *Argyrotaenia citrana* (Fernald) on pheromone trap caches of released males in a semi-enclosed courtyard, *Environ. Entomol.* 15, pp. 232–239, 1985.

David, C.T. Mechanisms of directional flight in wind, in *Mechanisms in Insect Olfaction,* Payne, T.L., M.C. Birch, and C.E.J. Kennedy, Eds., Clarendon Press, Oxford, U.K., 1986.

Delate, K.M., J.K. Grace, and C.H.M. Tome. Potential use of pathogenic fungi in baits to control the Formosan subterranean termite (Isoptera: Rhinotermitidae), *J. Appl. Entomol.* 11, pp. 429–433, 1995.

Dicke, M. Direct and indirect effects of plants on performance of beneficial organisms, in *Handbook of Pest Management,* Ruberson, J.R., Ed., Marcel Dekker, New York, pp. 105–153, 1998.

Elkington, J.S. and R.T Cardé. Odor dispersion, in *Chemical Ecology of Insects,* Bell, W.J. and R.T. Cardé, Eds., Chapman and Hall, London, pp. 73–88, 1984.

Färbert, P., U.T. Koch, A. Färbert, and R.T. Staten. Measuring pheromone concentrations in cotton fields with the EAG method, in *Insect Pheromone Research: New Directions,* Cardé, R.T. and A.K. Minks, Eds., Chapman and Hall, New York, pp. 347–358, 1997.

Fitzgerald, T.D. Trail following and recruitment: Response of eastern tent caterpillar *Malacosoma americanum* to 5-β-cholestane-2,4-dione and 5-β-cholestane-3-one, *J. Chem. Ecol.* 19, pp. 449–457, 1993.

Fletcher, B.S. and T.E. Bellas. Pheromones in Diptera, in *CRC Handbook of Natural Pesticides Vol IV. B,* Morgan, E.D. and N.B. Mandava, Eds., CRC Press, Boca Raton, FL, pp. 1–57, 1988.

Flint, H.M., J.R. Merkle, and A. Yamamoto. Aerial concentration of gossyplure, the sex pheromone of the pink bollworm (Lepidoptera: Gelechiidae), in cotton fields treated with long lasting dispensers, *Environ. Entomol.* 19, pp. 1845–1851, 1990.

Francke, W., G. Hindorf, and W. Reith. Alkyl-1,6-dioxaspiro[4,5]-decanes, a new class of pheromones, *Naturwissenschaften* 66, p. 618, 1979.

Free, J.B. *Pheromones of Social Bees,* Combstock Pub., Ithaca, New York, 1987.

Furlong, M.J., J.K. Pell, O.P. Choo, and S.A. Rahman. Field and laboratory evaluation of a sex pheromone trap for the autodissemination of a fungal entomopathogen *Zoophthora radicans* (Entomophthorales) by the diamondback moth *Plutella xylostella* (Lepidoptera: Yponomeutidae), *Bull. Ent. Res.* 85, pp. 331–337, 1995.

Gabel, B., and D. Thiery. Oviposition response of *Lobesia botrana* females to long-chain free fatty acids and esters from its eggs, *J. Chem. Ecol.* 22, pp. 161–171, 1996.

Gieselmann, M.J. and R.E. Rice. Use of pheromone traps, in *Armoured Scale Insects, Their Biology, Natural Enemies and Control*, Rosen, D., Ed., Elsevier, Amsterdam. 1990.

González, R. and M. Campos. A preliminary study on the use of trap-trees baited with ethylene for the integrated management of the olive beetle, *Phloetribus scarabaeoides* (Bern.) (Col., Scolytidae), *J. Appl. Entomol.* 119, pp. 601–605, 1995.

Hall, D.R. Use of host odor attractants for monitoring and control of tsetse flies. in *Behavior-Modifying Chemicals for Insect Management,* Ridgway, R.L., R.M. Silverstein, M.N. Inscoe, Eds. Marcel Dekker, New York, pp. 517–530, 1990.

Hallett, R.H., G. Gries, R. Gries, J.H. Borden, E. Czyewska, A.C. Oeschlager, H.D. Pierce, N.P.D. Angerilli, and A. Rauf. Field testing of aggregation pheromones of two asian palm weevils *Rhynchophorus ferrugineus* and *R. vulneratus. Naturwissenschaften* 80, pp. 328–331, 1993.

Haniotakis, G., M. Kozyrakis, T. Fitsakis, and A. Andronidaki. An effective mass trapping method for the control of *Dacus oleae* (Diptera: Tephritidae), *J. Econ. Entomol.* 84, pp. 564–569, 1991.

Hare, J.D., D.J.W. Morgan, and T. Nguyun. Increased parasitization of California red scale in the field after exposing its parasitoid, *Aphelinus melinus*, to a synthetic kairomone, *Entomol. Expl. Appl.* 82, pp. 73–81, 1997.

Haynes, K.F, W.G. Li, and T.C. Baker. Control of pink bollworm moth (Lepidoptera: Gelechiidae) with insecticides and pheromones (attracticide): lethal and sublethal effects, *J. Econ. Entomol.* 79, pp. 1466–1471, 1986.

Haynes, K.F., T.A. Miller, R.T. Staten, W.G. Li, and T.C. Baker. Pheromone trap for monitoring insecticide resistance in the pink bollworm moth *Pectinophora gossypiella* (Lepidoptera: Gelechiidae): New tool for resistance management, *Environ. Entomol.* 16, pp. 84–89, 1987.

Hefetz, A., I. Bar-Zakay, and B.A. Peleg. Application of scale sex pheromone for their control in citrus. IOBC SROP Working Group on The Use of Pheromones and Other Semiochemicals in Integrated Control, 10–15 September 1990, Granada, Spain, 1990.

Hoffmann, M.P., J.J. Kirkwyland, R.F. Smith, and R.F. Long. Field tests with kairomone-baited traps for cucumber beetles and corn rootworms in cucurbits. *Environ. Entomol.* 25, pp. 1173–1181, 1996.

Hoffmeister, T.S. and B.D. Roitberg. Counterespionage in an insect herbivore-parasitoid system, *Naturwissenschaften* 84, 117–119, 1997.

Howse, P.E., I.D.R. Stevens, and O.T. Jones, *Insect Pheromones and Their Use in Pest Management,* Chapman and Hall, London, 1998.

Hrdy, I., F. Kuldova, F. Kocourek, J. Berankova, and O. Pultar. Insect viruses and juvenoids disseminated by means of pheromone traps: a potential tool of integrated pest management in orchards, Abstracts XX Int. Congress of Entomology, August 1996, Firenze, 1996.

Inscoe, M.N., B.A. Leonhardt, and R.L. Ridgway. Commercial availability of insect pheromones and other attractants, in *Behavior-Modifying Chemicals for Insect Management,* Ridgway, R.L., R.M Silverstein, and M.N. Inscoe, Eds., Marcel Dekker, New York, pp. 631–715, 1990.

Izquierdo, J.I. *Helicoverpa armigera* (Hubner) (Lep., Noctuidae): relationship between captures in pheromone traps and egg counts in tomato and carnation crops, *J. Appl. Entomol.* 120, pp. 281–290, 1996.

Jackson, D.M., G.C. Brown, G.L. Nordin, and D.W. Johnson. Autodissemination of a baculovirus for management of tobacco budworms (Lepidoptera: Noctuidae) on tobacco, *J. Econ. Entomol.* 85, pp. 710–719, 1992.

James, D.G., R.J. Bartelt, and C.J. Moore. Mass-trapping of *Carpophilus* spp. (Coleoptera: Nitidulidae) in stone fruit orchards using synthetic aggregation pheromones and a coattractant: Development of a strategy for population supression, *J. Chem. Ecol.* 22, pp. 1541–1556, 1996.

Jang, E.B. and D.M. Light. Olfactory semiochemicals of Tephritids, in *Fruit Fly Pests: A World Assessment of their Biology and Management*, McPheron, B.A., and G.J. Steck, Eds. St. Lucie Press, Delray Beach, FL, pp. 73–90, 1996.

Jones, O.J. Pest monitoring, in *Insect Pheromones and Their Use in Pest Management*, Howse P.E., I.D.R. Stevens., and O.T. Jones, Eds., Chapman and Hall, London, pp. 280–299, 1998.

Jutsum, A.R. and R.F.S. Gordon. *Insect Pheromones in Plant Protection*, Wiley, Chichester, U.K. 1989.

Karg, G. and A.E. Sauer. Spatial distribution of pheromones in fields treated for mating disruption of the European grape wine moth *Lobesia botrana* measured with electroantennograms, *J. Chem. Ecol.* 21, pp. 1299–1314, 1995.

Karg, G. and D.M. Suckling. Polyethylene dispensers generate large scale temporal fluctuations in pheromone concentrations, *Environ. Entomol.* 26, pp. 896–905, 1997.

Karg, G., D.M. Suckling, and S.J. Bradley. Absorption and release of pheromone of *Epiphyas postvittana* (Lepidoptera: Tortricidae) by apple leaves, *J. Chem. Ecol.* 20, pp. 1825–1841, 1994.

Karlson, P. and M. Lüscher. "Pheromones": a new term for a class of biologically active substances, *Nature* 153, pp. 55–56. 1959.

Kerguelen, V. and R.T. Cardé. Manoeuvres of female *Brachymeria intermedia* flying toward host-related odours in a wind tunnel, *Physiol. Entomol.* 22, pp. 344–356. 1997.

Klein, M.G. and L.A. Lacey. An attractive trap for autodissemination of entomopathogenic fungi into populations of Japanese beetle, *Popillia japonica* (Coleoptera: Scarabaeidae), *Biocontr. Sci. Techn.*, 1998.

Knight, A.L. and Croft, B.A. Modelling and prediction technology, in *Tortricid Pests: Their Biology, Natural Enemies and Control*, World Crop Pests Volume 5, van der Geest, L.P.S and H.H. Evenhuis, Eds., Elsevier, Amsterdam, pp. 301–312, 1991.

Knipling, E.F. *The Basic Principles of Insect Population Suppression and Management*, USDA Agric. Handbook 512, 1979.

Leal, W., E.F. Vilela, M.S. Bento, T.M.C. Della Luci, and M. Ono. Long range sex pheromone of the longhorn beetle *Migdolus fryanus* Westwood: Identification and application, in *XX Int. Congress of Entomology*. Firenze, Italy, August 25–31, p. 542, 1997.

Leonhardt, B.A., R.T. Cunningham, W.A. Dickerson, V.C. Mastro, R.L. Ridgway, and C.P. Schwalbe. Dispenser design and performance criteria for insect attractants, in *Behaviormodifying Chemicals for Insect Management*, Ridgway R.L., R.M. Silverstein, and M.N. Inscoe, Eds. Marcel Dekker, New York, pp. 113–130, 1990.

Lewis, W.J. and W.R. Martin, Jr. Semiochemicals for use with parasitoids: Status and future, *J. Chem. Ecol.* 16, pp. 3067–3089, 1990.

Mafra-Neto, A. and T.C. Baker. Timed, metered sprays of pheromone disrupt mating of *Cadra cautella* (Lepidoptera: Pyralidae), *J. Agric. Entomol.* 13, pp. 149–168, 1996.

Mafra Neto, A. and R.T. Cardé. Dissection of the pheromone modulated flight of moths using single-pulse response as a template, *Experientia* 52, pp. 373–379, 1996.

Maganga, M., G. Gries, and R. Gries. Repellency of various oils and pine oil consituents to house flies (Diptera: Muscidae), *Environ. Entomol.* 25, pp. 1182–1187, 1996.

Mason, L.J. and R.K. Jansson. Disruption of sex pheromone communication in *Cylas formicarius* (Coleoptera: Apionidae) as a potential means of control, *Florida Entomologist* 74, pp. 469–472, 1991.

Mayer, M.S. and J.R. McLaughlin. *Handbook of Insect Pheromones and Sex Attractants*, CRC Press, Boca Raton, FL, 1991.

McBrien, H.L., G.J.R. Judd, and J.H. Borden. *Campylomma verbasci* (Heteroptera: Miridae): Pheromone-based seasonal flight patterns and prediction of nymphal densities in apple orchards, *J. Econ. Entomol.* 87, pp. 1224–1229, 1994.

McBrien, H.L., G.J.R. Judd, and J.H. Borden. Population supression of *Campylomma verbasci* (Heteroptera: Miridae) by atmospheric permeation with synthetic sex pheromone, *J. Econ. Entomol.* 90, pp. 801–808, 1997.

McNair, C., C. Bedard, G. Gries, and R. Gries. Reciprocal recognition of host and nonhost volatiles by conifer and Roseacean-feeding tortricids, *Int. Soc. Chem. Ecol.*, Vancouver, 1997.

McVeigh, L.J. and B.W. Bettany. The development of a lure and kill technique for control of the Egyptian cotton leafworm (*Spodoptera littoralis*), in *Behaviour of Moths and Molecules, IOBC-SROP Working Group on the Use of Pheromones and Other Semiochemicals in Integrated Control*, Arn, H., Ed., 8–12 September 1986, Neustadt an der Weinstrasse, Germany, pp. 59–60, 1986.

Metcalf, R.L. and E.R. Metcalf. *Plant Kairomones in Insect Ecology and Control*, Chapman and Hall, New York, p. 168, 1992.

Miller, J.R. and R.S. Cowles. Stimulo-deterrent diversion: a concept and its possible applications to onion maggot control, *J. Chem. Ecol.* 16, pp. 3197–3212, 1990.

Miyatake, T., S. Moriya, T. Kohama, and Y. Shimoji. Dispersal potential of male *Cylas formicarius* (Coleoptera: Brentidae) over land and water, *Environ. Entomol.* 26, pp. 272–276, 1997.

Mookherjee, J., R.A. Wilson, K.R. Schrankel, I. Katz, and J.F. Butler. Semio activity of flavor and fragrance molecules on various insect species, in *Bioactive Volatile Compounds from Plants*, Teranishi, R., R.G. Buttery and H. Sugusawa, Eds., American Chemical Society, Washington, D.C., pp. 35–48, 1993.

Mordue, A.J., A. Blackwell, B.S. Hansson, L.-J. Wadhams, and J.A.Pickett. Behavioural and electrophysiological evaluation of the oviposition attractants for *Cule quinquefasciatus* Say (Diptera: Culicidae), *Experientia*, 48, pp. 1109–1111, 1992.

Mumcuoglu, K.Y., R. Galum, U. Bach, J. Miller, and S. Magadassi. Repellency of essential oils and their components to the human body louse, *Pediculus humanus humanus*, *Entomol. Exp. Appl.* 78, pp. 309–314, 1996.

Murchie, A.K., L.E. Smart, and I.H. Williams. Responses of *Dasineura brassicae* and its parasitoids *Platygaster subuliformis* and *Omphale clypealis* to field traps baited with organic isothiocyanates, *J. Chem. Ecol.* 23, pp. 917–926, 1997.

Murlis, J. and C.D. Jones. Fine scale structure of odour plumes in relation to insect orientation to distant pheromone and other attractant sources, *Physiol. Entomol.* 6, pp. 1–86, 1981.

Ngamine, M., M. Kinjo, H. Sugie, and Y. Tamaki. Mass trapping of the sugarcane wireworm *Melanotus okinawensis*, in Okinawa, *Japan Plant Protection Assoc. Symposium on Pheromones*, Tokyo, Oct. 18, 1988.

Nordlund, D.A. Semiochemicals: A review of the terminology, in *Semiochemicals: Their Role in Pest Control*, Nordlund, D.A., R.L. Jones, and W.J. Lewis, Eds., John Wiley & Sons, New York, pp. 13–23, 1981.

Ntiamoah, Y.A., J.H. Borden, H.D. Pierce, Jr. Identity and bioactivity of oviposition deterrents in pine oil for the onion maggot, *Delia antiqua*, *Entomol. Exp. Appl.* 79, pp. 219–226, 1996.

O'Callaghan, M. and T.A. Jackson. Adult grass grub dispersal of *Serratia entomophila*, *Proc. 46th NZ Plant Protection Conf.*, pp. 235–236, 1993.

Otieno, W.A., T.O. Onyango, M.M. Pile, B.R Laurence, G.W. Dawson, L.J. Wadhams, and J.A. Pickett. A field trial of the synthetic oviposition pheromone of *Culex quinquefasciatus* Say (Diptera: Culicidae) in Kenya, *Bull. Ent. Res.* 78, pp. 463–478, 1998.

Papaj, D.R. Use and avoidance of occupied hosts as a dynamic process in tephritid flies in *Insect-Plant Interactions*, Vol. V, Bernays, E.A., Ed., CRC Press, Boca Raton, FL, 1994.

Park, J.D. and H.G. Goh. Control of beet armyworm, *Spodoptera exigua* Hubner (Lepidoptera: Noctuidae), using synthetic sex pheromone. I. Control by mass trapping in *Allium fistulosum* field, *Korean J. Appl. Entomol.* 31, pp. 45–49, 1992.

Pell, J.K., E.D.M. Macaulay, and N. Wilding. A pheromone trap for dispersal of the pathogen *Zoophthora radicans* Brefeld (Entomophthorales) amongst populations of the diamondback moth *Plutella xylostella* L. (Lepidoptera: Yponomeutidae), *Biocontrol Sci. Tech.* 3, pp. 315–320, 1993.

Peña, A., C. Lozano, A.J. Sánchez-Raya, and M. Campos. Ethylene release under field conditions for the management of the olive bark beetle *Phloetribus scarabaeoides*, *J. Agric. Entomol.* 15, pp. 23–32, 1998.

Phelan, P.L. and T.C. Baker. An attracticide for control of *Amyelois transitella* (Lepidoptera: Pyralidae) in almonds, *J. Econ. Entomol.* 80, pp. 779–783, 1987.

Pickett, J.A., L.J. Wadhams, and C.M. Woodcock. Exploiting chemical ecology for sustainable pest control, *British Crop Protection Council Symposium* 63, pp. 353, 1995.

Pinninger, D.B., M.R. Stubbs, and J. Chambers. Evaluation of some food attractants for the detection of *Oryzaephilus surinamensis* (L.) and other storage pests, in *Proceedings of the third international working conference on stored-product entomology*, Kansas State University, Mannhattan, Kansas, pp. 640–650, 1984.

Prokopy, R.J. Evidence for a marking pheromone deterring repeated oviposition in apple maggot flies, *Environ. Entomol.* 1, pp. 326–332, 1972.

Riedl, H., A. Seaman, and F. Henrie. Monitoring susceptibility to azinphosmethyl in field populations of the codling moth (Lepidoptera: Tortricidae) with pheromone traps, *J. Econ. Entomol.* 78, pp. 693–699, 1985.

Roelofs, W.L., E.H. Glass, J. Tette, and A. Comeau. Sex pheromone trapping for red-banded leaf roller control: theoretical and actual, *J. Econ. Entomol.* 62, pp. 1162–1167, 1970.

Roland, J., K.E. Denford, and L. Jiminez. Borneol as an attractant for *Cyzenis albicans*, a tachnid parasitoid of the winter moth, *Operophthera brumata* L. (Lepidoptera: Geometridae), *Can. Entomol.* 127, pp. 413–421, 1995.

Rumbo, E.R., D.M. Suckling, and G. Karg. Measurements of airborne pheromone concentrations using EAG equipment: Interactions between environmental volatiles and pheromone, *J. Insect Physiol.* 41, pp. 465–471, 1995.

Salom, S.M., D.M. Grossman, Q.M. McClellan, and T.L. Payne. Effect of an inhibitor-based suppression tactic on abundance and distribution of southern pine beetle (Coleoptera: Scolytidae) and its natural enemies, *J. Econ. Entomol.* 88, pp. 1703–1716, 1995.

Sanders, C.J. Mechanisms of mating disruption in moths. in *Insect Pheromone Research: New Directions*, Cardé, R.T. and A.K. Minks, Eds., Chapman and Hall, New York, pp. 333–346, 1997.

Sanders, C.J. and D.B. Lyons. Development of an extensive pheromone trap monitoring system for forests pests. IOBC/WPRS Bull 16, pp 43–49, 1993.

Sant'Ana, J., R. Bruni, A.A. Abdul-Baki, J.R. Aldrich. Pheromone-induced movement of nymphs of the predator, *Podisus maculiventris* (Heteroptera: Pentatomidae), *Biol. Control* 10, pp. 123–128, 1997.

Sauer, A.E. and G. Karg. Variables affecting the pheromone concentration in vineyards treated for mating disruption of grape vine moth, *J. Chem. Ecol.* 24, pp. 289–302, 1998.

Sauer, A.E., G. Karg, U.T. Koch, R. Milli, and J.J. de Kramer. A portable EAG system for the measurement of pheromone concentrations in the field, *Chemical Senses* 17, pp. 543–588. 1992.

Schofield, S. and J. Brady. Effects of carbon dioxide, acetone and 1-octen-3-ol on the flight responses of the stable fly, *Stomoxys calcitrans*, in a wind tunnel, *Physiol. Entomol.* 22, pp. 380–386. 1997.

Schoonhoven, L.M. Host-marking pheromones in Lepidoptera, with special reference to two *Pieris* spp., *J. Chem. Ecol.* 16, pp. 3043–3052, 1990.

Schwalbe, C.P. and V.C. Mastro. Multispecific trapping techniques for exotic-pest detection, *Agriculture, Ecosystems and Environment* 21, pp. 43–51, 1988.

Shani, A. Role of pheromones in integrated pest management (Survey conducted in July–November 1990), *IOBC WPRS Bulletin* 16, pp. 359–372. 1993.

Shaw, P.W., V.M. Cruickshank, and D.M. Suckling. Geographic changes in leafrollers species composition in Nelson orchards, *N. Z. J. Zool.* 21, pp. 1–6, 1994.

Shorey, H.H., and R.G. Gerber. Use of puffers for disruption of sex pheromone communication among navel orangeworm moths (Lepidoptera: Pyralidae) in almonds, pistachios, and walnuts, *Environ. Entomol.* 25, pp. 1154–1157, 1996.

Smart, L.E., M.M. Blight, and A. Lane. Development of a monitoring trap for spring and summer pests of oilseed rape, *Proc. Brighton Crop Prot. Conf., Pests and Dis.* pp. 167–172, 1996.

Suckling, D.M. and N.P.D. Angerilli. Point source distribution affects pheromone spike frequency and communication disruption of *Epiphyas postvittana* (Lepidoptera:Tortricidae), *Environ. Entomol.* 25, pp. 101–128, 1996.

Suckling, D.M. and G.M. Burnip. Orientation and mating disruption in *Planotortrix octo* using pheromone and inhibitor, *Entomol. Exp. Appl.* 78, pp. 149–158, 1996.

Suckling, D.M. and G.M. Burnip. Orientation disruption of *Ctenopseustis herana* (Lepidoptera: Tortricidae), *J. Chem. Ecol.* 23, pp. 2425–2436, 1997.

Suckling D.M. R.M. Hill, A.H. Gourlay, and P. Witzgall. Sex attractant-based monitoring of a biological control agent of gorse, *Biocont. Sci. Tech.* 9, pp. 99–104, 1999.

Suckling, D.M., S.R. Green, A.R. Gibb, and G. Karg. Predicting atmospheric concentration of pheromone in treated apple orchards, *J. Chem. Ecol.,* 25, pp. 117–137, 1997.

Suckling, D.M., G. Karg, and S.J. Bradley. Factors affecting electroantennogram and behavioural responses of *Epiphyas postvittana* under low pheromone and inhibitor concentrations in orchards, *J. Econ. Entomol.* 87, pp. 1477–1487, 1994.

Suckling, D.M., G. Karg, and S.J. Bradley. Apple foliage enhances mating disruption of lightbrown apple moth, *J. Chem. Ecol.* 22, pp. 325–341, 1996.

Suckling, D.M. and P. Shaw. Conditions that favor mating disruption of *Epiphyas postvittana* (Lepidoptera: Tortricidae), *Environ. Entomol.* 21, 949–956, 1992.

Suckling, D.M., C.H. Wearing, W.P. Thomas, D.R. Penman, and R.B. Chapman. Pheromone use in insecticide resistance surveys of lightbrown apple moths (Lepidoptera: Tortricidae), *J. Econ. Entomol.* 78, pp. 204–207, 1985.

Teulon, D.J., D.R. Penman, and P.M.J. Ramakers. Volatile chemicals for thrips (Thysanoptera: Thripidae) host finding and applications for thrips pest management, *J. Econ. Entomol.* 86, pp. 1405–1415, 1993.

Thiery, D. and B. Gabel. Inter-specific avoidance of egg-associated semiochemicals in four tortricids, *Experientia* 49, pp. 998–1001, 1993.

Trematerra, P. Survey of pheromone uses in stored-products pest control, *Zeitschrift für Angewandte Zoologie* 76, pp. 129–142, 1989.

Trumble, J.T. Integrating pheromones into vegetable crop production, in *Insect Pheromone Research: New Directions*, Cardé, R.T. and A.K. Minks, Eds., Chapman and Hall, New York, pp. 397–410, 1997.

Uchijima, Z. Concentration and diffusion of pheromone, Japan Plant Protection Assoc. Symposium on Pheromones, Tokyo, Oct. 18, 1988. (In Japanese)

Vale, G.A., D.R. Hall, and A.J.E. Gough. The olfactory responses of tsetse flies, *Glossina* spp. (Diptera: Gossinidae) to phenol and urine in the field. *Bull. Entomol. Res.* 78, pp. 293–300, 1988.

van der Pers, J.N.C. and A.K. Minks. Measuring pheromone dispersion in the field with single sensillum recording technique, in *Insect Pheromone Research: New Directions*, Cardé, R.T. and A.K. Minks, Eds., Chapman and Hall, New York, pp. 359–371, 1997.

van der Pers, J.N.C. and A.K. Minks. Pheromone monitoring in the field using single sensillum technique, *Entomol. Expl. Appl.* 68, pp. 237–245, 1993.

Vargas, R.I., R.J. Prokopy, J.J. Duan, C. Albrecht, and X.L. Qing. Captures of wild Mediterranean and Oriental fruit flies (Diptera: Tephritidae) in Jackson and McPhail traps baited with coffee juice, *J. Econ. Entomol.* 90, pp. 165–169. 1997.

Vega, F.E., P.F. Dowd, and R.J. Bartelt. Dissemination of microbial agents using an autoinoculating device and several insect species as vectors, *Biol. Contr.* 5, pp. 545–552, 1995.

Vickers, N.J. and T.C. Baker. Latencies of behavioral response to interception of filaments of sex pheromone and clean air influence flight track shape in *Heliothis virescens* (F.) males, *J. Comp. Physiol.*, 178, pp. 831–847, 1996.

Wall, C. Monitoring and spray timing, in *Insect Pheromones in Plant Protection*, Jutsum , A.R. and R.F.S. Gordon, Eds., John Wiley & Sons, New York, pp. 39–66, 1989.

Weatherston, I. Principles of design of controlled-release formulations, in *Behavior-Modifying Chemicals for Insect Management*, Ridgway, R.L., R.M. Silverstein, and M.N. Inscoe, Eds., Marcel Dekker, New York, pp. 93–112, 1990.

Weston, R.J., A.D. Woolhouse, E.B. Spurr, R.J. Harris, and D.M. Suckling. Synthesis and use of spirochetals and other venom constituents as potential wasp (Hymenoptera: Vespidae) attractants, *J. Chem. Ecol.* 23, pp. 553–568, 1997.

Wheeler, W. and R.M. Duffield. Pheromones of Hymenoptera and Isoptera, in *CRC Handbook of Natural Pesticides Vol. IV: Pheromones, Part B*, Morgan, E.D. and N.B. Mandava, Eds., CRC Press, Boca Raton, FL, pp. 59–206, 1988.

Wilson, I.M., J.H. Borden, R. Gries, and G. Gries. Green leaf volatiles as antiaggregants for the mountain pine beetle, *Dendroctonus ponderosae* Hopkins (Coleoptera: Scolytidae), *J. Chem. Ecol.* 22, pp. 1861–1875, 1996.

Witzgall, P., M. Bengtsson, G. Karg, A-C. Bäckmann, L. Streinz, P.A. Kirsch, Z. Blum, and J. Löfqvist. Behavioral observations and measurements of aerial pheromone concentrations in mating disruption trials against pea moth *Cydia nigricana* F. (Lepidoptera, Tortricidae), *J. Chem. Ecol.* 22, pp. 191–206, 1996.

Wratten, S.D. and H.F. van Emden. Habitat management for enhanced activity of natural enemies of insect pests, in *Ecology and Integrated Farming Systems*, Glen, D.M., M.P. Greaves, and H. Anderson, Eds., John Wiley & Sons, London, pp. 117–144, 1995.

CHAPTER 4

Botanical Insecticides, Soaps, and Oils

Richard A. Weinzierl

CONTENTS

4.1 INTRODUCTION

Botanical (=plant-derived) insecticides, soaps, and oils comprise only a very small portion of the total volume of insecticides used annually on a worldwide basis. Nonetheless, they remain important in insect pest management for at least three reasons. (1) They sometimes provide the most effective control of insect pests that have become resistant to other insecticides. (2) Most are short-lived in the environment, and they pose relatively low risks to nontarget organisms, including the beneficial predators and parasites that help to regulate pests, the higher level predators in food chains, and the human consumers of treated crops. (3) They are naturally occurring or derived or manufactured with minimal technology, so they are sometimes accepted by organic certification programs and by certain consumer groups; they also may be more readily available than synthetic insecticides in some developing countries. Because they do not leave persistent toxic residues, botanical insecticides, insecticidal soaps, and oils may be more compatible with biological control efforts than many synthetic insecticides, but by most definitions, botanical insecticides are not biological control agents.

This review covers first the most common botanical insecticides, then insecticidal soaps (salts of fatty acids), and finally insecticidal formulations of petroleum and vegetable oils. For each individual insecticidal compound or group, the text focuses on the nature and origin of the component chemical(s), the history of its use, its mode of action, toxicity, and persistence, and its uses and limitations in insect pest management, primarily in North America. In this chapter, as in a similar review prepared for lay readers (Weinzierl and Henn 1991), the innumerable historical reports of insecticidal effectiveness against one or a few pests are not cited individually. Instead, the collective nature of such reports is summarized without citation, and only publications that provide key overviews are cited.

4.2   BOTANICAL INSECTICIDES

4.2.1   General Overview

Botanical insecticides are either naturally occurring plant materials or products derived rather simply from such plant materials. In their simplest form, botanical insecticides may be crude preparations of plant parts ground to produce a dust or powder that may be used full-strength or diluted in a carrier such as clay, talc, or diatomaceous earth (which itself is also insecticidal); such preparations include dusts from pyrethrum daisy flowers, cubé roots (rotenone), sabadilla seeds, ryania stems, or neem leaves, fruits, or bark. Only slightly more sophisticated are water extracts or organic solvent extracts of insecticidal components of plants. These extracts or resins are then prepared as liquid concentrates or applied to talcs or clays to use as insecticidal dusts. Pyrethrins, rotenone, neem, and citronella and other essential oils commonly are formulated as extracts or liquid concentrates. The most processed forms of botanical insecticides are purified insecticidal compounds that are isolated from plant materials by a series of extractions and distillations. Nicotine and limonene are distilled from plant materials or their extracts.

Crude botanical insecticides have been used for several centuries and were known in tribal or traditional cultures around the world before being introduced to Europe and the U.S. Those with long histories of traditional use include neem in India, rotenone in east Asia and South America, pyrethrum in Persia (Iran), and sabadilla in Central and South America.

From the late 1800s to the 1940s, botanical insecticides were used extensively on several high value fruit and vegetable crops in the U.S. Nicotine-based insecticides were important before 1900, whereas pyrethrum, rotenone, sabadilla, and ryania became more popular in the 1930s and 1940s. Beginning in the 1940s and 1950s, the discovery and increasing development of synthetic insecticides led to virtual abandonment of botanical insecticides in commercial agriculture in much of the developed world. Newer synthetic organic insecticides were less expensive, more effective, and longer lasting. In the U.S., the only botanical insecticides that remained in widespread use after 1950 were the pyrethrins (used in household and industrial sprays and aerosols) and nicotine (used primarily in greenhouses). Home gardeners continued to apply rotenone on a small scale, but the use of sabadilla and ryania all but ended. For all botanical insecticides, availability became limited and costs increased.

In the 1980s and 1990s, interest in the use of botanical insecticides increased significantly, though not dramatically. This interest was driven primarily by concerns over the longer-lasting synthetic insecticides' environmental impacts and their residues on food. Despite some appealing traits, however, botanical insecticides continue to fill only minor roles, primarily because most are very expensive in comparison with synthetic insecticides. Their availability is often limited, and the bioactivity of some products may vary among seemingly identical preparations. In the U.S., botanical insecticides are regulated by the U.S. Environmental Protection Agency in much the same way as synthetic insecticides (though some may qualify for "reduced risk" status or be exempt from the requirement to meet residue tolerances). The costs of required tests of environmental fate and of human and environmental toxicity are great in relation to potential sales for most botanical insecticides. As a result, their future availability is uncertain.

This review covers the pyrethrins, rotenone, sabadilla, ryania, nicotine, neem, and (together) limonene and linalool. Essential oils such as citronella and pennyroyal are discussed briefly, as is diatomaceous earth. This chapter does not cover the avermectins or spinosins, compounds derived from soil microorganisms. Reviews of the nature and uses of these compounds have been written by Campbell (1989), Clark et al. (1994), DeAmicis et al. (1997), Hale and Portwood (1996), Lasota and Dybas (1991), and Sparks et al. (1996). Also beyond the scope of this chapter are the many plants and constituent compounds that are known to have insecticidal or repellent characteristics but have not yet been produced for commercial use. Jacobson and Crosby (1971), Grainger and Ahmed (1988), Jacobson (1989), Berenbaum (1989), and Hedin (1991, 1997) have reviewed this subject.

Botanical insecticides, though "natural" in origin, are *not* nontoxic or "nonchemical." Some of the plant-derived compounds discussed below (particularly rotenone and nicotine) are as toxic or more toxic to humans than many common synthetic insecticides. In general, they are toxic to pest and beneficial insects alike, and if they are used repeatedly, botanical insecticides can disrupt natural biotic control of insect pests by their natural enemies. Their limited persistence in the environment helps to minimize their adverse effects, but plant-derived toxins certainly should not be viewed as harmless.

4.2.2   Pyrethrum and the Pyrethrins

Pyrethrum is a powder produced by grinding the dried flowers of the pyrethrum daisy, *Chrysanthemum cinerariefolium*, and a related species, *C. coccineum*. Crude pyrethrum powders were first used introduced to Europe around 1800, and they were in use worldwide by around 1850 (Casida 1973; Matsumura 1985; Casida and Quistad 1995). Pyrethrum's insecticidal activity is provided by six constituent esters known as pyrethrins. They are pyrethrin I, the pyrethrolone ester of chrysanthemic acid; pyrethrin II, the pyrethrolone ester of pyrethric acid; cinerin I, the cinerolone ester of chrysanthemic acid; cinerin II, the cinerolone ester of pyrethric acid; jasmolin I, the jasmolone ester of chrysanthemic acid; and jasmolin II, the jasmolone ester of pyrethric acid (Casida 1973; Hayes 1982; Matsumura 1985). Pyrethrin I and II are present in greatest amounts; jasmolin I and II are present in the lowest concentrations (Worthing and Walker 1987).

Pyrethroids are synthetic insecticides with core chemical structures that resemble natural pyrethrins. To these core structures are added radicals that are less reactive, yielding a more stable molecule. The side chains of several pyrethroids contain chlorine, bromine, or fluorine, and in general these compounds are more persistent in the environment (lasting as effective residues for a few to several days) and at least somewhat more toxic to humans and other nontarget organisms than are natural pyrethrins. This review does not cover the pyrethroid insecticides.

Although pyrethrum daisies were once grown in eastern Europe, the Mideast, and Japan, commercial production is now greatest in Kenya (along with Ecuador), at least in part because concentrations of pyrethrins in Kenyan flowers average 1.3% and may reach 3% in some strains; these levels are significantly greater than those found in flowers from Japan or eastern Europe (Casida 1973; Matsumura 1985). Crude, powdered pyethrum is still used as an insecticide, but extracts of the component pyrethrins in petroleum ether, acetone, glacial acetic acid, ethylene dichloride, or methanol are more effective. Final concentrates of such extracts may contain over 90% pyrethrins but more often are comprised of approximately 25 to 50% pyrethrins. At high concentrations, pyrethrins are less stable. For all except research purposes, extracts contain a blend of all six esters, because separating and purifying the distinct compounds is too expensive to be practical in the preparation of commercial insecticides.

Pyrethrins poison insects and mammals in a similar manner; they interfere with nerve transmission by slowing or preventing the shutting of sodium channels in nerve axons (Bloomquist 1996). The result in insects is hyperactivity and convulsions; whole body tremors occur in mammals. The mode of action and resulting symptoms of poisoning are generally similar for pyrethrins and synthetic organochlorine insecticides. Although plant-derived pyrethrins are very toxic and fast-acting against insects, they are not very toxic to mammals by oral or dermal routes (Table 4.1), at least in comparison with other insecticides (reviewed by Hayes 1982). When ingested, they are not readily absorbed from the digestive tract, and they are readily hydrolyzed in the acidic conditions of the gut and the liver. As a result, pyrethrins are more toxic to mammals via inhalation than ingestion, because inha-

Table 4.1 A Summary of Characteristics of the Major Botanical Insecticides

Botanical insecticide	Source plant(s)	Mode of action & toxicity*	Uses
Pyrethrum/ Pyrethrins	Flowers of the pyrethrum daisy, *Chrysanthemum cinerariaefolium*	Interferes with Na and K ion movement in nerve axons. Mammalian oral and dermal LD_{50}s greater than 1000; some allergic reactions can occur in humans and other mammals.	Many. On pets and humans to control fleas, ticks, lice. Used with synergists as aerosol "bombs" in homes and food plants. Breaks down very rapidly. Mixed with more stable botanicals for field and garden uses.
Rotenone	Roots of *Derris, Lonchocarpus,* other tropical legumes	Disrupts energy metabolism in mitochondria. Mammalian oral LD_{50}s range from 25–3000, dermal >1000. More acutely toxic to mammals and more persistent than many botanicals. Some chronic toxicity suspected. Extremely toxic to fish.	In gardens and orchards against many insects, especially beetles. Persists at effective levels for 3 to 5 days or more. Used purposefully as a fish poison.
Sabadilla	Seeds of the tropical lily *Schoenocaulon officinale* and European *Veratrum album*	Interferes with Na and K ion movement in nerve axons. Mammalian oral LD_{50}s near 4000. Irritates skin and mucous membranes; potent inducer of sneezing.	In vegetables and fruits, particularly against squash bug, harlequin bug, and citrus thrips. Breaks down very rapidly.
Ryania	Woody stems of *Ryania speciosa* (S. American woody shrub)	Activates calcium ion release channels and causes paralysis in muscles of insects and vertebrates. Mammalian oral LD_{50} near 1000; dermal near 4000. More persistent than rotenone but less potent.	In fruit and field crops, particularly against caterpillars and thrips. Often combined with rotenone and pyrethrins in commercial mixtures for garden use.
Nicotine	Tobacco, other *Nicotiana* species, also *Duboisia, Anabasis, Asclepias, Equisetum,* and *Lycopodium*	Mimics the neurotransmitter acetylcholine and overstimulates receptor cells to cause convulsions and paralysis. Mammalian oral LD_{50}s range from 3–188; dermal 50 or lower. **Nicotine insecticides are very toxic to humans.**	Mostly in greenhouses and organic gardens. Free nicotine fumigations target aphids, thrips, and mites. Nicotine sulfate in nonalkaline solutions may last 24 to 48 hours and give limited residual protection.
Neem/ azadirachtins	Leaves, bark, and seeds of neem (*Azadirachta indica*) and Chinaberry (*Melia azedarach*)	Biochemical nature of feeding deterrence, repellence, and growth regulation effects are not well described. Mammalian oral LD_{50} greater than 13,000; used medicinally in humans.	On many crops and landscape plants; especially against soft-bodied and sedentary pests. Very short persistence on treated plants.

Table 4.1 (continued)   A Summary of Characteristics of the Major Botanical Insecticides

Botanical insecticide	Source plant(s)	Mode of action & toxicity*	Uses
Limonene/ Linalool	Citrus oils (linalool is also present in many other plants)	Limonene: mammalian oral LD_{50} > 5000; linalool: mammalian oral LD_{50} > 2400; dermal > 3500. Limonene causes spontaneous stimulation of sensory nerves; biochemical nature of modes of action are not well described for either. GRAS, but chronic toxicity suspected for limonene.	Mostly in pet shampoos, dips, and sprays to kill fleas and ticks. Synergized by PBO. Very short persistence on treated surfaces.

* LD_{50} estimates are expressed in mg of toxin per kg of body weight for test animals. LD_{50} = dose estimated to kill 50% of the test animal population; higher numbers indicate lower toxicity. Sources: Hayes (1982); Worthing and Walker (1987); Ware (1988). See text for more details for all columns.

lation provides a more direct route to the blood (Hayes 1982). Pyrethrins are highly toxic to mammals when administered intravenously (Verschoyle and Barnes 1972; reviewed by Hayes 1982). Because pyrethrins are extremely unstable in sunlight, moisture, and air, they rarely pose safety threats to humans or other nontarget organisms except during handling and application, and they are the major active ingredient in louse control insecticides approved for topical use directly on humans. Pyrethrum and to a lesser extent the extracts that contain pyrethrins can, however, cause irritation and allergic reactions in humans (Barthel 1973). Pyrethrins are moderately to highly toxic to fish, but their labile nature in the environment greatly reduces their potential hazard (Pilmore 1973). Because the synthetic pyrethroids are much more stable, their toxicity to fish is a much greater threat.

In attempts to counteract their instability in the environment, pyrethrins usually are combined with antioxidants to extend their persistence slightly, and for long-term storage they are held in the dark at −25°C. Insecticidal formulations usually contain synergists to slow pyrethrin detoxification by oxidative enzymes in target pests. Common synergists include piperonyl butoxide (PBO) and N-octyl bicyclo-heptene dicarboximide (MGK 264). Synergists often are added to pyrethrin formulations at a ratio of 2:1 to 10:1 (synergist: insecticide).

Many companies currently formulate and market pyrethrins for commercial and homeowner uses. Pyrethrum and pyrethrins are used most commonly in ectoparasiticides for humans and pets (particularly against lice and fleas), aerosols for fly control in livestock and poultry facilities, and in closed spaces such as greenhouses, grain storages, and food processing plants. Pyrethrins are also combined with slower acting botanical insecticides such as rotenone or ryania in garden insecticide mixtures. Applications of pyrethrins directly to pets and humans are considered "safe" specifically because the pyrethrins are low in acute oral and dermal toxicity to mammals. In these uses and where pest control is needed around dairies or food plants, the instability of pyrethrins is desirable so that unwanted residues do not persist on treated individuals or in food products. Rapid breakdown (in only a few hours) makes pyrethrins ineffective in most outdoor applications, especially where some degree of residual activity is needed to protect crops or landscape plants.

4.2.3 Rotenone

Rotenone is an insecticidal compound present in the roots of plants in the genera *Lonchocarpus* in South America, *Derris* in Asia, and several other legumes in the tropics. It is one of six related compounds that are extracted from the roots in acetone, ether, or other organic solvents. Matsumura (1985) and Hayes (1982) reviewed rotenone's chemical structure and characteristics.

Shepard (1951) reported that rotenone's use as a fish poison was recorded by the mid-1700s and that such use was probably ancient, but that its insecticidal characteristics were first recognized in the 1800s. Commercial rotenone was once produced from Malaysian *Derris*, but the source for most current production is Peruvian *Lonchocarpus*, often called cubé root. Extraction in organic solvents produces resins that range in active concentrations of 2 to 40%; rotenone and other rotenoids are present in these extracts. For most commercial insecticidal preparations, this resin is used to make liquid concentrates or applied to inert dusts or other carriers to produce an insecticidal dust formulation. Alternatively, *Lonchocarpus* or *Derris* roots can be dried, powdered, and mixed directly with a carrier to produce a dust formulation.

Rotenone disrupts energy metabolism in cell mitochondria, either by inhibiting the electron transport system or by uncoupling the transport system from ATP production. Synthetic insecticides also known to act in one of these two ways include hydramethylnon, sulfuramid, and pyridaben. Reviews of the mode of action of rotenone and the synthetic insecticides that act similarly have been prepared by Haley (1978), Hayes (1982), Matsumura (1985), Hollingshaus (1987), Schnellmann and Manning (1990), and Hollingworth et al. (1994). In insects, rotenone is converted to metabolites that are highly toxic; in mammals, metabolism results in detoxication (Ware 1988).

Rotenone is moderately toxic to mammals, and although estimates of toxicity vary greatly among published reports (reviewed by Hayes 1982), oral and dermal LD$_{50}$ values of 60 mg/kg and ~1000 mg/kg, respectively, are common (see Table 4.1). Such estimates illustrate that rotenone's acute toxicity to mammals is similar to that of many common synthetic compounds. It is also particularly toxic to fish, and its use as a fish poison (to "clean up" unwanted species and allow restocking) exceeds its uses for insect control. Insects poisoned by rotenone cease feeding and lose locomotor function; paralysis eventually leads to death, often several hours to a few days after exposure.

The potential chronic toxicity of rotenone is of greater interest than the chronic toxicities of most other botanical insecticides because rotenone residues may persist for 3 to 5 days or longer on treated surfaces (considerably longer than most other botanical insecticides persist), and human consumption of residues on treated crops is possible. Chronic exposure to rotenone can lead to kidney and liver damage, and some trials of chronic toxicity in rodents indicate that rotenone may be carcinogenic (reviewed by Hayes 1982). The U.S. Environmental Protection Agency proposed (and ruled) in 1989 (Anon. 1989) that rotenone's exemption from residue tolerance regulations be ended and that substantial new data be generated on its toxicity.

Among the botanical insecticides, rotenone has had more approved (labeled) uses in the U.S. than any compound other than the pyrethrins. It is often used against beetles that feed on plant foliage or fruits, and its common targets include the Colorado potato beetle, *Leptinotarsa decemlineata* (Say), a pest that has developed resistance to many synthetic insecticides, the plum curculio, *Conotrachelus nenuphar* (Herbst), in orchards, and cucumber beetles, *Diabrotica* and *Acalymma* species (Weinzierl 1998). Applications to plant foliage often remain effective for 3 to 5 days or longer. Mixing rotenone with alkaline materials such as soaps or lime greatly reduces its persistence and effectiveness. Rotenone is often sold in mixtures with pyrethrins, ryania, copper, or sulfur, and it is used widely by organic gardeners.

4.2.4 Sabadilla

Sabadilla is derived from the ripe seeds of *Schoenocaulon officinale*, also known as cevadilla or caustic barley, a tropical lily that grows in Central and South America. This species and several others (including the "false hellebores," *Veratrum* spp., in the family Melanthaceae) produce veratrine alkaloids that are insecticidal. The veratrine alkaloids comprise approximately 0.3% of the weight of aged sabadilla seeds; of these alkaloids, cevadine and veratridine are the most active insecticidally. Other alkaloids present in the seed and in insecticidal extracts include sabadinine, sabadiline, and sabadine (Hayes 1982). Allen et al. (1944) evaluated the relative effectiveness of the principal alkaloids of sabadilla seed.

Allen et al. (1944) briefly reviewed the history of sabadilla and noted that Spanish explorers and colonists had recorded that "the Indians of New Spain" (Central and South America) used sabadilla to combat the insects associated with wounds and sores, and that a history of such use well in advance of the Spanish presence in the region was very likely. Peru and Venezuela currently supply most of the sabadilla used in the Western Hemisphere; European white hellebore, *Veratrum album*, is used to produce similar insecticidal preparations in Europe.

To prepare insecticides from sabadilla seeds, they are aged, heated, or alkali-treated to form or activate the insecticidal alkaloids. The activated seeds may be ground and used as crude insecticidal dusts, and "Red Devil" sabadilla dust, an example of such a preparation, was once a well-known product. In addition, the active alkaloids have been extracted in kerosene and used in spray emulsions or applied to inert carrier dusts. Such extracts must be stored in the dark to prevent rapid breakdown (Allen et al. 1944).

The mode of action of sabadilla is similar to that of the pyrethrins, as it affects the voltage-dependent sodium channels of nerve axons (Ohta et al. 1973; Levi et al. 1980; Hayes 1982; Liebowitz et al. 1986; Garber and Miller 1987; and Bloomquist 1996). Veratridine causes an increase in the duration of the action potential, repetitive firing, and a depolarization of the nerve membrane potential (Bloomquist 1996). In insects, the result of this nerve poisoning can be immediate death or several days of paralysis before death. Sabadilla is synergized by PBO and MGK 264. In mammals, sabadilla dust is very low in toxicity (see Table 4.1), even though its constituent veratrine alkaloids, in purified form, are very toxic. Sabadilla dusts are, however, very irritating to the skin and mucous membranes of humans, and they are powerful sneeze-inducers. Ingestion of small amounts may cause headache, nausea, vomiting, diarrhea, and cramps; ingestion of high doses causes convulsions, cardiac paralysis, and respiratory failure (Hayes 1982 and references summarized therein).

Sabadilla was used historically for the control of insects on crops, animals, and humans (Allen et al. 1944). Since the advent of synthetic insecticides, sabadilla's uses have declined to the point that organic gardeners currently provide the major market for sabadilla products. Pests for which sabadilla is considered to be particularly effective include such "true bugs" (Hemiptera) as the squash bug, *Anasa tristis* (DeGeer); the chinch bug, *Blissus leucopterus* (Say), and the stink bugs, including the Harlequin bug, *Murgantia histrionica* (Hahns). Its action against these and other insects can result from both contact exposure and ingestion. The insecticidal alkaloids in sabadilla degrade rapidly in air and sunlight, resulting in only very short residual control after application. Sabadilla is highly toxic to honeybees (Atkins and Anderson 1954); this is particularly noteworthy where dust formulations might be applied to blooming cucurbits where pollinators are active at the same time the most common target pest, the squash bug, causes crop damage.

4.2.5 Ryania

Insecticidal preparations of ryania are derived from the woody stem tissue of the shrub *Ryania speciosa* (family Flacourtiaceae), a plant that is native to South America. As is true for other botanical insecticides, a mixture of components is present in extracts or powders of this plant material. Jeffries et al. (1992) identified 11 component compounds with insecticidal activity; they found the most abundant and insecticidally active constituents of these alkaloids (ryanoids) to be ryanodine and 9,21-dehydroryanodine.

Ryania has been used in the U.S. since the 1940s (Pepper and Carruth 1945; Reed and Filmer 1950). Ware (1988) states that most current production is from plants grown in Trinidad. Most commercial formulations are crude dusts (50% ryania powder), though the constituent alkaloids can be extracted in water, alcohol, acetone, ether, or chloroform to produce liquid or wettable powder formulations.

Ryanodine affects the calcium 2+ ion release channel in muscle; reports and reviews of its toxic action have been prepared by Goblet and Mounier (1981), Pessah et al. (1985), Fill and Cornado (1988), Jeffries et al. (1992), Coats (1994), Lehmberg and Casida (1994), and Bloomquist (1996). The result of poisoning in insects and in mammals is a sustained contraction of skeletal muscle without depolarization of the muscle membrane, then eventual paralysis (Bloomquist 1996). Ryania is effectively synergized by PBO.

The relative toxicity of ryania to mammals differs greatly between crude preparations of insecticidal dusts and the more purified extracts of the active ryanoids themselves. Ware (1988) notes that ryania's LD_{50} values in rats are 750 and 4000 mg/kg, respectively, reflecting relatively low mammalian toxicity. However, ryanodine and 9,21-dehydroryanodine are much more toxic to mammals (Pessah et al. 1985; Bloomquist 1996). Human ingestion of large doses of ryania causes weakness, deep and slow respiration, vomiting, diarrhea, and tremors; convulsions and coma precede death in fatal doses. Exposure to the more potent ryanoids in extracts causes symptoms similar to those of organophosphate poisoning. (Depending on exposure, organophosphate poisoning symptoms may include sweating, headache, twitching, muscle cramps, confusion, tightness in the chest, blurred vision, vomiting, evacuation of the bowels and bladder, convulsions, respiratory failure, coma, and death.)

Ryania's toxicity to insects can result from contact or ingestion. It is used most often for control of caterpillar pests of fruits and foliage, and the codling moth, *Cydia pomonella* (L.) in apples and pears, the citrus thrips, *Scirtothrips citri* (Moulton), in citrus, and the European corn borer, *Ostrinia nubilalis* (Hübner), in corn are among the most common targets of ryania used by organic farmers. It is also sold in combination products with rotenone and pyrethrins. Because of its high cost, it is rarely used except by home gardeners and organic producers. It can, however, perform as well as synthetic insecticides in some uses (as reported by Day et al. 1995, for example). Like rotenone, ryania persists longer in the field after application than most other plant-derived insecticides, with residues giving some degree of residual control for up to 3 to 5 days after application to plant surfaces.

4.2.6 Nicotine

Nicotine, nornicotine, and anabasine are related alkaloids derived from tobacco and other plant species in which they may comprise 2 to 8% of the dry weight of leaves. Hayes (1982) noted that nicotine is usually derived from *Nicotiana tabacum* but also occurs in *Nicotiana rustica* and *Duboisia*, another solanaceous genus, and in *Asclepias* (Asclepidaceae), *Equisetum* (Equisitaceae), and *Lycopodium* (Lycopodiaceae). Nornicotine is prevalent in *Nicotiana sylvestris* and *Duboisia hopwoodii*, and anabasine is present in *Nicotiana glauca* but is usually extracted from *Anabasis aphylla* (Chenopodiaceae). One or more of these compounds may be present in insecticide products marketed as nicotine.

Beinhart (1951) reviewed the production of nicotine and noted that water extracts of tobacco were used in England to kill garden insects as early as 1690, and that by the 1890s the principal active ingredient in such extracts was known to be nicotine. Nicotine may be removed from tobacco by water or hydrocarbon solvents or by dry or steam distillation; Ware (1988) reports that steam distillation is the most common current method for the preparation of commercial insecticides. Free, basic nicotine is very unstable and short-lived in the environment, and it is used most often for greenhouse fumigation. Its combination with acids forms salts (many of which do not crystallize because they are hygroscopic), and most are somewhat longer-lived in the environment. Nicotine sulfate, one such salt, is commonly sold for use as a spray intended to leave an effective residue on plants for one day or more. Salts that form crystals (nicotine benzoate, oxalate, salicylate, and tartrate) and fixed nicotines (water-insoluble salts such as nicotine tannate and nicotine bentonite) are yet more stable and in the past the fixed nicotines were used in baits and in residual ectoparasiticides for livestock.

Nicotine poisons insects and mammals by a similar mode of action — it is an acetylcholine mimic that binds to postsynaptic receptors (Yamamoto 1970;

Matsumura 1985; Ware 1988). Its breakdown is not catalyzed by acetylcholinesterase, so its presence causes repeated stimulation of the receptor. Overall symptoms of nicotine poisoning resemble those of poisoning by organophosphate or carbamate insecticides. Free, basic nicotine and nicotine sulfate are very toxic to humans as well as insects. Noteworthy is that the dermal toxicity of nicotine poses real dangers to users. Oral and dermal LD_{50} values have been estimated in the range of 50 to 60 mg/kg, though a great range in toxicity estimates have been published (Hayes 1982; Ware 1988; Coats 1994).

Nicotine is used most commonly as a fumigant and as a contact spray in greenhouses to kill soft-bodied, sucking and rasping pests such as aphids, thrips, and mites. When nicotine sulfate is combined with alkaline water or soap solutions, free nicotine is released (and degrades) rapidly. Nonalkaline solutions of nicotine sulfate liberate free nicotine more slowly (over 24 to 48 hours), and have some value as residual sprays against leaf-eating pests. Many "natural" pest control recommendations suggest the preparation of tobacco "teas" from tobacco products sold for smoking or chewing. These home-made preparations can be effective against soft-bodied insects on house plants or in greenhouses, but if they are strong enough to be effective against pest insects, they also are strong enough to poison humans as a result of oral or dermal exposure.

4.2.7 Neem

Neem insecticides are derived from the tropical and subtropical *Azadirachta indica* (=*Melia azidarachta*), a tree in the family Meliaceae (the mahogany family). This species, also known as margosa and Indian lilac, is native to southern and southeastern Asia and is now grown in portions of Africa, the Americas, and Australia (Jacobson 1986; Saxena 1989; Schmutterer 1990). Azadirachtin, the principal active ingredient in neem extracts, is one of several liminoids extracted from neem seeds. It is a steroid-like tetranortriterpenoid (structure determined by Kraus et al. 1985; illustrated in Schmutterer 1990). Novel insecticidal compounds are still being identified from *A. indica* and from the related chinaberry tree, *Melia azedarach*, now grown in the U.S. (Lee et al. 1991).

Neem-based insecticides are relatively new to North America, with significant interest in their use originating in the 1970s. However, neem's value as an insect deterrent and an insecticide has been known in India for centuries (Radwanski 1977 a, b, and c). Insecticidal products include teas or dusts made from leaves and bark, extracts prepared from fruits, seeds, or seed kernels, and oils pressed from the seeds. Although all plant parts show some characteristics of insect repellence or suppression, seeds or seed kernels provide the greatest amounts of insecticidal liminoids (Schmutterer 1990). Extraction in methanol is a common method of obtaining azidarachtin from seeds.

The numerous reported effects of neem on insects include repellence, feeding deterrence, oviposition deterrence, reduced growth and development, and interference with reproduction (Schmutterer 1990), but the biochemical or molecular mode of action for such effects remain unknown. Neem and its principal component, azidarachtin, are extremely low in mammalian toxicity (Schmutterer 1990; see

Table 4.1), and most forms are nonirritating to skin and mucous membranes. Neem extracts have been used medicinally for centuries in Asia and India to lower blood pressure, reduce inflammation, and reduce fevers. Exposure to seed dusts causes dermal or respiratory tract irritation in some individuals.

Neem's broad activity against plant-eating insects and its virtual nontoxicity to mammals makes it an extremely appealing insecticide, and various products registered (approved) by the U.S. Environmental Protection Agency are currently marketed for use as sprays, dusts, or soil-applied systemics. Warthen (1989) and Schmutterer (1990) indexed research on target pests and reviewed neem's uses. It is generally considered to be most effective against the soft-bodied, immature stages of plant pests, including whiteflies, thrips, mealybugs, and various caterpillars, and it is low in toxicity to insect predators and parasites that do not feed on plant tissues. However, many field tests reported in Extension publications and *Arthropod Management Tests* (formerly *Insecticide and Acaracide Tests*, Entomological Society of America, Lanham, MD) in the 1980s and 1990s indicate that neem's field performance in many pest control uses has failed to equal that of standard synthetic insecticides, perhaps because of its limited persistence and limited effectiveness in the presence of sunlight and rainfall.

4.2.8 Limonene and Linalool

Citrus extracts and two components of those extracts are recent additions to the list of commercial botanical insecticides. Limonene (particularly *d*-limonene, *p*-mentha-1,8-diene) is a monoterpenoid that comprises up to 90% of crude citrus oil and is easily extracted from it by steam distillation. The monoterpenoids in general are major components of the fragrant, volatile essential oils of such plants as mint, pine, cedar, citrus, and eucalyptus. Other components of these essential oils include ketones, aldehydes, esters, and various alcohols, many of which are also biologically active in insects and mammals. Linalool, a terpene alcohol, is present in citrus peel and in more than 200 other herbs, flowers, fruits, and woods.

Limonene and/or linalool are reported to act as neurotoxins (Coats et al. 1991), insect growth regulators (Karr and Coats 1992), and repellents and fumigants (Karr and Coats 1988). Limonene affects the sensory nerves of the peripheral nervous system, but it is not an inhibitor of acetylcholinesterase. It causes spontaneous stimulation of sensory nerves and subsequent signaling to motor nerves that results in muscle twitching, convulsions, and then paralysis in insects. Synergism by PBO is necessary to prevent recovery of adult insects such as fleas. Little is published on the mode of action of linalool, however it too is synergized by PBO.

Both limonene and linalool were deemed GRAS (generally regarded as safe) by the U.S. Food and Drug Administration in 1965, and they are used extensively at low concentrations as flavorings and scents in foods, cosmetics, soaps, and perfumes. They were considered safe for these uses because they are very low in acute oral and dermal toxicity to mammals (Table 4.1). At high concentrations, however, limonene and linalool injure mammals, and their chronic toxicity has been called into question. When applied topically, limonene is irritating to skin, eyes, and mucous membranes; it acts as a vasodilator and skin sensitizer. It has been shown to promote tumor formation in mice (Roe and Field 1965), and Coats (1994) noted other chronic effects.

Limonene and linalool evaporate rapidly from treated surfaces (animal skin, plant leaves, or household surfaces), so they provide little or no residual control of insect pests. Consequently, their use is limited to applications that directly contact target pests. Most formulations include PBO to synergize the primary active ingredients. Both limonene and linalool are common ingredients in flea dips and shampoos, where they appear to be safer to dogs and cats than crude citrus extracts. Even so, these compounds and especially crude citrus oils can injure treated animals (Powers 1985; Hooser et al. 1986). Citrus oil extracts, limonene, and linalool also have been added to insecticidal soaps for applications to plants to enhance their contact control of aphids and mites. In all uses, the extremely labile nature of the citrus compounds renders them inactive within a few hours. Karr and Coats (1988) found the effectiveness of limonene to be limited, and they described it to be "overall, a poor candidate for general employment as an insecticide."

4.2.9   Other Essential Oils

Several additional essential oils, including oils of cedar, lavendar, eucalyptus, pennyroyal, and citronella are used as insect repellents on pets and humans, primarily to discourage fleas and mosquitoes. With the exception of oil of pennyroyal, these essential oils are thought to pose little or no risk to pets or people, though they may cause skin irritation. Their modes of action are poorly understood, but it is assumed that most interfere with insect perception of chemical cues (odors) that help them to locate their mammalian hosts.

Oil of pennyroyal contains pulegone, a very toxic compound that can cause death in humans at an oral dose of as low as one tablespoon. At lower doses it can cause abortion, liver damage, and renal failure. Although the dermal toxicity of pennyroyal is fairly low, cats are sometimes victims of poisoning after applications of pennyroyal to their coats, presumably because they ingest it as a result of grooming.

Citronella, derived from lemongrass (*Cymbopogon winterianus* Jowitt), is sold mainly in the form of candles to be burned outdoors to repel mosquitoes from the immediate surroundings, and it is included in some mosquito repellent lotions. Despite its widespread use, relatively little scientific data support citronella's effectiveness.

4.2.10   Diatomaceous Earth

Diatomaceous earth is not usually considered to be a botanical insecticide, but it is comprised of the mineralized cell walls of one-celled aquatic plants called diatoms. Over millions of years, as diatoms died and settled to the bottoms of bodies of water, sediments rich in diatom "shells" were formed. These sediments are now dug or mined to yield diatomaceous earth. (Diatomaceous earth that is destined for use in swimming pool filters is ground to different particle sizes and characteristics than that prepared for insecticidal use. As a result, diatomaceous earth sold for water filtration is not an effective insecticide.)

The insecticidal activity of a range of chemically inert dusts, including diatomaceous earth, results from their abrasiveness to the insect cuticle and/or their sorption of cuticular fats (Ebeling 1971). Fats and oils in the insect cuticle make it nearly waterproof and prevent water loss. Sorptive dusts such as the silica aerogels (ammonium fluosilicate) adsorb fats to disrupt the cuticle's waterproofing ability; abrasive dusts such as diatomaceous earth actually cut or scratch the cuticle. Although diatomaceous earth does adsorb cuticular fats, its function is primarily as an abrasive. Where inert dusts are effective as insecticides, dehydration causes the insect's death.

Diatomaceous earth is virtually nontoxic orally to mammals, and its dermal toxicity is limited to its ability to adsorb oils and dry the skin. It is, however, a serious respiratory irritant, and applicators must wear a particle respirator while using it.

Moderately effective uses of diatomaceous earth include its historic application as a dust on livestock animals or humans to reduce infestations of lice, fleas, or some mites and as a dust added to stored grains or seeds to reduce losses to stored-product insects. Many other claims for the effectiveness of diatomaceous earth are unsubstantiated, and uses that target insects in moist environments are unlikely to be successful. In very dry conditions (grain storages, for example), diatomaceous earth can remain effective for several weeks after application, but on plant surfaces, adsorption of moisture and oils to diatomaceous earth particles rapidly reduces their abrasiveness and effectiveness.

4.3 INSECTICIDAL SOAPS (FATTY ACID SALTS)

Insecticidal soaps, though not classed as botanical insecticides, are produced from oils that are usually of plant origin. Insecticidal soaps (and all soaps in general) are made from the salts of fatty acids, and fatty acids are the principal components of the fats and oils found in plants and animals. Fatty acids and their salts (soaps) have been known for centuries to be insecticidal, and they were used widely for insect control in the U.S. and other developed countries through the early 1900s. Renewed interest in their use developed in the 1970s and 1980s for much the same reason as botanical insecticides were reexamined — concern over the environmental impacts and persistence of synthetic insecticides.

The salts of certain fatty acids seem to be most effective as insecticides (Tattersfield and Gimingham 1927; Puritch 1978), and most commercial insecticidal soaps now on the market contain potassium oleate (the potassium salt of oleic acid). The nature of the toxicity of insecticidal soaps has been attributed to caustic action of the base released by hydrolysis (Shepard 1951), entry into tracheae and subsequent lysis of hemocytes and body walls (Dills and Menusan 1935), and disruption of the nervous system (Richards and Weygandt 1945). More recent studies suggest that fatty acids disrupt cell membranes and uncouple oxidative phosphorylation (Samson and Dahl 1955; Lehninger and Remmert 1959; Ahmed and Scholfield 1961a,b); Scholfield 1963; Puritch 1975). Some toxicity due to a disruptive influence on the insect cuticle is presumed as well.

The mammalian toxicity of insecticidal soaps is basically the same as that of any other soap or detergent. Ingestion causes vomiting and general gastric upset, but no serious systemic consequences occur. Insecticidal soaps may irritate or dry the skin, and they are irritating to the eyes and mucous membranes, but these effects are no different from those of most soaps used for washing. In addition to insecticidal soaps — in which the active ingredient is the soap or fatty acid — insecticides may be ingredients in soaps used in dip or shampoo products for pets or in louse shampoos for humans. In such products, the dermal toxicity of the primary active ingredient may be increased slightly by the wetting action of the soap or shampoo. These products are not "insecticidal soaps" as discussed here.

Soaps are contact insecticides; to be effective they must be sprayed directly onto pests or the pests must encounter the spray while it is still wet. Dried residues on plant surfaces have no residual effect against insects or related pests. Soaps can be used effectively to kill soft-bodied pests such as aphids, thrips, scale crawlers, whitefly and leafhopper nymphs, and mites. They are less effective against the adults of hard-bodied insects, because these insects' cuticles are thicker and more resistant to penetration and because they are mobile and able to escape the spray as it is applied. Insecticidal soaps are often used on ornamental landscape plants and on house plants, though some plants are injured by the wax-dissolving nature of the spray. Soap sprays do not show any selectivity against pest insects versus predators and parasites (though adult stages of beneficial insects may be mobile enough to escape spray contact), and such predators as lady beetle and lacewing larvae are killed if they are present on treated plants.

In addition to commercial insecticidal soaps, many common household soaps and detergents are insecticidal when applied as a 1 to 2% solution in water. Dishwashing liquids and laundry detergents are, however, designed to dissolve grease, and they may damage or kill plants by dissolving the waxy cuticle on leaf and stem surfaces. Detergents are generally more likely to cause plant injury than soaps are.

4.4 INSECTICIDAL OILS

Petroleum oils and vegetable oils are used as insecticides. The first uses of petroleum oils as sprays applied to plants date back to the 1870s or earlier, when kerosene emulsions were used, and phytotoxicity (plant injury) was a common problem. In the 1920s, "fast-breaking" oil and water emulsions were used (deOng 1927); the water portion of these sprays quickly dripped from plant parts while the oil remained on the plant surface. "Dormant" oils, applied at approximately 2% by volume, were in widespread use on fruit trees by the 1930s (Quaintance 1933).

The early use of petroleum oils was limited to the dormant season because these oils injured plants when foliage was present. However, the phytotoxicity of oils was reduced by selecting oils with specific physical characteristics. Chapman et al. (1962) and Chapman (1967) described the use and benefits of "narrow-range" oils with specific distillation temperatures, distillation ranges, and unsulfonated residue levels. In brief, Davidson et al. (1991) described the ideal spray oils to be characterized by traits that include the following:

- A 50% distillation temperature of 415 to 440°F
- A 10 to 90% distillation range of 60 to 80°F (or less)
- An unsulfonated residue rating of 92 or higher.

Oils with distillation temperatures lower than 390°F have low pesticidal activity, and phytotoxic potential increases as distillation temperatures approach 455 to 460°F. The distillation range lists the temperature required to distill 10 to 90% of the oil; a wider range reflects less uniformity in molecular composition. The unsulfonated residue (UR) rating indicates the percentage of the oil that is free of phytotoxic compounds that can remain after distillation; a 92% rating is the lower limit for current narrow-range oils. The petroleum oils currently used as dormant and summer oils on a range of plants include Gavicide Super 90, Volck Supreme (though it is not truly a narrow-range oil), Orchex 796, and SunSpray Ultra-Fine (Davidson 1991). These oils are clear to light-colored, odorless, stable, and formulated with a nonionic emulsifier so that they readily mix with water.

In the 1920s, vegetable oils were tested in a broad range of conditions for insect control. Cottonseed oil, linseed oil, and castor oil were tested, and although cottonseed oil was more effective against certain pests than petroleum oil, it was also more damaging to citrus, and its development was not pursued (deOng 1927). Butler and Henneberry (1990) found cottonseed, corn, peanut, safflower, and soybean oils to be toxic to several soft-bodied insect pests.

Petroleum oils and vegetable oils block the spiracles of the insects that are present at the time of application, killing those insects as a result of suffocation. Oils also prevent gas exchange through the egg membrane, and insect eggs are common targets of oil applications. The fatty acids in vegetable oils are also likely to act in the same fashion as described above for soaps, possibly disrupting cell membranes and uncoupling oxidative phosphorylation (Samson and Dahl 1955; Lehninger and Remmert 1959; Ahmed and Scholfield 1961a,b; Scholfield 1963; Puritch 1975). Because oils kill only those insects that are present when they are applied, they most often affect relatively immobile, soft-bodied pests such as aphids, scales, leafhopper nymphs, caterpillars, and mites, as well as the eggs of several arthropod taxa. Oils are toxic to beneficial species as well as plant-feeding pests, and although adults of lady beetles, lacewings, and Hemipteran predators escape oil applications, immature stages of these predators are often killed.

Petroleum and vegetable oils used as insecticides are relatively nontoxic (in comparison with other insecticides) to mammals, and there are no data that suggest any chronic effects from low-dose exposures. However, petroleum oils are in general corrosive to membranes that line the digestive and respiratory tracts, and spills on skin can cause irritation and dermatitis.

Oils are applied in early spring to the majority of the apple orchards in the U.S. to kill San Jose scale, *Quadraspidiotis perniciosus* (Comstock), eggs of the European red mite, *Panonychus ulmi* (Koch), and eggs of the rosy apple aphid, *Dysaphis plantaginea* (Passerini). Oils are especially important in the control of these pests in apples because the target pests are resistant to several organophosphates, carbamates, or pyrethroids. In addition, oil sprays applied early in the season have almost no adverse effects on beneficial insects and mites. Oils are used on nursery crops at a similar time (as buds begin to open) to control a range of aphid, mite, and scale pests (Davidson et al. 1991; Johnson 1994). Applications to field crops and vegetables are less common, but Davidson et al. (1991) list susceptible pests in several such crops. Stylet oils (narrow-range petroleum oils) are used to prevent aphid and leafhopper transmission of viruses and related plant pathogens; these oils are thought to "clog" the insect's stylet (feeding tube) and prevent the introduction of virus particles into plants. Stylet oils are most effective if applied weekly or more often and if plant coverage is very thorough.

Plant injury from oil applications is most common where older, heavier oils are used; injury results from the oil penetrating and blocking stomata and lenticels (Davidson 1991). Plant injury can be reduced by *not* using oils if (1) plants are under moisture or disease stress; (2) temperatures are less than 30°F or greater than 90°F; and (3) relative humidity is less than 20%. Exceeding recommended rates of application also increases the likelihood of plant injury. Dormant season oils are commonly applied at 1 to 2% concentrations by volume; summer sprays rarely contain more than 1% by volume.

4.5 SUMMARY

The botanical insecticides, soaps, and oils, as noted in the introduction to this chapter, represent a very small portion of the total volume of insecticides used annually on a worldwide basis. Even so, they are important in insect pest management. These "alternative" insecticides ...

- are a diverse group of chemicals with different modes of action, some of which are similar to those of synthetic insecticides.
- include compounds that are relatively nontoxic to mammals and compounds that are moderately to highly toxic.
- are characterized by short to very short persistence in the environment. Rapid breakdown in the environment is both an environmental benefit and a pest control shortcoming.
- do not replace synthetic insecticides, but do supplement them and contribute to integrated pest management in several crop systems.

Botanical insecticides, soaps, and oils are *not* nontoxic to humans, and they are not "nonchemical" means of insect control. They are, however, less toxic and less persistent than many synthetic insecticides, and as a result they are less likely to cause environmental damage or reach human consumers of treated crops. They are, in general, more costly than synthetic insecticides, and to be effective despite their short persistence, well timed or repeated applications may be needed.

REFERENCES

Ahmed, K., and P.G. Scholfield. 1961a. Studies on fatty acid oxidation. 7. The effects of fatty acids on the phosphate metabolism of slice and mitochondrial preparations of rat liver. Biochem. J. 81: 37.

Ahmed, K., and P.G. Scholfield. 1961b. Studies on fatty acid oxidation. 8. The effects of fatty acids on metabolism of rat brain cortex in vitro. Biochem. J. 81: 45.

Allen, T.C., R.J. Dicke, and H.H. Harris. 1944. Sabadilla, *Schoenocaulon* spp., with reference to its toxicity to houseflies. J. Econ. Entomol. 37: 400–408.

Allen, T.C., P. Link, M. Ikawa, and L. Brunn. 1945. The relative effectiveness of the principal alkaloids of sabadilla seed. J. Econ. Entomol. 38: 293–296.

Anon. 1989. Guidance for the Reregistration of Pesticide Products Containing Rotenone as the Active Ingredient. Office of Pesticide Programs, United States Environmental Protection Agency, Washington, D.C. 116 pp.

Atkins, E.L., Jr., and L.D. Anderson. 1954. Toxicity of pesticide dusts to honeybees. J. Econ. Entomol. 47: 969–972.

Barthel, W.F., 1973. Toxicity of pyrethrum and its constituents to mammals. Pp. 123–142, in: Casida, J.E. (ed.), Pyrethrum: The Natural Insecticide. Academic Press, New York. 329 pp.

Beinhart, E.G. 1951. Production and use of nicotine. Pp. 773–779, in: 1950–1951 Yearbook of Agriculture. United States Department of Agriculture, Washington, D.C.

Berenbaum, M.R. 1989. North American ethnobotanicals as sources of novel plant-based insecticides. Pp. 11–24, in: Arnason, J.T., B.J.R. Philogene, and P. Morand (eds.), Insecticides of Plant Origin. ACS Symposium Series 387, American Chemical Society, Washington, D.C.

Bloomquist, J.R. 1996. Ion channels as targets for insecticides. Annu. Rev. Entomol. 41: 163–190.

Butler, G.D., and T.J. Henneberry. 1990. Pest control on vegetables and cotton with household cooking oils and liquid detergents. Southwestern Entomol. 15: 123–131.

Campbell, W.C. (ed.). 1989. Ivermectin and Abamectin. Springer-Verlag, New York. 363 pp.

Casida, J.E. 1973. Pyrethrum: The Natural Insecticide. Academic Press, New York.

Casida, J.E., and G.B. Quistad. 1995. Pyrethrum Flowers: Production, Chemistry, Toxicology, and Uses. Oxford Univ. Press.

Chapman, P.J. 1967. Petroleum oils for the control of orchard pests. N.Y. State Agr. Exp. Sta. Bull. 814: 1–22.

Chapman, P.J., S.E. Lienk, A.W. Avens, and R.W. White. 1962. Selection of a plant spray oil combining full pesticidal efficiency with minimal plant injury hazards. J. Econ. Entomol. 55: 737–744.

Clark, J.M., J.G. Scott, F. Campos, and J.R. Bloomquist. 1994. Resistance to avermectins: extent, mechanisms, and management implications. Annu. Rev. Entomol. 40: 1–30.

Coats, J.R. 1994. Risks from natural versus synthetic insecticides. Annu. Rev. Entomol. 39: 489–515.

Coats, J.R., L.L. Karr, and C.D. Drew. 1991. Toxicity and neurotoxic effects of monoterpenoids in insects and an earthworm. Pp. 305–316, in: Hedin, P.A. (ed.), Naturally Occurring Pest Bioregulators. American Chemical Society, Washington, D.C. 456 pp.

Davidson, N.A., J.E. Dibble, M.L. Flint, P.J. Marer, and A. Guye. 1991. Managing Insects and Mites with Spray Oils. Publ. 3347, University of California, Oakland, 47 pp.

Day, M.L., H.W. Hogmire, and M.W. Brown. 1995. Biology and management of rose leafhopper (Homoptera: Cicadellidae) on apple in West Virginia. J. Econ. Entomol. 88: 1012–1016.

DeAmicis, C.V., J.E. Dripps, C.J. Hatton, and L.L. Karr. 1997. Physical and biological properties of the spinosyns: novel macrolide pest-control agents from fermentation. Pp. 144–154, in: Hedin, P.A. (ed.), Phytochemicals for Pest Control. American Chemical Society, Washington, D.C.

deOng, E.R., H. Knight, and J.C. Chamberlain. 1927. A preliminary study of petroleum oil as an insecticide for citrus trees. Hilgardia 2: 351–384.

Dills, L.E., and H. Menusan, Jr. 1935. A study of some fatty acids and their soaps as contact insecticides. Contrib. Boyce Thompson Inst. 7: 63–82.

Ebeling, W. 1971. Sorptive dusts for pest control. Annu. Rev. Entomol. 16: 123–158.

Fill, M., and R. Coronado. 1988. Ryanodine receptor channel of sarcoplasmic reticulum. Trends Neurosci. 11: 453–457.

Garber, S.S., and C. Miller. 1987. Single Na+ channels activated by veratridine and batrachotoxin. J. Gen. Physiol. 89: 459–480.

Goblet, C., and Y. Mounier. 1981. Effects of ryanodine on the ionic currents and the calcium conductance in crab muscle fibers. J. Pharmacol. Exp. Ther. 219: 526–533.

Grainger, M., and S. Ahmed. 1988. Handbook of Plants with Pest Control Properties. John Wiley and Sons, New York. 470 pp.

Hale, K.A., and D.E. Portwood. 1996. The aerobic soil degradation of spinosad — a novel natural insect control agent. J. Environ. Sci. and Health 31: 477–484.

Haley, J.T. 1978. A review of the literature of rotenone 1,2,12,12a-tetra hydro-8,9-dimethoxy-(2-(1-methylethenyl)-1-benzopyrano[3,5-β]fluoro[2,3-H][1]-benzopyran-6(6h)-one. J. Environ. Pathol. Toxicol 1: 315–337.

Hayes, W.J. 1982. Pesticides derived from plants and other organisms. Pp. 75–111, in: W.J. Hayes (ed.), Pesticides Studied in Man. Williamson and Williamson, Baltimore.

Hedin, P.A. (ed.). 1991. Naturally Occurring Pest Bioregulators. American Chemical Society, Washington, D.C. 456 pp.

Hedin, P.A. (ed.). 1997. Phytochemicals for Pest Control. American Chemical Society, Washington, D.C.

Hollingshaus, J. 1987. Inhibition of mitochondrial electron transport by hydramethylnon: a new amidinohydrazone insecticide. Pestic. Biochem. Physiol. 27: 61–70.

Hollingworth, R., K. Ahmmadsahib, G. Gedelhak, and J. McLaughlin. 1994. New inhibitors of complex 1 of the mitochondrial electron transport chain with activity as pesticides. Biochem. Soc. Trans. (London) 22: 230–233.

Hooser, S.B., V.R. Beasley, and J.I. Everitt. 1986. Effects of an insecticidal dip containing d-limonene in the cat. J. Am. Vet. Med. Assoc. 189: 905–908.

Jacobson, M. 1986. The neem tree: natural resistance par excellence. Pp. 220–232, in: Green, M.B., and P.A. Hedin (eds.), Natural Resistance of Plants to Pests. American Chemical Society., Washington, D.C. 243 pp.

Jacobson, M. 1989. Botanical pesticides, past, present, and future. Pp. 1–10, in: Arnason, J.T., B.J.R. Philogene, and P. Morand (eds.), Insecticides of Plant Origin. American Chemical Society, Washington, D.C.

Jacobson, M., and D.G. Crosby. 1971. Naturally Occurring Insecticides. Marcel Dekker, New York.

Jeffries, P.A., R.F. Toia, B. Brannigan, J. Pessah, and J.E. Casida. 1992. Ryania insecticide: analysis and biological activity of 10 natural ryanoids. J. Agric. Food Chem. 40: 142–146.

Johnson, W.T. 1994. Oils as pesticides for ornamental plants. Pp. 557–581, in: Leslie, A.R. (ed.), Integrated Pest Management for Turf and Ornamentals. Lewis Publishers, Boca Raton, FL, 660 pp.

Karr, L.L., and J.R. Coats. 1988. Insecticidal properties of d-limonene. J. Pestic. Sci. 13: 287–290.

Karr, L.L., and J.R. Coats. 1992. Effects of four monoterpenoids on growth and reproduction of the German cockroach (Blattodea: Blatellidae). J. Econ. Entomol. 85: 424–429.

Kraus, W., M. Bokel, A. Klenk, and H. Ponhl. 1985. The structure of azadirachtin and 22,23-dihdro-23-B-methoxyazadirachtin. Tetrahedron Lett. 26: 6435–6438.

Lasota, J.A., and R.A. Dybas. 1991. Avermectins, a novel class of compounds: implications for use in arthropod pest control. Annu. Rev. Entomol. 36: 91–117.

Lee, S.M., J.A. Klocke, M.A. Barnby, R.B. Yamasaki, and M.F. Balandrin. 1991. Insecticidal constituents of *Azidarachta indica* and *Melia azedarach* (Meliaceae). Pp. 293–304, in: Hedin, P.A. (ed.), Naturally Occurring Pest Bioregulators. American Chemical Society, Washington, D.C. 339 pp.

Lehmberg, E., and J.E. Casida. 1994. Similarity of insect and mammalian ryanodine binding sites. Pestic. Biochem. Physiol. 48: 145–152.

Lehninger, A.L., and L.F. Remmert. 1959. An endogenous uncoupling and swelling agent in liver mitochondria and its enzymatic formation. J. Biol. Chem. 234: 2459.

Levi, G., V. Gallo, and M. Raiteri. 1980. A reevaluation of veratridine as a tool for studying the depolarization-induced release of neurotransmitters from nerve endings. Neurochem. Res. 5: 281–295.

Liebowitz, M.D., J.B. Sutro, and B. Hille. 1986. Voltage-dependent gating of veratridine-modified Na channels. J. Gen. Physiol. 87: 25–46.

Matsumura, F. 1985. Toxicology of Insecticides (2nd ed.). Plenum Press, New York.

Ohta, M., T. Narahashi, and T. Keeler. 1973. Effects of veratrum alkaloids on membrane potential and conductance of squid and crayfish giant axons. J. Pharmacol. Exp. Ther. 184: 143–154.

Pepper, B.P., and L.A. Carruth. 1945. A new plant insecticide for control of the European corn borer. J. Econ. Entomol. 38: 59–66.

Pessah, I.N., A.L. Waterhouse, and J.E. Casida. 1985. The calcium-ryanodine receptor complex of skeletal and cardiac muscle. Biochem. Biophys. Res. Commun. 128: 449–556.

Pilmore, R.E. 1973. Toxicity of pyrethrum to fish and wildlife. Pp. 143–165, in: Casida, J.E. (ed.), Pyrethrum: The Natural Insecticide. Academic Press, New York. 329 pp.

Powers, K.A. 1985. Toxicological effects of linalool: a review. Vet. and Human Toxicol. 27: 484–486.

Puritch, G.S. 1975. The toxic effects of fatty acids and their salts on the balsam woolly aphid, *Adelges piceae* (Ratz). Can. J. For. Res. 5:515–522.

Puritch, G.S. 1978. Biocidal effects of fatty acid salts on various forest insect pests. AOCS Monograph No. 5: 105–112.

Quaintance, A.L. 1933. Lubricating oil sprays for use on dormant fruit trees. U.S. Dept. Agric. Farmers' Bull. 1676: 1–18.

Radwanski, S. 1977a. Neem tree 1. Commercial potential, characteristics and distribution. World Crops and Livestock 29: 62–65.

Radwanski, S. 1977b. Neem tree 2. Uses and potential uses. World Crops and Livestock 29: 111–113.

Radwanski, S. 1977c. Neem tree 3. Further uses and potential uses. World Crops and Livestock 29: 167–168.

Reed, J. P. and R.S. Filmer. 1950. Activation of ryania dusts by piperonyl cyclonene and N-propyl isome. J. Econ. Entomol. 43: 161–164.

Richards, A.G., Jr., and J.L. Weygandt. 1945. The selective penetration of fat solvents into the nervous system of mosquito larvae. J. New York Entomol. Soc. 53: 153.

Roe, F.J., and W.E. Field. 1965. Chronic toxicity of essential oils and certain other products of natural origin. Food and Cosmetic Toxicol. 3: 311–324.

Samson, F.E., Jr., and N. Dahl. 1955. Coma induced by short chain fatty acids. Am. Physiol. Soc. Fed. Proc. 14: 129.

Saxena, R.C. 1989. Insecticides from neem. Pp. 110–135, in: J.T. Arnason, B.J.R. Philogene, and P. Morrand, (eds.), Insecticides of Plant Origin. American Chemical Society, Washington, D.C. 213 pp.

Schmutterer, H. 1990. Properties and potential of natural pesticides from the neem tree, *Azidarachta indica*. Annu. Rev. Entomol. 35: 271–297.

Schnellmann, R., and R. Manning. 1990. Perfluorooctane sulfonamide: a structurally novel uncoupler of oxidative phosphorylation. Biochim. Biophys. Acta 1016: 344–348.

Scholfield, P.G. 1963. Fatty acids and their analogs. Pp. 153–172, in: Hochster, R.M., and J.H. Quastel (eds.), Metabolic Inhibitors (Vol. 1). Academic Press, New York.

Shepard, H.H. 1951. The Chemistry and Action of Insecticides. McGraw-Hill, New York. 504 pp.

Sparks, T.C., H.A. Kirst, J.S. Mynderse, G.D. Thompson, J.R. Turner, O.K. Jantz, M.B. Hertlein, L.L. Larson, P.J. Baker, M.C. Broughton, J.D. Busacca, L.C. Creemer, M.L. Huber, J.W. Martin, W.M. Natasukasa, J.W. Paschal, and T.V. Worden. 1996. Chemistry and biology of the spinosyns: components of spinosad (Tracer), the first entry into DowElanco's naturalyte class of insect control products. 1996 Proceedings Beltwide Cotton Conferences, Nashville, TN, January 9–12, 1996; vol. 2: 692–695. National Cotton Council, Memphis.

Tattersfield, F., and C.T. Gimingham. 1927. Studies on contact insecticides. Part VI. The insecticidal action of fatty acids, their methyl esters and sodium and ammonium salts. Ann. Appl. Biol. 14: 331–358.

Verschoyle, R.D., and J.M. Barnes. 1972. Toxicity of natural and synthetic pyrethrins to rats. Pestic. Biochem. Physiol. 2: 308–311.

Ware, G.W. 1988. Complete Guide to Pest control (2nd ed.). Thompson Publ., Fresno, CA.

Warthen, J.D., Jr. 1989. Neem (*Azidarachta indica* A. juss): organisms affected and reference list update. Proc. Entomol. Soc. Wash. 91:367–388.

Weinzierl, R. 1998. Insect management for commercial vegetable crops. Pp. 181–208, in: K. Steffey (ed.), 1998 Illinois Agricultural Pest Management Handbook. University of Illinois at Urbana-Champaign.

Weinzierl, R., and T. Henn. 1991. Alternatives in insect management: biological and biorational approaches. Regional Extension Publ. NCR 401, Cooperative Extension Service, University of Illinois at Urbana-Champaign. 73 pp.

Worthing, C.R., and S.B. Walker (eds.). 1987. The Pesticide Manual, A World Compendium (8th ed.). The British Crop Protection Council, Thornton Heath, U.K.

Yamamoto, I. 1970. Mode of action of pyrethroids, nicotinoids, and rotenoids. Annu. Rev. Entomol. 15: 257–272.

CHAPTER 5

Insect Growth Regulators

Nancy E. Beckage

CONTENTS

5.1 OVERVIEW

Insect growth regulators (IGRs) are insecticides that mimic the action of hormones on the growth and development of insect pests. The advantages of using IGRs instead of classic insecticides in pest control include their reduced toxicity to the environment and their specificity for insects. Moreover, because many IGRs act specifically on target pests such as Lepidoptera they show minimal toxicity for beneficial parasites and predators.

Williams (1967) was the first to suggest that insect hormones might be utilized as insect-specific environmentally benign pesticides, and Staal (1982) presented a review of these strategies. Mimics of ecdysteroids (molting hormones) and juvenile hormones are two classes of hormone-based pesticides that have been developed for commercial use in insect pest control. Azadirachtin, extracted from neem seeds, also appears to disrupt growth and molting in a large number of insect species. These three classes of compounds will be discussed in this review.

Insect growth and development are regulated by a combination of hormones, including juvenile hormone, ecdysteroids, and several neuropeptides including eclosion hormone and ecdysis triggering hormone (see Nijhout, 1994; Riddiford, 1994;

Horodyski, 1996; Zitnan et al., 1996). During early larval life, the JH titer is maintained at a high level, and growth is interrupted by periodic molts that follow the release of 20-hydroxyecdysone by the prothoracic glands. The production of ecdysteroids occurs in response to the release of prothoracicotropic hormone from neurosecretory cells localized in the brain and retrocerebral complex. Following the production of a new cuticle during molting, eclosion hormone is released, followed by release of ecdysis-triggering hormone, and the insect sheds the exuvium. In Lepidoptera, the JH titer descends to nondetectable levels in the last larval instar, providing a signal for release of a small prewandering ecdysteroid peak and a switch from larval to pupal commitment. A larger prepupal peak of ecdysteroids, which occurs in the presence of a low titer of JH, then stimulates synthesis of pupal cuticle and the insect makes the transition to the pupal stage. Following pupation, ecdysteroid is again produced to cause formation of adult structures, and the pupal cuticle is shed at adult eclosion. JH then plays a prominent role in the reproduction of many species.

Due to the prominence of hormones in the insect life cycle, the administration of hormone agonists and antagonists has disruptive effects on growth and metamorphosis. One disadvantage of using some hormonally based IGRs, however, is the narrow window of sensitivity during which they must be administered to have any discernable effect on development. This can be avoided by the use of compounds with a longer *in vivo* half-life, so that the concentration of compound within the animal remains high during the sensitive period. Also, some stages of insects may be sensitive to an IGR, whereas other stage(s) are refractory. There may be a long lag period between administration of the compound and the observed induction of disruptive effects. Thus, it is sometimes difficult to achieve rapid knockdown of an insect population with IGR-based pesticides. However, the rapidity with which an insect responds varies with the compound. Juvenile hormone agonists may be utilized for control in the long term to disrupt metamorphosis, whereas ecdysone agonists act very quickly (within 24 hours) to trigger an unsuccessful molt. Azadirachtin frequently has an immediate antifeedant effect that later disrupts molting or reproduction.

Aside from acting as pesticides, hormonally based IGRs are important research tools for the study of hormone action (Oberlander et al., 1995). They offer new insights into how hormones regulate insect growth and development.

5.2 ECDYSTEROID AGONISTS

In recent years several nonsteroidal bisacylhydrazine ecdysone agonists have been synthesized by Rohm and Haas (Figure 5.1, compounds 1–5). These chemicals are much more potent than the native hormone 20-hydroxyecdysone in inducing molting. They also reduce feeding and weight gain. In lepidopteran insects, a lethal molt is induced following administration of the ecdysone agonist, and the animal dies trapped within the exuvial cuticle. Feeding stops 4–16 hours after ingestion of toxic doses of the agonist, and molting is initiated in the absence of an ecdysteroid increase, as Wing et al. (1988) demonstrated using isolated abdomens that lacked

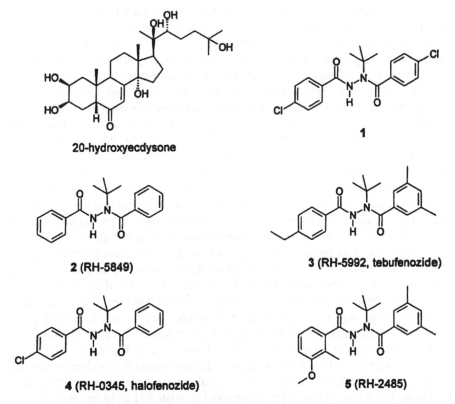

20-hydroxyecdysone

1

2 (RH-5849)

3 (RH-5992, tebufenozide)

4 (RH-0345, halofenozide)

5 (RH-2485)

Figure 5.1 Structures of 20-hydroxyecdysone and the bisacylhydrazine ecdysteroid agonists. Reprinted with permission from Dhadialla et al. (1998).

prothoracic glands. Usually the animal dies in the slipped head capsule stage following onset of apolysis (Dhadialla et al., 1998). However, as described by Oberlander et al. (1995), supernumerary larval molts may also occur (Gadenne et al., 1990; Musynska-Pytel et al., 1992) when the JH titer is high.

Ecdysteriod agonists have been shown to act in many lepidopterans including *Manduca sexta* (Wing et al., 1988), *Plodia interpunctella* (Silhacek et al., 1990), *Spodoptera littoralis* (Smagghe and Degheele, 1992), *Spodoptera litura* (Tateishi et al., 1993), *Spodoptera exempta* (Smagghe and Degheele, 1994a), *Spodoptera exigua* (Smagghe and Degheele, 1994b), and *Pieris brassicae* (Darvis et al., 1992). RH-5992 (tebufenozide) is more toxic to lepidopteran larvae than RH-5849 (Dhadialla et al., 1998). The compound RH-0345 has a different activity profile and acts on scarabid beetle larvae, cutworms, and webworms (Dhadialla et al., 1998). RH-2485 shows promise because it is more potent than tebufenozide against lepidopteran pests of corn (Trisyono and Chippendale, 1997), and cotton (Ishaaya et al., 1995) (see Dhadialla et al., 1998 for review).

Lepidopteran, dipteran, and coleopteran larvae are the primary insects affected by these ecdysone agonists. All larval stages are affected, but the effect induced depends on when during an instar the insect ingests or is treated with the compound.

An immediate lethal molt is induced if treatment occurs early in an instar, but if the insect is treated toward the end of an instar, first a normal molt will occur, which is then followed by the lethal molt (Dhadialla et al., 1998). In adult stage insects, egg production and spermatogenesis may be deleteriously affected by exposure to bisacylhydrazines (Smagghe and Degheele, 1994a; Carpenter and Chandler, 1994).

These compounds appear to interact with the ecdysteroid receptor complex (Wing, 1988) and thereby induce their effects. Tebufenozide competes with tritiated ponasterone A for binding to ecdysteroid receptors, which are EcR/*ultraspiracle* heterodimers, indicating that the compound acts as an ecdysteroid mimic at the molecular level. Experiments using cells (e.g., *Drosphila* Kc cells — see Wing, 1988; or *Chironomus* cells — see Spindler-Barth et al., 1991) or tissues cultured *in vitro* indicate that the bisacylhydrazines have the same mode of action as 20-hydroxyecdysone, but the effects are much longer lasting compared to 20-HE (Retnakaran et al., 1995). Even though these ecdysone agonists are not steroids, they act similar to 20-HE in causing wing disc eversion (Silhacek et al., 1990; Smagghe et al., 1996) and other physiological responses, such as initiation of spermatogenesis (Friedlander and Brown, 1995) or adult development in diapausing pupae (Sielezniew and Cymborowski, 1997) that are normally induced by 20-HE.

The high binding affinity of tebufenozide and RH-2485 to proteins in nuclear extracts of lepidopteran cells is correlated with their selective action on lepidopteran insects (Dhadialla et al., 1998). By contrast, the ecdysteroid receptors of coleopteran insects bind RH 5992 with low affinity (Dhadialla and Tzertzinis, 1997). This difference thereby explains the specificity of this compound for lepidopteran insects.

The tomato moth *Lacanobia oleacea* rapidly metabolizes ingested 20-hydroxy-ecdysone, but is susceptible to the ecdysteroid agonists RH-5849 and RH-5992 and undergoes a lethal larval molt (Blackford and Dinan, 1997). This insect feeds on a variety of weeds containing high concentrations of phytoecdysteroids in addition to tomato. Blackford and Dinan (1997) propose that the phytoecdysteroids may be rapidly detoxified by conjugation with long-chain fatty acids and excreted in an inactive form, while the ecdysteroid agonists exhibit enhanced *in vivo* stability.

In *Spodoptera littoralis,* application of RH-5849 delays the onset of wandering in a dose-dependent manner (Pszczolkowski and Kuszczak, 1996). These results suggest that not only an increase, but also a decrease in ecdysteroid levels appears to be vital for wandering to be initiated at the normal time.

Due to the selectivity of tebufenoxide and RH-2485 for Lepidoptera, the chemicals can be used without the risk of direct deleterious effects on beneficial species, although there may be indirect effects on parasitoids due to the death of the host. Fifth instar *Manduca sexta* larvae that are injected with 10 µg RH-5849 in DMSO slip their head capsules and later have few or no emerging *Cotesia congregata* parasitoids (N.E. Beckage and F.F. Tan, unpubl. data). Brown (1994) analyzed the effects of tebufenoxide on two parasitoids, *Ascogaster quadridentata,* an endoparasitoid, and *Hyssopus* sp., an ectoparasitoid, and assessed its effect on the host–parasitoid interaction. In general, ectoparasitoids are less affected by hormonally active IGRs compared to endoparasitoids, which are constantly exposed *in vivo* to the IGR.

Both tebufenoxide and RH-2485 are highly active in the field against a variety of vegetable, fruit, and ornamental pests (Dhadialla et al., 1998), indicating that the

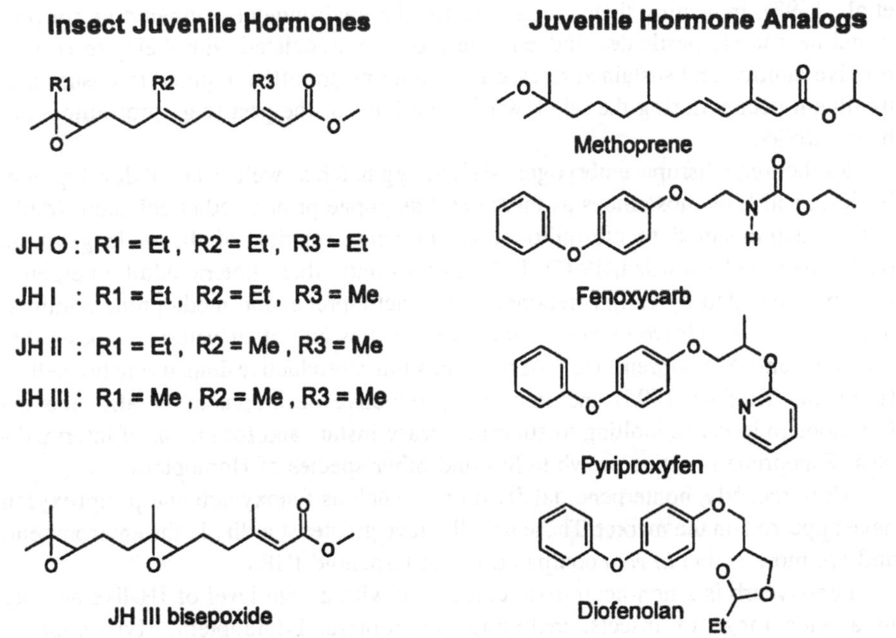

Insect Juvenile Hormones

JH O : R1 = Et , R2 = Et , R3 = Et

JH I : R1 = Et , R2 = Et , R3 = Me

JH II : R1 = Et , R2 = Me , R3 = Me

JH III : R1 = Me , R2 = Me , R3 = Me

JH III bisepoxide

Juvenile Hormone Analogs

Methoprene

Fenoxycarb

Pyriproxyfen

Diofenolan

Figure 5.2 Structures of the juvenile hormones and juvenile hormone agonists. Reprinted with permission from Dhadialla et al. (1998).

compounds have the potential for wide application. Importantly, the compounds exhibit minimal toxicity to vertebrates; thus they can be safely utilized in the field without adverse effects on human and animal health.

5.3 JUVENILE HORMONE AGONISTS

The agonists of juvenile hormone include compounds such as methoprene and hydroprene, which are terpenoids, and nonterpenoids such as fenoxycarb and pyriproxyfen (Figure 5.2). In order for metamorphosis to occur in holometabolous insects, JH must descend to nondetectable levels so that ecdysteroid is released in the absence of JH and the switch from larval to pupal commitment occurs (Riddiford, 1994). In hemimetabolous insects, the JH titer must decrease to permit molting to the adult form. If the JH titer is maintained at too high a level, due to administration of a JH agonist in the larval or nymphal stage, then molting to a supernumerary instar or an intermediate (larval–pupal, nymphal–adult, or pupal–adult) is induced. Such intermediates are nonviable. The supernumerary instars that form may have morphological anomalies, which likewise prevent the animal from completing a normal metamorphosis.

JH agonists are highly effective insect growth regulators that cause a wide range of developmental derangements in susceptible species, affecting embryogenesis, larval development, metamorphosis, and reproduction. The JH agonists have low acute toxicity for fish, birds, and mammals (Grenier and Grenier, 1993; Dhadialla et al., 1998), indicating their use is safe for the environment compared to conventional neurotoxic pesticides. Indeed, one problem associated with their use is their relative lability, and sustained release formulations are often required to ensure that treatment occurs during the window of sensitivity of the pest (e.g., mosquitoes) to be controlled.

Methoprene disrupts embryogenesis and egg hatch as well as larval development in fleas, and kills mosquitoes as larvae and as pupae prior to adult eclosion. Methoprene administered to mosquito pupae prevents rotation of the male genitalia (O'Donnell and Klowden, 1997). In houseflies and other Diptera, adult emergence may be prevented by earlier treatment with methoprene, and methoprene administered as a feed additive to cattle controls horn flies and other veterinary pests that breed in dung. Methoprene acts to terminate adult reproductive diapause in houseflies (Kim and Krafsur, 1995) and induce reproductive development. It also acts on Lepidoptera to cause molting to supernumerary instars and formation of intermediates. Kinoprene is active on whiteflies and other species of Homoptera.

More recently, nonterpenoidal JH mimics such as fenoxycarb and pyriproxyfen have appeared on the market. These usually have greater stability in the environment, and are more active *in vivo* compared to the terpenoid IGRs.

Fenoxycarb is a non-neurotoxic carbamate with a high level of JH-like activity in a wide range of insects, including Heteroptera, Lepidoptera, Hymenoptera, Coleoptera, Diptera, Dictyoptera, Isoptera, and Homoptera (see Grenier and Grenier, 1993 for review). The administration of fenoxycarb has been reported to cause sterility, kill insect eggs, interfere with metamorphosis, induce permanent larvae, cause formation of nonviable intermediates, and affect caste differentiation. Ultralow doses of fenoxycarb (100 picograms) induce permanent larvae in the silkworm *Bombyx mori* (Monconduit and Mauchamp, 1998), which have inactive prothoracic glands. Incubation of prothoracic glands in the presence of fenoxycarb reduces production of ecdysteroid by the glands (Dedos and Fugo, 1996). Larval growth and food consumption is reduced following the administration of fenoxycarb in *B. mori* (Leonardi et al., 1996). Fenoxycarb appears to qualitatively affect the biosynthesis of fatty acids by the fat body in the Eastern spruce budworm *Choristoneura fumiferana* (Mulye and Gordon, 1993).

Moreno et al. (1993a,b) investigated the effects of fenoxycarb on the parasitoid *Phanerotoma ocularis* and found signficant adverse effects on parasitoid development whether the compound was applied via the host (Moreno et al., 1993a) or directly to the parasitoid (Moreno et al., 1993b). Development of the tachinid *Pseudoperichaeta nigrolineata* is also disrupted by fenoxycarb applied to its host (Grenier and Plantevin, 1990). Other JH agonists such as methoprene may also deleteriously affect the emergence and metamorphosis of parasitoids (Beckage and Riddiford, 1982), so hormonally active IGRs may have some adverse effects on parasitoid populations but overall the effects seem less detrimental compared to the effects of conventional neurotoxic pesticides.

Pyriproxyfen is another potent JH agonist that is active in a wide range of arthropods, including ants (Vail and Williams, 1995; Vail et al., 1996), fleas (Palma et al., 1993), mole crickets (Parkman and Frank, 1998), mosquitoes (Kono et al., 1997), flies (Hargrove and Langley, 1993; Bull and Meola, 1993), whiteflies (Ishaaya et al., 1994; Ishaaya and Horowitz, 1995), scales (Peleg, 1988), cockroaches (Koehler and Patterson, 1991; Lim and Yap, 1996), and ticks (Teal et al., 1996; Donahue et al., 1997), as well as lepidopterans (Smagghe and Degheele, 1994b). As seen with other JH agonists, multiple effects are induced in a single species. The compound interferes with embryogenesis, disrupts the metamorphic molt, and causes morphological deformities in the desert locust, *Schistocerca gregaria* (Vennard et al., 1998). In the tobacco cutworm, *Spodoptera litura,* topical application of 0.3 ng pyroproxyfen to day 0 female pupae reduces the number of eggs oviposited, because the treated females lack an oviposition-stimulating factor, which is necessary for the deposition of eggs (Hatakoshi, 1992). Thus, reproduction may be inhibited through oviposition inhibition as well toxic effects on developing oocytes following treatment with pyriproxyfen.

Similar to methoprene, pyriproxyfen fed to cattle and other animals has an insecticidal effect on the subsequent emergence of horn flies, face flies, and houseflies from manure (Miller, 1989; Miller and Miller, 1994). Pyriproxyfen also disrupts development of horn flies following direct application to the fly (Bull and Meola, 1993).

In the Colorado potato beetle, pyriproxyfen inhibits expression of diapause protein 1, and induces expression of vitellogenin following application to short-day adults destined to diapause (De Kort et al., 1997). In last instar nymphs, the compound prevents metamorphosis at low doses. Pyriproxyfen also induces synthesis of vitellogenin in *Locusta migratoria* (Edwards et al., 1993) and inhibits the synthesis of larva-specific hemolymph proteins (De Kort and Koopmanschap, 1991).

Pyriproxyfen resistance has been generated in houseflies selected for 17 generations for resistance (Zhang and Shono, 1997; Zhang et al., 1997; Zhang et al., 1998). In these flies, cytochrome P450 monooxygenase enzymes in the fat body and gut play critical roles in the metabolism of pyriproxyfen (Zhang et al., 1998). Resistance to pyriproxyfen has also been reported in the homopteran *Bamesia tabaci* (Horowitz and Ishaaya, 1994). This is not surprising, given that resistance to other JH agonists such as methoprene can also be induced (Turner and Wilson, 1995).

5.4 AZADIRACHTIN

Azadirachtin is a tetranortriterpenoid present in seeds from the Indian neem tree, *Azadirachta indica.* This compound was originally isolated by Butterworth and Morgan (1968) and its structure was subsequently described (Zanno et al., 1975; Kraus et al., 1985). Azadirachtin has strong antifeedant and repellant activity (Ascher, 1993; Simmonds and Blaney, 1996) and has pleiotropic effects on growth, development, and reproduction (Schmutterer, 1990; Mordue (Luntz) and Blackwell, 1993). In contrast to the wealth of information we have about its effects on development, its biochemical effects at the cellular and molecular levels are still barely known.

A wide range (>200 species; Ascher, 1993) of both chewing and sucking phytophagous insects, and stored product pests have been shown to be affected by azadirachtin. Insects that are susceptible include aphids, lepidopterans, hemipterans, cockroches, beetles, and orthopterans. Azadirachtin is taken up systemically and translocated into the tissues of treated plants (Arpaia and Van Loon, 1993), in addition to affecting insects via the leaf surface or by direct contact with the target pest. Aquatic species such as mosquitoes are also affected (Mordue (Luntz) and Blackwell, 1993). While neem derivatives have been reported to provide broad-spectrum control of pest species, they appear less toxic to natural enemies of insect pests than to the pests themselves (Schmutterer, 1990; Banken and Stark, 1997).

The activity of azadirachtin is due to a complex combination of antifeedant and toxic properties that affect growth, molting, and reproduction (see Schmutterer, 1990, Ascher, 1993, and Mordue (Luntz) and Blackwell, 1993 for reviews). Aside from its insecticidal properties, it also exhibits nematocidal, antiviral, and antifungal properties (Mordue (Luntz) and Blackwell, 1993). Protozoa (e.g., malaria parasites) are also deleteriously affected (Jones et al., 1994) by azadirachtin.

Azadirachtin's antifeedant properties to some extent reflect its action on gustatory chemoreceptors and its activity in suppressing food consumption has been reported in numerous species (Mordue (Luntz) and Blackwell, 1993). However, topical application of azadiractin also has potent effects, indicating a nongustatory pathway also is involved in its mechanism of action. Second, azadirachtin affects ecdysteroid and juvenile hormone titers, resulting in severe growth and molting aberrations. The neuropeptides regulating ecdysteroid and JH production may be affected. Disruption of molting leads to formation of larval–pupal, nymphal–pupal, nymphal–adult, and pupal–adult intermediates. Formation of permanent larvae is also observed. Third, it has direct detrimental effects on many insect tissues such as endocrine glands, muscles, fat body, and gut epithelial cells. Effects on reproduction include effects on spermatogenesis in males, and fecundity and oviposition in females. Also, longevity of adult males or females may be reduced.

Azadirachtin is a growth retardant in *Periplaneta americana* and reduced ingestion of food in the immature stages results in smaller adults, which exhibit a decrease in the rate of reproduction relative to untreated control insects (Richter et al., 1997). Effects on egg viability have been reported in many species.

Reproductive effects have been noted in Orthoptera, Heteroptera, Homoptera, Hymenoptera, Coleoptera, Lepidoptera, and Diptera (Schmutterer, 1990). In females, azadirachtin treatment may result in total sterility. Degenerative changes are seen in the follicle cells as well as the oocytes themselves. Observations include a separation of the follicle from the oocyte, and lack of vitellogenin in both fat body and hemolymph as well as in developing oocytes. The effects of azadirachtin may be rescuable by treatment with juvenile hormone (Sayah et al., 1996). Azadirachtin inhibits the growth of oocytes of *Rhodnius prolixus* when administered in a blood meal, and phospholipid transfer from lipophorin to the developing oocytes is inhibited (Moreira et al., 1994).

In males, effects are seen on testes development and spermatogenesis, resulting in a decrease in fertility (Dorn, 1986; Shimizu, 1988). Injection of azadirachtin inhibits the growth of testes in the desert locust, *Schistocerca gregaria,* and sperm meiosis is halted at metaphase I (Linton et al., 1997). Dihydrozadirachtin binds sperm in the testes and reduces sperm motility in the locust *Schistocerca gregaria* (Nisbet et al., 1996).

In the African armyworm *Spodoptera exempta*, a highly damaging pest of cereal crops in Africa, azadirachtin treatment lowers food intake, efficiency of conversion of ingested food, and efficiency of conversion of digested food to body mass (Tanzubil, 1995). At 10 μg per insect, growth is drastically curtailed. In this experiment, azadirachtin was applied topically, confirming that the compound can interfere with feeding via a nonsensory mechanism.

Aside from affecting feeding, azadirachtin deleteriously affects the gut itself. Azadirachtin disrupts the ultrastructural organization of the epithelial cells lining the midgut of *Rhodnius prolixus* (Nogueira et al., 1997). Changes include clustering of the microvilli, disorganization of the extracellular membrane layers, and alterations in the basal portion of the epithelial cells. The midgut is also disrupted in the locusts *Schistocerca gregaria* and *Locusta migratoria,* and the effects are distinctly different from those induced by starvation (Nasiruddin and Mordue, 1993).

Salannin, which is also extracted from neem seeds, deters feeding, delays molting by increasing larval duration, causes larval and pupal mortalities, and decreases pupal weights of *Spodoptera litura* (Govinachari et al., 1996). Smaller pupae yield smaller adults with reduced fecundity.

Azadirachtin affects the ultrastructure of the ring gland of *Lucilia cuprina* (Meurant et al., 1994). The ring gland is comprised of the prothoracic gland, corpus cardiacum, and corpus allatum. Changes seen include crenulation of nuclear shape, clumping of the heterochromatin, and pyknosis. The observed effects likely contribute to a generalized disturbance of neuroendocrine function. Rembold et al. (1989) reported that azadirachtin targets the corpus cardiacum and causes disruption of its function. Ecdysis may be inhibited (Kubo and Klocke, 1982; Garcia and Rembold, 1984; Arpaia and Van Loon, 1993), suggesting that the production of eclosion hormone (Horodyscki, 1996) or ecdysis triggering hormone (Zitnan et al., 1996) may be affected. Levels of ecdysteroids and prothoracicotropic hormone are altered following treatment with azadirachtin (Barnby and Klocke, 1990). Thus, the observed developmental disturbances likely reflect changes caused in the endocrine glands of treated insects, and consequently hormone titers.

Azadirachtin, along with three other neem seed compounds, inhibits ecdysone 20-monooxygenase in a dose-dependent manner, indicating ecdysteroid metabolism is slowed (Mitchell et al., 1997) by these agents. Metabolism of hormones as well as their synthesis may be disrupted, contributing to alterations in the endocrine milieu.

Azadirachtin binds to the nuclei of Sf9 cells in culture, and binding is irreversible and saturable (Nisbet et al., 1997). Isolated cells in culture can therefore be used to study the mechanism of action of azadirachtin.

When tested in combination with gypsy moth NPV, azadirachtin plus NPV was significantly more effective in killing gypsy moth larvae compared to NPV alone or azadirachtin only (Cook et al., 1996). The combination of azadirachtin and virus was predicted to result in good foliage protection if used against gypsy moth larvae. However, the addition of azadirachtin to viral formulations might also result in less virus being produced within the larval cadaver and released into the environment because the treated larvae are smaller (Cook et al., 1996).

Aside from the IGRs discussed here, several other insect growth regulators are presently being utilized in pest control, including chitin synthesis inhibitors (e.g., benzoylphenyl ureas; Cohen, 1993) and other compounds such as cyromazine, which act on cuticle deposition (Venuela and Budia, 1994), Chitinases also offer opportunities for development of new biopesticides (Kramer and Muthukrishnan, 1997). Insect chitinase genes can be expressed in plants, thereby disrupting the feeding and growth of the insect feeding upon the plant, or expressed in insect baculoviruses, enhancing the insecticidal activity and speed of kill of the viruses (Gopalakrishnan et al., 1995). Insect neuropeptides also offer potential for development of new insecticides similar to the hormonally based IGRs that have already been discovered and utilized in pest control (Masler et al., 1993).

REFERENCES

Arpaia, S. and J.J.A. Van Loon. Effects of azadirachtin after systemic uptake into *Brassica oleracea* on larvae of *Pieris brassicae. Entomologia Experimentalis et Applicata* 66, 39–45, 1993.

Ascher, K.R.S. Nonconventional insecticidal effects of pesticides available from the neem tree, *Azadirachta indica. Archives of Insect Biochemistry and Physiology* 22, 433–449, 1993.

Banken, J.A.O. and J.D. Stark. Stage and age influence on the susceptibility of *Coiccinella septempunctata* (Coleoptera: Coccinellidae) after direct exposure to Neemix, a neem insecticide. *Journal of Economic Entomology* 90, 1102–1105, 1997.

Barnby, M.A. and J. Klocke. Effects of azadirachtin on levels of ecdysteroids and prothoracicotropic hormone-like activity in *Heliothis virescens* (Fabr.) larvae. *Journal of Insect Physiology* 36, 125–131, 1990.

Beckage, N.E. and L.M. Riddiford. Effects of methoprene and juvenile hormone on larval ecdysis, emergence, and metamorphosis of the endoparasitic wasp, *Apanteles congregatus. Journal of Insect Physiology* 28, 329–334, 1982.

Blackford, M. and L. Dinan. The tomato moth *Lacaobia oleracea* (Lepidoptera: Noctuidae) detoxifies ingested 20-hydroxyecdysoine, but is susceptible to the ecdysteroid agonists RH-5849 and RH-5992. *Insect Biochemistry and Molecular Biology* 27, 291–309, 1997.

Brown, J.J. Effects of a nonsteroidal ecdysone agonist, tebufenozide, on host/parasitoid interactions. *Archives of Insect Biochemistry and Physiology*, 235–248, 1994.

Bull, D.L. and R.W. Meola. Effect and fate of the insect growth regulator pyriproxyfen after application to the horn fly (Diptera: Muscidae). *Journal of Economic Entomology* 86, 1754–1760, 1993.

Butterworth, J.H. and E.D. Morgan. Isolation of a substance that suppresses feeding in locusts. *J. Chem. Soc., Chem. Commun.,* 23–24, 1968.

Carpenter, J.E. and L.D. Chandler. Effects of sublethal doses of two insect growth regulators on *Heliocoverpa zea* (Lepidoptera: Noctuidae) reproduction. *Journal of Entomological Science* 29, 425–435, 1994.

Cohen, E. Chitin synthesis and degradation as targets for pesticide action. *Archives of Insect Biochemistry and Physiology* 22, 245–261, 1993.

Cook, S.P., R.E. Webb, and K.E. Thorpe. Potential enhancement of the gypsy moth (Lepidoptera: Lymantriidae) nuclear polyhedrosis virus with the triterpene azadirachtin. *Environmental Entomology* 25, 1209–1214, 1966.

Darvis, B., L. Polgar, M.H. Tag El-Din, E. Katakin, and K.D. Wing. Developmental disturbances in different insect orders caused by an ecdysteroid agonist, RH 5849. *Journal of Economic Entomology* 85, 2107–2112, 1992.

Dedos, S.G. and H. Fugo. Effects of fenoxycarb on the secretory activity of the prothoracic glands in the fifth instar of the silkworm, *Bombyx mori. General and Comparative Endocrinology* 104, 213–224, 1996.

De Kort, C.A.D. and A.B. Koopmanschap. A juvenile hormone analogue affects the protein pattern of the haemolymph in last-instar larvae of *Locusta migratoria. Journal of Insect Physiology* 37, 87–93, 1991.

De Kort, C.A.D., A.B. Koopmanship, and A.M.V. Vermunt. Influence of pyriproxyfen on the expression of haemolymph protein genes in the Colorado potato beetle, *Leptinotarsa decemlineata. Journal of Insect Physiology* 43, 363–371, 1997.

Dhadialla, T.S. and G. Tzertzinis. Characterization and partial cloning of ecdysteroid receptor from a cotton boll weevil embryonic cell line. *Archives of Insect Biochemistry and Physiology* 35, 45–57, 1997.

Dhadialla, T.S., G.R. Carlson, and D.P. Le. New insecticides with ecdysteroidal and juvenile hormone activity. *Annual Review of Entomology* 43, 545–569, 1998.

Donahue, W.A., P.D. Teel, O.F. Strey, and R.W. Meola. Pyriproxyfen effects on newly engorged larvae and nymphs of the lone star tick (Acari: Ixodidae). *Journal of Medical Entomology* 34, 206–211, 1997.

Dorn, A. Effects of azadirachtin on reproduction and egg development of the heteropteran *Oncopeltus fasciatus* Dallus. *Journal of Applied Entomology* 102, 313–319, 1986.

Edwards, G.C., R.P. Braun, and G.R. Wyatt. Induction of vitellogenin synthesis in *Locusta migratoria* by the juvenile hormone analog pyriproxyfen. *Journal of Insect Physiology* 39, 609–614, 1993.

Friedlander, M. and J.J. Brown. Tebufenozide (Mimic), a non-ecdysteroidal ecdysone agonist, induces spermatogenesis reintiation in isolated abdomens of dispausing codling moth larvae (*Cydia pomonella*). *Journal of Insect Physiology* 41, 403–411, 1995.

Gadenne, C., L. Varjas, and B. Mauchamp. Effects of the non-steroidal ecdysone mimic RH-5849, on diapause and non-diapause larvae of the European corn borer, *Ostrinia nubilalis. Journal of Insect Physiology* 36, 555–559, 1990.

Garcia, H. and H. Rembold. Effects of azadirachtin on ecdysis in *Rhodnius prolixus. Journal of Insect Physiology* 30, 939–941, 1984.

Gopalakrishnan, B., S. Muthukrishnan, and K.J. Kramer. Baculovirus-mediated expression of a *Manduca sexta* chitinase gene: properties of the recombinant protein. *Insect Biochemistry and Molecular Biology* 25, 255–265, 1995.

Grenier, S., and A.M. Grenier. Fenoxycarb, a fairly new insect growth regulator: a review of its effects on insects. *Annals of Applied Biology* 122, 369–403, 1993.

Grenier, S. and G. Plantevin. Development modifications of the parasitoid *Pseudoperichaeta nigrolineata* (Dept. Tachinidae) by fenoxycarb, an insect growth regulator, applied onto its host *Ostrinia nubilalis* (Lep. Pyralidae). *Journal of Applied Biology* 110, 462–470, 1990.

Hargrove, J.W. and P.A. Langley. A field trial of pyriproxyfen-treated targets as an alternative method for controlling tsetse (Diptera: Glossinidae). *Bulletin of Entomological Research* 83, 361–368, 1993.

Hatakoshi, M. An inhibitory mechanism over oviposition in the tobacco cutworm, *Spodoptera litura* by juvenile hormone analogue pyriproxyfen. *Journal of Insect Physiology* 38, 793–801, 1992.

Horodyski, F. M. Neuroendocrine control of insect ecdysis by eclosion hormone. *Journal of Insect Physiology* 42, 917–24, 1996.

Horowitz, A.R. and I. Ishaaya. Managing resistance to insect growth regulators in the sweet-potato whitefly (Homoptera: Åleyrodidae). *Journal of Economic Entomology* 87, 866–871, 1994.

Ishaaya, I. and A.R. Horowitz. Pyriproxyfen, a novel insect growth regulator for controlling whiteflies: Mechanisms and resistance management. *Pesticide Science* 43, 227–232, 1995.

Ishaaya, I., A. De Cock, and D. Degheele. Pyriproxyfen, a potent suppressor of egg hatch and adult formation of the greenhouse whitefly (Homoptera: Aleyrodidae). *Journal of Economic Entomology* 87, 1185–1189, 1994.

Ishaaya, I., S. Yablonsski, and A.R. Horowitz. Comparative toxicology of two ecdysteroid agonists, RH-2485 and RH-5992, on susceptible and pyrethroid-resistant strains of the Egyptian cotton leafworm, *Spodoptera littoralis*. *Phytoparasitica* 23, 139–145, 1995.

Jones, I.W., A.A. Denholm, S.V. Ley, H. Lovell, A. Wood, and R.E. Sinden. Sexual development of malaria parasites is inhibited *in vitro* by the need extract azadirachtin, and its semi-synthetic derivatives. *FEMS Microbiology Letters* 120, 267–273, 1994.

Kim, Y. and E. S. Krafsur. *In vivo* and *in vitro* effects of 20-hydroxyecdysone and methoprene on diapause maintenance and reproductive development in *Musca autumnalis*. *Physiological Entomology* 20, 52–58, 1995.

Koehler, P.G. and R.S. Patterson. Incorporation of pyriproxyfen in a German cockroach (Dictyoptera: Blattellidae) management program. *Journal of Economic Entomology* 84, 917–921, 1991.

Kono, Y., K. Omata-Iwabuchi, and M. Takahashi. Changes in susceptiblity to pyriproxyfen, a JH mimic, during late larval and early pupal stages of *Culex pipiens molestus*. *Medical Entomology and Zoology* 48, 85–89, 1997.

Kramer, K.J. and S. Muthukrishnan. Insect chitinases: molecular biology and potential use as biopestides. *Insect Biochemistry and Molecular Biology* 27, 887–900, 1997.

Kraus, W., M. Bokel, A. Klenk, and H. Pohnl. The structure of azadiractin and 22,23-dihydro-23B-methoxyazadirachtin. *Tetrahedron Letters* 26, 6435–6438, 1985.

Kubo, I. and J.A. Klocke. Azadirachtin, insect ecdysis inhitor. *Agricultural and Biological Chemistry* 46, (7), 1951–1953, 1982.

Leonardi, M.G., S. Cappellozza, P. Ianne, L. Cappellozza, P. Parenti, and B. Giordana. Effects of the topical application of an insect growth regulator (fenoxycarb) on some physiological parameters in the fifth instar larvae of the silkworm *Bombyx mori*. *Comparative Biochemistry and Physiology* B 113, 361–365, 1996.

Lim, J.L. and H.H. Yap. Induction of wing twisting abnormalities and sterility on German cockroaches (Dictyoptera: Blattellidae) by a juvenoid pyriproxyfen. *Journal of Economic Entomology* 89, 1161–1165, 1996.

Linton, Y.M., A.J. Nisbet, and A.J. Mordue (Luntz). The effects of azadirachtin on the testes of the desert locust. *Journal of Insect Physiology* 11, 1077–1084, 1997.

Masler, E.P., T.J. Kelly, and J.J. Menn. Insect neuropeptides: discovery and application in insect management. *Archives of Insect Biochemistry and Physiology* 22, 87–111, 1993.

Meola, R., S. Pullen, and S. Meola. Toxicity and histopathology of the growth regulator pyriproxyfen to adults and eggs of the cat flea (Siphonaptera: Pulicidae). *Journal of Medical Entomology* 33, 670–679, 1996.

Meurant, K., C. Sernia, and H. Rembold. The effects of azadirachtin A on the morphology of the ring complex of *Lucilia cuprina* (Wied) larvae. *Cell and Tissue Research* 275, 247–254, 1994.

Miller, R.W. Evaluation of S-31183 for fly (Diptera: Muscidae) control as a feed-through compound for poultry, cattle, and swine. *Journal of Agricultural Entomology* 6, 77–81, 1989.

Miller, R.W. and J.A. Miller. Pyriproxyfen bolus for control of fly larvae. *Journal of Agricultural Entomology* 11, 39–44, 1994.

Mitchell, M.J., S.L. Smith, S. Johnson, and E.D. Morgan. Effects of the neem tree compounds azadirachtin, salannin, nimbin, and 6-desacetylnimbin on ecdysone 20-monooxygenase activity. *Archives of Insect Biochemistry and Physiology* 35, 199–209, 1997.

Monconduit, H. and B. Mauchamp. Effects of ultralow doses of fenoxycarb on juvenile hormone-regulated physiological parameters in the silkworm, *Bombyx mori*. *Archives of Insect Biochemistry and Physiology* 37, 178–189, 1998.

Mordue (Luntz), A.J., and A. Blackwell. Azadirachtin: an update. *Journal of Insect Physiology* 39, 903–924, 1993.

Moreira, M.F., E.S. Garcia, and H. Masuda. Inhibition by azadirachtin of phospholipid transfer from lipophorin to the oocytes in *Rhodnius prolixus*. *Archives of Insect Biochemistry and Physiology* 27, 287–299, 1994.

Moreno, J., N. Hawlitzky, and R. Jimenez. Effect of the juvenile hormone analogue fenoxycarb applied via the host on the parasitoid *Phanerotoma ocularis* Kohl Hym. Braconidae. *Journal of Insect Physiology* 39, 183–186, 1993a.

Moreno, J., N. Hawlitzky, and R. Jimenez. Morphological abnormalities induced by fenoxycarb on the pupa of *Phanerotoma ocularis* Kohl Hym. Braconidae. *Journal of Applied Entomology* 115, 170–175, 1993b.

Mulye, H. and R. Gordon. Effects of fenoxycarb, a juvenile hormone analogue, on lipid metabolism of the Eastern spruce budworm *Choristoneura fumiferana*. *Journal of Insect Physiology* 39: 721–727, 1993.

Muszynska-Pytel, M., M.A. Pszczolkowski, P. Mikolajczk, and B. Cymboroski. Strain-specificity of Galleria mellonella larvae to juvenilizing treatments. *Comparative Biochemistry and Physiology* 103A, 119–125, 1992.

Nasiruddin, M. and A.J. Mordue. The effect of azadirachtin on the midgut histology of the locusts, *Schistocerca gregaria* and *Locusta migratoria*. *Tissue & Cell* 25, 875–884, 1993.

Nijhout, H.F. *Insect Hormones*. Princeton University Press, Princeton, NJ, 267 pp., 1994.

Nisbet, A.J., A.J. Mordue (Luntz), R.B. Grossman, L. Jennens, S.V. Ley, and W. Mordue. 1997. Characterization of azadirachtin binding to Sf9 nuclei *in vitro*. *Archives of Insect Biochemistry and Physiology* 34, 461–473, 1997.

Nisbet, A.J., A.J. Mordue (Luntz), L.M. Williams, L. Hannah, L. Jennens, S.V. Ley, and W. Mordue. Autoradiographic localization of [22,23-^{3}H] dehydroazadirachtin binding sites in desert locust testes and effects of azadirachtin on sperm motility. *Tissue & Cell* 28, 725–729, 1996.

Nogueira, N.F.S., M. Gonzales, E.M. Garcia, and W. De Souza. Effect of azadirachtin A on the fine structure of the midgut of *Rhodnius prolixa*. *Journal of Invertebrate Pathology* 66, 58–63, 1977.

Oberlander, H., D.L. Silhacek, and P. Porcheron. Non-steroidal ecdysteroid agonists: Tools for the study of hormone action. *Archives of Insect Biochemistry and Physiology* 28, 209–223, 1995.

O'Donnell, P.P. and M.J. Klowden. Methoprene affects the rotation of the male terminalia of *Aedes aegypti* mosquitoes. *Journal of the American Mosquito Control Association* 13, 1–4, 1997.

Palma, K.G., S.M. Meola, and R. W. Meola. Mode of action of pyriproxyfen and methoprene on eggs of *Ctenocephalides*–Felis Siphonaptera Pulicidae. *Journal of Medical Entomology* 30, 421–426, 1993.

Parkman, J.P. and J.H. Frank. Development and reproduction of mole crickets (Orthoptera: Gryllotalpidae) after treatments with hydroprene and pyriproxyfen. *Journal of Economic Entomology* 91, 392–397, 1998.

Peleg, B.A. Effect of a new phenoxy juvenile hormone analog on California red scale (Homoptera: Diaspididae), Florida wax scale (Homoptera: Coccidae) and the ectoparasite *Aphytis holoxanthus* De Bache (Hymenoptera: Aphelinidae). *Journal of Economic Entomology* 81, 88–92, 1988.

Pszczolkowski, M.A. and B. Kuszczak. Effect of an ecdysone agonist, RH-5849, on wandering behavior in *Spodoptera littoralis*. *Comparative Biochemistry and Physiology* 113C, 359–367, 1996.

Rembold, H., B. Subrahmanyam, and T. Muller. Corpus cardiacum — a target for azadirachtin. *Experientia* 45, 361–363, 1989.

Retnakaran, A., K. Hiruma, S.R. Palli, and L.M. Riddiford. Molecular analysis of the mode of action of RH-5992, a lepidopteran-specific nonsteroidal ecdysteroid agonist. *Insect Biochemistry and Molecular Biology* 25, 109–117, 1995.

Richter, K., G.A. Boehm, and H. Kleeberg. Effect of NeemAzal, a natural azadirachtin-containing preparation, on *Periplaneta americana*. *Journal of Applied Entomology* 121, 59–64, 1997.

Riddiford, L.M. Cellular and molecular actions of juvenile hormone I. General considerations and premetamorphic actions. *Advances in Insect Physiology* 24, 213–274, 1994.

Sayah, F., C. Fayet, M. Idaomar, and A. Karlinsky. Effect of azadiractin on vitellogenesis of *Labidura riparia* (Insect Dermaptera). *Tissue & Cell* 28, 741–749, 1996.

Schmutterer, H. Properties and potential of natural pesticides from the neem tree. *Annual Review of Entomology* 35, 271–297, 1990.

Shimizu, T. Suppressive effects of azadirachtin on spermiogenesis of the diapausing cabbage armyworm, *Mamestra brassicae, in vitro*. *Entomologia Experimentalis et Applicata* 46, 197–199, 1988.

Sielezniew, M. and B. Cymboroski. Effects of ecdysteroid agonist RH-5849 on pupal diapause of the tobacco hornworm (*Manduca sexta*). *Archives of Insect Biochemistry and Physiology* 35, 191–197, 1997.

Silhacek, D.L., H. Oberlander, and P. Porcheron. Action of RH 5849, a non-steroidal ecdysteroid mimic, on *Plodia interpunctella* (Hübner) *in vivo* and *in vitro*. *Archives of Insect Biochemistry and Physiology* 15, 201–212, 1990.

Simmonds, M.S.J. and W.M. Blaney. Azadirachtin: Advances in understanding its activity as an antifeedant. *Entomologia Experimentalis et Applicata* 80, 23–26, 1996.

Smagghe, G. and D. Degheele. Effects of RH-5849, the first nonsteroidal ecdysone agonist, on larvae of *Spodoptera littoralis* (Boisd.) (Lepidoptera: Noctuidae). *Archives of Insect Biochemistry and Physiology* 21, 119–128, 1992.

Smagghe, G. and D. Degheele. Action of the nonsteroidal ecdysteroid mimic RH-5849 on larval development and adult reproduction of insects of different orders. *Invertebrate Reproduction and Development* 25, 227–236, 1994a.

Smagghe, G. and D. Degheele. Effects of the ecdysteriod agonists RH 5849 and RH 5992, alone and in combination with a juvenile hormone analogue, pyriproxyfen, on larvae of *Spodoptera exigua*. *Entomologia Experimentalis et Applicata* 72, 115–123, 1994b.

Smagghe, G., H. Eelen, E. Verschelde, K. Richter, and D. Degheele. Differential effects of nonsteroidal ecdysteroid agonists in Coleoptera and Lepidoptera: analysis of evagination and receptor binding in imaginal discs. *Insect Biochemistry and Molecular Biology* 26, 687–695, 1996.

Spindler-Barth, M., A. Turberg, and K.D. Spindler. On the action of RH5849, a nonsteroidal ecdysteroid agonist, on a cell line from *Chironomus tentans*. *Archives of Insect Biochemistry and Physiology* 16, 11–18, 1991.

Staal, G.B. Insect control with growth regulators interfering with the endocrine system. *Entomologia Experimentalis et Applicata* 31, 15–23, 1982.

Tanzubil, P.B. Effects of neem (*Azadirachta indica* A. Juss) extracts on food intake and utilization in the African armyworm, *Spodoptera exempta* (Walker). *Insect Science and its Application* 16, 167–170, 1995.

Tateishi, K., M. Kiguchi, and S. Takeda. New cuticle formation and mote inhibition by RH-5849 in the common cutworm, *Spodoptera litura* (Lepidoptera: Noctuidae). *Applied Entomology and Zoology* 28, 177–184, 1993.

Teal, P.D., W.A. Donahue, O.F. Strey, and R.W. Meola. Effects of pyriproxyfen on engorged females and newly oviposited eggs of the lone star tick (Acari: Ixodidae). *Journal of Medical Entomology* 33, 721–725, 1996.

Trisyono, A. and M. Chippendale. Effect of the nonsteroidal ecdysone agonists, methoxy-fenozide and tebufenozide, on the European corn borer (Lepidoptera: Pyralidae). *Journal of Economic Entomology*, 90, 1486–1492. 1997.

Turner, C. and T.G. Wilson. Molecular analysis of the methoprene-tolerant gene region of *Drosophila melanogaster Archives of Insect Biochemistry and Physiology* 30, 133–147, 1995.

Vail, K.M. and D.F. Williams. Pharoah ant (Hymenoptera: Formicidae) colony development after consumption of pyriproxyfen baits. *Journal of Economic Entomology* 88, 1695–1702, 1995.

Vail, K.M., D.F. Williams, and D.H. Oi. Perimeter treatments with two bait formulations of pyriproxyfen for control of Pharoah ants (Hymenoptera: Formicidae). *Journal of Economic Entomology* 89, 1501–1507, 1996.

Vennard, C., B. Nguama, R.J. Dillon, H. Ouchi, and A.K. Charnley. Effects of the juvenile hormone mimic pyriproxyfen on egg development, embryogenesis, larval development, and metamorphosis in the desert locust *Schistocerca gregaria*. *Journal of Economic Entomology* 91, 41–49, 1998.

Venuela, E. and F. Budia. Ultrastructure of *Ceratitis capitata* Wiedemann larval integument and changes induced by the IGI cyromazine. *Pesticide Biochemistry and Physiology* 48, 191–201, 1994.

Williams, C.M. The juvenile hormone: Its role in the endocrine control of molting, pupation, and adult development in the cecropia silkworm. *Biological Bulletin Woods Hole* 121:572–585.

Wilson, T.G. and J. Fabian. A *Drosophila melanogaster* mutant resistant to a chemical analog of juvenile hormone. *Developmental Biology* 118, 190–201, 1986.

Wing, K.D. RH-5849, a nonsteroidal ecdysone agonist: Effects on a *Drosophila* cell line. *Science* 241, 467–469, 1988.

Wing, K.D., R.A. Slawecki, and G.R. Carlson. RH-5849, a nonsteroidal ecdysone agonist: Effects on larval Lepidoptera. *Science* 241: 470–472, 1988.

Zanno, P.R., I. Miura, K. Nakanishi, and D.L. Elder. Structure of the insect phagorepellent azadirachtin. Application of PRFT/CQD carbon-13 nuclear magnetic resonance. *Journal of the American Chemical Society* 97, 1975–1977, 1975.

Zhang, L. and T. Shono. Toxicities of pyriproxyfen to susceptible and resistant strains of houseflies. *Applied Entomology and Zoology* 32, 373–378, 1997.

Zhang, L., K. Harada, and T. Shono. Genetic analysis of pyriproxyfen resistance in the housefly. *Musca domestica. Applied Entomology and Zoology* 32, 217–226, 1997.

Zhang, L., S. Kasai, and T. Shono. *In vitro* metabolism of pyriproxyfen by microsomes from susceptible and resistant housefly larvae. *Archives of Insect Biochemistry and Physiology* 37, 215–224, 1998.

Zitnan, D., T.G. Kingan, J.L. Hermesman, and M.E. Adams. Identification of ecdysis-trigger-ing hormone from an epitracheal system. *Science* 271, 88–91, 1996.

Physiological Approaches

CHAPTER **6**

Genetic Control of Insect Pests

Alan S. Robinson

CONTENTS

6.1 INTRODUCTION

The principles underlying the diverse genetic approaches proposed for the management of insect-related problems are based on an understanding of genes and chromosomes and their role in the interaction of the insect with its environment. The term *genetic control* is often used to collectively describe these approaches, but this term carries with it considerable ambiguity by the use of the word "control." Most entomologists would interpret the term as meaning a reduction of insect population numbers leading directly to the amelioration of the insect-related problem. However, it can also be interpreted as the manipulation of the insect genome to modulate the characteristic that makes the insect a pest. Genetic control has therefore both qualitative and quantitative aspects and it is in this wide sense that the term is interpreted in the present chapter. A second difficulty associated with the use of the word "control" concerns its temporal connotation with the suggestion that the procedure has to be implemented on a continuous basis. However, one form of genetic control has been shown to be very effective in the eradication of large insect populations over considerable geographic areas. Because of the possibility of achieving eradication, the discussion of control versus eradication has special relevance for genetic techniques. The release of radiation-sterilized insects can lead to population eradication, and for certain pests the use of this principle is an integral part of the conventional modern approach to insect management.

6.1.1 General Principles

Once the mechanics of Mendelian genetics and chromosome theory were fully interpreted, geneticists realized that certain concepts could be exploited to develop insect control techniques (Serebrovskii 1940). Implicit in this realization was the understanding that an insect, once genetically modified, could be released into the field, mate with the natural population, and cause a reduction in the pest status of the species. This perception came long before concerns relating to environmental protection and insecticide resistance initiated the drive for more biologically and socially acceptable forms of insect control techniques. Entomologists, not geneticists, originally focused attention on the search for agents that would sterilize insects and they eventually concluded that ionizing radiation could be the agent required (Knipling 1955,1960). Some 10 years later, there was an explosion of other ideas that formed the basis for current thinking on genetic control (Curtis 1968a,b; Whitten 1970, 1971a; Foster et al. 1972, Smith and von Borstel 1972; Whitten and Foster 1975). Theoretical analyses of the effectiveness of many of these mechanisms indicated their potential (Knipling and Klassen 1976). Genetic control has therefore a long pedigree, sufficiently long in fact that it has already been evaluated as a "growth industry or lead balloon?" (Curtis 1985). Several full texts on the subject have been published together with a series of Symposium proceedings organized by the International Atomic Energy Agency (IAEA 1993, 1988, 1982; Davidson 1974; Pal and Whitten 1974; Hoy and Mckelvey 1979; Steiner et al. 1982)

Genetic control techniques require the transmission, through at least one generation, of modified hereditary material and thus they require that mating occur between the released and wild insects and that fertilisation take place. This means that they are, by definition, species specific. An exception to this can be seen in the use of hybrid sterility in species complexes, where closely related species have not yet evolved effective premating isolation but where genetic differentiation is such that hybrids can be sterile (Potts 1944; Davidson 1969). Species targeted approach to insect control has gained much support in recent years and integrated pest management is generally based on this principle. Species specificity ensures virtually no deleterious effects on the ecosystem in general but requires that each species be targeted individually. This specificity is in stark contrast to the effects of many pesticides or even to some forms of biological control that are now coming under increasing criticism because of unexpected negative environmental effects (Howarth 1991).

The majority of genetic control techniques have the unique property of becoming more effective as the target population is reduced in numbers. However, they tend to be less effective at high population densities. This was elegantly shown in the first models used to describe the use of sterile insects (Knipling 1955). This contrasts sharply with the use of insecticides where net effectiveness decreases when populations become small. The reason for this contrasting effectiveness is that genetic control relies on an insect–insect interaction, e.g., mate seeking, whereas insecticides rely on a chemical–insect interaction. In the former case both components will actively seek each other out, whereas in the latter the "inert" component has still to be placed wherever the insect may be found.

Insects are very adept at developing resistance to chemical poisons, even to the new generation of microbial insecticides (Gould et al. 1992; Tabashnik 1994; Tabashnik et al. 1997). The current trend to incorporate insect toxin genes in plants is likely to meet the same constraint. It seems that the biochemical machinery of insects, coupled with their large numbers and relatively short generation time, can be very easily adapted to nullifying the effects of environmental poisons. The development of resistance to genetic control would require that the target insect be able to recognize and reject for mating the genetically modified insect; in other words a form of premating isolation mechanism would need to evolve. Theoretically, if the genetically modified insect retains the same mating behaviour as the target insect, there is no variation for natural selection to act on and hence resistance cannot develop even if the fitness of that mating is zero. In practice, however, laboratory rearing of insects can change many behavioural traits (Cayol in press) so that the possibility that resistance may develop has to be considered. There have been two published cases of resistance to genetic control (Hibino and Iwahashi 1991; McInnis et al. 1996). In both cases although there was behavioural evidence that wild females appeared to reject the radiation-sterilized males, there was no evidence that genetic selection was the cause as no attempt was made to genetically analyse the trait. The behavioural resistance in one of these populations has now disappeared (McInnis pers. comm.) and the status of the original observation could be questioned. Quality control of released insects has a major role to play in minimizing the chances that "resistance" can occur. If effective resistance developed in a wild population, it could in some cases be dealt with by establishing a new laboratory colony from the resistant field population.

Genetic control is a technology that lends itself very well to integration with other pest management procedures. For example, if transgenic Bt plants are being used to control the larval stages of plant-feeding insects, then genetically modified adult insects could be released to increase the pressure on the pest population. In other situations, integrated crop-protection measures are able to manage all of the insect pests present in a particular ecosystem with the exception of one that still requires pesticide application and this impacts negatively on the whole integrated approach. This key pest would be an ideal candidate for genetic control. Many genetic techniques can also be combined with the release of parasitoids (Knipling 1992; Mannion et al. 1995). The combinatorial approach to insect control can be well served by genetic control.

Presuppression strategies are essential for sterility mediated genetic control techniques because insect numbers in the field are at a level where it would be logistically impossible to produce the required number of insects for release. In certain situations genetically modified insects can be released at a time to coincide with a natural reduction in pest numbers, for example, at the end of a winter period.

Chemical pest control is generally undertaken in response to (a) the perception that an insect problem is present, (b) the reality that one is about to occur, or (c) the emergency of a new outbreak; in other words chemical control can be characterized as being reactive or even retroactive. The neutral environmental impact of genetic control opens the way to the development of a prophylactic approach to insect pest management where an area is protected from insect colonization by the permanent release of genetically modified insects. This approach would be inconceivable for pesticides or even for conventional biological control agents. The Los Angeles Basin area in California is now protected from medfly colonization by the permanent release of sterile males (Anon. 1996). This approach provides a more sound economic strategy to address the problem of repeated introductions of this exotic pest. Although prevention is in general better than cure, the economics of this approach will probably not be suitable for every situation.

Genetic control in most cases has to be viewed as an area-wide approach in which a crop, or an animal or human population is protected from insect attack over a large geographic area. It is not suitable for a field by field, or even a farm by farm approach as both the biology and the economics demand large-scale application. This means that effective genetic control programmes require considerable start-up funds and the large financial resources required is a major reason why these types of approaches have not been more attractive to funding agencies; it is far easier to obtain funding for ten small projects than one large one. However, a recent study (Enkerlin and Mumford 1997) has clearly shown that in the long term, area-wide approaches, including the SIT, have a much better return on investment than do conventional farmer by farmer approaches. A key element in area-wide economics is the mobilisation and organisation of the beneficiaries. In the long term, genetic control techniques will only be successful if they become commercially viable and are able to compete economically with other control methods. Commercial viability can be approached by introducing a levy for all the beneficiaries, but to be effective it requires that all farmers in the target area are participants of the scheme. This again is a major difference when compared with the purchase of insecticides or biological control agents by individual farmers where individual choices can be made.

The decision as to when and in which species genetic control techniques could and should be developed is complex and multifaceted. It involves consideration of the biology and pest status of the species, other methods available for control, and economic evaluation. There are two popular misconceptions relating to genetic control techniques: first, that they can be developed only in species that have a rich infrastructure of genetic information and second, that the use of sterilized males is only applicable in species in which the females mate only once. Neither of these statements is true. The number of times a female mates is irrelevant providing the sperm that is transferred from the sterilized male is competitive with sperm from a normal male. Although the acquisition of a reasonable genetic tool kit can be of enormous help and is essential for some approaches, the most spectacular success of genetic control against the screwworm was achieved "...without knowing how many chromosomes they had" (LaChance 1979 quoting R. C. Bushland). The simplest and so far the most effective genetic control technique, the sterile insect technique (SIT), can be developed with very limited genetic knowledge of the target species.

6.2 REQUIREMENTS FOR APPLICATION

Absence of detailed knowledge of the population dynamics, ecology, and behaviour of the target pest is a guarantee of failure for any genetic control technique. The level of knowledge required is much greater than for most other insect control strategies. Techniques employing sterility can be very sensitive to density-dependent processes that regulate natural populations and some data on the level of this type of regulation is essential. In a reciprocal manner, once sterility is being induced in a natural population and it can be correlated with changes in population density, the level of density-dependent regulation can be assessed. In this way the induction of sterility can be used as a tool by ecologists to further refine their population models.

6.2.1 Colonization, Mass Rearing, and Quality

All types of genetic control require the colonization and to some extent the mass rearing of the target species with individual species differing in the ease with which they accept these two processes. There is no "real science" of laboratory colonization for insects in terms of sampling frequency and sample size to ensure that a colony once established is representative of the original population. However, Mackauer (1976) has described some of the genetic aspects of insect colonization. Sampling is generally done with the philosophy "the more the better." Superimposed on this shaky beginning the colony will be subjected to selection that will inevitably occur during the long-term maintenance of a population in the laboratory. The move to large-scale mass rearing in preparation for release will exert another level of pressure on the population, and for operational programmes the economic factor in production costs becomes extremely important. All developmental stages of the insect have to be provided with an environment that not only enables them to reproduce in a predictable and efficient manner, but which also produces individuals with a certain level of quality at an acceptable economic price. These often opposing forces of quality versus quantity will always lead to a compromise, but a reduction in quality of released insects below a reasonable level will make any technique impractical.

The effects of laboratory colonization on many aspects of insect behaviour are incremental, heterogeneous, and to a certain degree unpredictable, and many ideas have been developed as to how quality can be monitored in the laboratory and how rearing systems can be adapted in an attempt to retain quality (Boller 1972; Chambers 1977; Huettel 1976; Ochieng-Odero 1994). Many quality parameters can be effectively monitored in the laboratory, for example, size, survival, etc., but the assessment of parameters related to behaviour would seem to be of little value when carried out under these conditions. As all genetic control techniques require the mating of the released insects with the wild population, any change in mating behavioural patterns will have an immediate detrimental effect on the efficiency of the technique, and this aspect of quality has to be monitored in a representative and meaningful way — probably in the open field or in field cages. Dispersal is another key behaviour that is critical for success.

In an operational programme it is essential to have a predicable supply of insects of known quality for a specified period. These are difficult requirements to meet for managers of rearing facilities and demand an industrial approach in terms of logistics and human resources.

6.2.2 Post-Production Processes

For any area-wide genetic control programme, large numbers of insects have to be prepared for release. This involves marking the insects so that they can be recognized in the field, sterilizing them if necessary, transport to the field area, and then their dispersal over the treatment area. These post-production processes are considerable and require just as much attention as does the production component. The processes have to be carried out within a defined and generally short time frame, and have to be simple, robust, economical, and cause little damage to the insect. Despite these constraints ingenious systems have been developed for many species. In general adult insects are released as they are mobile and less likely to be attacked by predators, being mobile they can also aid in the dispersal process and for large programmes they are usually released from aircraft. Aerial release is often much cheaper than ground release and ensures a much better distribution of insects at a relatively low cost.

6.2.3 Field Monitoring

A continuous evaluation of a field programme is essential both in terms of monitoring effectiveness and in making programme adjustments. Released flies must be clearly distinguishable from field insects in a rapid and secure way and methods must be available to monitor the wild and released population. The issue of marking is critical and current methods that rely on fluorescent dust are not optimal. The misclassification of a single fly as wild as opposed to released can have a major impact on a programme where eradication is the goal. The use of genetic transformation technology to introduce benign genetic markers will provide a high degree of security for the determination of the origin of a trapped insect. Real-time evaluation of the programme enables managers to make decisions as to where an increased or a decreased number of flies need to be released. The monitoring process also provides the key evidence relating to the quality of the flies being released.

If any form of sterility technique is being used for control, a measure of the population fertility before and during the programme is highly informative; unfortunately, this parameter is not always easy to monitor in the field. It is also the only direct evidence that the released insects have interacted with the wild population. Without this parameter, critics can always invoke other cause for population collapse or even eradication (Readshaw 1986). However, in the case critiqued by Readshaw (1986) this parameter was available and it could be correlated with the decrease in population size.

6.3 QUANTITATIVE AND QUALITATIVE APPROACHES

Insect problems are modulated by the number of insects and their virulence, and both these components can be targeted using genetic control. The number of insects can be reduced by increasing the genetic load in a population by a variety of approaches outlined below. Genetic load is a term coined by Muller (1950) that expresses the amount of genetic sterility in a population. The amount of genetic load required to cause a continuous reduction in the target population will depend on the degree of density-dependent regulation, the stage where it occurs, and the immigration of fertilized females into the treatment area (Prout 1978; Dietz 1976). The response of a population to an increase in genetic load can also enable ecologists to quantify the degree of density-dependent regulation and reproductive increase (Weidhaas et al. 1972). The imposition of a genetic load, when of sufficient size to generate a reduction in population size, will if continually applied lead to the eradication of the target population. This means that when the target population begins to decrease in size there is no way back and eradication is inevitable. The attainment of eradication constitutes a shift from a quantitative to a qualitative situation, at the trivial level from one insect to no insect.

Qualitative changes in the genomes of insects can alter their status from pestiferous to benign and vice versa. Genetic control theory offers several mechanisms by which this status can be manipulated. Chromosomal translocations (Curtis 1968b), compound chromosomes (Childress 1972), and cytoplasmic incompatibility (Curtis and Adak 1974) rely on some form of inter-population sterility to manipulate gene frequency, whereas meiotic drive (Foster and Whitten 1974) relies on non-Mendelian segregation leading to the unequal recovery of particular chromosomes (Sandler and Novitski 1957). All of the above systems are driven by a dynamic process that uses the motor of natural selection to introduce a particular genotype into a population; in theory the genotype can be driven to fixation. If a beneficial gene is absolutely linked to the genetic entity being driven into the population, it too will reach fixation. Different types of beneficial gene have been suggested as appropriate candidates for this approach including inability to diapause (Hogan 1966), temperature sensitivity (Smith 1971), insecticide susceptibility (Whitten 1970), inability to transmit a pathogen (Curtis and Graves 1984) and eye colour mutations (Foster et al. 1985a). The introduction of beneficial genes by simply overflooding a target population has also been proposed as a method to achieve qualitative change (Klassen et al. 1970).

All of the above theoretical approaches are subject to many constraints, both biological and operational, which have determined their acceptance as potential components in insect control programmes. A recent review highlights the pros and cons of the qualitative approach to vector-borne disease control (Pettigrew and O'Neill 1997). Experience has shown that the quantitative approach, being conceptually the simpler and in operation certainly so, has been the one most used in field application and to date is the only approach used for operational insect control.

6.4 MECHANISMS

The principles involved in the use of the various approaches have been well described elsewhere and do not need repetition (see references above). This section will simply summarize these principles and highlight the aspects that are relevant to the practical application of genetic control.

6.4.1 Dominant Lethality

Dominant lethality is the basis of the Sterile Insect Technique (SIT), undoubtedly the most successful application of genetic control of insects. Dominant lethality occurs when a haploid nucleus has been altered in such a way that when combined with a normal haploid nucleus the resulting zygote dies immediately or some time later (Muller 1927). Dominant lethals are easy to induce, common, and easy to score. As long ago as 1916 the sterilizing effect of ionizing radiation on insects was demonstrated (Runner 1916). Some time later, during studies on mustard gas, it was also shown that chemicals could produce the same effect (Auerbach and Robson 1942). LaChance (1967) produced an excellent review on this subject and synthesized the then current ideas on the use of radiation and chemicals to induce dominant lethal mutations in insects.

Dominant lethality in males is not sperm inactivation, if this were so, it could not be used for the SIT. It relies on genetic damage induced in the sperm being able to cause zygote lethality following fusion with the oocyte and it requires normal sperm function in terms of motility and fertilizing ability. The genetic basis of dominant lethality is well understood (LaChance 1967) and is the same for most insect species with the exception of the Hemiptera, Homoptera, and Lepidotera. These three orders of insect have an unusual chromosome structure (North and Holt 1970), which has major consequences for the development of genetic control procedures (see below). The dose–response relationship of ionizing radiation and the induction of dominant lethals in the different types of germ cells in males and females has been well described in *Drosophila* (Sankaranarayanan and Sobels 1976), and for each new species this relationship is important to determine. Mass rearing and release logistics often determine the developmental stage of the insect that has to be irradiated.

The chromosomal breaks induced by radiation and chemicals, although produced by different mechanisms, are the fundamental cause of dominant lethality. These breaks, although of no consequence to the haploid nucleus (sperm), cause chromosomal imbalance in the developing zygote through the breakage–fusion–bridge cycle (Curtis 1971) and lead to zygotic death. The time when the zygote dies depends on the amount of genetic damage inherited; the more the damage, the earlier the zygotes will die. For the SIT, full sterility throughout the life of the released insect in the field is required and sterility is traditionally measured by using egg hatch. However, dominant lethals can exert their effect at any time during development, and in theory a dose of radiation that guarantees that no fertile adults are produced following a mating between irradiated and a nonirradiated insect could be defined as the sterilizing dose and would indeed fulfil the requirements of the SIT. This latter dose would be much lower than the one causing zero egg hatch and would produce a much more competitive insect. The exponential component of dose response kinetics for dominant lethal induction at high levels of egg death requires an increasing amount of radiation for less biological effect. Notwithstanding this situation of diminishing returns, there is a strong reluctance on the part of SIT programme managers to use a lower dose of radiation that would lead to a low percentage of egg hatch but that would guarantee that no fertile adults develop. This reduced treatment would of course have to guarantee that the females that are released are fully sterile and that the commodity being protected could sustain a small amount of insect damage from the few larvae that would hatch but that would not develop to fertile adults.

In most SIT programmes both sexes are released and the response of both males and females to the sterilizing treatment has to be assessed. In some species the males are the more sensitive sex, e.g., the screw-worm, *Cochliomyia hominivorax* (LaChance and Crystal 1965), in other species the females are, e.g., the medfly, *Ceratitis capitata* (Hooper 1971). In the case where the male is the more sensitive sex it would be very advantageous to have a system for the removal of females so that a lower radiation dose could be given to the males.

As stated above both chemicals and ionizing radiation can cause dominant lethality and hence are potential candidates for use in SIT. In practice, ionizing radiation has been the agent of choice to produce competitive sterile insects. In the 1960s there was an extensive search for chemical alternatives to ionizing radiation without really much success in terms of practical use of the chemicals (Smith et al. 1964). However, in certain species, e.g. mosquitoes, chemical sterilization was preferred and was used in a fairly large field trial (Weidhaas et al. 1974). The emphasis on the use of chemosterilants in mosquitoes is probably due to two factors; first, adult mosquitoes are fairly fragile and are difficult to handle, thus a method for pupal treatment was preferred, and second, treatment of the pupae in their natural environment, water, with chemicals was much easier than the use of radiation. The major problem associated with the use of these chemicals is that they are mutagenic and environmental concerns, from the standpoint of both the treatment procedure and the release into the environment of large numbers of treated insects, are considerable.

6.4.2 Inherited Partial Sterility

Lepidoptera have chromosomes with diffuse centromeres, so-called holokinetic chromosomes (Bauer 1967), and this feature is shared with the Hemiptera and the Homoptera. All other insect species have a localized centromere. This phenomenon has a major impact on the interaction of these chromosomes with radiation. First, sterilizing doses of radiation are almost an order of magnitude higher for species with holokinetic chromosomes and second, if they are given substerilizing doses of radiation, their F1 progeny are more sterile than the parents. Proverbs (1962) was the first to demonstrate the inheritance of this type of partial sterility in the codling moth, *Laspeyresia pomonella*, and the mechanism by which it occurs is well understood (LaChance et al. 1970). The positive correlation of high radiation doses with reduced competitiveness encouraged the development in Lepidoptera of the use of inherited partial sterility for genetic control. In this technique the released insects are given a substerilizing dose of radiation to maximize competitiveness with their progeny being fully sterile. Mathematical models indicated the potential of the approach (Knipling 1970; Knipling and Klassen 1976).

There are three factors that must be taken into account when this type of genetic control is discussed. First, Lepidoptera transfer two types of sperm during mating, eupyrene and apyrene. The former are nucleate and effect fertilization, and both radiation and the partial sterility in the F1 generation can affect the transfer of these two sperm types by males. This can have a negative effect on the competitiveness of the insects (LaChance 1975), although the negative response to radiation is not shared by all species (North and Holt 1971). The mechanism by which inherited partial sterility can affect sperm transfer in the F1 generation is not known. Second, there is a distortion in the sex ratio in the progeny of irradiated males in favour of males, probably due to the expression of radiation-induced recessive lethals in the hemizygous F1 females (North 1975). However, the F1 females do have a higher level of fertility than the F1 males. If inherited partial sterility is therefore proposed, a radiation treatment should be identified which maximises the sex ratio distortion in favour of the male. Third, the two sexes differ in their sensitivity to the induction of partial sterility in the F1 generation following the same dose of radiation. Given the same substerilizing dose of radiation, progeny from irradiated females are less sterile than progeny from unirradiated males, but in both cases the F1 male is more sterile than the F1 female (North 1975). As female moths are in general more sensitive to radiation than male moths, radiation treatments can be designed that fully sterilize the female but leave the male with residual fertility leading to the production of an F1 generation that is almost completely sterile and composed mainly of males.

6.4.3 Autosomal Translocations and Compound Chromosomes

Both these types of chromosomal rearrangement can play a role in insect control because they generate sterility when individuals carrying them are mated to individuals with a wild-type chromosomal karyotype. They differ from the more classical hybrid sterility syndrome as they have to be induced, generally by irradiation, and isolated following a series of genetic crosses. The use of autosomal translocations was first proposed by Serebrovskii (1940 in Russian), but the concept was independently developed by Curtis (1968a) and Curtis and Hill (1968, 1971) and Curtis and Robinson (1971). Autosomal translocations are produced following the exchange of chromosome material between nonhomologous chromosomes (Robinson 1976), whereas compound chromosomes result from the exchange of chromosome arms between homologous chromosomes (Holm 1976). Homozygous autosomal translocations should be fully fertile when inbred, but they produce a hybrid with reduced fertility when mated to chromosomally wild-type individuals. Compound chromosome strains are characterized by a reduced fertility when inbred, but they cause full sterility when outcrossed to a wild-type strain (Foster et al. 1972). The principle of generating strains of insects that show complete reproductive isolation from each other had already been experimentally demonstrated in *Drosophila* (Kozhevnikov 1936).

The sterility generated by these two types of rearrangement is due to chromosomal imbalance. Autosomal homozygous translocations have the full complement of genetic material and are able to pass through all stages of cell division without any difficulty. However, translocation heterozygotes, even though they carry the full genetic complement, generate a proportion of gametes that do not. These functional unbalanced gametes will, following fertilisation, lead to the death of the zygote. The unbalanced gametes are produced as a consequence of the segregation of the translocation complex during meiosis (Robinson 1976). A characteristic of translocations already mentioned above is that the semisterility they produce is inherited and, in the case of autosomal translocations, by both sexes. With compound chromosomes, the chromosomal imbalance is such that all F1 zygotes die as eggs; there is no inherited sterility. An individual carrying compound chromosomes is in fact a genetically contrived sterile male.

A phenomenon that characterizes both these types of rearrangement is that of negative heterosis, i.e., the hybrid is less fit than either parent. Inherent to this type of fitness relationship is the property of frequency-dependent selection whereby natural selection will drive one of the chromosomal types to fixation. The inference of this is that there must be an unstable equilibrium on either side of which selection will act to cause fixation of one chromosomal type or the other. If the fitness of both parental strains is 1, this unstable equilibrium will be when the frequencies of the two chromosomal types are equal. For other fitness levels the equilibrium point will change (Whitten 1971a,b). The presence of an unstable equilibrium means that if a gene is tightly linked to one of the chromsomal types it can be driven to fixation if the frequency of the particular chromosomal type is above the equilibrium frequency. It is in this way that these types of rearrangement were recruited for insect control as a way to manipulate gene frequencies in natural populations (Curtis 1968b; Foster et al. 1972).The major reason this approach has not been successful is that both translocation homozygotes and compound chromosome stocks were shown to have fitness values far below that required to achieve realistic population replacement (Robinson and Curtis 1973; Fitz-Earle et al. 1973).

6.4.4 Male-Linked Translocations

Male-linked translocations are exchanges of genetic material between an auto-some and the chromosome involved in male determination (Roberts 1976). This chromosomal rearrangement is inherited from father to son and because of the segregation of the translocation complex during male meiosis, males carrying the translocation have reduced fertility (Laven et al. 1971). Male-linked translocations therefore induce inherited sterility in males but have no effect on female fertility. As they are semisterile they will always be eliminated from a population with the exception of the extreme case of fixation, i.e., when all the males in a population carry the translocation. They were proposed as genetic control agents because of their ability to introduce inherited partial sterility into populations. Because these rearrangements are male-linked they are in general easy to maintain as in many Diptera genetic recombination is extremely rare in males. However, in species where sex is determined by a male determining gene that is carried on an autosome the position of the breakpoint relative to the gene is crucial to their stability as in these species recombination occurs in both sexes.

Insect species differ markedly in the genetic mechanisms that determine sex, and the induction of male-linked translocations has to take into account this under-lying sex-determination mechanism (Robinson 1983). However, even in pest species with quite different sex-determination mechanisms, these rearrangements can be easily induced, e.g., *Ceratitis capitata* (Steffans 1983) and *Culex pipiens* (Laven et al. 1971). Although the use of male-linked translocations for direct control has been limited, they have been extensively used to develop genetic sexing strains for use in genetic control programmes (Robinson 1983).

6.4.5 Hybrid Sterility

Hybrid sterilty between different tsetse species was the first genetic principle to be proposed as a means of developing new ways to control insects (Potts 1944). The sterility generated when closely related species or geographically distinct popula-tions are crossed can be genetic (Davidson 1969), cytoplasmic (Laven 1967a) or possibly a combination of the two. Genetic divergence between the members of species complexes can be of a degree that generates sterility in the hybrids and in general the heterogametic sex in the F1 generation tends to be the most affected (Haldane 1922, and see review Orr 1997). In most cases hybrid male sterility is accompanied by residual fertility in the F1 females as in most insect species the males are the heterogametic sex. However, in Lepidoptera the opposite is the case. The fact that F1 females remain partially fertile enables gene exchange to occur between sibling species in areas where the two cryptic species are sympatric. F1 hybrid males can show two deleterious effects of hybridisation, namely sterility and inviability, and it appears that they can be ascribed to different genetic phenomena, with inviability being due to X-autosome imbalance and sterility being due to interaction of sex specific genes (Wu and Davis 1993). The fact that these two mechanisms can be found within the same genus (Gooding 1997; Rawlings 1985) indicates the different ways in which speciation can develop. In Drosophila, the genetics of hybrid sterility has been analysed to the degree that many specific gene–chromosome interactions have been identified as the cause of the low fitness of hybrids (Palopoli and Wu 1994). Hybrid inviability is generally not suitable for genetic control as the reproductive system of the males can be poorly developed leading to noninsemination of females following mating. This would allow the female to remate with a fertile male. To be of any use in insect genetic control, the mating behaviour and reproductive physiology of the hybrid males must be equivalent to those of the wild males. The residual female fertility in F1 females can also present problems for application of this approach if some way is not found to remove them before release. However, in *Anopheles gambiae* certain hybrid crosses produce only sterile males in the F1 generation (Curtis 1982).

The phenomenon of cytoplasmic incompatibility (CI) can be seen as another manifestation of hybrid sterility and it is often expressed when allopatric populations within the same species are crossed. It was first described in *Culex pipiens* (Laven 1967a) and the causative agent was identified as a bacterium of the Genus *Wolbachia* (Yen and Barr 1971), a rickettsial endosymbiont that is found in the reproductive organs. Bacteria of this Genus have now been found to be very widely distributed throughout the Class Insecta and elsewhere, where in addition to cytoplasmic incompatibility, they cause parthenogenesis induction (Stouthamer et al. 1993) and feminization of males (Rousset et al. 1992). Cytoplasmic incompatibility is inherited maternally (Laven 1967b) and can be uni- or bi-directional. In uni-directional incompatibility sperm from a male that is infected with *Wolbachia* will lead to the death of the zygote following fertilization of the egg from a noninfected female. The reciprocal cross, infected female with noninfected male, is fertile as are the other two homozygous crosses. In essence, rickettsia in males induce incompatibility while rickettsia in females restore compatibility. In bi-directional incompatibility, males and females carry different strains of *Wolbachia* so that all heterozygous crosses are incompatible. Antibiotic treatment, leading to elimination of the bacterium, destroys the incompatibility phenotype (Yen and Barr 1971). The "selfish" behaviour of *Wolbachia* enables it to spread rapidly through a naive wild population (Turelli and Hoffman 1991). Intra- and interspecific horizontal transfer of these types of organisms has been experimentally demonstrated (Boyle et al. 1993), but the significance of this mode of transmission in nature is unclear. It does, however, open the possibility that *Wolbachia* could be transferred to a naive host and so generate novel cases of incompatibility. For recent reviews of *Wolbachia* biology see Werren (1997) and Bourtzis and O'Neill (1998).

There are currently two ways in which the phenomenon of incompatibility can be exploited for genetic control. First, males infected with a CI-inducing *Wolbachia* could be released into a naive population and all matings between the released males and the wild females would be sterile; this assumes that there are, in fact, naive populations in the field. It is known, however, that there is a high degree of incompatibility polymorphism in natural populations (Barr 1980) that could seriously interfere with this approach. The release of incompatible males would be equivalent to the sterile insect technique. Second, if genes could be introduced into other maternally inherited factors such as endosymbionts, the ability of *Wolbachia* to spread through a naive field population could be used to "piggy back" specific genes into the target population (Beard et al. 1993). It has already been possible to introduce genes expressing antiparasite proteins into a symbiont of *Rhodnius prolixus*, the vector of Chagas disease (Durvasula et al. 1997). The use of cytoplasmic incompatibility to transport genetically manipulated, maternally inherited organelles or organisms will certainly require detailed studies of population dynamics and genetics before it can be used in insect control.

6.5 FIELD TRIALS

Field trials have been carried out with all of the techniques described above with varying results. The move from the laboratory to the field often revealed predictable difficulties, but occasionally new problems arose especially in the area of laboratory adaptation. In many cases the technique could not be adequately tested because of poor field quality of the flies. In the early days there was little attention paid to the quality of the laboratory material that was released into the field. It is not surprising that many field trials gave very disappointing results as the overall poor quality of the insect masked any beneficial effect that the was being exerted by the genetic technique being tested.

Two sets of field trials will be discussed in detail as they represent the most expansive tests carried out; unfortunately neither led to the establishment of operational programmes for reasons that were different for the two sets of trials but were unrelated to the techniques being evaluated. In other species, e.g., the house fly, quite extensive field trials have also been carried out (Wagoner et al. 1973; McDonald et al. 1983).

6.5.1 *Lucilia cuprina,* the Sheep Blowfly

The theoretical framework for the development of genetic methods for pest control was greatly stimulated by the extensive work carried out on this species in Australia. Very early on in the programme a decision was made to focus on the use of male-linked translocations as opposed to the use of the SIT and the concept of control as opposed to eradication was accepted. The programme was solidly based on the development of sheep blowfly genetics to a level where fine grained genetic analysis could be carried out. This species was probably the most intensively studied pest species from the genetic point of view and a marvelous collection of mutants and strains were assembled. During the course of the programme a series of informative field trials were carried out using compound chromosome strains and male-linked translocations.

The field trials carried out using compound chromosome strains in *L. cuprina* remain the only field experience with pest insects using this system. For almost 7 months about 1 million larvae carrying compound chromosomes were released into a 10 sq. kilometer valley. Genetic analysis in the year of the releases revealed that males from the released strain were mating with the wild females and that females from the released strain were ovipositing in sheep. At the end of the release

period, 90% of the field matings were between flies carrying compound chromosomes, and it was expected that the compound chromosome type would eliminate
the normal chromosome. However, the next year no compound chromosome individuals could be found in the field, indicating that the fitness of the released strain
was much inferior to that of the field strain (Foster et al. 1985b).

The field trials with male-linked translocations were much more successful.
Preliminary trials in Wee Jasper and Boorowa, N.S.W (Vogt et al. 1985) solved many
logistical problems associated with the rearing and release of these strains and the
significance of immigration was highlighted. During these trials the amount of
genetic load, i.e., sterility, present in the field population, was not translated into a
concomitant reduction in population size, suggesting that immigration of fertilized
females was taking place. To check the importance of immigration a third trial was
carried out on Flinders Island. For 39 weeks about 1.35 million males were released
each week from 15 sites on the island and a very high level of sterility (88%) was
induced in the population. The authors provided strong evidence that the high levels
of sterility had a major impact on the density levels both in the year of release and
in the following year. This field trial provides the most convincing evidence that
male-linked translocations used in the appropriate way can produce a real reduction
in pest population density with the added advantage that the genetic sterility persists
after the releases have been terminated (Foster 1986).

There were no more subsequent field trials of any genetic control technique for
the sheep blowfly. What is the legacy of this programme? The original decision not
to include the SIT as an option directed the programme to the development of more
sophisticated techniques requiring the development of complex genetic strains. This
required a considerable amount of basic genetics and the necessary time. It is also
true that key theoretical expectations proved to be inaccurate, leading to major
problems with strain construction, stability, and viability. The fact that the economics
of wool production and marketing went through a traumatic period during the
programme did nothing to help the continuation of the programme.

6.5.2 Mosquitoes

This group of insects, because of its major importance to human and animal
health, has been the target for the development of genetic control techniques. A
complete listing of all field trials including the species and techniques used is given
by Rai (1996), and Asman et al. (1981) in an earlier review summarized the status
of mosquito genetic control on a species by species basis. The SIT, cytoplasmic
incompatibility, hybrid sterility, male linked translocations, and autosomal translocations have all been tested in field trials. In contrast to many agricultural pests,
where the larval stage causes damage, mosquitoes are characterized by having as
the pestiferous stage the adult female responsible for disease transmission. This
means that female mosquitoes cannot be released and it requires the development
of a system to remove females before release (Seawright et al. 1978; Curtis 1978;
Robinson 1986). Other problems associated with mosquitoes are their fragility, short
adult lifespan, and relatively long aquatic larval stage, requiring that rearing, handling,

and release procedures be suitably adapted. Following the revival of interest in genetic control in the late 1960s, the first field trials using genetic control were carried out on mosquitoes. Laven (1967c, 1972), using both male-linked translocations and cytoplasmic incompatibility, demonstrated that these types of genetic phenomena could introduce sterility into natural populations of *Culex* mosquitoes. However, there is some discussion as to the effect that the attained sterility had on population density (Weidhaas and Seawright 1974). In some situations mosquitoes are very strongly regulated by density dependent factors, which can act as a buffer to any reduction in egg fertility. At about the same time a field trial was carried out with sterile hybrids generated by crossing cryptic species of the *Anopheles gambiae* complex (Davidson et al. 1970). The trial was a failure probably due to the fact that the hybrid males released were generated from a cross between two species and used against a third. While in the laboratory there was no premating isolation apparent, this was not so in the field.

The major attempt to develop genetic control techniques for mosquitoes was undertaken by a Unit sponsored by the World Health Organisation and the Indian Council of Medical Research in Delhi during the early 1970s. The objective of the Unit was "to determine the operational feasibility of genetic control techniques for the control or eradication of mosquito vectors, and the diseases they transmit," and the following vectors were targeted: *Culex pipiens fatigans*, *A. aegypti*, and *Anopheles stephensi* (Ramachandra Rao 1974). Extensive work was carried out on many genetic systems including sterile males, cytoplasmic incompatibility, meiotic drive, and male-linked and autosomal translocations, and attempts were made to combine several of these approaches (see special issue of the *Journal of Communicable Diseases* 1974). Extensive field trials of several of these systems following many years of development were abruptly halted by a very aggressive press campaign that basically accused the Unit of developing biological warfare strategies (*Nature* 1975) by carrying out extensive work on *A. aegypti*, which then was not a major disease vector in India. This species is the vector of yellow fever. In fact, far more research work and field experiments were carried out with *C.p. fatigans*, and work on *An. stephensi* was just beginning. The advantage of using *Aedes* and *Culex* initially was that there were naturally occurring genetic systems that could quickly be evaluated as to their efficacy. The press campaign resulted directly in the WHO withdrawing from the project, which together with the impossible position of the Indian scientists led to the closing of the Unit. The termination of this programme created the perception that genetic control attempts of mosquitoes had been tested and shown to be ineffective, and it led to the cessation of most work in this area although recently the interest in the use of SIT as a component of vector control in urban situations has been revived (*Nature* 1997).

Current approaches to malaria control are aimed at population replacement with refractory strains generated by transgenic technology. The lack of any proven mechanism to transport a gene construct into a population is a serious obstacle to this approach and will be much more difficult to achieve than the development in the laboratory of transgenic refractory mosquitoes. Regulatory restrictions will also be considerable when fertile transgenic mosquitoes are required to be released and initial feasibility studies will probably have to be done with sterilized individuals.

Field trials are the key factor for success. If they succeed, the subsequent operational programmes are also generally successful. If, however, they fail, the technology is immediately written off and future development is severely curtailed. The scale of the field trial has to be such that the potential of a technique can be effectively evaluated. Small, artificial "field" situations where in fact laboratory experiments are simply performed outside are not convincing and have little predictive value.

6.6 OPERATIONAL PROGRAMMES

Operational programmes have an area-wide approach, an economic perspective, an application character, and a defined end point, which does not always have to be eradication. They are also characterized by having a constituency that has a vested interest in the programme as the constituency members often pay for the programme. Within the field of genetic control the only technology that has achieved this status so far is the SIT. This is probably because it is the simplest of the technologies proposed and only requires the transfer of irradiated sperm from the released male to the wild female to be successful. Major international SIT programmes for two key agricultural pests have not only made a tremendous impact on agricultural production and trade in the target areas but have led to the export of the technology to other areas and even to the export of sterile flies. (The transport of sterile flies from a major production source to another area where they are needed opens the door to the commercial application of this technique). Programmes for two key pests, the Mediterranean fruit fly, *Ceratitis capitata*, and the screwworm, *Cochliomyia hominivorax,* are almost identical in principle and very similar in implementation, but they deal with two quite different pest situations. The medfly is predominantly a quarantine pest, although in many areas it does cause severe damage, whereas the screwworm causes huge losses through direct damage. However, for both species the goal of the programme was the same, eradication of the insect from the targeted area. Recently, Krafsur (1998 in press) has cogently argued the case for area-wide SIT and effectively rebutted the criticisms that are often leveled at this technique.

6.6.1 New World Screwworm, *Cochliomyia hominivorax*

The success of the SIT for the control and subsequent eradication of the screwworm from large areas of North and Central America has been well documented (Graham 1985). The programme began in the late 1950s in the southern states of America and the last endemic case of screwworm in the U.S. was recorded in 1982. The final goal of the programme is to eradicate the pest from Central America and prevent reinvasion by maintaining a sterile fly release barrier at the Isthmus of Panama. The programme is well on the way to achieving these goals with eradication having now been achieved in Mexico, Guatemala, Belize, El Salvador, and Honduras (Wyss 1998 in press). There are no technical reasons why this programme should not reach the goals that were set and in so doing provide a sustainable and environmentally acceptable solution to the problem of screwworm in Central America.

The programme has not been without its problems and its critics. In 1972–1976 there was an outbreak that has never been fully explained, although poor programme implementation was the most likely cause coupled with reduced surveillance by farmers (Krafsur 1985). This outbreak was seized upon by critics of the approach in an attempt to decry the effectiveness of the technique itself (Richardson et al. 1982). It was suggested that the target population had developed premating barriers and that in fact there were incipient speciation processes operating. These suggestions and the data on which they were based were taken seriously (LaChance et al. 1982), but any evidence for the development of, or selection for, a new mating type in the field could not be identified. The fact that the programme has subsequently demonstrated its effectiveness throughout Central America is an eloquent demonstration of the power of the technique. It has also been suggested that climate was the major factor responsible for the disappearance of the pest (Readshaw 1986), but this argument has been adequately dealt with (Krafsur et al. 1986) and the fact that many Central American countries have subsequently declared eradication would suggest that something more than weather is influencing screwworm numbers.

The accidental introduction of the screwworm into Libya in March 1988 represented a major threat to wildlife and agricultural production in North Africa and it was decided to try to eradicate the outbreak using sterile males. Sterile pupae were transported from the Mexico–U.S. Commission production plant at Tuxtla Gutierrez in Mexico, emerged in Libya, and released over a treatment area of 40,000 km². Between 1990 and 1992, 1300 million sterile insects were shipped from Mexico to Libya for release and the last case of screwworm was reported in April 1991 (Lindquist et al. 1992; Vargas-Teran et al. 1994; FAO 1992). Here again the suggestion was made that weather was the factor regulating and eventually eradicating the population, but a thorough analysis of the biological and climatological data indicated the key role that the sterile insects played in population suppression and eventual eradication (Krafsur and Lindquist 1996). It would have been an abdication of responsibility to have done nothing and simply hoped that the infestation would die a natural death.

The success of screwworm SIT remains the paradigm for genetic control and its level of success has been approached for other programmes but never bettered. The programme is ongoing and the goal remains the establishment of a barrier zone at the Isthmus of Panama. There is also now a programme now being formulated to eradicate this pest from the island of Jamaica using flies from the Tuxtla Gutierrez facility (Grant et al. 1998 in press).

6.6.2 Mediterranean Fruit Fly, *Ceratitis capitata*

Fruit flies and especially the medfly have been the target for many genetic control techniques, especially the SIT. Fruit flies are of great economic importance mainly from a quarantine aspect as they prevent free trade in many agricultural commodities. The Mediterranen fruit fly, medfly, is a particularly notorious quarantine pest because of the wide range of pests that it attacks (Liquido et al. 1991). The medfly is an Old World species with its ancestral home probably being East Africa, but it has been spread over most of the world now as a consequence of trade and man's activities. It has therefore the status of an introduced pest in most of the areas where it is of economic importance. The SIT, because it can lead to eradication on an area-wide basis, is the ideal tool to deal with a quarantine pest of this type and the concept of "fly free" areas (Malavasi et al. 1994) out of which agricultural produce can be exported has been the key factor pushing the development of this technology. The eradication of the medfly from Chile (Lobos and Machuca 1998, in press) using the SIT as an indispensable component, has opened up a U.S.$ 400 million trade for the country.

The largest programme to control and eradicate this pest was initiated in Mexico in the 1970s. The aim was to prevent the invasion of this pest, which had become established in Guatemala, into Mexico. The presence of the medfly in Mexico would have threatened its multimillion dollar agricultural export trade with the U.S. The success of the programme in achieving its original goal is well documented (Orozco et al. 1994) and the programme goals have now been extended to included large parts of Guatemala (Villasenor et al. 1998 in press). The technology and philosophy developed for this programme have found supporters in many other parts of the world and for several other species of fruit flies, e.g., the melon fly, *Bactrocera cucurbitae* (Ito and Kakinohana 1995), the Queensland fruit fly, *B. tryoni* (Fisher 1994), the Oriental fruit fly, *B. dorsalis* (Shiga 1989), and *Anastrepha* species. The use of this technology for three species of *Anastrepha* is now the major component of a very large eradication project in Mexico (Rull et al. 1996).

The repeated introductions of medfly into California over the past 10 years (Dowell and Penrose 1995) have required emergency actions costing millions of U.S. dollars. The public opposition to aerial bait-spraying to eradicate these out-breaks has encouraged the authorities in the use of the SIT. Initially this was used in a reactive mode following the detection of outbreaks, and sterile flies were purchased where possible and released over the designated area. This emergency type response to a recurring problem proved to be an ineffective way of dealing with the problem. It was difficult to operate and to budget so it was decided to initiate a prophylactic approach in which sterile medflies are released over high-risk areas on a continuous basis. The programme has been running in this way from 1996 with no major outbreaks of medfly occurring (Dowell 1998 in press).

Following on from the success of medfly SIT in Mexico, other countries where this pest is a serious permanent problem have recently embraced this effective pest control methodology, e.g., Argentina (De Longo et al. 1998 in press), Portugal (Pereira et al. 1998 in press), Israel (Gomes et al. 1998 in press) and South Africa (Barnes and Eyles 1998 in press). There is no doubt that the use of SIT for fruit fly control has been and will continue to provide a viable option for the area-wide control and suppression of this important group of agricultural pests.

6.6.3 Other Operational Programmes of Note

(1) The onion fly (*Delia antiqua*) is the sole insect pest of onions in temperate regions of the world. A small commercial SIT programme started in 1981 is currently operating in the Netherlands for the control of this pest. This is probably the only truly commercially run programme of its kind in the world (Loosjes 1998 in press). About 400 million flies are produced annually and are used for the control of the pest on some 2600 ha of onion, representing about one sixth of the Dutch onion

crop. The programme is technically very successful but suffers from poor grower uptake and in some way the selfish behaviour of a minority of growers who try to benefit from sterile flies released on their neighbours fields. This problem illustrates the need for an area-wide philosophy when implementing this sort of programme and the need that all potential beneficiaries participate in the programme. The cost of the programme to the farmer was initially less than chemical control; this created some suspicion in the mind of the farmers, and when the price was increased so did farmer uptake. The programme has been operating since 1981 and there is no technical reason why it could not be expanded to cover the whole of the Dutch onion crop given the right political and social support.

(2) Tsetse flies of the Genus *Glossina* represent a major threat to agricultural development in sub-Saharan agriculture as they transmit protozoan parasites of the Genus Trypanosoma, which cause a debilitating sickness in livestock (nagana) and sleeping sickness in humans. For an extensive recent review of tsetse control see Rogers et al. (1994). Tsetse flies of both sexes are obligate blood feeders, and are larviparous with females, producing a single third instar larvae every 9 to 10 days. Because of this unusual mode of reproduction natural populations of tsetse are usually at a low density and are an ideal target for any genetic control technique that induces sterility in the population (Knipling 1979). As already indicated the first genetic control attempts of any insect were conducted with this species in the 1940s (Vanderplank 1944, 1947, 1948) where hybrid sterility between the three subspecies of *G. morsitans* group was used. Over 100,000 field collected *G. m. centralis* pupae were released into an isolated population of *G. swynnertoni* over a 7 month period. The sterility generated led to the replacement of the latter species by the former. Due to the arid conditions the *G. m. centralis* population rapidly disappeared and the area became tsetse free. The success of this approach before any significant mass-rearing technology for this group of insects had been developed indicated their susceptibility to any form of applied genetic load. When *in vitro* rearing for tsetse was developed (Feldmann 1994), this opened the way to efficient mass rearing and therefore the use of the SIT. Large-scale field trials using the release of sterilized males have been very successfully carried out in Nigeria and Burkino Faso (Offori 1993). However, in both cases reinvasion occurred when the programmes were terminated and the tsetse free areas were recolonized.

Recently the same technique has been used with stunning success to eradicate *G. austeni* from Unguja Island, Zululand, Republic of Tanzania. The aerial releases of sterile males followed an effective presuppression of the wild population with animal insecticidal pour-ons. The island has now been declared tsetse free and is free to develop its agriculture without the threat of trypanosomosis (Msangi et al. 1998 in press).

6.7 CONCLUDING REMARKS

The potential of genetic control for insect pests has yet to be fully exploited in terms of operational feasibility, despite the theories having been with us for about 30 years. There are many reasons for this, among the most important being the size with which these programme have to be implemented. This tends to alarm both scientists and administrators, but programmes of a limited size will not be successful. There have also been some spectacular failures of genetic control that have not helped promote the technology. Academia generally takes a dim view of these sorts of approaches as they cannot adequately be experimented with once they reach the stage of application. In this mode the programme simply has to implemented and there is very little room for diversion. This does not mean that the programmes are unscientific or are not carried out within a strict scientific discipline. It remains a fact, however, that this type of big science fills many scientists with a feeling of unease even though most of these programmes are industry driven and partially financed and consequently do not divert money away from other more research-oriented approaches.

Commercialization is another aspect that is difficult to integrate with genetic control. The development of commercially based genetic control, which provides a product that competes successfully on the open market with conventional control will have to be the way if genetic control wants to fully capitalize on its potential.

ACKNOWLEDGMENTS

I thank Dr. J. Hendrichs for many discussions on this subject and for reviewing the text.

REFERENCES

Anon. California Dept. of Food and Agriculture. Report to the Legislature: Medierranean Fruit Fly Preventative Release Program. March 1996.

Asman, S.M., P.T. McDonald, and T. Prout. Field Studies of Genetic Control Systems for Mosquitoes. *Annual Review of Entomology* 26, 289–318,1981.

Auerbach, C. and J.M. Robson. Experiments on the Action of Mustard Gas in Drosophila. Production of Sterility and Mutation. *Report to the Ministry of Supply*, W 3979, 1942.

Barnes, B. and D.K. Eyles. Feasibility of Eradicating *Ceratitis* spp. Fruit Flies from the Western Cape of South Africa by the Sterile Insect Technique. In *Proceedings of FAO/IAEA International Conference on Area-Wide Control of Insect Pests Integrating the Sterile Insect and Related Nuclear and other Techniques.* In Press 1998.

Barr, A.R. Cytoplasmic Incompatibility in Natural Populations of a Mosquito, *Culex pipiens*, L. *Nature* 283, 71–72, 1980.

Bauer, H. Die Kinetische Organisation der Lepidopteran-Chromosomen. *Chromosoma* 22, 101–125, 1967.

Beard, C.B., S.L. O'Neill, R.B. Tesh, F.F. Richards, and S. Aksoy. Modification of Arthropod Vector Competence via Symbiotic Bacteria. *Parasitology Today* 9, 129–133, 1993.

Boller, E.F. Behavioural aspects of mass-rearing insects. *Entomophaga* 17, 9–25, 1972.

Bourtzis, K. and S. O'Neill. Wolbachia Infections and Arthropod Reproduction. *BioScience* 48, 287–293, 1998

Boyle, L., S.L. O'Neill, H.M. Robertson, and T.L. Karr. Interspecific and Intraspecific Horizontal Transfer of *Wolbachia* in *Drosophila*. *Science* 260, 1796–1799, 1993.

Cayol, J.P. Changes in Sexual Behaviour and Some Life Traits of Tephritid Species Caused by Mass-Rearing Processes. In *Fruit Flies (Diptera: Tephritidae) Phylogeny and Evolution of Behaviour,* Ed. M. Aluha. In press.

Chambers, D.L. Quality Control in Mass Rearing. *Annual Review of Entomology* 22, 289–308, 1977.

Childress, D. Changing Population Structure Through the Use of Compound Chromosomes. *Genetics* 72, 183–186, 1972.

Curtis, C.F. Genetic Control of Insect Pests: Growth Industry or Lead Balloon? *Biological Journal of the Linnean Society* 26, 359–374, 1985.

Curtis, C.F. Genetic Sex Separation in *Anopheles arabiensis* and the Production of Sterile Hybrids. *Bulletin of the World Health Organisation* 45, 453–456, 1978.

Curtis, C.F. Induced Sterility in Insects. *Advances in Reproductive Physiology* 5, 120–165, 1971.

Curtis, C.F. The Mechanism of Hybrid Male Sterility from Crosses in the *Anopheles gambiae* and *Glossina morsitans* complexes. *Recent Developments in the Genetics of Insect Disease Vectors* Eds. Steiner, W.W.M., W. Tabashnik, K.S. Rai, and S. K. Narang. Stipes Publishing Company, New York, 1982, 290–312.

Curtis, C.F. A Possible Genetic Method for the Control of Insect Pests with Special Reference to Tsetse Flies (*Glossina* spp). *Bulletin of Entomological Research* 67, 509–523, 1968a.

Curtis, C.F. Possible Use of Translocations to Fix Desirable Genes in Insect Pest Populations. *Nature* 218, 368–369, 1968b.

Curtis, C.F. and T. Adak. Population Replacement in *Culex fatigans* by Means of Cytoplasmic Incompatibility. *Bulletin of the World Health Organisation* 51, 249–255, 1974.

Curtis, C.F. and P. Graves. Genetic Variation in the Ability of Insects to Transmit Filariae, Trypanosomes and Malarial Parasites. *Current Topics in Vector Research* 1, 31–62, 1984.

Curtis, C.F. and W.G. Hill. Theoretical and Practical Studies on a Possible Genetic Method for Tsetse Fly Control. *Isotopes and Radiation in Entomology*. IAEA, Vienna, 1968, 243–247.

Curtis, C.F. and W.G. Hill. Theoretical Studies on the Use of Translocations for the Control of Tsetse Flies and Other Disease Vectors. *Theoretical Population Biology* 2, 71–90, 1971.

Curtis, C.F. and A.S. Robinson. Computer Simulation of the Use of Double Translocations for Pest Control. *Genetics* 69, 97–113, 1971.

Davidson, G. *Genetic Control of Insect Pests*. Academic Press, London, 1974, 158pp.

Davidson, G. The Potential Use of Sterile Hybrid Males for the Eradication of Member Species of the *Anopheles gambiae* Complex. *Bulletin of the World Health Organisation* 40, 221–228, 1969.

Davidson, G., J.A. Odetoyinbo, B. Colussa, and J. Coz. A Field Attempt to Assess the Mating Competitiveness of Sterile Males Produced by Crossing 2 Member Species of the *Anopheles gambiae* Complex. *Bulletin of the World Health Organisation* 42, 55–67, 1970.

De Longo, O., A. Colombo, and P. Gomez-Riera. Use of Massive SIT for the Control of Medfly (*Ceratitis capitata* Weid.) Strain SEIB 6-96 in Mendoza, Argentina. *Proceedings of FAO/IAEA International Conference on Area-Wide Control of Insect Pests Integrating the Sterile Insect and Related Nuclear and other Techniques.* In Press 1998.

Dietz, K. The Effect of Immigration on Genetic Control. *Theoretical Population Biology* 9, 58–67, 1976.

Dowell, R.V. Mediterranean Fruit Fly Preventative Release Programme in Southern California. *Proceedings of FAO/IAEA International Conference on Area-Wide Control of Insect Pests Integrating the Sterile Insect and Related Nuclear and other Techniques.* In Press 1998.

Dowell, R.V. and R. Penrose. Mediterranean Fruit Fly Eradication in California 1994–1995. *The Mediterranean Fruit fly in California: Defining Critical Research.* College of Natural and Agricultural Sciences, University of California, Riverside, 161–185, 1995.

Durvasula, R., A. Gumbs, A. Panackal, O. Kruglov, S. Aksoy, R.B. Merryfield, F.F. Richards, and C.B. Beard. Prevention of Insect Borne Disease: An Approach Using Transgenic Symbiotic Bacteria. *Proceedings of the National Academy of Science* 94, 3274–3278, 1997.

Enkerlin W. and J. Mumford. Economic Evaluation of Three Alternative Methods for Control of Mediterranean Fruit Fly (Diptera:Tephritidae) in Israel, Palestinian Territories and Jordan. *Journal of Economic Entomology* 90, 1066–1072, 1997.

FAO. The New World Screwworm Eradication Programme, North Africa 1988–1992, 199.

Feldmann, U. Guidelines for the Rearing of Tsetse Flies Using the Membrane Feeding Technique. *Techniques of Insect Rearing for the Development of Integrated Pest and Vector Management Strategies,* Ed. J.P.R. Ochieng-Odero, ICIPE Science Press, 1994, 49–71.

Fisher, K. The Eradication of the Queensland Fruit Fly, *Bactrocera tryoni,* from Western Australia. *Fruit Flies and the Sterile Insect Technique,* Eds. Calkins, C.O. , W. Klassen, and P. Liedo, CRC Press, Boca Raton, FL, 1994, 237–246.

Fitz-Earle, M., D.G. Holm, and D.T. Suzuki. Genetic Control of Insect Populations: 1. Cage Studies of Chromosome Replacement by Compound Autosomes in *Drosophila melanogaster. Genetics* 74, 461–475, 1973.

Foster, G.G. *Evaluation of the Feasibility of Genetic Control of Sheep Blowfly.* CSIRO, Division of Entomology, 1986, 117pp.

Foster, G.G. and M. J. Whitten. The Development of Genetic Methods of Controlling the Australian Sheep Blowfly, *Lucilia cuprina. The Use of Genetics in Insect Control,* Eds. Pal, R., and M.J. Whitten, Elsevier/North Holland, Amsterdam, 1974.

Foster, G.G., W.G. Vogt, and T.L. Woodburn. Genetic Analysis of Field Trials of Sex-linked Translocation Strains for Genetic Control of the Australian Sheep Blowfly, *Lucilia cuprina. Australian Journal of Biological Science* 38, 275–293, 1985a.

Foster, G.G., R.H. Maddern, R.A. Helman, and E.M. Reed. Field Trial of a Compound Chromosome Strain for Genetic Control of the Sheep Blowfly, *Lucilia cuprina. Theoretical and Applied Genetics* 70, 13–21, 1985b.

Foster, G.G., M.J. Whitten, T. Prout, and R.Gill. Chromosome Rearrangements of the Control of Insect Pests. *Science* 176, 875–880, 1972.

Gomes, P.J., E. Ravins, and M. Bahdousheh. Area-Wide Control of Medfly in the Lower Jordan Rift Valley. *Proceedings of FAO/IAEA International Conference on Area-Wide Control of Insect Pests Integrating the Sterile Insect and Related Nuclear and other Techniques.* In Press 1998.

Gooding, R.H. Genetic Analysis of Hybrid Sterility in Crosses of the Tsetse Flies *Glossina palpalis palpalis* and *Glossina palpalis gambiensis* (Diptera: Glossinidae). *Canadian Journal of Zoology* 75, 1109–1117, 1997.

Gould, F., A. Martinez-Ramirez, A. Anderson, J. Ferre, F.J. Silva, and W.J. Moar. Broad-spectrum Resistance to *Bacillus thuringiensis* toxins in *Heliothis virescens. Proceedings of the National Academy of Sciences* 89, 7986–7990, 1992.

Graham, O.H. Symposium on Eradication of the Screwworm from the United States and Mexico. *Miscellaneous Publications of the Entomological Society of America* 62, 1985.

Grant, G.H., W. Snow, and A. Vargas. A Screwworm Eradication Programme for Jamaica and Other Caribbean Nations. In *Proceedings of FAO/IAEA International Conference on Area-Wide Control of Insect Pests Integrating the Sterile Insect and Related Nuclear and other Techniques.* In Press 1998.

Haldane, J.B.S. Sex Ratio and Unisexual Sterility in Animal Species. *Journal of Genetics* 12, 101–109, 1922.

Hibino, Y. and O. Iwahashi. Appearance of Wild Females Unreceptive to Sterilized Males on Okinawa Island in the Eradication Program of the Melon Fly, *Dacus cucurbitae* Coquillet (Diptera: Tephritidae). *Applied Entomology and Zoology* 26, 265–270, 1991.

Hogan, T.W. Physiological Differences Between Races of Telogryllus commodus (Walker) (Orthoptera:Gryllidae) Related to a Proposed Genetic Approach to Control. *Australian Journal of Zoology* 14, 245–251, 1966.

Holm, D.G. Compound Autosomes. *The Genetics and Biology of Drosophila* Vol. 1B. Eds. Ashburner, M. and E. Novitski. Academic Press, London, 1976, 529–562.

Hooper, G.H.S. Gamma Sterilization of the Mediterranean Fruit fly. *Sterility Principle for Insect Control or Eradication.* IAEA, Vienna, 1971, 87–95.

Howarth, F.G. Environmental Impacts of Classical Biological Control. *Annual Review of Entomology* 36, 485–509, 1991.

Hoy, M. and J.J. McKelvey, eds. *Genetics in Relation to Insect Management*, Rockefeller Foundation, New York, 1979, 179 pp.

Huettel, M.D. Monitoring the Quality of Laboratory-Reared Insects: A Biological and Behavioural Perspective. *Environmental Entomology* 5, 807–814, 1976.

IAEA. *Management of Insect Pests: Nuclear and Related Molecular and Genetic Techniques*, IAEA, Vienna, STI/PUB/909, 1993.

IAEA. *Modern Insect Control: Nuclear Techniques and Biotechnology*, IAEA, Vienna, STI/PUB/763, 1988.

IAEA. *Sterile Insect Technique and Radiation in Insect Control*, IAEA, Vienna, STI/PUB/595, 1982.

Ito, Y. and H. Kakinohana. Eradication of the Melon Fly (Diptera:Tephritidae) from the Ryukyu Archipelago with the Sterile Insect Technique: Possible Reasons for Success. *The Mediterranean Fruit Fly in California: Defining Critical Research.* College of Natural and Agricultural Sciences, University of California, Riverside, 1995, 215–231. *Journal of Communicable Diseases* 6, 1974.

Klassen, W., E.F. Knipling, and J.U. McGuire. The Potential for Insect-Population Suppression by Dominant Conditional Lethal Traits. *Annals of the Entomological Society of America* 63, 238–255, 1970.

Knipling, E.F. The Basic Principles of Insect Population Suppression and management. United States Department of Agriculture Handbook No. 512, 659 pp., 1979.

Knipling, E.F. Possibilities of Insect Control or Eradication Through the Use of Sexually Sterile Males. *Journal of Economic Entomology* 48, 459–469, 1955.

Knipling, E.F. Principles of Insect Parasitism Analysed from New Perspectives: Practical Implications for Regulating Insect Populations by Biological Means. United States Department Of Agriculture, *Agricultural Handbook* 693, 1992.

Knipling, E.F. Suppression of Pest Lepidoptera by Releasing Partially Sterile Males. A Theoretical Appraisal. *BioScience* 20, 465–470, 1970.

Knipling, E.F. Use of Insects for Their Own Destruction. *Journal of Economic Entomology* 53 415–420, 1960.

Knipling, E.F. and W. Klassen. Relative Efficiency of Various Genetic Mechanisms for Suppression of Insect Populations. USDA, ARS, *Technical Bulletin* No. 1533, 1976.

Kozhevnikov, B.T. Experimentally Produced Karyotypical Isolation. *Biological Zhurnal* 5, 741–752, 1938.

Krafsur, E.S. Screwworm Flies (Diptera: Calliphoridae): Analysis of Sterile Mating Frequencies and Covariates. *Bulletin of the Entomological Society of America* 44, 36–40, 1985.

Krafsur, E.S. Sterile Insect Technique for Suppressing and Eradicating Insect Populations: 55 Years and Counting. In press.

Krafsur, E.S. and D.A. Lindquist. Did the Sterile Insect Technique or Weather Eradicate Screwworms (Diptera:Calliphoridae) from Libya? *Journal of Medical Entomology* 33, 877–887, 1996.

Krafsur, E.S., H. Townson, G. Davidson, and C. Curtis. Screwworm Eradication *is* What it Seems. *Nature* 323, 495–496, 1986.

LaChance, L.E. Genetic Strategies Affecting the Success of the Sterile Insect Release Method. *Genetics in Relation to Insect Pest Management*, Eds. Hoy, M. A. and J.J. McKelvey, Rockefeller Foundation, New York, 1979, pp. 8–18.

LaChance, L.E. Induced Sterilty in Irradiated Diptera and Lepidoptera. Sperm Transfer and Dominant Lethal mutations. *Sterility Principle for Insect Control 1974*. IAEA-SM-186/41, 401–411, 1975.

LaChance, L.E. The Induction of Dominant Lethal Mutations in Insects by Ionizing Radiation and Chemicals, as Related to the Sterile Male Technique of Insect Control. *Genetics of Insect Vectors of Disease*, Eds. Wright, J.W. and R. Pal, Elsevier, Amsterdam, 1967, 617–650.

LaChance, L.E. and M.M. Crystal. Induction of Dominant Lethal Mutations in Insect Oocytes and Sperm by Gamma rays and an Alkylating Agent: Dose response and Joint Action Studies. *Genetics* 51, 699–708, 1965.

LaChance, L.E., A.C. Bartlett, R.A. Bram, R.J. Gagne, O.H. Graham, D.O. McInnis, C.J. Whitten, and J.A. Seawright. Mating Types in Screwworm Populations? *Science* 218, 1142–1145, 1982.

LaChance, L.E., M. Degrugillier, and A.P. Leverich. Cytogenetics of Inherited Sterility in Three Generations of the Large Milkweed Bug as Related to Holokinetic Chromosomes. *Chromosoma* 29, 20–41, 1970.

Laven, H.E. A Possible Model for Speciation by Cytoplasmic Isolation in the *Culex* pipiens complex. *Bulletin of the World Health Organisation* 37, 263–266, 1967a.

Laven, H.E. Speciation and Evolution in *Culex pipiens*. *Genetics of Insect Vectors of Disease*. Eds. Wright, J.W. and R. Pal, Elsevier, Amsterdam, 251–275, 1967b.

Laven, H.E. Eradication of *Culex pipiens fatigans* through Cytoplasmic Incompatibility. *Nature* 216, 383–384, 1967c.

Laven, H.E., J. Cousserans, and G. Guile. Eradicating Mosquitoes Using Translocations: A First Field Experiment. *Nature* 236, 4456–457, 1972.

Laven, H.E., E. Jost, H. Meyer, and R. Selinger. Semisterility for Insect Control. *Sterility Principle for Insect Control or Eradication*. IAEA-SM-138/16, 415–424, 1971.

Lindquist, D.A., M. Abusowa, and M.J.R. Hall. The New World Screwworm in Libya: A Review of its Introduction and Eradication. *Medical and Veterinary Entomology* 6, 2–8, 1992.

Liquido, N., L.E. Shinoda, and R.T. Cunningham, Host Plants of the Mediterranean Fruit Fly (Diptera:Tephritidae): An Annotated World Review. *Miscellaneous Publications of the Entomological Society of America* 77, 1–52, 1991.

Lobos, C. and J. Machuca. Eradication of Medfly from Chile and Joint Programme in Southern Peru. *Proceedings of FAO/IAEA International Conference on Area-Wide Control of Insect Pests Integrating the Sterile Insect and Related Nuclear and other Techniques*. In Press 1998.

Loosjes, T. The Sterile Insect Technique for Commercial Control of the Onion Fly. *Proceedings of FAO/IAEA International Conference on Area-Wide Control of Insect Pests Integrating the Sterile Insect and Related Nuclear and other Techniques*. In Press 1998.

Mackauer, M. Genetic Problems in the Production of Biological Control Agents. *Annual Review of Entomology* 21, 369–385, 1976.

Malavasi, A., G. Rohwer, and D.S. Campbell. Fruit Fly Free Areas: Strategies to Develop Them. *Fruit Flies and the Sterile Insect Technique,* Eds. Calkins, C.O. , W. Klassen, and P. Liedo. CRC Press, Boca Raton, FL, 1994, 165–180.

Mannion, C.M., J.E. Carpenter, and H.R. Gross. Integration of Inherited Sterility and a Parasitoid, *Archytas marmoratus* (Diptera: Tachinidae), for Managing *Helicoverpa zea* (Lepidoptera: Noctuidae): Acceptability and Suitability of Hosts. *Environmental Entomology* 24, 1679–1684, 1995.

McDonald, I.C., O.A. Johnson, C. Nickel, P. Everson, and D. Birkmeyer. House Fly (Diptera:Muscidae) Genetics: Field Studies with Males from a Male-Producing Strain. *Annals of the Entomological Society of America* 76, 333–338, 1983.

McInnis, D.O., D.R. Lance, and C.G. Jackson. Behavioural Resistance to the Sterile Insect technique by Mediterranean Fruit Fly (Diptera: Tephritidae) in Hawaii. *Annals Entomological Society of America* 89, 739–744, 1996.

Msangi, A.M., K.M. Saleh, N. Kiwia, W.A. Mussa, F. Mramba, K.G. Juma, V.A. Dyck, M.J.B. Vreysen, A.G. Parker, U. Feldmann, Z-R Zhu, and H. Pan. Success in Zanzibar: Eradication of Tsetse. *Proceedings of FAO/IAEA International Conference on Area-Wide Control of Insect Pests Integrating the Sterile Insect and Related Nuclear and other Techniques.* In Press 1998.

Muller, H. J. Artificial Transmutation of the Gene. *Science* 66, 84–88, 1927.

Muller, H. J. Our Load of Mutations. *American Journal of Human Genetics* 2, 111–176, 1950.

Nature. Oh, New Delhi; Oh, Geneva. 256, 355–357, 1975.

Nature. Consortium Aims to Revive Sterile-Mosquito Project. 389, 488, 1997.

North, D.T. Inherited Sterility in Lepidoptera. *Annual Review of Entomology* 20, 167–182, 1975.

North, D.T. and G.C. Holt. Population Control of Lepidoptera: The Genetic and Physiological Basis. *The Manitoba Entomologist* 4, 53–69,1970.

North, D.T. and G.C. Holt. Radiation Studies of Sperm Transfer in Relation to Competitiveness and Oviposition in the Cabbage Looper and Corn Earworm. *Application of Induced Sterility for Control of Lepidopterous Populations.* IAEA, Vienna, 1971, 87–97.

Ochieng-Odero, J.P.R. Does Adaptation Occur in Insect Rearing Systems, or Is It a Case of Selection, Acclimatization and Domestication? *Insect Science and Application* 15, 1–7, 1994.

Offori, E.D. Tsetse Sterile Insect Technique Programmes in Africa: Review and Analysis of Future Prospects. In *Management of Insect Pests: Nuclear and Related Molecular and Genetic Techniques* IAEA, 345–357, 1993.

Orozco, D., W.S. Enkerlin, and J. Reyes. The Moscamed Program: Practical Achievements and Contribution to Science. *Fruit Flies and the Sterile Insect Technique,* Eds. Calkins, C.O., W. Klassen, and P. Liedo. CRC Press, Boca Raton, FL, 1994, pp. 209–222.

Orr, H.A. Haldane's Rule. *Annual Review of Ecology and Systematics* 28, 195–218, 1997.

Pal, R. and M.J. Whitten. *The Use of Genetics in Insect Control,* Elsevier, North Holland, Amsterdam, 1974.

Palopoli, M.F. and C.-I. Wu. Genetics of Hybrid Male Sterility between Drosophila Sibling Species: A Complex Web of Epistasis is Revealed in Interspecific Studies. *Genetics* 138, 329–341, 1994.

Pereira, R., A. Barbosa, N. Silva, J. Caldeira, L. Dantas, and J. Pachero. Madeira-Med Programme, A Sterile Insect Technique Program for Control of the Mediterranean Fruit Fly in Madeira, Portugal. *Proceedings of FAO/IAEA International Conference on Area-Wide Control of Insect Pests Integrating the Sterile Insect and Related Nuclear and other Techniques.* In Press 1998.

Pettigrew, M.M. and S.L. O'Neill. Control of Vector-Borne Disease by Genetic Manipulation of Insect Populations: Technological Requirements and Research Priorities. *Australian Journal of Entomology* 36, 309–317, 1997.

Potts, W.H. Tsetse Hybrids. *Nature* 154, 606–607, 1944.

Prout, T. The Joint Effects of the Release of Sterile Males and Immigration of Fertilized Females on a Density Regulated Population. *Theoretical Population Biology* 13, 40–71, 1978.

Proverbs, M.D. Progress on the Use of Sexual Sterility for the Control of the Codling Moth, *Carpocapsa pomonella* (L.) (Lepidoptera:Olethreutidae). *Proceedings of the Entomological Society of Ontario* 92, 5–11, 1962.

Rai, K.S. Genetic Control of Vectors. *The Biology of Disease Vectors,* Eds. Beaty, B.J. and W.C. Marquardt, University Press of Colorado, Boulder, 1996, 564–574.

Ramachandra Rao, T. Research on Genetic Control of Mosquitoes in India: review of the Work of the WHO/ICMR Research Unit, New Delhi. *Journal of Communicable Diseases* 6, 57–72, 1974.

Rawlings, P. The Genetics of Hybrid Sterility between Subspecies of the Complex of *Glossina morsitans* Westwood (Diptera: Glossinidae). *Bulletin of Entomological Research* 75, 689–699, 1985.

Readshaw, J.L. Screwworm Eradication: A Grand Delusion. *Nature* 320, 407–410, 1986.

Richardson, R.H., J.R. Ellison, and W.W. Averhoff. Autocidal Control of Screwworms in North America. *Science* 215, 361–370, 1982.

Roberts, P.A. The Genetics of Chromosome Aberration. *The Genetics and Biology of Drosophila, Vol 1a,* Eds. Ashburner, M. and E. Novitski, Academic Press, London, 1976, 68–174.

Robinson, A.S. Genetic Sexing in *Anopheles stephensi* using Dieldrin Resistance. *Journal of the American Mosquito Control Association* 2, 93–95, 1986.

Robinson, A.S. Progress in the Use of Chromosomal Translocations for the Control of Insect Pests. *Biological Reviews* 51, 1–24, 1976.

Robinson, A.S. Sex Ratio Manipulation in Relation to Insect Control. *Annual Review of Genetics* 17, 191–214, 1983.

Robinson, A.S. and C.F. Curtis. Controlled Crosses and cage Experiments with a Translocation in *Drosophila. Genetica* 44, 129–137, 1973.

Rogers, D.J., C. Hendrickx, and J.H.W. Slingenbergh. Tsetse Flies and Their Control. *Review of the Scientific and Technical Office of the International Epizootics* 13, 1075–1124, 1994.

Rousset, F., D. Bouchon, B. Pintereau, P. Juchault, and M. Solignac. *Wolbachia* Endosymbionts Responsible for Various Alterations in Sexuality of Arthropods. *Proceedings of the Royal Society of London, Series B* 250, 91–98, 1992.

Rull, J.A., J.F. Reyes and W. Enkerlin. The Mexican National Fruit Fly Eradication Campaign: Largest Fruit Fly Industrial Complex in the World. *Fruit Fly Pests: A World Assessment of their Biology and Management,* Eds. McPheron, B.A. and G. Steck. St Lucie Press, Delray Beach, FL, 1996, 561–563.

Runner, G.A. Effect of Roentgen Rays on the Tobacco, or Cigarette Beetle and the Results of Experiments with a New Roentgen Tube. *Journal of Agricultural Research* 6, 383–388, 1916.

Sandler, L. and E. Novitski. Meiotic Drive as an Evolutionary Force. *American Naturalist* 91, 105–110, 1957.

Sankaranarayanan, K. and F.H. Sobels. Radiation Genetics. *The Genetics and Biology of Drosophila, Vol 1c.* Eds. Ashburner, M. and E. Novitski. Academic Press, London, 1976, 1089–1250.

Seawright, J.A., P.E. Kaiser, D.A. Dame, and C.S. Lofgren. Genetic Method for the Preferential Elimination of Females of *Anopheles albimanus. Science* 200, 1303–1314, 1978.

Serebrovskii, A. S. On the Possibility of a New Method for the Control of Insect Pests. *Zoologicheskii Zhurnal* 19, 618–630, 1940.

Shiga, M. Current Programme in Japan. *Fruit Flies: Their Biology, Natural Enemies and Control,* Eds. Robinson, A.S. and G.H. Hooper, Elsevier, Amsterdam, 1989, 375–386.

Smith, C.N., G.C. LaBrecque, and A.B. Borkovec. Insect Chemosterilants. *Annual Review of Entomology* 9, 269–284, 1964.

Smith, R.H. Induced Conditional Lethal Mutations for the Control of Insect Populations, in *Sterility Principle for Insect Control or Eradication.* IAEA, 1971.

Smith, R.H. and R.C. von Borstel. Genetic Control of Insect Populations. *Science* 178, 1164–1174, 1972.

Steffans, R.J. Methodology for Translocation Production and Stability of Translocations in the Mediterranean Fruit Fly, *Ceratitis capitata* Wied. (Dipt., Tephritidae). *Zeitschrift fur Angewandte Entomologie* 95, 181–188, 1983.

Steiner, W.W.M., W. Tabashnik, K.S. Rai, and S. K. Narang. *Recent Developments in the Genetics of Insect Disease Vectors.* Stipes Publishing Company, New York, 1982.

Stouthamer, R., J.A.J. Breeuwer, R.F. Luck, and J.H. Werren. Molecular Identification of Parthenogenesis Associated Micro-organisms. *Nature* 361, 66–68, 1993.

Tabashnik, B.E. Evolution of Resistance to *Bacillus thuringiensis. Annual Review of Entomology* 39, 47–79, 1994.

Tabashnik, B.E., Y-B. Liu, T. Malvar, D.G. Heckel, L. Masson, V. Ballester, F. Granero, J.L. Mensua, and J. Ferre. Global Variation in the Genetic and Biochemical Basis of Diamondback Moth Resistance to *Bacillus thuringiensis. Proceedings of the National Academy of Science* 94, 12780–12785, 1997.

Turelli, M. and A.A. Hoffmann. Rapid Spread of an Inherited Incompatibility Factor in California *Drosophila. Nature* 353, 440–442, 1991.

Vanderplank, F.L. Experiments in Cross Breeding Tsetse Flies (Glossina Species). *Annals of Tropical Medical Parasitology* 42, 131–152, 1948.

Vanderplank, F.L. Experiments in the Hybridisation of Tsetse Flies (*Glossina,* Diptera) and the Possibility of a New Method of Control. *Transactions of the Royal Entomological Society, London* 98, 1–18, 1947

Vanderplank, F.L. Hybridisation between *Glossina* species and Suggested New Method for Control of Certain Species of Tsetse. *Nature* 154, 607–608, 1944.

Vargas-Tehran, M., B.S. Hursey, and E.P. Cunningham. Eradication of the Screwworm from Libya Using the Sterile Insect Technique. *Parasitology Today* 10, 119–122, 1994.

Villasenor, A., J. Carillo, J. Zavala, J. Stewart, C. Lira, and J. Reyes. Current Progress in the Medfly Program Mexico-Guatemala. In *Proceedings of FAO/IAEA International Conference on Area-Wide Control of Insect Pests Integrating the Sterile Insect and Related Nuclear and other Techniques.* In Press 1998.

Vogt, W.G., T.L. Woodburn, and G.G. Foster. Ecological Analysis of Field Trials Conducted to Assess the Potential of Sex-linked Translocation Strains for Genetic Control of the Sheep Blowfly, *Lucilia cuprina* (Wiedemann). *Australian Journal of Biological Science* 38, 259–273, 1985.

Wagoner, D.E., P.B.Morgan, G.C. LaBrecque, and A.O. Johnson. Genetic Manipulation Used Against a Field Population of House Flies, 1. Males Bearing a Heterozygous Translocation. *Environmental Entomology* 2, 128–134, 1973.

Weidhaas, D.E. and J.A. Seawright. Comments on the Article "Genetic Control of Mosquitoes" (Laven 1974). Proceedings of *Tall Timbers Conference on Ecological Animal Control by Habitat Management* 5, 211–220, 1974.

Weidhaas, D.E., S.G. Breeland, C.S. Lofgren, D.A. Dame, and R. Kaiser. Release of Chemos-terilized Males for the Control of *Anopheles albimanus* in El Salvador. IV. Dynamics of the Test Population. *American Journal of Tropical Medicine and Hygiene* 23, 298–308, 1974.

Weidhaas, D.E., G.C. LaBrecque, C.S. Lofgren, and C.H. Schmidt. Insect Sterility in Insect Population Dynamics Research. *Bulletin of the World Health Organisation* 47, 309–315, 1972.

Werren, J.H. Biology of *Wolbachia*. *Annual Review of Entomology* 42, 587–609, 1997.

Whitten, M.J. Genetics of Pests in Their Management, in *Concepts of Pest Management,* Eds. Rabb, R.L. and F.E. Guthrie, N.C. State University, Raleigh, 1970, 119–135.

Whitten, M.J. Insect Control by Genetic Manipulation of Natural Populations. *Science* 172, 682–684, 1971a.

Whitten, M.J. Use of Chromosome Rearrangements for Mosquito Control. *Sterility Principle for Insect Control or Eradication.* IAEA, Vienna, 399–413, 1971b.

Whitten, M.J. and G.G. Foster. Genetical Methods of Pest Control. *Annual Review of Entomology* 20, 461–476, 1975.

Wu, C.-I. and A.W. Davis. Evolution of Postmating Reproductive Isolation: The Composite Nature of Haldane's Rule and its Genetic Basis. *American Naturalist* 142, 187–212, 1993.

Wyss, J.H. Screwworm Eradication in the Americas: Overview. *Proceedings of FAO/IAEA International Conference on Area-Wide Control of Insect Pests Integrating the Sterile Insect and Related Nuclear and other Techniques.* In Press 1998.

Yen, J.H. and A.R. Barr. New Hypothesis of the Cause of Cytoplasmic Incompatibility in *Culex pipiens. Nature* 232, 657–658, 1971.

CHAPTER 7

Plant Resistance to Insects

C. Michael Smith

CONTENTS

1-56670-479-0/00/$0.00+$.50
© 2000 by CRC Press LLC

7.1 INTRODUCTION

7.1.1 Terminology

Plants with constitutive insect resistance possess genetically inherited qualities that result in a plant of one cultivar being less damaged than a susceptible plant lacking these qualities (Painter, 1951). Plant resistance to insects is a relative property, based on the comparative reaction of resistant and susceptible plants, grown under similar conditions, to the pest insect. Pseudoresistance can occur in susceptible plants due to fluctuations in plant age, moisture content, insect population density, temperature, photoperiod, soil chemistry, or soil moisture. Associational resistance occurs when a normally susceptible plant is grown in association with a resistant plant and derives protection from insect predation (Alfaro, 1995; Ampong-Nyarko et al., 1994; Letourneau, 1986). A unique type of associational resistance results from insects feeding on plants infected by *Neotyphodium* (formerly *Acremonium*) endophytes, which produce alkaloids that have negative effects on insect feeding and growth (Breen, 1994; Clement et al., 1994).

Induced insect resistance may also occur when a plant's defensive system is stimulated by external physical or chemical stimuli (Kogan and Paxton, 1983), eliciting the accumulation of increased levels of endogenous plant metabolites (Baldwin, 1994). Induced resistance to insects exists over a broad range of plant taxa, including Brassicaceae (Agrawal, 1998; Bodnaryk and Rymerson, 1994; Palaniswamy and Lamb, 1993; Siemens and Mitchellolds, 1996), Chenopodiaceae (Mutikainen et al., 1996), Compositae (Roseland and Grosz, 1997), Graminae (Bentur and Kalode, 1996; Gianoli and Niemeyer, 1997), Leguminoseae (Wheeler and Slansky, 1991), Malvaceae (McAuslane et al., 1997; Thaler and Karban, 1997), Pinaceae (Alfaro, 1995; Jung et al., 1994), Salicaceae (Zvereva et al., 1997), and Solanaceae (Bronner et al., 1991, Stout and Duffey, 1996; Westphal et al., 1991).

7.1.2 History

Pest insect-resistant plants have been recognized for many years as a sound approach to crop protection in the U.S. Two early examples of resistant cultivars are wheat cultivars found to have resistance to the Hessian fly, *Mayetiola destructor* (Say), in New York in 1788 and apple cultivars that were resistant to the woolly apple aphid, *Eriosoma lanigerum* (Hausmann) in the early 1900s (Painter, 1951). The most famous example of the successful use of plant resistance to insects was when the distinguished 19th century entomologist Charles Valentine Riley imported American grape rootstocks to France in the late 1800s to save the French wine industry from destruction by the grape phylloxera, *Phylloxera vitifoliae* (Fitch).

Today hundreds of insect-resistant crop cultivars are grown globally (Smith, 1989). Many of these are major cereal grain food crops developed by cooperative research efforts between plant breeders and entomologists at International Agricultural Research Centers, Provincial or State Agricultural Experiment Stations, and national Department of Agriculture laboratories. These efforts have led to a detailed understanding of the type and genetic nature of insect resistance in several crop plants, and have significantly improved the major food production areas of the world during the past 40 years (Maxwell and Jennings, 1980; Smith, 1989).

In one of the earliest comprehensive reviews of plant resistance to insects, Snelling (1941) identified over 150 publications dealing with plant resistance to insects in the U.S. from 1931 until 1940. Since then numerous reviews have chronicled the progress and accomplishments of scientists conducting research on plant resistance to insects (Beck, 1965; Green and Hedin, 1986; Harris, 1980; Hedin, 1978, 1983; Maxwell et al., 1972; Painter, 1958).

The first book on the subject of plant resistance to insects, *Plant Resistance to Insect Pests*, was written by Reginald Painter (1951), who is considered the founder of organized plant resistance to insects research in the U.S.. In Russia, Chesnokov (1953) published the book *Methods of Investigating Plant Resistance to Pests*, the first comprehensive review of techniques to evaluate plants for resistance to insects.

In recent years intensified research in plant resistance has led to the publication of several additional texts on the subject. These include Lara (1979), *Principios de Resistancia de Plantas a Insectos*; Maxwell and Jennings (1980), *Breeding Plants Resistant to Insects*; Panda (1979), *Principles of Host-Plant Resistance to Insects*; Panda and Kush (1995), *Host-Plant Resistance to Insects*; Russell (1978), *Plant Breeding for Pest and Disease Resistance*; Smith (1989), *Plant Resistance to Insects — A Fundamental Approach*; and Smith et al. (1994), *Techniques for Evaluating Insect Resistance in Crop Plants*.

7.1.3 Economic Benefits

Insect-resistant cultivars provide a substantial economic return on economic investment. Insect-resistant cultivars of alfalfa, corn, and wheat produced in the midwestern U.S. during the 1960s provided a 300% return on every dollar invested in research (Luginbill, 1969). Wheat cultivars developed with resistance to the Hessian fly provided a 120-fold greater return on investment than pesticides (Painter, 1968). More recently, Hessian fly resistance developed in Moroccan bread wheats provided a 9:1 return on investment of research (Azzam et al., 1997).

The current value of insect-resistant cultivars, due to reduced insect damage and reduced costs of insecticide applications, varies with economic conditions. Teetes et al. (1986) estimated the annual value of grain sorghum cultivars resistant to the greenbug, *Schizaphis graminum* Rondani, in Texas to be approximately $30 million. The estimated value of Kansas grain sorghum cultivars with resistance to the greenbug or the chinch bug, *Blissus leucopterous* (Say), is $45 million per year (Anonymous, 1995). The economic value of genetic resistance in wheat to all major worldwide arthropod pests amounts to just over $250 million per year (Smith et al., 1998). The rice cultivar, IR36, which contains multiple insect resistance, has provided $1 billion of additional annual income to rice producers and processors in South and Southeast Asia (Khush and Brar, 1991).

Cultivars of corn, cotton, and potatoes containing the insect-specific toxin gene from the bacteria *Bacillus thuringiensis* (Bt) have begun to be produced in U. S. agriculture, and will be introduced into Asian crop production before the end of the century. The value of Bt cotton production in the U. S. state of Mississippi alone is estimated to be $400 million per year, as a result of reduced applications of conventional insecticides (Dr. Johnnie Jenkins, personal communication).

The effects of insect-resistant cultivars are cumulative. The longer insect-resistant plant genes are employed and effective, the greater the benefits of their use. Tenfold reductions in pest insect populations and 50% increases in crop yield are not unusual where insect-resistant cultivars have been introduced and maintained in several rice production systems in South and Southeast Asia (Panda, 1979; Waibel, 1987; IRRI, 1984).

7.1.4 Environmental Benefits

Schalk and Ratcliffe (1976) estimated that production of insect-resistant cultivars eliminated the annual application of over 300,000 tons of insecticides in the U.S. If this trend has remained constant since then, insect-resistant cultivars have helped avoid the application of more than 6 million tons of insecticides. Improved cultivars of cotton, sorghum, corn, and vegetables have contributed greatly to this statistic (Cuthbert and Jones, 1978; Cuthbert and Fery, 1979; George and Wilson, 1983; Jones et al., 1986; Teetes et al., 1986; Wiseman et al., 1975).

7.2 CATEGORIES OF RESISTANCE

Three categories or modalities of plant resistance to insects were first described by Painter (1951), to classify plant-pest insect interactions. They include antibiosis, antixenosis and tolerance. Antibiosis and antixenosis resistance categories describe the reaction of an insect to a plant, while tolerance resistance describes the reaction of a plant to insect infestation and damage.

7.2.1 Antibiosis and Antixenosis

Antibiosis describes a plant trait that adversely affects the biology of an insect or mite when the plant is used for food. Antixenosis, known previously as nonpreference, describes a plant trait that limits a plant from serving as a host to an insect, resulting in an adverse affect on the behavior of the insect when it feeds or oviposits on a plant or uses it for shelter.

Antibiotic and antixenotic effects manifested in insects may occur because of either the presence of detrimental chemical and morphological plant factors. Morphological factors include trichomes, both glandular (Hawthorne et al., 1992; Heinz and Zalom, 1995; Kreitner and Sorensen, 1979; Nihoul, 1994; Steffens and Walters, 1991; Yoshida et al., 1995) and nonglandular (Baur et al., 1991; Elden, 1997; Gannon and Bach, 1996; Oghiakhe et al., 1995; Palaniswamy and Bodnaryk, 1994; Park et al., 1994; Quiring et al., 1992; Ramalho et al., 1984), surface waxes (Bodnaryk, 1992; Bergman et al., 1991; Stoner, 1990; Yang et al., 1993), tightly packed vascular bundles (Brewer et al., 1986; Cohen et al., 1996; Mutikainen et al., 1996), or high fiber content (Beeghly et al., 1997; Bergvinson, 1994; Davis et al., 1995).

Detrimental phytochemical factors include toxins (Barbour and Kennedy, 1991; Barria et al., 1992; Barry et al., 1994; Reichardt et al., 1991), feeding and oviposition deterrents (Hattori et al., 1992; Huang and Renwick, 1993; Schoonhoven et al., 1992), repellents (Snyder et al., 1993), high concentrations of digestibility reducing substances such as lignin and silica (Ukwungwu and Obebiyi, 1985; Rojanaridpiched et al., 1984; Muller et al., 1960; Blum, 1968). Conversely, resistance may also be due the absence of essential nutrients (Cole, 1997; Febvay et al., 1988).

The ingestion of allelochemicals from resistant plants by insects does not necessarily result in a decreased activity of insect detoxication enzymes and associated enhanced insect mortality. In some cases, ingestion of resistant plant allelochemicals synergizes toxicity (Rose et al., 1988). However, in some cases allelochemicals do not synergize toxicity (Kennedy, 1984). In other cases, allelochemicals from insect-resistant plants have no effect on insecticidal toxicity (Kennedy and Farrar, 1987).

Determining whether the antibiosis or antixenosis (or both) categories of resistance are involved in insect resistance depends on the particular point in the sequence of insect host finding, location, and acceptance viewed by the researcher (Visser, 1983). Antixenotic resistance functions by altering the olfactory (Dickens et al., 1993; Lapis and Borden, 1993; Seifelnasr, 1991), visual (Fiori and Craig, 1987; Green et al., 1994; Shifriss, 1981), tactile (Mitchell et al., 1973), and gustatory (Roessingh et al., 1992) plant cues used by an insect to successfully locate a host plant, feed on it and/or use it as a habitat for reproduction. Antibiosis resistance works by causing insect mortality or delayed development after contact with or ingestion of plant tissues containing the morphological or allelochemical defenses described previously.

7.2.2 Tolerance

Tolerance describes properties that enable a resistant plant to yield more biomass than a susceptible plant, due to the ability to withstand or recover from insect damage caused by insect populations equal to those on plants of a susceptible cultivar. Essentially, tolerant plants can outgrow an insect infestation or recover and add new growth after the destruction or removal of damaged tissues. Tolerance is well documented in recent research on maize (Anglade et al., 1996; Kumar and Mihm, 1995), sorghum (Vandenberg et al., 1994), rice (Nguessan et al., 1994), turfgrass (Crutchfield and Potter, 1995), and cassava (Leru and Tertuliano, 1993), and oilseed crops (Brandt and Lamb, 1994). For additional information, readers are referred to reviews by Reese et al. (1994), Smith (1989), and Velusamy and Heinrichs (1986).

7.3 IDENTIFYING AND INCORPORATING INSECT RESISTANCE GENES

7.3.1 Conventional Genes

Sources of potential insect-resistant germplasm are available for evaluation in numerous international, national, and private seed collections. The International

Plant Genetic Resources Institute (IPGRI), Rome, Italy, (formerly the International Board of Plant Genetic Resources), in conjunction with several international research centers that comprise the Consultative Group for International Agricultural Research (CGIAR), maintains a database of the number, location, and condition of all existing major world crop plant germplasm (IPGRI, 1997). The mandate of IPGRI is to advance the conservation and use of plant genetic resources for the benefit of present and future generations. IPGRI is a convening center for the CGIAR Genetic Resources Program, and is linked to the Food and Agriculture Organization of the United Nations. IPGRI, FAO, CGIAR, and national germplasm collections such as the U. S. National Plant Germplasm System work together. These organizations have a common goal to collect, preserve, and maintain germplasm of the major food crops of the world with as much genetic diversity as possible, in order to guard against the occurrence of outbreaks of disease and insect pests in crop cultivars with limited genetic diversity. The U. S. National Plant Germplasm System is comprised of more than 350,000 crop accessions and is the largest supplier of germplasm to the world.

Agricultural researchers are continually concerned that germplasm centers should enhance their efforts to collect and preserve wild crop species (Hargrove et al., 1985; National Research Council, 1991). This is not an easy task, however, as global germplasm preservation efforts are jeopardized by slash and burn agricultural practices, population expansion, and timber and mining activities in many parts of the world. The governments of many countries are also reluctant to allow the collection and exchange of germplasm, because of fears that businesses in developed countries will use these genetic resources for profit (Plucknett et al., 1987). The 1996 Global Plan of Action for the Conservation and Sustainable Utilization of Plant Genetic Resources for Food and Agriculture was a plan developed and launched by 150 governments, with the help of IPGRI, to promote the active conservation and use of plant genetic resources (IPGRI, 1997). Bretting and Duvick (1997) extensively reviewed the need to conserve plant genetic resources in both static (*ex situ*) and dynamic (*in situ*) conditions.

With decreasing amounts of wild germplasm available for use in many crop plant species, it is more necessary than ever to better preserve existing global crop plant germplasm collections. Additional efforts are now necessary to increase the diversity and amount of collections and to make efforts to collect new genetic materials that can be incorporated into domestic crop plant species and further broaden the genetic composition of these species. Activity by plant resistance researchers in both areas is expressly needed. Few collections have been thoroughly evaluated under controlled conditions for resistance to the major pests of each crop. There are many opportunities available for close interdisciplinary research between entomologists and plant breeders to conduct these studies.

7.3.2 Transgenes

Insect pest management systems now have an additional type of insect resistance gene from a non-plant source. Genes from the bacteria *Bacillus thuringiensis* (Bt), encoding various delta–endotoxin insecticidal proteins have effective and specific insecticidal effects against economically important species of Coleoptera and Lepidoptera. The Bt genes are expressed in transgenic maize (Armstrong et al., 1993; Koziel et al., 1993; Williams et al., 1997), cotton (Benedict et al., 1996; Jenkins et al., 1997), poplar (Kleiner et al., 1995; Robison et al., 1994), potato (Ebora et al., 1994; Gatehouse et al., 1997), and tomato (Rhim et al., 1996). These cultivars are currently marketed and produced in Asia, Australia, Europe, and the U. S. Transgenic eggplant (Jelenkovic et al., 1998), persimmon (Tao et al., 1997), and rice (Ghareyazie et al., 1997) have also been constructed and are being developed for commercial production.

Other proteins toxic to insects have also been identified. These include the carbohydrate-binding proteins lectins (Marconi et al., 1993); proteinase inhibitors from maize, potato, rice, and tomato (Heath et al., 1997); proteinase inhibitors from insects (Kanost et al., 1989); chymotrypsin and trypsin inhibitors from cowpea and sweet potato (Hoffmann et al., 1992; Lombardiboccia et al., 1991; Yeh et al., 1997; Zhu et al., 1994); and alpha-amylase inhibitors from common bean (Fory et al., 1996; Ishimoto and Kitamura, 1993). Transgenes encoding several of these inhibitors have been transferred into plants such as bean (Ishimoto et al., 1996; Schroeder et al., 1995), cotton (Thomas et al., 1995a), poplar (Klopfenstein et al., 1993; Leple et al., 1995), potato (Benchekroun et al., 1995), rice (Duan et al., 1996; Xu et al., 1996), strawberry (Graham et al., 1997) and tobacco (Hilder et al., 1987; Masoud et al., 1993; Sane et al., 1997; Thomas et al., 1995b).

Conventional plant resistance is often a complex mixture of plant physical and chemical factors, which often results in substantial pest insect mortality. In contrast, transgenes have thus far been expressed at high levels to impart high insect mortality, which more than likely will result in the development of virulent, resistance-breaking insect biotypes. Deploying them with moderate levels of conventional insect resistance (Daly and Wellings, 1996) will most likely enhance the effectiveness of transgenes.

Initial research results have demonstrated that conventional genes and transgenes can be combined for enhanced and more stable insect resistance. Davis et al. (1995) produced the first maize hybrids with fall armyworm resistance derived from both a Bt transgene and a conventional maize resistance gene. Similar results were reported by Sachs et al. (1996), who demonstrated increased and more durable resistance in cotton to the tobacco budworm, *Heliothis virescens* (F.), after transforming a high-terpenoid content cotton cultivar with the CryIA (b) insecticidal Bt protein. Mu et al. (unpublished) have produced rice hybrids containing both Bt constructs and potato protease inhibitors with moderate levels of stable resistance to the pink stem borer, *Sesamia inferens* (Walker).

7.3.3 Conventional Breeding and Selection of Insect-Resistant Plants

Since humans began to domesticate and produce crops, they have enhanced the processes of natural plant adaptation and selection by selecting seeds with some degree of resistance to abiotic and biotic stresses, including insects. Plant breeding as a discipline of agricultural research has, in comparison, created resistant cultivars for only about 60 years. This research has been accomplished by identifying traits in resistant donor plants and transferring them to existing susceptible cultivars using conventional breeding techniques or, more recently, using gene transfer techniques.

The genetic control of insect resistance is normally determined by evaluating the segregating F_2 progeny from crosses between resistant and susceptible parents, or from diallel crosses (Ajala, 1993) involving several resistant and susceptible parents. In addition to the level of resistance in progeny *per se*, standard measures of the genetic expression of resistance involve determination of the inheritance of genes from resistant plants as well as the general and specific combining ability of genes transferred for resistance.

Many different methods are used in conventional plant breeding to develop insect-resistant cultivars. Mass selection (Sanford and Ladd, 1983), pure line selection, and recurrent selection (Dhillon and Wehner, 1991) are used routinely for incorporating insect resistance into crop plants. See Smith (1989) for an extensive review of insect resistance via recurrent selection. These methods can be used in both cross- and self-pollinated plants. In self-pollinated crops, backcross breeding (Wiseman and Bondari, 1995), bulk breeding and pedigree breeding (Khush, 1980) have also been used to add insect resistance to agronomically desirable cultivars.

7.3.4   Molecular Marker-Assisted Breeding

DNA marker technology has been established as a tool for crop improvement, but its utility depends on the crop in which it is being applied (Mohan et al., 1997; Staub et al., 1996). Lee (1995) extensively reviewed the existent use of DNA markers to overcome some of the weaknesses of traditional plant breeding. Unlike the morphological markers traditionally used in conventional plant breeding, DNA markers have the advantages of revealing neutral sites of variation in DNA sequences, are much more numerous than morphological markers, and they have no disruptive effect on plant physiology (Jones et al., 1997). Marker-assisted selection of plant traits is especially more efficient than phenotypic selection in larger populations of lower heritabilities (Hospital et al., 1997). Plant resistance research teams have begun to use DNA markers to select insect-resistant plants. The first such markers used were restriction fragment length polymorphisms (RFLPs) derived from cloned DNA fragments. With RFLP analysis, high-density genetic maps are being constructed to map insect resistance genes in cowpea (Myers et al., 1996), rice (Fukuta et al., 1998; Hirabayashi and Ogawa, 1995; Ishii et al., 1994; Mohan et al., 1994), mungbean (Young et al., 1992), barley (Nieto-Lopez and Blake, 1994), and wheat (Chen et al., 1996; Gill et al., 1987; Ma et al., 1993) (Table 7.1).

Randomly amplified polymorphic DNA (RAPD) markers have also been used to show allelic variation between plant genotypes for insect resistance. RAPD markers are short DNA sequences approximately 10 nucleotides long, which, when used to amplify genomic DNA in the polymerase chain reaction, amplify homologous sequences. The differences in sequences of resistant and susceptible plant DNA result in differential primer binding sites, which in turn permit the visualization of polymorphisms between the two types of DNA. RAPD markers have been used to detect insect resistance in wheat (Dweiket et al., 1994, 1997) and rice (Nair, 1995;

Table 7.1 Crop Plants Exhibiting Arthropod Resistance Linked to a DNA Marker

Plant	Arthroopd	Reference(s)
Apple	Rosy leaf curling aphid	Roche et al., 1997
Barley	Russian wheat aphid	Nieto-Lopez and Blake, 1994
Cowpea	Cowpea aphid	Myers et al., 1996
Maize	Corn earworm	Byrne et al., 1996
	European corn borer	Shon et al., 1993
	Southwestern corn borer	Khairallah et al., 1997
	Sugarcane borer	Bohn et al., 1996
Mungbean	Bruchid weevil	Young et al., 1992
Potato	Colorado potato beetle	Bonierbale et al., 1994; Yencho et al., 1996
Rice	Brown planthopper	Hirabayashi and Ogawa, 1995; Huang et al., 1997; Ishii et al., 1994
	Gall midge	Mohan et al., 1994; Nair et al., 1995, 1996
Tomato	Tobacco hornworm	Maliepaard et al., 1995; Mutschler et al., 1996
Wheat	Hessian fly	Dweikat et al., 1994, 1997; Gill et al., 1987; Ma et al., 1993; Seo et al., 1997
	Wheat curl mite	Chen et al., 1996

Nair et al., 1996). Both RFLP and RAPD markers are linked to genes expressing insect resistance in apple (Roche et al., 1997).

The markers described above are linked to the expression of major genes. Some insect resistance, like many other plant traits, is often the result of the action of several minor genes and is expressed in segregating populations as a continuum between resistance and susceptibility. Quantitative trait loci (QTL) statistical analyses can be used to define the RFLP map location of QTLs, contributing to the expression of minor gene resistance to insects. QTL analysis has been used to map insect resistance genes in maize (Bohn et al., 1996; Byrne et al., 1996; Khairallah et al., 1997; Lee et al., 1997; Schon et al., 1993), potato (Bonierbale et al., 1994; Yencho et al., 1996), rice (Huang et al., 1997), and tomato (Maliepaard et al., 1995; Mutschler et al., 1996). Comparisons are already beginning to be made between the advantages and disadvantages of different types of DNA markers used in marker assisted selection (Powell et al., 1996).

Since these genes have shown to be linked with an RFLP marker, their future selection can be based on the genotype of the RFLP marker, rather than the plant phenotype. This process of marker-assisted selection of plants based on RFLP genotype, before the phenotypic trait for resistance is expressed, holds promise for greatly accelerating the rate of development of arthropod-resistant crops (Paterson et al., 1991).

7.4 METHODS FOR ASSESSING RESISTANCE

Entomologists, plant breeders, and related plant scientists are continuously in need of more accurate and more efficient techniques with which to assess the resistance or susceptibility of plant germplasm. The technique used depends on the pest insect damage being evaluated and the age and stage of plant tissue being

Table 7.2   Insects Successfully Dispensed Using a Mechanical Innoculator

Insect	Reference(s)
Chinch bug, *Blissus leucopterous* (Say)	Harvey et al., 1985
Corn earworm, *Heliothis zea* (Boddie)	Mihm, 1982
Corn leaf aphid, *Rhopalosiphum maidis* (Fitch)	Harvey et al., 1985
English grain aphid, *Sitobion avenae* (Fabricius)	Harvey et al., 1985
European corn borer, *Ostrinia nubilalis* (Hubner)	Guthrie et al., 1984
Fall armyworm, *Spodoptera frugiperda* (J. E. Smith)	Mihm, 1983a; Pantoja et al., 1986
Green peach aphid, *Myzus persicae* (Sulzer)	Harvey et al., 1985
Greenbug, *Schizaphis graminum* (Rondani)	Harvey et al., 1985
Pea aphid, *Acyrthosiphon pisum* (Harris)	Harvey et al., 1985
Southwestern corn borer, *Diatraea grandiosella* Dyar	Davis, 1985; Mihm, 1983b

Modified from Smith, C. M., *Plant Resistance to Insects — A Fundamental Approach*. John Wiley & Sons, New York, 1989, p. 286. With permission.

damaged. Smith et al. (1994) developed a comprehensive review of existing techniques for assessing the effects of plant resistance on both plants and insects. The following discussion describes the major considerations for the use and development of such techniques.

The routine use of artificial diets to produce most of the pest Lepidoptera of the major world food crops (Davis and Guthrie, 1992; Singh and Moore, 1985), coupled with the development of mechanical insect rearing and plant infestation techniques, have allowed major increases in the quantity of germplasm that can be evaluated for insect resistance (Davis, 1985; Davis et al., 1985; Mihm, 1982; Mihm, 1983a,b). The larval plant innoculator, a major technological development in plant resistance to insects research, dispenses predetermined numbers of insects onto plants in sterilized corn grit medium (Mihm et al., 1978; Wiseman et al., 1980). This device is routinely used to make rapid, accurate placement of several species of insects onto test plants (Table 7.2.). Standardized damage rating scales are used to evaluate most major crop plants for insect resistance (Davis, 1985; Smith et al., 1994; Tingey, 1986). Measurements of insect damage to plants are usually more useful than measurements of insect growth or population development on plants, because reduced insect damage to plants and the resulting increases in yield or quality are the ultimate goals of most crop improvement programs.

Greenhouse experiments allow large-scale evaluation of seedling plants in a relatively short period of time. Identification of seedling-resistant plants also allows crosses involving these plants to be made in the same growing season and reduces the time required to develop resistant cultivars. However, plants resistant as seedlings may be susceptible in later growth stages (see Section 7.5, Biotic and Abiotic Factors Affecting the Expression of Resistance), necessitating field verification of resistance in mature plants. If resistance is evaluated in field studies where plants cannot be artificially infested, planting dates should be adjusted to coincide with the expected time of peak insect abundance. Two or three separate plantings at different dates may be necessary in order to have one planting that best coincides with the insect population peak. Spreader rows of a susceptible variety or related crop species have also been used very effectively to attract pest insects into field plantings.

Phenotypic plant chemical or morphological characters thought to mediate insect resistance can be monitored during the selection process to provide a rapid determination of potentially resistant plants. However, the demonstration of allelochemicals or morphological differences between resistant and susceptible plants does not always conclusively demonstrate that these factors mediate insect resistance. This process removes the variation due to the test insect until a later stage of study, when results can be confirmed in replicated field experiments.

Both physical and allelochemical resistance factors have been used to monitor for insect resistance (Andersson et al., 1980; Cole, 1987; Hamilton-Kemp et al., 1988; Kitch et al., 1985; Robinson et al., 1982). Several methods have been developed to alter the configuration of plant tissues, in order to determine the factors that mediate resistance. For in-depth reviews of these methods, readers are referred to Smith et al. (1994).

Measurements of insect population growth rate, insect development, and insect behavior have all been used to supplement basic information about plant measurements of resistance, and are used to determine the existence of antibiosis, antixenosis, and/or tolerance. Nutritional indices developed by Walbauer (1964, 1968) provide highly accurate measurements of insect consumption, digestion, and utilization of plant tissues. These measurements have been used to access the foliar insect resistance of cotton (Montandon et al., 1987), maize (Manuwoto and Scriber, 1982), potato (Cantelo et al., 1987), and soybean (Reynolds and Smith, 1985). For additional information, see the review of Van Loon (1991).

An electronic feeding monitor (McLean and Kinsey, 1966) passes a small electrical current across the insect and plant, both of which are wired to a recording device such as a strip chart recorder or oscilloscope. Insect feeding activity is detected when insect stylets penetrate the plant tissue at various depths, causing a change in the electrical conductance by the plant tissues. These changes are converted electronically and displayed as electronic penetration graphs. Differences in the type of graph produced during insect feeding indicate the frequency of feeding and differences in food source (plant xylem or phloem). Electronic penetration graphs have been used to study the resistance of several plants to different species of pest aphids and planthoppers (Holbrook, 1980; Kennedy et al., 1978; Nielson and Don 1974; Shanks and Chase, 1976; Khan and Saxena, 1984; Velusamy and Heinrichs, 1986). Tarn and Adams (1982) reviewed the history, development, and use of this technique.

Plant tolerance is assessed by comparing the production of plant biomass (yield) in insect-infested and noninfested plants of the same cultivar (Smith, 1989). Yield differences between the two plant groups are then used to calculate percent yield loss of each cultivar evaluated, based on the ratio: yield of infested plants/yield of noninfested plants.

A tolerance evaluation involves preparing replicated plantings that include the different cultivars being evaluated and a susceptible control cultivar, caging all plants in each replicate, and infesting caged plants in one half of each replicate with insect populations at or above the economic injury level for that insect. Plants remain infested until susceptible controls exhibit marked growth reduction or until the pest insect has completed at least one generation of development. Volumetric or plant biomass production measurements are then taken to calculate percent yield loss.

In an extensive review of methods to assess tolerance to aphids, Reese et al. (1994) determined that measuring tolerance as the slope described by graphing the relationship between weights of infested and control plants gave more accurate assessments than methods that consider only ratios of the two plant weight variables. More recent research (Ma et al., 1998; Deol et al., 1998) has determined that tolerance can also be accurately assessed from leaf chlorophyll loss measurements.

7.5 BIOTIC AND ABIOTIC FACTORS AFFECTING THE EXPRESSION OF RESISTANCE

The expression of plant resistance to insects is affected by variation in insect, plants, and the environment (Heinrichs, 1988; Smith, 1989). Plant tissue age affects the expression of insect resistance in maize (Kumar and Asino, 1993; Videla et al., 1992; Wiseman and Snook, 1995), oil seed crops (McCloskey and Isman, 1995; Nault et al., 1992), tree crops (Bingaman and Hart, 1993), vegetables (de Kogel et al., 1997; Diawara et al., 1994; Nihoul, 1994; Vaughn and Hoy, 1993) and wheat (Hein, 1992). In several cases, younger, more succulent leaves of resistant plants are more palatable to insects than older, more mature leaves (de Kogel et al., 1997a; Reynolds and Smith, 1985; Rodriguez et al., 1983). However, Laska et al. (1986), demonstrated that young leaves of a sweet pepper cultivar are more resistant to greenhouse whitefly, *Trialeurodes vaporariorum* (Westwood), feeding damage than older leaves. Even the plant that test insects are fed prior to germplasm evaluation can influence the degree of resistance expressed (Schotzko and Smith, 1991). These findings emphasize the need to standardize the plant tissue age expressing the greatest degree of insect resistance as well as the most critical stages in the growth of the target plant or life cycle of the pest insect.

The quality of light under which plants are grown also conditions the expression of insect resistance. This general phenomenon has been demonstrated in legume and solanaceous crops (Elden and Kenworthy, 1995; de Kogel, 1997b; Nkansah-poku and Hodgson, 1995). There is a direct relationship between increased intensity of light used to grow resistant plants and the expression of specific allelochemicals that mediate insect resistance (Ahman and Johansson, 1994; Bergvinson et al., 1995; Deahl et al., 1991; Jansen and Stamp, 1997). Light quality, in addition to intensity, also conditions insect resistance, as evidenced by the fact that plants grown under increased amounts of short-wave ultraviolet light exhibit higher levels of insect resistance (McCloud and Berenbaum, 1994).

Plants grown at abnormally high or low temperatures often exhibit a diminished expression of resistance. This relationship exists in insect-resistant wheat (Ratanatham and Gallun, 1986), sorghum (Wood and Starks, 1972), and tomato (Nihoul, 1993) grown at high temperatures and in insect-resistant alfalfa clones grown at low temperatures (Karner and Manglitz, 1985).

Soil nutrients play an important role in determining actual insect resistance in plants. Annan et al. (1997) determined that high levels of phosphorous increased aphid resistance in cowpea. Similar results have been detected in pearl millet (Leuck, 1972). Increasing the amount of potassium fertilizer enhances insect resistance in alfalfa and sorghum (Kindler and Staples, 1970; Schwessing and Wilde, 1979). Increased amounts of nitrogen fertilizer generally have an opposite effect (Annan et al., 1997), creating a super-optimal nutrition source for insects. Increasing the rate of nitrogen fertilization decreases the glandular trichome production in insect-resistant tomato, as well as the toxic methyl-ketone, 2-tridecanone produced by the trichomes (Barbour et al., 1991).

Soil-moisture changes also affect the expression of insect resistance. Jenkins et al. (1997a) observed that resistant cultivars of soybean plants grown in high soil-moisture conditions were less resistant to Mexican bean beetle, *Epilachna varivestis* (Mulsant), than plants grown under a normal moisture regime.

7.6 PLANT-INSECT GENE FOR GENE INTERACTIONS

The genetics and inheritance of many different crop plant genes resistant to insects have been documented in several reviews (Gatehouse et al., 1994; Khush and Brar, 1991; Singh, 1986). Both the expression and durability of these genes depend on the category of resistance, the pest insect genotype, and the interaction between the cultivar, the pest, and the environment. Insect biotypes are strains of the pest insect that mutate to express virulence genes that overcome resistance, often in response to high levels of antibiosis (vertical gene) resistance. The concepts of vertical and horizontal (several minor) resistance genes originated in research describing the effects of plants genes expressing pathogen resistance.

Biotypes form in much the same way that pest insects develop resistance to insecticides, by the selection of individuals with behavioral or physiological mechanisms that enable them to survive exposure to the toxin. This change involves genetic selection, mutation, or recombination in the pest population.

Eighteen arthropods exhibit biotypes with the ability to overcome genetic plant resistance to insects (Table 7.3). Nine of the existing biotypes are aphid species, in which parthenogenic reproduction contributes greatly to their successful development. Four of the existing biotypes are sexually dimorphic Diptera with high reproductive potentials. The brown planthopper, *Nilaparvata lugens* Stal, green leafhopper, *Nephotettix virescens* (Distant), and rice green leafhopper, *Nephotettix cincticeps* Uhler, occur continuously on large rice monocultures in much of Asia. For additional general information on aphid biotypes, see Webster and Inayatulluh (1985) and Ratcliffe et al. (1994).

The loss of resistance caused by genetic changes in the pest is commonly related to the gene-for-gene selection of virulence genes in the pest insect that corresponds to cultivar genes for resistance. The gene-for-gene hypothesis is well documented in the interactions between genes of the gall midge, *Orseolia oryzae* Wood Mason, and rice (Kumar et al., 1994; Tomar and Prasad, 1992) the Hessian fly and wheat (Ratcliffe and Hatchett, 1997), and the greenbug and sorghum (Puterka and Peters, 1995).

Tolerance resistance does not exert sufficient selection pressure on pest insects to evolve virulence genes (Heinrichs et al., 1984). However, agricultural producers often

Table 7.3 Arthropods Developing Biotypes in Response to Plant Resistance

Crop	Insect	Number of biotypes	Reference(s)
Alfalfa	Pea aphid	4	Auclair, 1978; Frazer, 1972
	Spotted alfalfa aphid	6	Nielson and Lehman, 1980
Apple	Wooly apple aphid	2	Sen Gupta and Miles, 1975
	Rosy leaf curling aphid	3	Alston and Briggs, 1977
	Apple maggot fly	2	Prokopy et al., 1988
Corn	Corn leaf aphid	5	Painter and Pathak, 1962; Singh and Painter, 1964; Wilde and Feese, 1973
Grape	Grape phylloxera	2	Fergusson-Kolmes and Dennehy, 1993; Hawthorne and Via, 1994
Raspberry	Raspberry aphid	4	Briggs, 1965; Keep and Knight, 1967
Rice	Green rice leafhopper	2	Sato and Sogawa, 1981
	Green leafhopper	3	Heinrichs and Rapusas, 1985; Takita and Hashim, 1985
	Brown planthopper	4	Verma et al., 1979
	Rice gall midge	4	Heinrich and Pathak, 1981
Sorghum	Greenbug	11	Harvey and Hackerott, 1969; Harvey et al., 1991, 1997; Kindler and Spomer, 1986; Puterka et al., 1982; Porter et al., 1982; Teetes et al., 1975; Wood, 1961
Vegetables	Cabbage aphid	2-4	Dunn and Kempton, 1972; Lammerink, 1968
	Sweetpotato whitefly	2	Brown et al., 1995
Wheat	Wheat curl mite	5	Harvey et al., 1995
	English grain aphid	3	Lowe, 1981
	Hessian fly	16	Ratcliffe et al., 1994

Modified from Smith, C. M., *Plant Resistance to Insects — A Fundamental Approach*. John Wiley & Sons, New York, 1989, p. 286. With permission.

prefer cultivars with antibiosis or antixenosis resistance, which reduces pest insect populations. In contrast to the use of high levels of antibiosis resistance, Kennedy et al. (1987) demonstrated that moderate levels of both antibiosis and antixenosis have substantial value in reducing population levels of migratory pest Lepidoptera.

Bt-based plant resistance to insects expressed as a single strong (vertical) resistance gene functions in the same manner as conventional plant antibiosis genes (Llewellyn et al., 1994), and the Bt toxin causes high mortality among insects feeding on these cultivars. However, laboratory research with insect pests of both stored grain and field crops suggests that this level of gene expression will lead to the rapid development of pest insect biotypes virulent to Bt plants (Huang et al., 1997; Johnson et al., 1990; McGaughey, 1985; McGaughey and Beeman, 1988; Miller et al., 1990; Moar et al., 1995; Ramachandran et al., 1998; Stone et al., 1989).

Since monogenic resistance is generally more vulnerable to biotype development than polygenic resistance, various tactics to delay the development of Bt-virulent biotypes have been proposed. These include adjusting the level of toxin expression, pyramiding multiple toxin genes, seed mixtures of Bt and non-Bt plants, and "patchwork planting" of Bt and non-Bt cultivars (Alstad and Andow, 1995; Gould, 1994;

Gould et al., 1991; McGaughey and Johnson, 1992; Roush, 1997; Wigley et al., 1994). Several of these strategies are similar to those devised for deploying conventional antibiosis insect resistant plant genes (Gallun and Khush, 1980; Smith, 1989). Currently, however, all transgenic crops produced in the U. S. are marketed using a high-dose strategy, which relies on the maximum expression of various Bt constructs (Daly and Wellings, 1996; Roush, 1997).

7.7   PLANT RESISTANCE AS THE FOUNDATION OF INTEGRATED INSECT PEST MANAGEMENT

Conventional plant genes in the major food and fiber crops of the world have been used to develop many insect-resistant cultivars during the past 30 years. Pertinent examples exist in maize (Mihm, 1997), rice (Heinrichs, 1994), and wheat (Smith, 1989). Presently, insect-resistant cultivars are integral components of insect pest management programs in world agricultural systems. These cultivars interact synergistically with biological, chemical, and cultural control methods, and reduce the spread of plant diseases vectored by pest insects and related arthropods (Harvey et al., 1994; Kennedy et al., 1976; Maramorosch, 1980).

Plant resistance increases the effectiveness of insect biological control agents by synergizing the interactions between insect-resistant barley, maize, sorghum, and wheat, and the parasitoids of insect pests attacking these crops (Isenhour and Wiseman, 1987; Reed et al., 1991; Riggin et al., 1992; Starks et al., 1972). Larvae of the tobacco budworm suffer similar increased mortality when exposed to transgenic maize plants containing the Bt toxin and the fungus *Nomuraea rileyi* (Johnson et al., 1997). Maize cultivars with conventional gene resistance to the fall armyworm, *Spodoptera frugiperda* (J. E. Smith), or the corn earworm, *Heliothis zea* (Boddie), are more effective when used in combination with applications of nuclear polyhedrosis virus (Hamm and Wiseman, 1986; Wiseman and Hamm, 1993).

Limitations to the effective amount of synergism that can occur between resistant cultivars and biological control agents have been determined. The frego bract cotton character that imparts resistance to the boll weevil also increases weevil susceptibility to parasitism (McGovern and Cross, 1976). However, frego bract plants suffer enhanced susceptibility to *Lygus* spp. plant feeding bugs (Jenkins et al., 1971). Some sources of insect-resistant potato, tomato, and soybean contain levels of toxic allelochemicals that have negative effects on beneficial insects (Barbour et al., 1993; Duffey, 1986; Kauffman and Flanders, 1986; Orr and Boethel, 1985; Powell and Lambert, 1984; Yanes and Boethel, 1983), entomophathic fungi (Gallardo et al., 1990), and insect viruses (Felton and Duffey, 1990).

High trichome density in insect-resistant cotton and tomato have been shown to be detrimental to beneficial insects (Stipanovic, 1983; Treacy et al., 1985). However, moderate levels of plant trichome density in insect-resistant cultivars of cucumber, potato, and wheat effectively synergize the actions of parasites and predators on these crops (Lampert et al., 1983; van Lentern, 1991; Obrycki et al., 1983). Bottrell et al. (1998) reviewed the differences in the effects of plant resistance factors on biological control agents. Their results suggested that a better understanding of the evolution of crop plants, pests, and pest biological control agents is needed to better determine how plant resistance and biological control can be combined for more durable insect pest management.

Insect-resistant cultivars also complement the effects of variation in time of planting and trap crops. Antixenotic cotton cultivars grown in combination with early-maturing cotton cultivars that trap boll weevils allow a 20% reduction in insecticide application (Burris et al., 1983). Rice trap crops planted 20 days ahead of the main crop, a brown plant hopper–resistant cultivar, attract the hopper population earlier and serve as reservoirs for natural enemies (Heinrichs et al., 1984).

The integration of resistant cultivars with insecticides is also well documented. Cotton cultivars exhibiting the frego bract and okra (thin) leaf traits allow greater than 30% penetration of insecticides into the cotton foliage canopy, increasing the efficiency and decreasing the amount of insecticide required for control (Jenkins et al., 1971). Plant resistance in carrots to the carrot fly, *Psilia rosae* (F.), and in *Brassica* spp. to the turnip fly, *Delia floralis* (Fallen), reduces insecticide use by 50 to 80% (Ellis, 1990; Taksdal, 1992). Insect-resistant rice or sorghum cultivars require much less insecticide to maintain net crop yield and value (Heinrichs et al., 1984; Teetes et al., 1986; van den Berg et al., 1994a). Some insect-resistant cultivars of rice (Kalode, 1980; Reissig et al., 1981), sorghum (Kishore, 1984), vegetables (Cuthbert and Fery, 1979), and wheat (Buntin et al., 1992) have been developed that derive no synergistic benefit from insecticides. As with biological control, some negative interactions between insect-resistant cultivars and insecticidal control also exist. Enhanced detoxication of insecticides occurs when pest insects are fed foliage containing high levels of allelochemicals that mediate insect resistance in Solanaceous crops (Ghidiu et al., 1990; Kennedy, 1984).

In addition to the synergism documented above, insect-resistant cultivars also have advantages over these biological, cultural, and insecticidal control methods. As described previously, resistant cultivars are compatible with insecticide use, but in many cases biological control is not. Insecticides applied at recommended rates are not specific and often kill beneficial insects. Resistant cultivars, especially those with moderate levels of resistance, affect only the target pest insect and generally do not kill beneficial organisms, depending on the category and mechanism of resistance as mentioned above. The effects of insect-resistant cultivars are density independent, operating at all levels of pest population abundance, but biological control organisms depend on the sustained density of their hosts or prey insects to remain effective (Panda and Khush, 1995).

Transgenic insect-resistant cotton, maize, and potato cultivars with Bt-based resistance have been marketed in the U. S. for only a few years on a small portion of the total hectarage of each crop. However, their use will increase during the next decade. Although the initial field performance of transgenic (Bt) crops is impressive, Daly and Wellings (1996) have compared the various aspects of both conventional and transgenic plant resistance to insects (Table 7.4). As discussed in previous sections, conventional resistance may be expressed as antibiosis, antixenosis, tolerance, or a combination of these, and mediated by plant allelochemicals and/or plant

Table 7.4   A Comparison of Natural and Engineered Plant Resistance to Insects

Category	Natural plant resistance	Engineered plant resistance
Mechanisms	Antibiosis, antixenosis, tolerance	Antibiosis
Basis	Diverse chemical and physical	Chemical – antimetabolic
Pest Mortality	Variable	High
Expression	Variable	Constitutive
Tritrophic Interactions	Complex	Possibly simple
Management	May be required	Required

From Daly, J. C. and P. W. Wellings. Ecological Constraints to the Deployment of Arthropod Resistant Crop Plants: A Cautionary Tale, In: *Frontiers of Population Ecology*, Floyd, R. B., A.W. Shepard, and P.J. De Barro, Eds., CSIRO Publishing, Melbourne, FL, 1996. With Permission.

physical factors or both. Transgenic resistance is only antibiotic, due to a toxin. The two types of resistance are also expressed in very different ways. Finally, conventional resistance is expressed at different plant-growth stages and in different plant tissues, while the current transgenic resistant cultivars exhibit high levels of Bt toxin expressed at any plant developmental stage. As a result, the utility of crops with high levels of Bt-based insect resistance on large areas of crop production with small area of pest refugia, is as yet an unproven plant resistance tactic.

7.8   CONCLUSIONS

The cooperative efforts of biochemists, entomologists, geneticists, molecular biologists, and plant breeders to identify, quantify, and develop insect-resistant crop cultivars during the past several decades are some of the most significant accomplishments of modern agricultural research. These efforts have utilized the genetic diversity in wild and closely related species of world crop plants to identify genes that express resistance to the major arthropod pests of world agriculture.

The current world economic value of this resistance is several hundred million dollars per year. The ecological value of insect resistance has greatly decreased world pesticide usage, contributing to a healthier environment for humans, livestock, and wildlife. Agricultural producers have benefited from crops with arthropod resistance through decreased production costs. Consumer benefits derived from insect-resistant crops include safer and more economically produced food.

Although many arthropod-resistant cultivars have been developed, research and development must continue, in order to maintain the benefits of this resistance in global food production. Crops developed using either conventional plant genes or transgenes must be monitored for the occurrence of virulence genes in newly developing resistance-breaking biotypes. Where possible, accurate and efficient techniques based on molecular genetic markers must be adapted or developed and implemented to monitor biotypes, such as those developed by Gould et al. (1997). The need to identify biotypes of pests infesting transgenic crops expressing high levels of resistance is critical. There is also an acute need for actual field data to develop functional gene-release strategies that slow or avoid the development of biotypes, especially for highly polyphagous pests exposed to transgene toxins in several different crops.

New and improved insect infestation techniques and devices that safely and efficiently place test insects onto plants, such as the mechanical innoculator, will also be essential to future progress. The development and refinement of standardized rating scales to determine insect damage to more crops will greatly facilitate the development of insect-resistant cultivars in several additional crop plant species. There is also a need for a more complete knowledge of plant nutrient composition, in order to design artificial diets that more accurately represent an insect's host plant, so that the true contributions of plant allelochemicals to insect resistance can be ascertained.

Whether developing new resistant cultivars or improving existing cultivars, new resistance genes must continue to be identified, from both conventional and transgenic sources. Significant fractions of the world germplasm collections remain to be evaluated for resistance to many pest insects. Major initiatives to translate the entire maize and rice genomes are progressing. Molecular genetic information gained from these efforts and from the use of new DNA technologies (Kopp, 1998; Lutz, 1977; Schena et al., 1995) will accelerate the rate of major advancements in the molecular genetics of plant resistance research. It is most likely that several plant genes governing plant-insect interactions will be sequenced. Eventually, it will be possible to predict the plant-insect resistance genes necessary to achieve an economically significant level of management of a given pest insect. In the interim, however, efforts must be made to merge the benefits of proven conventional plant genes with those of transgenes for durable insect-resistant crop plants. The problems of nontarget insect susceptibility and the potential for development of biotypes will be present in the resistant cultivars developed, whether by conventional or transgenic means.

With the world population expected to exceed 10 billion people before 2040, it is essential that global food production be increased to meet that need. Arthropod-resistant crops should continue to be integral components of that food production system, because of their proven economic and environmental benefits. A continual supply of safe food produced with insect-resistant crop cultivars will depend heavily on 21st century plant resistance research teams that develop durable insect-resistant gene products. The combination of improved curation and maintenance of germplasm collections and rapidly emerging new molecular genetic technologies will provide many opportunities for interdisciplinary research efforts to identify and develop new sources of insect resistance.

ACKNOWLEDGMENTS

The author wishes to express sincere thanks to Dr. Nilsa Bosque Pérez, Department of Plant Soil, and Entomological Sciences, University of Idaho, and Dr. Kimberly Stoner, Connecticut Agricultural Experiment Station, for their insightful reviews of the manuscript.

REFERENCES

Agrawal, A. A. Induced responses to herbivory and increased plant performance. *Science* 279, 1201–1202, 1998.

Ahman, I., and M. Johansson. Effect of light on DIMBOA-glucoside concentration in wheat (*Triticum aestivum* L). *Ann. Appl. Biol.* 124, 569–574, 1994.

Ajala, S. O. Population cross diallel among maize genotypes with varying levels of resistance to the spotted stem borer *Chilo partellus* (Swinhoe). *Maydica* 38, 39–45, 1993.

Alfaro, R. I. An induced defense reaction in white spruce to attack by the white pine weevil, *Pissodes strobi. Can. J. Forest Res.* 25, 1725–1730, 1995.

Alstad, D. N., and D. A. Andow. Managing the evolution of insect resistance to transgenic plants. *Science* 268, 1894–1896, 1995.

Alston, F. H., and J. B. Briggs. Resistance genes in apple and biotypes of *Dysaphis devecta. Ann. Appl. Biol.* 87, 75–81, 1977.

Ampong-Nyarko, K., K. V. S. Reddy, R. A. Nyangor, and K. N. Saxena. Reduction of insect pest attack on sorghum and cowpea by intercropping. *Entomol. Exp. Appl.* 70, 179–184, 1994.

Andersson, B. A., R. T. Holman, L. Lundgren, and G. Stenhagen. Capillary gas chromatograms of leaf volatiles. A possible aid to breeders for pest and disease resistance. *J. Agric. Food Chem.* 28, 985–989, 1980.

Anglade, P., B. Gouesnard, A. Boyat, and A. Panouille. Effects of multitrait recurrent selection for European corn borer tolerance and for agronomic traits in FS12 maize synthetic. *Maydica* 41, 97–104, 1996.

Annan, I. B., K. Ampong-Nyarko, W. M. Tingey, and G. A. Schaefers. Interactions of fertilizer, cultivar selection, and infestation by cowpea aphid (Aphididae) on growth and yield of cowpeas. *Intl. J. Pest Manage.* 43, 307–312, 1997.

Anonymous. *Agriculture and the Kansas Economy — Examples of Potential Economic Enhancement.* Informal Report to the Kansas Legislature. Kansas Agricultural Experiment Station/Kansas Cooperative Extension Service, Kansas State University, Manhattan, Kansas, 1995.

Armstrong, C. L., G. B. Parker, J. C. Pershing, S. M. Brown, P. R. Sanders, D. R. Duncan, T. Stone, D. A. Dean, D. L. Deboer, J. Hart, A. R. Howe, F. M. Morrish, M. E. Pajeau, W. L. Petersen, B. J. Reich, S. J. Sate, S. R. Sims, S. Stehling, R. Rodriguez, C. G. Santino, W. Schuler, L. J. Tarochione, and M. E. Fromm. Field evaluation of European corn borer control in progeny of 173 transgenic corn events expressing an insecticidal protein from Bacillus thuringiensis. *Crop Sci.* 35, 550–557, 1995.

Auclair, J. L. Biotypes of the pea aphid *Acyrthrosiphon pisum* in relation to host plants and chemically defined diets. *Entomol. Exp. Appl.* 24, 12–16, 1978.

Azzam, A., S. Azzam, S. Lhaloui, A. Amri, M. El Bouhssini, and M. Moussaoui. Economic returns to research in Hessian fly (Diptera: Cecidomyidae) resistant bread-wheat varieties in Morocco. *J. Econ. Entomol.* 90, 1–5, 1997.

Baldwin, I. T. Chemical Changes Rapidly Induced by Folivory. In: *Insect-Plant Interactions V,* Bernays, E., Ed., CRC Press, Boca Raton, FL. 1994.

Barbour, J. D., and G. G. Kennedy. Role of steroidal glycoalkaloid alpha tomatine in host-plant resistance of tomato to Colorado potato beetle. *J. Chem. Ecol.* 17, 989–1005, 1991.

Barbour, J. D., R. R. Farrar, and G. G. Kennedy. Interaction of fertilizer regime with host-plant resistance in tomato. *Entomol. Exp. Appl.* 60, 289–300, 1991.

Barbour, J. D., R. R. Farrar, and G. G. Kennedy. Interaction of *Manduca sexta* resistance in tomato with insect predators of *Helicoverpa zea. Entomol. Exp. Appl.* 68, 143–155, 1993.

Barria, B. N., S. V. Copaja, and H. M. Niemeyer. Occurrence of DIBOA in wild *Hordeum* species and its relation to aphid resistance. *Phytochemistry* 31, 89–91, 1992.

Barry, D., L. L. Darrah, and D. Alfaro. Relation of European corn borer (Lepidoptera: Pyralidae) leaf-feeding resistance and DIMBOA content in maize. *Environ. Entomol.* 23, 177–182, 1994.

Baur, R., S. Binder, and G. Benz. Nonglandular leaf trichomes as short-term inducible defense of the grey alder, *Alnus-incana* (L.), against the chrysomelid beetle, *Agelastica alni* L. *Oecologia* 87, 219–226, 1991.

Beck, S. D. Resistance of plants to insects. *Annu. Rev. Entomol.* 10, 107–232, 1965.

Beeghly, H. H., J. G. Coors, and M. Lee. Plant fiber composition and resistance to European corn borer in four maize populations. *Maydica* 42, 297–303, 1997.

Benchekroun, A., D. Michaud, B. Nguyen-Quoc, S. Overney, Y. Desjardins, and S. Yelle. Synthesis of active oryzacystatin I in transgenic potato plants. *Plant Cell Rep.* 14, 585–588, 1995.

Benedict, J. H., E. S. Sachs, D. W. Altman, W. R. Deaton, R. J. Kohel, D. R. Ring, and S. A. Berberich. Field performance of cottons expressing transgenic CryIA insecticidal proteins for resistance to *Heliothis virescens* and *Helicoverpa zea* (Lepidoptera: Noctuidae). *J. Econ. Entomol.* 89, 230–238, 1996.

Bentur, J. S., and M. B. Kalode. Hypersensitive reaction and induced resistance in rice against the Asian rice gall midge, *Orseolia oryzae*. *Entomol. Exp. Appl.* 78, 77–81, 1996.

Bergman, D. K., J. W. Dillwith, A. A. Zarrabi, J. L. Caddel, and R. C. Berberet. Epicuticular lipids of alfalfa relative to its susceptibility to spotted alfalfa aphids (Homoptera, Aphididae). *Environ. Entomol.* 20, 781–785, 1991.

Bergvinson, D. J., J. T. Arnason, and L. N. Pietrzak. Localization and quantification of cell wall phenolics in European corn borer resistant and susceptible maize inbreds. *Can. J. Bot.* 72, 243–1249, 1994.

Bergvinson, D. J., J. S. Larsen, and J. T. Arnason. Effect of light on changes in maize resistance against the European corn borer, *Ostrinia nubilalis* (Hubner). *Can. Entomol.* 127, 111–122, 1995.

Bingaman, B. R., and E. R. Hart. Clonal and leaf age variation in *Populus* phenolic glycosides, implications for host selection by *Chrysomela scripta* (Coleoptera, Chrysomelidae). *Environ. Entomol.* 22, 397–403, 1993.

Blum, A. Anatomical phenomena in seedlings of sorghum varieties resistant to the shoot fly (*Atherigona varia soccata*). *Crop Sci.* 8, 388–390, 1968.

Board on Agriculture National Research Council. *Managing Global Resources. The U.S. National Plant Germplasm System/Committee on Managing Global Genetic Resources: Agricultural Imperatives.* National Academy Press, Washington, D.C., 1990, p. 171.

Bodnaryk, R. P. Leaf epicuticular wax, an antixenotic factor in Brassicaceae that affects the rate and pattern of feeding of flea beetles, *Phyllotreta cruciferae* (Goeze). *Can. J. Plant Sci.* 72, 1295–1303, 1992.

Bodnaryk, R. P., and R. T. Rymerson. Effect of wounding and jasmonates on the physicochemical properties and flea beetle defence responses of canola seedlings, *Brassica napus* L. *Can J. Plant Sci.* 74, 899–907, 1994.

Bohn, M., M. M. Khairallah, D. González de Leoñ, D. A. Hoisington, H. F. Utz, J. A. Deutsch, D. C. Jewell, J. A. Mihm, and A. E. Melchinger. QTL mapping in tropical maize. 1. Genomic regions affecting leaf feeding resistance to sugarcane borer and other traits. *Crop Sci.* 36, 1352–1361, 1996.

Bonierbale, M. W., R. L. Plaisted, O. Pineda, and S. D. Tanksley. Qtl analysis of trichomemediated insect resistance in potato. *Theor. Appl. Genet.* 87, 973–987, 1994.

Brandt, R. N., and R. J. Lamb. Importance of tolerance and growth rate in the resistance of oilseed rapes and mustards to flea beetles, *Phyllotreta cruciferae* (Goeze) (Coleoptera: Chrysomelidae). *Can. J. Plant Sci.* 74, 169–176, 1994.

Breen, J. P. *Acremonium* endophyte interactions with enhanced plant resistance to insects. *Annu. Rev. Entomol.* 39, 401–423, 1994.

Bretting, P. K., and D. N. Duvick. Dynamic conservation of plant genetic resources. *Adv. Agron.* 61, 1–51, 1997.

Brewer, G. J., E. L. Sorensen, E. K. Horber, and G. L. Kreitner. Alfalfa stem anatomy and potato leafhopper (Homoptera: Cicadellidae) resistance. *J. Econ. Entomol.* 79, 1249–1253, 1986.

Briggs, J. B. The distribution, abundance, and genetic relationships of four strains of the Rubus aphid (*Amphorophora rubi*) in relation to raspberry breeding. *J. Hort. Sci.* 49, 109–117, 1965.

Bronner, R., E. Westphal, and F. Dreger. Enhanced peroxidase activity associated with the hypersensitive response of *Solanum dulcamara* to the gall mite *Aceri cladophthirus* (Acari, Eriophyoidea). *Can. J. Bot.* 69, 2192–2196, 1991.

Brown, J. K., D. R. Frohlich, and R. C. Rosell. The sweet potato or silverleaf whiteflies: Biotypes of *Bemisia tabaci* or a species complex? *Annu. Rev. Entomol.* 40, 511–534, 1995.

Buntin, G. D., S. L. Ott, and J. W. Johnson. Integration of plant resistance, insecticides, and planting date for management of the Hessian fly (Diptera,Cecidomyiidae) in winter wheat. *J. Econ. Entomol.* 85, 530–538, 1992.

Burris, E., D. F. Clower, J. E. Jones and S.L. Anthony. Controlling boll weevils with trap cropping, resistant cotton. *La. Agric.* 26, 22–24, 1983.

Byrne, P. F., M. D. McMullen, M. E. Snook, T. A. Musket, J. M. Theuri, N. W. Widstrom, B. R. Wiseman, and E. H. Coe. Quantitative trait loci and metabolic pathways: Genetic control of the concentration of maysin, a corn earworm resistance factor, in maize silks. *Proc. Natl. Acad. Sci. U.S.A.* 93, 8820–8825, 1996.

Cantelo, W. W., L. W. Douglass, L. L. Sanford, S. L. Sinden, and K. L. Deahl. Measuring resistance to the Colorado potato beetle (Coleoptera: Chrysomelidae) in potato. *J. Entomol. Sci.* 22, 245–252, 1987.

Chen, Q., R. L. Conner, and A. Laroche. Molecular characterization of *Haynaldia villosa* chromatin in wheat lines carrying resistance to wheat curl mite colonization. *Theor. Appl. Genet.* 93, 679–684, 1996.

Chesnokov, P. G. *Methods of Investigating Plant Resistance to Pests.* National Science Foundation, Washington, D.C., 1953, p. 107. (Israel Program for Scientific Translation.)

Clement, S. L., W. J. Kaiser, and H. Eischenseer. *Acremonium* endophytes in germplasms of major grasses and their utilization for insect resistance. In *Biotechnology of Endophytic Fungi of Grasses,* Bacon, C. W. and White, J. F., Eds., CRC, Boca Raton, FL, 1994.

Cohen, A. C., T. J. Henneberry, and C. C. Chu. Geometric relationships between whitefly feeding behavior and vascular bundle arangements. *Entomol. Exp. Appl.* 78, 135–142, 1996.

Cole, R. A. Intensity of radicle fluorescence as related to the resistance of seedlings of lettuce to the lettuce root aphid and carrot to the carrot fly. *Ann. Appl. Biol.* 111, 629–639, 1987.

Cole, R. A. The relative importance of glucosinolates and amino acids to the development of two aphid pests *Brevicoryne brassicae* and *Myzus persicae* on wild and cultivated brassica species. *Entomol. Exp. Appl.* 85, 121–133, 1997.

Crutchfield, B. A., and D. A. Potter. Tolerance of cool-season turfgrasses to feeding by Japanese beetle and southern masked chafer (Coleoptera: Scarabaeidae) grubs. *J. Econ. Entomol.* 88, 1380–1387, 1995.

Cuthbert, F. P., and R. L. Fery. Value of plant resistance for reducing cowpea curculio damage to the southern pea (*Vigna unguiculata* (L.) Walp.). *J. Am. Soc. Hort. Sci.* 104, 199–201, 1979.

Cuthbert, F. P., Jr., and A. Jones. Insect resistance as an adjunct or alternative to insecticides for control of sweet potato soil insects. *J. Am. Soc. Hort. Sci.* 103, 443–445, 1978.

Daly, J. C., and P. W. Wellings. Ecological Constraints to the Deployment of Arthropod Resistant Crop Plants: A Cautionary Tale. In: *Frontiers of Population Ecology*, Floyd, R. B., Shepard, A. W., and De Barro, P. J., Eds., CSIRO Publishing, Melbourne, FL, 1996.

Davis, F. M. Entomological techniques and methodologies used in research programmes on plant resistance to insects. *Ins. Sci. Applic.* 6, 391–400, 1985.

Davis, F. M., and W. D. Guthrie. Rearing Lepidoptera for Plant Resistance Research. In: *Advances in Insect Rearing for Research and Pest Management*, Anderson, T. E., and Leppla, N. C., Eds., Oxford & IBH Publishing, New Delhi, 1992.

Davis, F. M., G. T. Baker, and W. P. Williams. Anatomical characteristics of maize resistant to leaf feeding by southwestern corn borer (Lepidoptera: Pyralidae) and fall armyworm (Lepidoptera: Noctuidae). *J. Agric. Entomol.* 12, 55–65, 1995.

Davis, F. M., T. G. Oswalt, and S. S. Ng. Improved oviposition and egg collection system for the fall armyworm (Lepidoptera: Noctuidae). *J. Econ. Entomol.* 78, 725–729, 1985.

Deahl, K. L., W. W. Cantelo, S. L. Sinden, and L. L. Sanford. The effect of light intensity on Colorado potato beetle resistance and foliar glycoalkaloid concentration of four *Solanum chacoense* clones. *Am. Potato J.* 68, 659–666, 1991.

de Kogel, W. J., A. Balkemaboomstra, M. Vanderhoek, S. Zijlstra, and C. Mollema. Resistance to western flower thrips in greenhouse cucumber: Effect of leaf position and plant age on thrips reproduction. *Euphytica* 94, 63–67, 1997a.

de Kogel, W. J., M. van der Hoek, M. T. A. Dik, B. Gebala, F. R. van Dijken, and C. Mollema. Seasonal variation in resistance of chrysanthemum cultivars to *Frankliniella occidentalis* (Thysanoptera: Thripidae). *Euphytica* 94, 283–288, 1997b.

Deol, G. S., J. C. Reese, and B. S. Gill. A rapid, nondestructive, technique for assessing chlorophyll loss from greenbug (Homoptera: Aphididae) feeding damage on sorghum leaves. *J. Kansas Entomol. Soc.* 70, 305–312, 1997.

Dhillon, N. P. S., and T. C. Wehner. Host-plant resistance to insects in cucurbits — germplasm resources, genetics and breeding. *Trop. Pest Manage.* 37, 421–428, 1991.

Diawara, M. M., J. T. Trumble, C. F. Quiros, K. K. White, and C. Adams. Plant age and seasonal variations in genotypic resistance of celery to beet armyworm (Lepidoptera, Noctuidae). *J. Econ. Entomol.* 87, 514–522, 1994.

Dickens, J. C., G. D. Prestwich, C. S. Ng, and J. H. Visser. Selectively fluorinated analogs reveal differential olfactory reception and inactivation of green leaf volatiles in insects. *J. Chem. Ecol.* 19, 1981–1991, 1993.

Duan, J. J., and R. J. Prokopy. Visual and odor stimuli influencing effectiveness of sticky spheres for trapping apple maggot flies *Rhagoletis pomonella* (Walsh) (Diptera Trephritidae). *J. Appl. Entomol.* 113, 271–279, 1992.

Duan, X., X. Li, Q. Xue, M. Abo-El-Saad, D. Xu, and R. Wu. Transgenic rice plants harboring an introduced potato proteinase inhibitor II gene are insect resistant. *Nature Biotechnology* 14, 494–498, 1996.

Duffey, S. S. Plant Glandular Trichomes: Their Partial Role in Defense Against Insects. In: *Insect and the Plant Surface*, Juniper, B. E. and Southwood, T. R. E., Eds., Edward Arnold Ltd., London, 1986.

Dunn, J. A., and D. P. H. Kempton. Resistance to attack by *Brevicoryne brassicae* among plants of Brussels sprouts. *Ann. Appl. Biol.* 72, 1–11, 1972.

Dweikat, I., H. Ohm, S. MacKenzie, F. Patterson, S. Cambron, and R. Ratcliffe. Association of a DNA marker with Hessian fly resistance gene *H9* in wheat. *Theor. Appl. Genet.* 89, 964–968, 1994.

Dweikat, I., H. Ohm, F. Patterson, and S. Cambron. Identification of RAPD markers for 11 Hessian fly resistance genes in wheat. *Theor. Appl. Genet.* 94, 419–423, 1997.

Ebora, R. V., M. M. Ebora, and M. B. Sticklen. Transgenic potato expressing the *Bacillus thuringiensis* CryIA(c) gene effects on the survival and food consumption of *Phthorimea operculella* (Lepidoptera, Gelechiidae) and *Ostrinia nubilalis* (Lepidoptera, Nochuidae). *J. Econ. Entomol.* 87, 1122–1127, 1994.

Elden, T. C. Influence of soybean lines isogenic for pubescence type on two spotted spider mite (Acarina: Tetranychidae) development and feeding damage. *J. Entomol. Sci.* 32, 296–302, 1997.

Elden, T. C., and W. J. Kenworthy. Physiological responses of an insect-resistant soybean line to light and nutrient stress. *J. Econ. Entomol.* 88, 430–436, 1995.

Ellis, P. R. The role of host plant resistance to pests in organic and low input agriculture. *Organic and Low Input Agriculture. BCPC Monograph* 45, 93–102, 1990.

Febvay, G., J. Bonnin, Y. Rahbe, R. Bournoville, S. Delrot, and J. L. Bonnemain. Resistance of different lucerne cultivars to the pea aphid, *Acyrthrosiphon pisum*: influence of phloem composition on aphid fecundity. *Entomol. Exp. Appl.* 48, 127–134, 1988.

Felton, G. W., and S. S. Duffey. Inactivation of baculovirus by quinones formed in insect damaged plant tissues. *J. Chem. Ecol.* 16, 1221–1236, 1990.

Fergusson-Kolmes, L. A., and T. J. Dennehy. Differences in host utilization by populations of North American grape phylloxera (Homoptera, Phylloxeridae). *J. Econ. Entomol.* 86, 1502–1511, 1993.

Fiori, B. J., and D. W. Craig. Relationship between color intensity of leaf supernatants from resistant and susceptible birch trees and rate of oviposition by the birch leafminer (Hymenoptera: Tenthredinidae). *J. Econ. Entomol.* 80, 1331–1333, 1987.

Fory, L. F., F. Finardi, C. M. Quintero, T. C. Osborn, C. Cardona, M. J. Chrispeels, and J. E. Mayer. Alpha-amylase inhibitors in resistance of common beans to the Mexican bean weevil and the bean weevil (Coleoptera: Bruchidae). *J. Econ. Entomol.* 89, 204–210, 1996.

Frazer, B. D. Population dynamics and recognition of biotypes in the pea aphid (Homoptera: Aphididae). *Can. Entomol.* 10.1729–1733, 1972.

Fukuta, Y., K. Tamura, M. Hirae, and S. Oya. Genetic analysis of resistance to green rice leafhopper (*Nephotettix cincticeps* Uhler) in rice parental line, Norin-PL6, using RFLP markers. *Breeding Science* 48, 243–249, 1998.

Gallardo, F., D. J. Boethel, J. R. Fuxa, and A. Richter. Susceptibility of *Heliothis zea* (Boddie) larvae to *Nomuraea ridleyi* (Farlow) Samson. Effects of α-tomatine at the third trophic level. *J. Chem. Ecol.* 16, 1751–1759, 1990.

Gallun, R. L., and G. S. Khush. Genetic Factors Affecting Expression and Stability of Resistance. In: *Breeding Plants Resistant to Insects,* Maxwell, F. G. and Jennings, P. R., Eds., John Wiley & Sons, New York, 1980.

Gannon, A. J., and C. E. Bach. Effects of soybean trichome density on Mexican bean beetle (Coleoptera: Coccinellidae) development and feeding preference. *Environ. Entomol.* 25, 1077–1082, 1996.

Gatehouse, A. M. R., D. Boulter, and V. A. Hilder. Potential of plant-derived genes in the genetic manipulation of crops for insect resistance. *Plant Genetic Manipulation for Crop Protection* 7, 155–181, 1994.

Gatehouse, A. M. R., G. M. Davison, C. A. Newell, A. Merryweather, W. D. O. Hamilton, E. P. J. Burgess, R. J. C. Gilbert, and J. A. Gatehouse. Transgenic potato plants with enhanced resistance to the tomato moth, *Lacanobia oleracea*: growth room trials. *Mol. Breed.* 3, 49–63, 1997.

George, B. W., and F. D. Wilson. Pink bollworm (Lepidoptera: Gelechiidae) effects of natural infestation on upland and pima cottons untreated and treated with insecticides. *J. Econ. Entomol.* 76, 1152–1155, 1983.

Ghareyazie, B., F. Alinia, C. A. Menguito, L. G. Rubia, J. M. de Palma, E. A. Liwanag, M. B. Cohen, G. S. Kush, and J. Bennett. Enhanced resistance to two stem borers in an aromatic rice containing a synthetic *cryIA(b)* gene. *Mol. Breeding* 3, 401–414, 1997.

Ghidiu, G.M., C. Carter, and C. A. Silcox. The effect of host plant on Colorado potato beetle (*Coleoptera: Chrysomelidae*) susceptibility to pyrethroid insecticides. *Pestic. Sci.* 28, 259–270, 1990.

Gianoli, E., and H. M. Niemeyer. Characteristics of hydroxamic acid induction in wheat triggered by aphid infestation. *J. Chem. Ecol.* 23, 2695–2705, 1997.

Gill, B. S., J. H. Hatchett, and W. J. Raupp. Chromosomal mapping of Hessian fly-resistance gene *H13* in the D genome of wheat. *J. Hered.* 78, 97–100, 1987.

Gould, F. Potential and problems with high-dose strategies for pesticidal engineered crops. *Biocontrol Sci. Technol.* 4, 451–461, 1994.

Gould, F. Sustainability of transgenic insecticidal cultivars: Integrating pest genetics and ecology. *Annu. Rev. Entomol.* 43, 701–726, 1998.

Gould, F., A. Anderson, A. Jones, D. Sumerford, D. J. Heckel, J. Lopez, S. Micinski, R. Leonard, and M. Laster. Initial frequency of alleles for resistance to *Bacillus thuringiensis* toxins in field populations of *Heliothis virescens*. *Proc. Natl. Acad. Sci. U.S.A.* 94, 3519–3523, 1997.

Gould, F., G. G. Kennedy, and M. T. Johnson. Effects of natural enemies on the rate of herbivore adaptation to resistant host plants. *Entomol. Exp. Appl.* 58, 1–14, 1991.

Graham, J., S. C. Gordon, and R. J. McNicol. The effect of the CpTi gene in strawberry against attack by vine weevil (*Otiorhynchus sulcatus* F., Coleoptera: Curculionidae). *Ann. Appl. Biol.* 131, 133–139, 1997.

Green, M. B., and P. A. Hedin, Eds., *Natural Resistance of Plants to Pests. Role of Allelochemicals.* ACS Symposium Series 296. American Chemical Society, Washington, D.C., 1986, p. 243.

Green, T. A., R. J. Prokopy, and D. W. Hosmer. Distance of response to host tree models by female apple maggot flies, *Rhagoletis pomonella* (Walsh) (Diptera: Trephritidae): Interaction of visual and olfactory stimuli. *J. Chem. Ecol.* 20, 2393–2413, 1994.

Guthrie, W. D., and J. L. Jarvis. Plant damage and survival of European corn borer (Lepidoptera: Pyralidae) larvae reared for 22 years on resistant and susceptible inbred lines of maize. *J. Kansas Entom. Soc.* 63, 193–195, 1990.

Guthrie, W. D., J. L. Jarvis, and J. C. Robbins. Damage from infesting maize plants with European corn borer egg masses and larvae. *J. Agric. Entomol.* 1, 6–16, 1984.

Hamilton-Kemp, T. R., R. A. Andersen, J. G. Rodriguez, J. H. Loughrin, and C. G. Patterson. Strawberry foliage headspace vapor components at periods of susceptibility and resistance to *Tetranychus urticae* Koch. *J. Chem. Ecol.* 14, 789–796, 1988.

Hamm, J. J., and B. R. Wiseman. Plant resistance and nuclear polyhedrosis virus for suppression of the fall armyworm (Lepidoptera: Noctuidae). *Florida Entomol.* 69, 549–559, 1986.

Hargrove, T. R., V. L. Cabanilla, and W. R. Coffman. Changes in rice breeding in 10 Asian countries: 1965–84. Diffusion of genetic materials, breeding objectives, and cytoplasm. *International Rice Research Institute Research Paper Series* 111, 1–18, 1985.

Harris, M. K., Ed., *Biology and Breeding for Resistance to Arthropods and Pathogens in Agricultural Plants.* Texas Agric. Exp. Sta. Publ. MP-1451, 1980, p. 605.

Harvey, T. L., and H. L. Hackerott. Recognition of a greenbug biotype injurious to sorghum. *J. Econ. Entomol.* 62, 776–779, 1969.

Harvey, T. L., K. D. Kofoid, T. J. Martin, and P. E. Sloderbeck. A new greenbug biotype virulent to E-biotype resistant sorghum. *Crop Sci.* 31, 1689–1691, 1991.

Harvey, T. L., H. L. Hackerott, T. J. Martin, and W. D. Stegmeier. Mechanical insect dispenser for infesting plants with greenbugs (Homoptera: Aphididae). *J. Econ. Entomol.* 78, 489–492, 1985.

Harvey, T. L., T. J. Martin, and D. L. Seifers. Importance of plant resistance to insect and mite vectors in controlling virus diseases of plants: resistance to the wheat curl mite (Acari: Eriophyidae). *J. Agric. Entomol.* 11, 271–277, 1994.

Harvey, T. L., T. J. Martin, and D. L. Seifers. Survival of five wheat curl mite *Aceria tosichella* Keifer (Acari: Eriophyidae) strains on mite resistant wheat. *Exp. Appl. Acarol.* 19, 459–463, 1995.

Harvey, T. L., G. E. Wilde, and K. D. Kofoid. Designation of a new greenbug biotype, biotype K, injurious to resistant sorghum. *Crop Sci.* 37, 989–991, 1997.

Hattori, M., Y. Sakagami, and S. Marumo. Oviposition deterrents for the limabean pod borer, *Etiella zinckenella* (Treitschke) (Lepidoptera, Pyralidae) from *Populus nigra* L. c.v. *italica* leaves. *Appl. Entomol. Zool.* 27, 195–204, 1992.

Hawthorne, D. J., and S. Via. Variation in performance on two grape cultivars within and among populations of grape phylloxera from wild and cultivated habitats. *Entomol. Exp. Appl.* 70, 63–76, 1994.

Hawthorne, D. J., J. A. Shapiro, W. M. Tingey, and M. A. Mutschler. Trichome-borne and artificially applied acylsugars of wild tomato deter feeding and oviposition of the leafminer *Liriomyza trifolii. Entomol. Exp. Appl.* 65, 65–73, 1992.

Heath, R. L., G. McDonald, J. T. Christeller, M. Lee, K. Bateman, J. West, R. vanHeeswijck, and M. A. Anderson. Proteinase inhibitors from *Nicotiana alata* enhance plant resistance to insect pests. *J. Insect Physiol.* 43, 833–842, 1997.

Hedin, P. A., Ed., *Plant Resistance to Insects.* ACS Symposium Series 62. American Chemical Society, Washington, D.C., 1978, p. 286.

Hedin, P. A., Ed., *Plant Resistance to Insects.* ACS Symposium Series 208, American Chemical Society, Washington, D.C., 1983, p. 374.

Hein, G. L. Influence of plant growth stage on Russian wheat aphid, *Diuraphis noxia* (Homoptera, Aphididae), reproduction and damage symptom expression. *J. Kansas Entomol. Soc.* 65, 369–376, 1992.

Heinrichs, E. A. Host Plant Resistance. In: *Biology and Management of Rice Insects,* Heinrichs, E. A., Ed., Wiley Eastern Limited, New Delhi, 1994.

Heinrichs, E. A., and P. K. Pathak. Resistance to the rice gall midge, *Orseolia oryzae* in rice. *Insect Sci. Applic.* 1, 123–132, 1981.

Heinrichs, E. A., and H. R. Rapusas. Cross-virulence of *Nephotettix virescens* (Homoptera: Cicadellidae) biotypes among some rice cultivars with the same major-resistance gene. *Environ. Entomol.* 14, 696–700, 1985.

Heinrichs, E. A., L. T. Fabellar, R. P. Basilio, T.-C. Wen, and F. Medrano. Susceptibility of rice planthoppers *Nilaparvata lugens* and *Sogatella frucifera* (Homoptera: Delphacidae) to insecticides as influenced by level of resistance in the host plant. *Environ. Entomol.* 13, 455–458, 1984.

Heinz, K. M., and F. G. Zalom. Variation in trichome-based resistance to *Bemisia argentifolii* (Homoptera: Aleyrodidae) oviposition on tomato. *J. Econ. Entomol.* 88, 1494–1502, 1995.

Hilder, V. A., A. M. R. Gatehouse, S. E. Sheerman, R. F. Barker, and D. Boulter. A novel mechanism of insect resistance engineered into tobacco. *Nature* 330, 160–163, 1987.

Hirabayashi, H., and T. Ogawa. RFLP mapping of Bph-1 (Brown planthopper resistance gene) in rice. *Breeding Sci.* 45, 369–371, 1995.

Hoffman, M. P., F. G. Zalom, L. T. Wilson, J. M. Smilanick, L. D. Malyj, J. Kiser, V. A. Hilder, and W. M. Barnes. Field evaluation of transgenic tobacco containing genes encoding *Bacillus thuringensis* delta-endotoxin or cowpea trypsin inhibitor-efficacy against *Helicoverpa zea* (Lepidoptera, Noctuidae). *J. Econ. Entomol.* 85, 2516–2522, 1992.

Holbrook, F.R. An index of acceptability to green peach aphids for *Solanum* germplasm and for a suspected non-host plant. *American Potato Journal* 57, 1–6, 1980.

Hospital, F., L. Moreau, F. Lacoudre, A. Charcosset, and A. Gallais. More on the efficiency of marker-assisted selection. *Theor. Appl. Genet.* 95, 1181–1189, 1997.

Huang, F., R. A. Higgins, and L. L. Buschman. Baseline susceptibility and changes in susceptibility to *Bacillus thuringiensis* subsp. *kurstaki* under selection pressure in European corn borer (Lepidoptera: Pyralidae). *J. Econ. Entomol.* 90, 1137–1143, 1997.

Huang, N., A. Parco, T. Mew, G. Magpantay, S. Mccouch, E. Guiderdoni, J. C. Xu, P. Subudhi, E. R. Angeles, and G. S. Khush. RFLP mapping of isozymes, RAPD and QTLs for grain shape, brown planthopper resistance in a doubled haploid rice population. *Mol. Breeding* 3, 105–113, 1997.

Huang, X. P., and J. A. A. Renwick. Differential selection of host plants by two *Pieris* species — The role of oviposition stimulants and deterrents. *Entomol. Exp. Appl.* 68, 59–69, 1993.

Inayatullah, C., J. A. Webster, and W. S. Fargo. Index for measuring plant resistance to insects. *Southwest. Entomologist* 109, 146–152, 1990.

International Rice Research Institute, Entomology. In: *A Decade of Cooperation and Collaboration Between Sukamandi (AAARD) and IRRI 1972–1982*. International Rice Research Institute, Los Banos, Philippines, 1984.

IPGRI, *Annual Report 1996*. International Plant Genetic Resources Institute, Rome, Italy, 1997, p. 128.

Isenhour, D.J., and B. R. Wiseman, Foliage consumption and development of the fall armyworm (Lepidoptera: Noctuidae) as affected by the interactions of a parasitoid, *Capoletis sonorensis* (Hymenoptera: Ichneumonidae), and resistant corn genotypes. *Environ. Entomol.* 16, 1181–1184, 1987.

Ishii, T., D. S. Brar, D. S. Multani, and G. S. Khush. Molecular tagging of genes for brown planthopper resistance and earliness introgressed from *Oryza australiensis* into cultivated rice, *O. sativa. Genome* 37, 217–221, 1994.

Ishimoto, M., and K. Kitamura. Specific inhibitory activity and inheritance of an alpha-amylase inhibitor in a wild common bean accession resistant to the Mexican bean weevil. *Jpn. J. Breed.* 43, 69–73, 1993.

Ishimoto, M., T. Sato, M. J. Chrispeels, and K. Kitamura. Bruchid resistance of transgenic azuki bean expressing seed alpha-amylase inhibitor of common bean. *Entomol. Exp. Appl.* 79, 309–315, 1996.

Jansen, M. P. T., and N. E. Stamp. Effects of light availability on host plant chemistry and the consequences for behavior and growth of an insect herbivore. *Entomol. Exp. Appl.* 82, 319–333, 1997.

Jelenkovic, G., S. Billings, Q. Chen, J. Lashomb, G. Hamilton, and G. Ghidiu. Transformation of eggplant with synthetic *crylIIA* gene produces a high level of resistance to the Colorado potato beetle. *J. Am. Soc. Hort. Sci.* 123, 19–25, 1998.

Jenkins, E. B., R. B. Hammond, S. K. St. Martin, and R. L. Cooper. Effect of soil moisture and soybean growth stage on resistance to Mexican bean beetle (Coleoptera: Coccinellidae). *J. Econ. Entomol.* 90, 697–703, 1997a.

Jenkins, J. N., J. C. McCarty, R. E. Buehler, J. Kiser, C. Williams, and T. Wofford. Resistance of cotton with delta-endotoxin genes from *Bacillus thuringensis* var. *kurstaki* on selected lepidopteran insects. *Agron. J.* 89, 768–780, 1997b.

Jenkins, J. N., W. L. Parrot, J. E. Jones, W. D. Caldwell, D. T. Bowman, J. M. Brand, A. Coco, J. G. Marhsall, D. J. Boquet, R. Hutchinson, W. Aguillard, and D. F. Clower. Gumbo 500: An improved open-canopy cotton. *La. Agric. Exp. Sta. Circ.* 114, 1–14, 1971.

Johnson, D. E., G. L. Brookhart, K. J. Kramer, B. D. Barnett, and W. H. McGaughey. Resistance to *Bacillus thuringensis* by the Indian meal moth, *Plodia interpunctella*: Comparison of midgut proteinases from susceptible and resistant larvae. *J. Invert. Path.* 55, 235–244, 1990.

Johnson, M. T., F. Gould, and G. G. Kennedy. Effect of an entomopathogen on adaptation of *Heliothis virescens* populations to transgenic host plants. *Entomol. Exp. Appl.* 83, 121–135, 1997.

Jones, J. E., D. James, F. E. Sistler, and S. J. Stringer. Spray penetration of cotton canopies as affected by leaf and bract isolines. *La. Agric.* 29, 15–17, 1986.

Jones, N., H. Ougham, and H. Thomas. Markers and mapping: we are all geneticists now. *New Phytologist* 137, 165–177, 1997.

Jung, P., M. Rohde, and J. Lunderstadt. Induced resistance in the phloem of the European larch *Larix decidua* (Mill) after attack by the great larch bark beetle *Ips cembrae* (Heer) (Col, Scolytidae). *J. Appl. Entomol.* 117, 427–433, 1994.

Kalode, M. B. The Rice Gall Midge-Varietal Resistance and Chemical Control. In: *Rice Improvement in China and Other Asian Countries.* International Rice Research Institute and Chinese Academy of Agricultural Sciences, 1980.

Kanost, M. R., V. P. Sarvamangala, and M. A. Wells. Primary structure of a member of the serpine superfamily of proteinase inhibitors from an insect *Manduca sexta. J. Biol. Chem.* 264, 965–972, 1989.

Karban, R., and G. M. English-Loeb. A "vaccination" of Willamette spider mites (Acari: Tetranychidae) to prevent large populations of Pacific spider mites on grapevines. *J. Econ. Entomol.* 83, 2252–2257, 1990.

Karner, M. A., and G. R. Manglitz. Effects of temperature and alfalfa cultivar on pea aphid (Homoptera: Aphididae) fecundity and feeding activity of convergent lady beetle (Coleoptera: Coccinellidae). *J. Kansas Entomol. Soc.* 58, 131–136, 1985.

Kauffman, W. C., and R. V. Flanders. Effects of variably resistant soybean and lima bean cultivars on *Pediobius foveolatus* (Hymenoptera: Eulophidae), a parasitoid of the Mexican bean beetle, *Epilachna varivestis* (Coleoptera: Coccinellidae). *Environ. Entomol.* 14, 678–682, 1985.

Keep, E., and R.L. Knight. A new gene from *Rubus occidentalis* L. for resistance to strains 1, 2, and 3 of the rubus aphid, *Amphorophora rubi* Kalt. *Euphytica* 16, 209–214, 1967.

Kennedy, G. G. Host plant resistance and the spread of plant viruses. *Environ. Entomol.* 5, 827–832, 1976.

Kennedy, G. G. 2-tridecanone, tomatoes and *Heliothis zea*: Potential incompatility of plant antibiosis with insecticidal control. *Entomol. Exp. Appl.* 35, 305–311, 1984.

Kennedy, G. G., and R. R. Farrar, Jr. Response of insecticide-resistant and susceptible Colorado potato beetles, *Leptinotarsa decemlineata* to 2-tridecanone and resistant tomato foliage: The absence of cross resistance. *Entomol. Exp. Appl.* 45, 187–192, 1987.

Kennedy, G. G., F. Gould, O. M. B. Deponti, and R. E. Stinner. Ecological, agricultural, genetic, and commercial considerations in the deployment of insect-resistant germplasm. *Environ. Entomol.* 16, 327–338, 1987.

Kennedy, G. G., D. L. McLean and M. G. Kinsey. Probing behavior of *Aphis gossyppi* on resistant and susceptible muskmelon. *J. Econ. Entomol.* 71, 13–16, 1978.

Khairallah, M., M. Bohn, D. C. Jewell, J. A. Deutsch, J. Mihm, D. Hoisington, A. Melchinger, and D. González-de-León. Location and Effect of Quantitiative Trait Loci for South-western Corn Borer and Sugarcane Borer Resistance in Tropical Maize. In: *Insect Resistant Maize: Recent Advances and Utilization; Proceedings of an International Symposium, International Maize and Wheat Improvement Center (CIMMYT), Mexico, D. F., 1994,* Mihm, J. A., Ed., CIMMYT, El Batan, Mexico, 1997.

Khan, Z. R., and R. C. Saxena. Electronically recorded waveforms associated with the feeding behavior of *Sogatella furcifera* (Homoptera: Delphacidae) on susceptible and resistant rice varieties. *J. Econ. Entomol.* 77, 1479–1482, 1984.

Khush, G. S. Breeding Rice for Multiple Disease and Insect Resistance. In: *Rice Improvement in China and Other Asian Countries.* International Rice Research Institute. Los Banos, Philippines, 219–238, 1980.

Khush, G. S., and D. S. Brar. Genetics of Resistance to Insects in Crop Plants. *Adv. Agron.* 45: 45, 223–274, 1991.

Kindler, S. D., and S. M. Spomer. Biotypic status of six greenbug (Homoptera: Aphididae) isolates. *Environ. Entomol.* 15, 567–572, 1986.

Kindler, S. D., and R. Staples. Nutrients and the reaction of two alfalfa clones to the spotted alfalfa aphid. *J. Econ. Entomol.* 63, 938–940, 1970.

Kishore, P. Integration of host-plant resistance and chemical control for the management of sorghum stem-borer. *Indian J. Agric. Sci.* 54, 131–133, 1984.

Kitch, L. W., R. E. Shade, W. E. Nyquist, and J. D. Axtell. Inheritance of density of erect glandular trichomes in the genus *Medicago. Crop Sci.* 25, 607–611, 1985.

Kleiner, K. W., D. D. Ellis, B. H. Mccown, and K. F. Raffa. Field evaluation of transgenic poplar expressing a *Bacillus thuringensis* cry1A(a) d-endotoxin gene against forest tent caterpillar (Lepidoptera: Lasiocampidae) and gypsy moth (Lepidoptera: Lymantriidae) following winter dormancy. *Environ. Entomol.* 24, 1358–1364, 1995.

Klopfenstein, N. B., H. S. McNabb, E. R. Hart, R. B. Hall, R. D. Hanna, S. A. Heuchelin, K. K. Allen, N.-Q. Shi, and R. W. Thornburg. Transformation of *Populus* hybrids to study and improve pest resistance. *Silvae Genet.* 42, 2–3, 1993.

Kogan, M., and J. Paxton. Natural inducers of plant resistance to insects. In: *Plant Resistance to Insects,* ACS Symposium Series 208, Hedin, P. A. Ed., American Chemical Society, Washington, D.C., 1983.

Kopp, M. U., A. J. de Mello, and A. Manz. Chemical amplification: Continuous-flow PCR on a chip. *Science* 280, 1046–1048, 1998.

Koziel, M. G., G. L. Beland, C. Bowman, N.B. Carozzi, R. Crenshaw, L. Crossland, J. Dawson, N. Desai, M. Hill, S. Kadwell, K. Launis, K. Lewis, D. Maddox, K. McPherson, M. R. Meghji, E. Merlin, R. Rhodes, G. W. Warren, M. Wrightand, and S. V. Evola. Field performance of elite transgenic maize plants expressing an insecticidal protein derived from *Bacillus thuringiensis. Bio/Technology* 11, 194–200, 1993.

Kreitner, G. L., and E. L. Sorensen. Glandular trichomes on *Medicago* species. *Crop Sci.* 19, 380–384, 1979.

Kumar, H. Resistance in maize to *Chilo partellus* (Lepidoptera, Pyralidae) in relation to crop phenology, larval rearing medium, and larval development stages. *J. Econ. Entomol.* 86, 886–890, 1993.

Kumar, H., and G. O. Asino. Resistance of maize to *Chilo partellus* (Lepidoptera, Pyralidae) — effect of plant phenology. *J. Econ. Entomol.* 86, 969–973, 1993.

Kumar, H., and J. A. Mihm. Antibiosis and tolerance to fall armyworm, *Spodoptera frugiperda* (J. E. Smith), southwestern corn borer, *Diatraea grandiosella* Dyar and sugarcane borer, *Diatraea saccharalis* Fabricius, in selected maize hybrids and varieties. *Maydica* 40, 245–251, 1995.

Kumar, A., S. S. Baghel, and M. N. Shrivastava. Inheritance and allelic relations of genes governing resistance to biotype 1 of rice-gall midge (*Orseolia oryzae*). *Indian J. Agr. Sci.* 64, 504–506, 1994.

Lammerink, J. A new biotype of cabbage aphid, *Brevicoryne brassicae* L. on aphid resistant rape (*Brassica napus* L.). *N. Z. J. Agric. Res.* 11, 341–344, 1968.

Lampert, E. P., D. L. Haynes, A. J. Sawyer, D.P. Jokinen, S. G. Wellso, R.L. Gallun, and J. J. Roberts. Effects of regional releases of resistant wheats on the population dynamics of the cereal leaf beetle (Coleoptera: Chrysomelidae). *Ann. Entomol. Soc. Am.* 76, 972–980, 1983.

Lapis, E. B., and J. H. Borden. Olfactory discrimination by *Heteropsylla cubana* (Homoptera, Psyllidae) between susceptible and resistant species of *Leucaena* (Leguminosae). *J. Chem. Ecol.* 19, 83–90, 1993.

Lara, F. M., *Principios de Resistancia de Plantas a Insectos*. (Portuguese). Piracicaba, Livro-ceres, Brazil, 1979, p. 207.

Laska, P., J. Betlach, and M. Havrankova. Variable resistance in sweet pepper, *Capsicum annuum*, to glasshouse whitefly, *Trialeurodes vaporariorum* (Homoptera, Aleyrodidae). *Acta Ent. Bohemoslov.* 83, 347–353, 1986.

Lee, E. A., P. F. Byrne, M. D. McMullen, M. E. Snook, B. R. Wiseman, N. W. Widstrom, and E. H. Coe. Genetic mechanisms underlying apimaysin and maysin synthesis and corn earworm antibiosis in maize (*Zea mays* L.) *Genetics* 149, 1997–2006, 1998.

Lee, M. DNA markers and plant breeding programs. *Adv. Agron.* 55, 265–344, 1995.

Leple, J. C., M. Bonadebottino, S. Augustin, G. Pilate, V. D. Letan, A. Delplanque, D. Cornu, and L. Jouanin. Toxicity to *Chrysomela tremulae* (Coleoptera: Chrysomelidae) of trans-genic poplars expressing a cysteine proteinase inhibitor. *Mol. Breeding* 1, 319–328, 1995.

Leru, B., and M. Tertuliano. Tolerance of different host-plants to the cassava mealybug *Phenacoccus manihoti* Matile-Ferrero (Homoptera, Pseudococcidae). *Intl. J. Pest Manage.* 39, 379–384, 1993.

Letourneau, D. K. Associational resistance in squash monocultures and polycultures in tropical Mexico. *Environ. Entomol.* 15, 285–292, 1986.

Leuck, D. B. Induced fall armyworm resistance in pearl millet. *J. Econ. Entomol.* 65, 1608–1611, 1972.

Llewellyn, D., Y. Cousins, A. Mathews, L. Hartweck, and B. Lyon. Expression of *Bacillus thuringensis* insecticidal protein genes in transgenic crop plants. *Agr. Ecosyst. Environ.* 49, 85–93, 1994.

Lombardiboccia, G., M. Carbonaro, and E. Carnovale. Trypsin and chymotrypsin inhibitors from a wild species and a domestic species of cowpea (*Vigna unguiculata*). *Food Sci. Technol.-Lebensm. Wiss.* 24, 370–372, 1991.

Lowe, H. J. B. Resistance and susceptibility to colour forms of the aphid *Sitobion avenae* in spring and winter wheats (*Triticum aestivum*). *Ann. App. Biol.* 99, 87–98, 1981.

Luginbill, P., Jr., Developing resistant plants — the ideal method of controlling insects. *U.S. Dep. Agric. ARS Prod. Res. Rep.* 111, 1–14, 1969.

Lutz, D. Microfluidic widgets. *Am. Sci.* 85, 321–322, 1997.

Ma, R. Z., J. C. Reese, W. C. Black IV, and P. Bramel-Cox. Chlorophyll loss of a greenbug-susceptible sorghum due to pectinases and pectin fragments. *J. Kansas Entomol. Soc.* (In Press), 1998.

Ma, Z.Q., B. S. Gill, M. E. Sorrells, and S. D. Tanksley. RFLP markers linked to 2 Hessian fly-resistance genes in wheat (*Triticum aestivum* L. from *Triticum tauschii* (Coss) Schmal. *Theor. Appl. Genet.* 85, 750–754, 1993.

Maliepaard, C., N. Bas, S. van Heusden, J. Kos, G. Pet, R. Verkerk, R. Vrielink, P. Zabel, and P. Lindhout. Mapping of QTLs for glandular trichome densities and *Trialeurodes vaporariorum* (greenhouse whitefly) resistance in an F_2 from *Lycopersicon esculentum* × *Lycopersicon hirsutum f. glabratum. Heredity* 75, 425–433, 1995.

Manuwoto, S., and J. M. Scriber. Consumption and utilization of three maize genotypes by the southern armyworm. *J. Econ. Entomol.* 75, 163–167, 1982.

Maramorosch, K. Insects and Plant Pathogens. In: *Breeding Plants Resistant to Insects.* Maxwell, F. G. and Jennings, P. R., Eds., John Wiley and Sons, New York, 1980.

Marconi, E., N. Q. Ng., and E. Carnovale. Protease inhibitors and lectins in cowpea. *Food Chem.* 47, 37–40, 1993.

Masoud, S. A., L. B. Johnson, F. F. White, and G. R. Reeck. Expression of a cysteine proteinase inhibitor (oryzacystatin-I) in transgenic tobacco plants. *Plant Mol. Biol.* 21, 655–663, 1993.

Maxwell, F. G., and P. R. Jennings, Eds. *Breeding Plants Resistant to Insects.* John Wiley and Sons, New York, 1980, p. 683.

Maxwell, F. G., J. N. Jenkins, and W. L. Parrott. Resistance of plants to insects. *Adv. Agron.* 24, 187–265, 1972.

McAuslane, H. T. Alborn, and J. P. Toth. Systemic induction of terpenoid aldehydes in cotton pigment glands by feeding of larval *Spodoptera exigua. J. Chem. Ecol.* 23, 2861–2879, 1997.

McCloskey, C., and M. B. Isman. Plant growth stage effects on the feeding and growth responses of the bertha armyworm, *Mamestra configurata* to canola and mustard foliage. *Entomol. Exp. Appl.* 74, 55–61, 1995.

McCloud, E. S., and M. R. Berenbaum. Stratospheric ozone depletion and plant-insect interactions: Effects of UVB radiation on foliage quality of *Citrus jambhiri* for *Trichoplusia ni. J. Chem. Ecol.* 20, 525–539, 1994.

McGaughey, W. H. Insect resistance to the biological insecticide *Bacillus thuringiensis. Science.* 229, 193–195, 1985.

McGaughey, W. H., and R. W. Beeman. Resistance to *Bacillus thuringiensis* in colonies of Indian meal moth and almond moth (Lepidoptera: Pyralidae). *J. Econ. Entomol.* 81, 28–33, 1989.

McGaughey, W. H., and D. E. Johnson. Managing insect resistance to *Bacillus thuringiensis* toxins. *Science* 258, 1451–1455, 1992.

McGovern, W.L., and W. H. Cross. Effects of two cotton varieties on levels of boll weevil parasitism (Coleoptera: Curculionidae). *Entomophaga* 21, 123–125, 1976.

McLean, D. L., and M. G. Kinsey. A technique for electronically recording aphid feeding and salivation. *Nature* 202, 1358–1359, 1964.

Mihm, J. A. Efficient mass-rearing and infestation techniques to screen for host plant resistance to fall armyworm, *Spodoptera frugiperda.* Centro Internacional de Mejoramiento de Maiz y Trigo, El Batan, Mexico. 1983a.

Mihm, J. A. Efficient mass-rearing and infestation techniques to screen for host plant resistance to maize stem borers, *Diatraea* sp. Centro Internacional de Mejoramiento de Maiz y Trigo, El Batan, Mexico. 1983b.

Mihm, J. A. Techniques for efficient mass rearing and infestation in screening for host plant resistance to corn earworm, *Heliothis zea.* Centro Internacional de Mejoramiento de Maiz y Trigo. El Batan, Mexico. 1982.

Mihm, J. A., Ed. *Insect Resistant Maize: Recent Advances and Utilization,* Proceedings of an International Symposium, International Maize and Wheat Improvement Center (CIM-MYT), Mexico, D. F., 1994, CIMMYT, El Batan,Mexico,1997, p. 302.

Mihm, J. A., F. B. Peairs, and A. Ortega. New procedures for efficient mass production and artificial infestation with lepidopterous pest of maize. *CIMMYT Review,* 1978, 138 pp.

Miller, D. L., U. Rahardja, and M. E. Whalon. Development of a strain of Colorado potato beetle resistant to the delta-endotoxin of B.t. *Pest Resistance Manage.* 2, 25, 1990.

Mitchell, H. C., W. H. Cross, W. L. McGovern, and E. M. Dawson. Behavior of the boll weevil on frego bract cotton. *J. Econ. Entomol.* 66, 677–680, 1973.

Moar, W. J., M. Pustzi-Carey, H. Van Faassen, D. Bosch, R.Frutos, C. Rang, K. Luo, and M.J. Adang. Development of *Bacillus thuringiensis* Cry1C resistance by *Spodoptera exigua* (Hubner) (Lepidoptera: Noctuidae). *Appl. Environ. Microbiol.* 61, 2086–2092, 1995.

Mohan, M., S. Nair, J. S. Bentur, U. P. Rao, and J. Bennett. RFLP and RAPD mapping of the rice *Gm2* gene confers resistance to biotype 1 of gall midge (*Orseolia oryzae*). *Theor. Appl. Genet.* 87, 782–788, 1994.

Mohan, M., S. Nair, A. Bhagwat, T. G. Krishna, M. Yano, M. C. R. Bhatia, and T. Sasaki. Genome mapping, molecular markers and marker-assisted selection in crop plants. *Mol. Breeding* 3, 87–103, 1997.

Montandon, R., R. D. Stipanovic, H. J. Williams, W. L. Sterling, and S. B. Vinson. Nutritional indices and excretion of gossypol by *Alabama argillacea* (Hubner) and *Heliothis virescens* (F.) (Lepidoptera: Noctuidae) fed glanded and glandless cotyledonary cotton leaves. *J. Econ. Entomol.* 80, 32–36, 1987.

Muller, B. S., R. J. Robinson, J. A. Johnson, E. T. Jones, and B. W. X. Ponnaiya. Studies on the relation between silica in wheat plants and resistance to Hessian fly attack. *J. Econ. Entomol.* 53, 945–949, 1960.

Mutikainen, P., M. Walls, and J. Ovaska. Herbivore-induced resistance in *Betula pendula*: The vole of plant vascular architecture. *Oecologia.* 108, 723–727, 1996.

Mutschler, M. A., R. W. Doerge, S.-C. Liv, J. P. Kuai, B. E. Liedl, and J. A. Shapiro. QTL analysis of pest resistance in the wild tomato *Lycopersicon pennellii*: QTL's controlling acylsugar level and composition. *Theor. Appl. Genet.* 92, 709–718, 1996.

Myers, G. O., C. A. Fatokun, and N. D. Young. RFLP mapping of an aphid resistance gene in cowpea (*Vigna unguiculata* L. Walp). *Euphytica* 91, 181–187, 1996.

Nair, S., J. S. Bentur, U. P. Rao, and M. Mohan. DNA markers tightly linked to a gall midge resistance gene (*Gm2*) are potentially useful for marker-aided selection in rice breeding. *Theor. Appl. Genet.* 91, 68–73, 1995.

Nair, S., A. Kumar, M.N. Srivastava, and M. Mohan. PCR-based DNA markers linked to a gall midge resistance gene, *Gm4t,* has potential for marker-aided selection in rice. *Theor. Appl. Genet.* 92, 660–665, 1996.

National Research Council. *Managing Global Resources. The U. S. National Plant Germplasm System/Committee on Managing Global Genetic Resources: Agricultural Imperatives.* National Academy Press, Washington, D.C., 1991, p. 171.

Nault, B. A., J. N. All, and H. R. Boerma. Influence of soybean planting date and leaf age on resistance to corn earworm (Lepidoptera, Noctuidae). *Environ. Entomol.* 21, 264–268, 1992.

Nguessan, F. K., S. S. Quisenberry, and T. P. Croughan. Evaluation of rice anther culture lines for tolerance to the rice water weevil (Coleoptera, Curculionidae). *Environ. Entomol.* 23, 331–336, 1994.

Nielson, M. W., and H. Don. Probing behavior of biotypes of the spotted alfalfa aphid on resistant and susceptible alfalfa clones. *Entomol. Exp. Appl.* 17, 477–486, 1974.

Nielson, M. W., and W. F. Lehman. Breeding Approaches in Alfalfa. In: *Breeding Plants Resistant to Insects,* Maxwell, F. G. and Jennings, P. R., Eds., John Wiley & Sons, New York, 1980.

Nieto-Lopez, R. M., and T. K. Blake. Russian wheat aphid resistance in barley: Inheritance and linked molecular markers. *Crop Sci.* 34, 655–659, 1994.

Nihoul, P. Do light intensity, temperature and photoperiod affect the entrapment of mites on glandular hairs of cultivated tomatoes. *Exp. Appl. Acarol.* 17, 709–718, 1993.

Nihoul, P. Phenology of glandular trichomes related to entrapment of *Phytoseiulus ersimilis* A.-H. in the glasshouse tomato. *J. Hort. Sci.* 69, 783–789, 1994.

Nkansah-poku, J., and C. J. Hodgson. Interaction between aphid resistant cowpea cultivars, three clones of cowpea aphid, and the effect of two light intensity regimes on this interaction. *Int. J. Pest Manage.* 41, 161–165, 1995.

Obrycki, J. J., M. J. Tauber, and W. M. Tingey. Predator and parasitoid interaction with aphid-resistant potatoes to reduce aphid densities: A two-year field study. *J. Econ. Entomol.* 76, 456–462, 1983.

Oghiakhe, S. Effect of pubescence in cowpea resistance to the legume pod borer *Maruca testulalis* (Lepidoptera: Pyralidae). *Crop Prot.* 14, 379–387, 1995.

Orr, D. B., and D. J. Boethel. Comparative development of *Copidosoma truncatellum* (Hymenoptera: Eucyrtidae) and its host, *Pseudoplusia includens* (Lepidoptera: Noctuidae), on resistant and susceptible soybean genotypes. *Environ. Entomol.* 14, 612–616, 1985.

Painter, R. H. *Crops that Resist Insects Provide a Way to Increase World Food Supply.* Kansas Agric. Exp. Stn. Bull. 1968.

Painter, R. H., *Insect Resistance in Crop Plants.* University of Kansas Press, Lawrence, KS, 1951, p. 520.

Painter, R. H. Resistance of plants to insects. *Annu. Rev. Entomol.* 3, 267–290, 1958.

Painter, R. H., and M. D. Pathak. The distinguishing features and significance of the four biotypes of the corn leaf aphid *Rhopalosiphum maidis* (Fitch). *Proc. XI Int. Cong. Entomol.* 2, 110–115, 1962.

Palaniswamy, P., and R. P. Bodnaryk. A wild *Brassica* from Sicily provides trichome-based resistance against flea beetles, *Phyllotreta cruciferae* (Goeze) (Coleoptera: Chrysomelidae). *Can. Entomol.* 126, 1119–1130, 1994.

Palaniswamy, P., and R. J. Lamb. Wound-induced antixenotic resistance to flea beetles, *Phyllotreta cruciferae* (Goeze) (Coleoptera, Chrysomelidae), in crucifers. *Can. Entomol.* 125, 903–912, 1993.

Panda, N. *Principles of Host-Plant Resistance to Insect Pests.* Allanheld, Osmun & Co. and Universe Books, New York, 1979, p. 386.

Panda, N., and G. S. Kush. *Host-Plant Resistance to Insects.* CAB/International Rice Research Institute, Wallingford, Oxon, U.K., 1995, p. 431.

Pantoja, A., C. M. Smith, and J. F. Robinson. Evaluation of rice germ plasm for resistance to the fall armyworm (Lepidoptera: Noctuidae). *J. Econ. Entomol.* 79, 1319–1323, 1986.

Park, S. J., P. R. Timmins, D. T. Quiring, and P.Y. Jul. Inheritance of leaf area and hooked trichome density of the first trifoliolate leaf in common bean (*Phaseolus vulgaris* L). *Can. J. Plant Sci.* 74, 235–240, 1994.

Plucknett, D. L., N. J. H. Smith, J. T. Williams, and N. M. Anishetty. *Gene Banks and the World's Food.* Princeton University Press, Princeton, NJ, 1987, p. 247.

Porter, K. B., G. L. Peterson, and O. Vise. A new greenbug biotype. *Crop Sci.* 22, 847–850, 1982.

Powell, J. E., and L. Lambert. Effects of three resistant soybean genotypes on development of *Microplitis croceipes* and leaf consumption by its *Heliothis* spp. hosts. *J. Agric. Entomol.* 1, 169–176, 1984.

Powell, W., M. Morgante, C. Andre, M. Hanafey, J. Vogel, S. Tingey, and A. Rafalski. The comparison of RFLP, RAPD, AFLP and SSR (microsatellite) markers for germplasm analysis. *Mol. Breeding.* 2, 225–238, 1996.

Prokopy, R. J., S. R. Diehl, and S. S. Cooley. Behavioral evidence for host races in *Rhagoletis pomonella* flies. *Oecologia* 76, 138–147, 1988.

Puterka, G. J., and D. C. Peters. Genetics of greenbug (Homoptera: Aphididae) virulence to resistance in sorghum. *J. Econ. Entomol.* 88, 421–429, 1995.

Puterka, G. J., D. C. Peters, D. L. Kerns, J. E. Slosser, L. Bush, D. W. Worrall, and R. W. McNew. Designation of two new greenbug (Homoptera: Aphididae) biotypes G and H. *J. Econ. Entomol.* 81, 1754–1759, 1988.

Quiring D. T., P. R. Timmins, and S. J. Park. Effect of variations in hooked trichome densities of *Phaseolus vulgaris* on longevity of *Liriomyza trifolii* (Diptera, Agromyzidae) adults. *Environ. Entomol.* 21, 1357–1361, 1992.

Ramachandran, S., G. D. Buntin, J. N. All, B. E. Tabashnik, P. L. Raymer, M. J. Adang, D. A. Pullman, and C. N. Stewart, Jr. Survival, development and oviposition of resistant diamondback moth (Lepidoptera: Plutellidae) on transgenic canola producing a *Bacillus thuringiensis* toxin. *J. Econ. Entomol.* 91, 1239–1244, 1998.

Ramalho, F. S., W. L. Parrott, J. N. Jenkins, and J. C. McCarty, Jr. Effects of cotton leaf trichomes on the mobility of newly hatched tobacco budworms (Lepidoptera: Noctuidae). *J. Econ. Entomol.* 77, 619–621, 1984.

Ratanatham, S., and R. L. Gallun. Resistance to Hessian fly (Diptera: Cecidomyidae) in wheat as affected by temperature and larval density. *Environ. Entomol.* 15, 305–310, 1986.

Ratcliffe, R. H., and J. H. Hatchett. Biology and Genetics of the Hessian Fly and Resistance in Wheat. In: *New Developments in Entomology,* Bobdari, K., Ed., Research Signpost, Scientific Information Guild, Trivandrum, 1997.

Ratcliffe, R. H., G. C. Safranski, F. L. Patterson, H. W. Ohm, and P. L. Taylor. Biotype status of Hessian fly (Diptera: Cecidomyidae) populations from the eastern United States and their responses to 14 Hessian fly resistance genes. *J. Econ. Entomol.* 87, 1113–1121, 1994.

Reed, D. K., J. A. Webster, B. C. Jones, and J. D. Burd. Tritrophic relationships of Russian wheat aphid (Homoptera: Aphididae), a hymenopterous parasitoid (*Diaeretiella rapae* McIntosh), and resistant and susceptible small grains. *Biol. Con.* 1, 35–41, 1991.

Reese, J. C., J. R. Schwenke, P. S. Lamont, and D. D. Zehr. Importance and quantification of plant tolerance in crop pest management programs for aphids: Greenbug resistance in sorghum. *J. Agr. Entomol.* 11, 255–270, 1994.

Reichardt, P. B., T. P. Clausen, and J. P. Bryant. Role of phenol glycosides in plant-herbivore interactions. *Handbook of Natural Toxins,* 6, 313–333, 1991.

Reissig, W. H., E. A. Heinrichs, L. Antonio, M. M. Salac, A. C. Santiago, and A. M. Tenorio. Management of pest insects of rice in farmers' fields in the Philippines. *Prot. Ecol.* 3, 203–218, 1981.

Reynolds, G. W., and C. M. Smith. Effects of leaf position, leaf wounding, and plant age of two soybean genotypes on soybean looper (Lepidoptera: Noctuidae) growth. *Environ. Entomol.* 14, 475–478, 1985.

Rhim, S. L., H. J. Cho, B. D. Kim, W. Schnetter, and K. Geider. Development of insect resistance in tomato plants expressing the delta-endotoxin gene of *Bacillus thuringensis* subsp. *tenebrionis. Mol. Breeding* 1, 229–236, 1995.

Riggin, T. M., B. R. Wiseman, D. J. Isenhour, and K. E. Espelie. Incidence of fall armyworm (Lepidoptera: Noctuidae) parasitoids on resistant and susceptible corn genotypes. *Environ. Entomol.* 21, 888–895, 1992.

Robinson, J. F., J. A. Klun, W. D. Guthrie, and T. A. Brindley. European corn borer leaf feeding resistance: A simplified technique for determing relative differences in concentrations of 6-methoxybenzoxazolinone (Lepidoptera: Pyralidae). *J. Kans. Ent. Soc.* 55, 297–301, 1982 .

Roche, P., F. H. Alston, C. Maliepaard, K.M. Evans, R. Vrielink, F. Dunemann, T. Markussen, S. Tartarini, L. M. Brown, C. Ryder, and G. J. King. RFLP and RAPD markers linked to the rosy leaf curling aphid resistance gene (Sd(1)) in apple. *Theor. Appl. Genet.* 94, 528–533, 1997.

Rodriguez, J. G., D. A. Reicosky, and C. G. Patterson. Soybean and mite interactions: Effects of cultivar and plant growth stage. *J. Kansas Entomol. Soc.* 56, 320–326, 1983.

Roessingh, P., E. Stadler, G. R. Fenwick, L. A. Lewis, J. K. Nielsen, J. Hurter, and T. Ramp. Oviposition and tarsal chemoreceptors of the cabbage root fly are stimulated by glucosinolates and host plant extracts. *Entomol. Exp. Appl.* 65, 267–282, 1992.

Rojanaridpiched, C., V. E. Gracen, H. L. Everett, J. C. Coors, B. F. Pugh, and P. Bouthyette. Multiple factor resistance in maize to European corn borer. *Maydica* 29, 305–315, 1984.

Rose, R. L., T. C. Sparks, and C. M. Smith. Insecticide toxicity to larvae of *Pseudoplusia includens* (Walker) and *Anticarsia gemmatalis* (Hubner) (Lepidoptera) as influenced by feeding on resistant soybean (PI227687) leaves and coumestrol. *J. Econ. Entomol.* 81, 1288–1294, 1988.

Roseland, C. R., and T. J. Grosz. Induced responses of common annual sunflower, *Helianthus annuus* L., from geographically diverse populations and deterrence to feeding by sunflower beetle. *J. Chem. Ecol.* 23, 517–542, 1997.

Roush, R. T. Bt-transgenic crops: just another pretty insecticide or a chance for a new start in resistance management? *Pesticide Science* 51, 328–334, 1997.

Russell, G. E., *Plant Breeding for Pest and Disease Resistance.* Butterworth Publishers, Boston, 1978, p. 485.

Sachs, E. S., J. H. Benedict, J. F. Taylor, D. M. Stelly, S. K. Davis, and D. W. Altman. Pyramiding CryIA(b) insecticidal protein and terpenoids in cotton to resist tobacco budworm (Lepidoptera: Noctuidae). *Environ. Entomol.* 25, 1257–1266, 1996.

Sane, V. A., P. Nath, Aminuddin, and P. V. Sane. Development of insect-resistant transgenic plants using plant genes: Expression of cowpea trypsin inhibitor in transgenic tobacco plants. *Curr. Sci.* 72, 741–747, 1997.

Sanford, L. L., and T. L. Ladd, Jr. Selection for resistance to potato leafhopper in potatoes. III. Comparison of two selection procedures. *Am. Potato J.* 60, 653–659, 1983.

Sato, A., and K. Sogawa. Biotypic variations in the green rice leafhopper, *Nephotettix virescens* Uhler (Homoptera: Deltocephalidae). *Appl. Ent. Zool.* 16, 55–57, 1981.

Schalk, J. M., and R. H. Ratcliffe. Evaluation of ARS program on alternative methods of insect control: Host plant resistance to insects. *Bull. Entomol. Soc. Am.* 22, 7–10, 1976.

Schena, M., D. Shalon, R. W. Davis, and P. O. Brown. Quantitative monitoring of gene expression patterns with a complementary DNA microarray. *Science* 270, 467–470, 1995.

Schön, C. C., M. Lee, A. E. Mechinger, W. D. Guthrie, and W. L. Woodman. Mapping and characterization of quantitative trait loci affecting resistance against second-generation European corn borer in maize with the aid of RFLPs. *Heredity* 70, 648–659, 1993.

Schoonhoven, L. M., W. M. Blaney, and M. S. J. Simmonds. Sensory Coding of Feeding Deterrents in Phytophagous Insects. *Insect-Plant Interactions IV,* Bernays E., Ed., CRC Press, Boca Raton, FL, 1992.

Schotzko, D. J., and C. M. Smith. Effects of preconditioning host plants on population development of Russian wheat aphids (Homoptera: Aphididae). *J. Econ. Entomol.* 84, 1083–1087, 1991.

Schroeder, H. E., S. Gollasch, A. Moore, L. M. Tabe, S. Craig, D. C. Hardie, M. J. Chrispeels, D. Spencer, and T. J. V. Higgins. Bean alpha-amylase inhibitor confers resistance to the pea weevil (*Bruchus pisorum*) in transgenic peas (*Pisum sativum* L). *Plant Physiol.* 107, 1233–1239, 1995.

Schwessing, F. C., and G. Wilde. Temperature and plant nutrient effects on resistance of seedling sorghum to the greenbug. *J. Econ. Entomol.* 72, 20–23, 1979.

Seifelnasr, Y. E. Influence of olfactory stimulants on resistance and susceptibility of pearl millet, *Pennisetum americanum* to the rice weevil, *Sitophilus oryzae. Entomol. Exp. Appl.* 59, 163–168, 1991.

Sen Gupta, G. C., and P. W. Miles. Studies on the susceptibility of varieties of apple to the feeding of two strains of wooly aphids (Homoptera) in relation to the chemical content of the tissues of the host. *Aust. J. Agric. Res.* 26, 157–168, 1975.

Seo, Y. W., J. W. Johnson, and R. L. Jarret. 1997. A molecular marker associated with the H21 Hessian fly resistance gene in wheat. *Mol. Breeding* 3, 177–181, 1997.

Shanks, C. H. Jr., and D. Chase. Electrical measurement of feeding by the strawberry aphid on susceptible and resistant strawberries and nonhost plants. *Ann. Entomol. Soc. Am.* 69, 784–786, 1976.

Shifriss, O. Do *Curcurbita* plants with silvery leaves escape virus infection? *Curcurbit Gen. Coop. Rep.* 4, 42–45, 1981.

Siemens, D. H., and T. Mitchellolds. Glucosinolates and herbivory by specialists (Coleoptera: Chrysomelidae, Lepidoptera: Plutellidae): Consequences of concentration and induced resistance. *Environ. Entomol.* 25, 1344–1353, 1996.

Singh, D. P. *Breeding for Resistance to Diseases and Insect Pests.* Springer-Verlag, New York, 1986, p. 222.

Singh, P., and R. F. Moore., Eds., *Handbook of Insect Rearing.* Vol. 1. Elsevier Science Publishing Co., New York, 1985, p. 481.

Singh, S. R., and R. H. Painter. Effect of temperature and host plants on progeny production of four biotypes of corn leaf aphid. *J. Econ. Entomol.* 75, 348–350, 1964.

Smith, C. M., *Plant Resistance to Insects — A Fundamental Approach.* John Wiley & Sons, New York, 1989, p. 286.

Smith, C. M., Z. R. Khan, and M. D. Pathak, *Techniques for Evaluating Insect Resistance in Crop Plants.* Lewis Publishers, Boca Raton, FL, 1994, p. 320.

Smith, C. M., S. S. Quisenberry, and F. du Toit. The Value of Conserved Wheat Germplasm Possessing Arthropod Resistance, in *Global Plant Genetic Resources for Insect Resistant Crops,* Clement, S. L. and Quisenberry, S. S., Eds., CRC Press, Boca Raton, FL, 1998.

Snelling, R. O. Resistance of plants to insect attack. *Bot. Rev.* 7, 543–586, 1941.

Snyder, J. C., Z. H. Guo, R. Thacker, J. P. Goodman, and J. Stpyrek. 2,3-dihydrofarnesoic acid, a unique terpene from trichomes of *Lycopersicon hirsutum,* repels spider mites. *J. Chem. Ecol.* 19, 2981–2997, 1993.

Starks, K. J., R. Muniappan, and R. D. Eikenbary. Interaction between plant resistance and parasitism against the greenbug on barley and sorghum. *Ann. Entomol. Soc. Am.* 65, 650–655, 1972.

Staub, J. E., F. C. Serquen, and M. Gupta. Genetic markers, map construction, and their application in plant breeding. *Hortscience* 31, 729–741, 1996.

Stettens, J. C., and D. S. Walters. Biochemical Aspects of Glandular Trichome-Mediated Insect Resistance in the Solanaceae. In: *Naturally Occurring Pest Bioregulators,* Hedin, P. A., Ed., ACS Symposium Series 449. American Chemical Society, Washington, D.C., 1991.

Stipanovic, D. D. Function and Chemistry of Plant Trichomes and Glands in Insect Resistance: Protective Chemicals in Plant Epidermal Glands and Appendages. In: *Plant Resistance to Insects.* Hedin, P. A., Ed., ACS Symposium Series 208. American Chemical Society, Washington, D.C., 1983.

Stone, T. B., S. R. Sims, and P. G. Marrone. Selection of tobacco budworm for resistance to genetically engineered *Pseudomonas fluorescens* containing the δ-endotoxin of *Bacillus thuringiensis* subsp. *kurstaki. J. Invertebr. Pathol.* 53, 228–234, 1989.

Stoner, K. A. Glossy leaf wax and host-plant resistance to insects in *Brassica oleracea* L. under natural infestations. *Environ. Entomol.* 19, 730–739, 1990.

Stout, M. J., and S. S. Duffey. Characterization of induced resistance in tomato plants. *Entomol. Exp. Appl.* 79, 273–283, 1996.

Takita, T., and H. Hashim. Relationship between laboratory-developed biotypes of green leafhopper and resistant varieties of rice in Malaysia. *Jpn. Agric. Res. Quart.* 19, 219–223, 1985.

Taksdal, G. The complementary effects of plant resistance and reduced pesticide dosage in field experiments to control the turnip root fly, *Delia floralis*, in swedes. *Ann. Appl. Biol.* 120, 117–125, 1992.

Tao, R., A. M. Dandekar, S. L. Uratsu, P. V. Vail, and J. S. Tebbets. Engineering genetic resistance against insects in Japanese persimmon using the cryIA© gene of *Bacillus thuringensis. J. Am. Soc. Hort. Sci.* 122, 764–771, 1997.

Tarn, T. R., and J. B. Adams. Aphid Probing and Feeding, Electronic Monitoring, and Plant Breeding. In: *Pathogens, Vectors, and Plant Diseases Approaches to Control,* Harris K. F. and Maramorosch, K., Eds., Academic Press, New York, 1982.

Teetes, G. L., M. I. Becerra, and G. C. Peterson. Sorghum midge (Diptera: Cecidomyidae) management with resistant sorghum and insecticide. *J. Econ. Entomol.* 79, 1091–1095, 1986.

Teetes, G. L., C. A. Schaefer, J. R. Gipson, R. C. McIntyre, and E. E. Latham. Greenbug resistance to organophosphorous insecticides on the Texas high plains. *J. Econ. Entomol.* 68, 214–216, 1975.

Thaler, J. S., and R. Karban. A phylogenetic reconstruction of constitutive and induced resistance in *Gossypium. Am. Nat.* 149, 1139–1146, 1997.

Thomas, J. C., D. G. Adams, V. D. Keppenne, C. C. Wasmann, J. K Brown, M. R. Kanost, and H. J. Bohnert. Protease inhibitors of *Manduca sexta* expressed in transgenic cotton. *Plant Cell Rep.* 14, 758–762, 1995a.

Thomas, J. C., D. G. Adams, V. D. Keppenne, C. C. Wasmann, J. K. Brown, M. R. Kanost, and H. J. Bohnert. *Manduca sexta* encoded protease inhibitors expressed in *Nicotiana tabacum* provide protection against insects. *Plant Physiol. Biochem.* 33, 611–614, 1995b.

Tingey, W. M. Potato Glandular Trichomes Defensive Activity Against Insect Attack. In: *Naturally Occurring Pest Bioregulators,* Hedin, P. A., Ed., ACS Symposium Series 449, American Chemical Society, Washington, D.C., 1991.

Tingey, W. M. Techniques for Evaluating Plant Resistance to Insects. In: *Plant-Insect Interactions*, Miller, J. A. and Miller, T. A. Eds., Springer-Verlag, New York, 1986.

Tomar, J. B., and S. C. Prasad. Genetic analysis of resistance to gall midge (*Orseolia oryzae* Wood Mason) in rice. *Plant Breed.* 109, 159–167, 1992.

Treacy, M. F., G. R. Zummo, and J. H. Benedict. Interactions of host-plant resistance in cotton with predators and parasites. *Agric. Ecosystems and Environ.* 13, 151–157, 1985.

Ukwungwu, M. N., and J. A. Obebiyi. Incidence of *Chilo Zacconius* Bleszynski on some rice varieties in relation to plant characters. *Insect Sci. Applic.* 6, 653–656, 1985.

van den Berg, J., G. D. J. van Rensburg, and M. C. van der Westhuizen. Host-plant resistance and chemical control of *Chilo partellus* (Swinhoe) and *Busseola fusca* (Fuller) in an integrated pest management system on grain sorghum. *Crop Prot.* 13, 308–310, 1994a.

van den Berg, J., W. G. Wenzel, and M. C. van der Westhuizen. Tolerance and recovery resistance of grain sorghum genotypes artificially infested with *Busseola fusca* (Fuller) (Lepidoptera, Noctuidae). *Insect Sci. Appl.* 15, 61–65, 1994b.

van Lentern, J. C. Biological Control in a Tritrophic System Approach. In: *Aphid–Plant Interactions: Populations to Molecules*. Peters, D. C. and Webster, J. A., Eds., Oklahoma State University Press, Stillwater, 1991.

Van Loon, J. J. A. Measuring Food Utilization in Plant-Feeding Insects: Toward a Metabolic and Dynamic Approach. In: *Insect Plant Interactions III,* Bernays, E., Ed., CRC Press, Boca Raton, FL, 1991.

Vaughn, T. T., and C. W. Hoy. Effects of leaf age, injury, morphology, and cultivars on feeding behavior of *Phyllotreta cruciferae* (Coleoptera, Chrysomelidae). *Environ. Entomol.* 22, 418–424, 1993.

Velusamy, R., and E. A. Heinrichs. Tolerance in crop plants to insect pests. *Insect Sci. Applic.* 7, 689–696, 1986.

Verma, S. K., P. K. Pathak, B. N. Singh, and M. N. Lal. Indian biotypes of the brown planthopper. *Int. Rice Res. Newsl.* 4, 7, 1979.

Videla, G. W., F. M. Davis, W. P. Williams, and S. S. Ng. Fall armyworm (Lepidoptera, Noctuidae) larval growth and survivorship on susceptible and resistant corn at different vegetative growth stages. *J. Econ. Entomol.* 85, 2486–2491, 1992.

Visser, J. H. Differential Sensory Perceptions of Plant Compounds by Insects. In: *Plant Resistance to Insects,* Hedin, P. A., Ed., ACS Symposium Series 208, American Chemical Society, Washington, D.C., 1983.

Waibel, H., *The Economics of Integrated Pest Control in Irrigated Rice. A Case Study from the Philippines.* Springer-Verlag, Berlin, 1987, p. 196.

Walbauer, G. P. The consumption, digestion, and utilization of solanaceous and non-solanaceous plants by larvae of the tobacco budworm, *Protoparce sexta* (Johan) (Lepidoptera: Sphingidae). *Entomol. Exp. Appl.* 7, 253–257, 1964.

Walbauer, G. P. The Consumption and Utilization of Food by Insects. In: *Advances in Insect Physiology,* V, Beament, J. W. L., Treherne, J. E., and Wigglesworth, V. B., Eds., Academic Press, New York, 1968.

Webster, J.A., and C. Inayatullah. Aphid biotypes in relation to plant resistance: A selected bibliography. *Southwest. Entomol.* 10, 116–125, 1985.

Westphal, E., F. Dreger, and R. Bronner. Induced resistance in *Solanum dulcamara* triggered by the gall mite *Aceria cladophthirus* (Acari, Eriophyoidea). *Exp. Appl. Acarol.* 12, 111–118, 1991.

Wheeler, G. S., and F. Slansky. Effect of constitutive and herbivore-induced extractables from susceptible and resistant soybean foliage on nonpest and pest noctuid caterpillars. *J. Econ. Entomol.* 84, 1068–1079, 1991.

Wigley, P. J., C. N. Chilcott, and A. H. Broadwell. Conservation of *Bacillus thuringensis* efficacy in New Zealand through the planned deployment of Bt genes in transgenic crops. *Biocontrol Sci. Technol.* 4, 527–534, 1994.

Wilde, G., and H. Fcesc. A new corn leaf aphid biotype and its effect on some cereal and small grains. *J. Econ. Entomol.* 66, 570–571, 1973.

Williams, W. P., F. M. Davis, J. L. Overman, and P. M. Buckley. Enhancing inherent fall armyworm resistance of corn with *Bacillus thuringensis* genes. *Fla. Entomol.* (In Press), 1998.

Williams, W. P., J. B. Sagers, J. A. Hanten, F. M. Davis, and P. M. Buckley. Transgenic corn evaluated for resistance to fall armyworm and southwestern corn borer. *Crop Sci.* 37, 957–962, 1997.

Wiseman, B. R., and K. Bondari. Inheritance of resistance in maize silks to the corn earworm. *Entomol. Exp. Appl.* 77, 315–321, 1995.

Wiseman, B. R., and J. J. Hamm. Nuclear polyhedrosis virus and resistant corn silks enhance mortality of corn earworm (Lepidoptera, Noctuidae) larvae. *Biol.Control* 3, 337–342, 1993.

Wiseman, B. R., and M. E. Snook. Effect of corn silk age on flavone content and development of corn earworm (Lepidoptera: Noctuidae) larvae. *J. Econ. Entomol.* 88, 1795–1800, 1995.

Wiseman, B.R., F. M. Davis, and J. E. Campbell. Mechanical infestation device used in fall armyworm plant resistance programs. *Fla. Entomol.* 63, 425–432, 1980.

Wiseman. B. R., E. A. Harrell, and W. W. McMillian. Continuation of tests of resistant sweet corn hybrid plus insecticides to reduce losses from corn earworm. *Environ. Entomol.* 2, 919–920, 1975.

Wood, E. A., Jr. Biological studies of a new greenbug biotype. *J. Econ. Entomol.* 54, 1171–1173, 1961.

Wood, E. A., Jr., and K. J. Starks. Effect of temperature and host plant interaction on the biology of three biotypes of the greenbug. *Environ. Entomol.* 1, 230–234, 1972.

Xu, D., Q. Xue, D. McElroy, Y. Mawal, V. A. Hilder, and R. Wu. Constitutive expression of a cowpea trypsin inhibitor gene, *CpTi*, in transgenic rice plants confers resistance to two major rice insect pests. *Mol. Breeding* 2, 167–173, 1996.

Yanes, J., Jr., and D. J. Boethel. Effect of a resistant soybean genotype on the development of the soybean looper (Lepidoptera: Noctuidae) and an introduced parasitoid, *Microplitis demolitor* Wilkinson (Hymenoptera: Braconidae). *Environ. Entomol.* 12, 1270–1274, 1983.

Yang, G., B. R. Wiseman, D. J. Isenhour, and K. E. Espelie. Chemical and ultrastructural analysis of corn cuticular lipids and their effect on feeding by fall armyworm larvae. *J. Chem. Ecol.* 19, 2055–2074, 1993.

Yeh, K.-W., M.-I. Lin, S.-J. Tuan, Y.-M. Chen, C.-Y. Lin, and S.-S. Kao. Sweet potato (*Ipomoea batatas*) trypsin inhibitors expressed in transgenic tobacco plants confer resistance against *Spodoptera litura*. *Plant Cell Reports* 16, 696–699, 1997.

Yencho, G. C., M. W. Bonierbale, W. M. Tingey, R. L. Plaisted, and S. D. Tanksley. Molecular markers locate genes for resistance to the Colorado potato beetle, *Leptinotarsa decemlineata*, in hybrid *Solanum tuberosum x S. berthaultii* potato progenies. *Entomol. Exp. Appl.* 81, 141–154, 1996.

Yoshida, M., S. E. Cowgill, and J. A. Wightman. Mechanism of resistance to *Helicoverpa armigera* (Lepidoptera: Noctuidae) in chickpea: The role of oxalic acid in leaf exudate as an antibiotic factor. *J. Econ. Entomol.* 88, 1783–1786, 1995.

Young, N. D., L. Kumar, D. Menancio-Hautea, D. Danesh, N. S. Talekar, S. Shanmugasundarum, and D.-H. Kim. RFLP mapping of a major bruchid resistance gene in mungbean (*Vigna radiata*, L. Wilczek). *Theor. Appl. Genet.* 84, 839–844, 1992.

Zhu, K. Y., J. E. Huesing, R. E. Shade, and L. L. Murdock. Cowpea trypsin inhibitor and resistance to cowpea weevil (Coleoptera: Bruchidae) in cowpea variety TVu-2027. *Environ. Entomol.* 23, 987–991, 1994.

Zvereva, E. L., M. V. Kozlov, P. Niemela, and E. Haukioja. Delayed induced resistance and increase in leaf fluctuating asymmetry as responses of *Salix borealis* to insect herbivory. *Oecologia* 109, 368–373, 1997.

Biotechnology

CHAPTER 8

Genetic Engineering of Plants for Insect Resistance

John A. Gatehouse and Angharad M.R. Gatehouse

CONTENTS

1-56670-479-0/00/$0.00+$.50
© 2000 by CRC Press LLC

8.1 INTRODUCTION

Since the wide-scale mechanisation of agriculture, and the revolution in plant breeding that has brought high-yielding crop varieties, the developed world has been largely protected from the scourge of food shortages. Yet the problem has not gone away, for people living in the developing countries are still experiencing food shortage, both in short-term events like the many well-publicised famines, and perhaps more seriously in long-term chronic shortages of both calories and essential nutrients. The world population is still increasing and is projected to reach 9 to 10 billion over the next four decades. Thus an immediate priority for agriculture is to achieve maximum production of food and other products.

Unfortunately, as has been all too clearly shown in both the developed and undeveloped worlds, the price for achieving maximum production can be too high, with irreversible depletion or destruction of the natural environment making certain agricultural practices unsustainable in the longer term. One of these practices is the indiscriminate use of pesticides to combat insect and other pests. While pesticides are very effective in dealing with the immediate problem of insect attack on crops, and have been responsible for dramatic yield increases in crops that are subject to serious pest problems, in the longer term severe drawbacks have become apparent. Nonspecific pesticides are harmful to nontarget organisms that would normally act to keep the pest population in check. They are toxic to beneficial insects, that act as predators or parasites to the pest species, and they have a harmful effect on higher animals that also act as predators for crop pests. The effects of pesticide residues working their way up the food chain to poison the well-loved predator species at the top of the chain is well known.

Many pesticides, particularly those based on organophosphates, are also toxic to humans. Further, to clearly demonstrate that overreliance on pesticides is non-sustainable, many insect pests have become resistant to pesticides. The selection pressure on the pest is very high, and thus resistance can appear within just a few generations. In the absence of the predators (killed by pesticide) that would normally keep it in check, a pest species can become an even greater problem than it was before the pesticide was introduced, as has been the case with rice brown planthopper (*Nilaparvata lugens*) through much of southeast Asia. Unfortunately, practices that are unsustainable in the long term may be commercially attractive in the short term, and thus indiscriminate use of nonspecific pesticides continues, especially where agriculture is less well regulated. Such short-term thinking is endemic in modern agriculture and has led to a gulf being opened between the agricultural industry (and most farmers) and the broad coalition of humanitarian interests grouped under the term "environmentalists."

In response to much criticism, the agrochemical industry has been actively looking for less damaging ways to control insect pests, and has introduced a number of less harmful pesticides. In addition, alternative strategies for pest control have been pursued, such as biological control, and the use of varieties with inherent resistance. From a commercial point of view, however, these strategies do not offer such high levels of return as the pesticides they are meant to replace, or at least supplement. From the farmer's point of view, the requirements of the alternative strategies are more difficult to implement and do not offer the same security that the old indiscriminate pesticides did. Also, despite integrated pest management strategies combining the use of chemicals, resistant germplasm, and the modifying of planting, harvesting, and handling practices, yield losses due to insects have actually increased slightly for most crops over the last two decades (Duck and Evola 1997). All these factors taken together have resulted in the worst excesses of pesticide usage being checked, but not in the changes necessary to move to true sustainability.

In this context, the emergence of technologies that have allowed plants to be stably transformed with foreign genes has been timely, and after some initial suspicion, genetic engineering of crops for insect resistance has now been adopted both by the agricultural industry and by government agencies with some enthusiasm. The technology allows the extension of the "gene pool" available to a particular crop species, and thus engineered inherent resistance to pests based on resistance genes from other plant species, or on resistance genes from species in other kingdoms, or even on entirely novel resistance genes becomes possible. Pesticide usage can be eliminated, or at least dramatically decreased, with concomitant economic and environmental benefits. Genetically engineered, insect-resistant seed can be sold as a high-value commodity, and thus both farmers and the agricultural industry are able to maximise their profits. Nor is this all in the future; insect resistance has been one of the major "success stories" of the application of plant genetic engineering to agriculture, and genetically engineered insect-tolerant corn, potato, and cotton plants expressing a gene encoding the bacterial endotoxin from *Bacillus thuringiensis* are now a commercial reality, at least in the U.S.

Despite these potential benefits, there has also been a good deal of public scepticism (at least in Europe) about genetically engineered crops in general, and insect-resistant crops specifically. The practical concerns focus around two questions: "are genetically engineered crops safe for humans?" and "are genetically engineered crops safe for the environment?" Both these questions are valid and must be addressed. In this review, as well as considering strategies for producing insect-resistant transgenic crops, the best ways of deploying these crops to meet the goal of sustainability, and to address public concerns about their use in agriculture, will be considered.

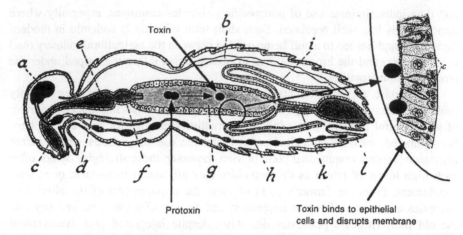

Figure 8.1   Mechanism of toxicity of *Bacillus thuringiensis* (Bt) δ-endotoxins toward insects. The insect ingests the crystalline protein deposits from Bt spores, which pass through the mouth (c) and foregut (e) and dissolve. Protoxin molecules are activated by proteolysis in the midgut (g) by insect digestive proteases. Cleavage of the protoxin generates an active toxin molecule (N-terminal region of protoxin), which binds to specific receptor glycoproteins on the surface of the epithelial cells lining the gut via domain II of the toxin protein. The bound toxin then causes ion channels to form in the membrane of the gut epithelial cells, by insertion of domain I of the protein into the membrane. Free passage of ions causes death and lysis of the gut epithelial cells, and disintegration of the gut lining, leading to death.

8.2   INSECT-RESISTANT TRANSGENIC PLANTS EXPRESSING *BACILLUS THURINGIENSIS* TOXINS

The production of transgenic plants that express the insecticidal toxins produced by different strains of the soil bacterium *Bacillus thuringiensis* (Bt) has been extensively reviewed (e.g., Koziel et al. 1993; Peferoen 1997).

Spores of Bt contain a crystalline protoxin protein encoded by a gene (*cry*) carried on a plasmid within the bacterium. On ingestion of spores by the insect, the crystals dissolve and the protoxin is cleaved by digestive proteinases in the insect gut to generate active *Bt* toxin molecules (Choma et al. 1990) (Figure 8.1). The active toxin molecule binds to a specific glycoprotein receptor that is situated in the cell membranes of gut cells lining the insect midgut, and then inserts itself into the gut cell membrane (Liang et al. 1995). The bound toxin interacts with the cell membrane, inserting part of the molecule to form a channel in the cell membrane that allows the free passage of ions (Knowles and Dow 1993). The toxin-created channels destroy the imbalance in ion concentrations that has been established across these membranes (which can be very considerable, since in many lepidopteran larvae the gut pH is approximately 10.5–11), resulting in the death and lysis of the cells lining the gut (Manthavan et al. 1989). Death of the insect rapidly follows, and the carcass forms a substrate for the growth of *B. thuringiensis* from the spores. The bacteria eventually sporulate, releasing fresh spores into the soil to repeat the cycle.

Bt toxins form an extensive range of preformed "natural" insecticides. Different strains of Bt contain plasmids encoding toxins with different sequences, and different

Table 8.1 Summary of Bt Crystal Protein Gene Family (Adapted from Peferoen 1997)

Designation of sub-family	Previous designation	Polypeptide Mr	Pesticidal activity
Cry1α-K	CryI, CryV (Cry1I), various	129–138,000 80–81,000 (Cry1I)	Lepidoptera
Cry2A	CryII	69–71,000	Lepidoptera (Diptera; Cry2A1)
Cry3α-C	CryIII	72–74,000	Coleoptera
Cry4α-B	CryIV	126–135,000	Diptera
Cry5α-B	CryV	139–154,000	Nematoda (Coleoptera; Cry5B)
Cry6α-B	CryVI	44–53,000	Nematoda
Cry7A	CryIIIC	127,000	Coleoptera
Cry8α-C	CryIIIE-G	128–130,000	Coleoptera
Cry9α-C	CryIG,X,H	127–130,000	Lepidoptera
Cry10A	CryIVC	75,000	Diptera
Cry11α-B	CryIVD	72,000	Diptera
Cry12A	CryVB	140,000	Nematoda
Cry13A	CryVC	89,000	Nematoda
Cry14A	CryVD	133,000	Coleoptera
Cry15A		38,000	Lepidoptera
Cyt1A, Cyt2A	CytA, CytB	27,000–29,000	Cytolytic proteins

specificities of action against insects; in general, a particular toxin shows a high level of specificity and is only effective against a limited range of closely related species. The different Bt toxins found in nature have been classified into types designated Cry1, Cry2, etc., on the basis of broad specificity and sequence homology of the proteins, as summarised in Table 8.1, and further subclassified into toxin types designated Cry 1A, Cry1B, etc., and individual toxin sequences designated Cry1Aa1, Cry1Ab1, etc. Active research into isolating further Bt toxin types is still under way to extend the range of insects that these toxins are active against. Broadly, Cry1, Cry2, and Cry9 toxins are active against Lepidoptera, Cry3, Cry7, and Cry8 toxins are active against Coleoptera; and Cry4, Cry10, and Cry11 toxins are active against Diptera. Cry1 toxins are the most common type. No Bt toxins with high levels of toxicity toward homoptera have yet been identified.

Bt preparations have been used for many years as an "organic" insecticide that is sprayed onto plant tissues (Peferoen 1997). However, the utility of Bt as a conventional insecticide is limited by instability of the protein when exposed to uv light and poor retention on plant surfaces in wet weather. The high level of toxicity of the Bt toxin protein, and the ease of isolating its encoding gene from bacterial plasmids, made it an obvious choice for initial experiments attempting to produce insect-resistant transgenic plants.

8.2.1 Genetic Engineering of Plants to Express Bt Toxins

Whereas the isolation of genes encoding Bt toxins was an easy task, subsequent engineering of transgenic plants that expressed these toxins proved much less straightforward. In fact, considerable modification to the Bt toxin genes has proved necessary in order to obtain adequate expression to confer insect resistance on transgenic plants. The necessary modifications have fallen into two classes: alterations to the protein sequence of the Bt toxins and alterations to the gene sequences.

8.2.1.1 Changes to Protein Sequence

As described above, Bt toxin genes encode an inactive protoxin molecule, which is activated by proteolytic cleavage in the insect gut. When different toxin genes are compared, the N-terminal regions of the encoded proteins (approximately 600 amino acids) are found to show significant sequence homologies, whereas the C-terminal regions are much more variable in both sequence and length. The N-terminal regions are resistant to proteolytic cleavage (Höfte and Whiteley 1989) and form the toxic part of the protoxin; they contain a highly conserved sequence of amino acids at the C-terminus of the processed, active toxin (Höfte et al. 1986), which seems to act as a processing site for protoxin activation. The C-terminal region of the protoxin appears to function in forming the crystalline structures observed for protoxin deposits in bacterial spores.

The structure of the processed Bt toxin protein, as produced by proteolysis of the crystalline protoxin, contains three domains (Li et al. 1991; Grochulski et al. 1995). The first (N-terminal) domain contains approximately 250 amino acids and forms a helical bundle with six α-helices surrounding a central α-helix. This part of the molecule is responsible for pore formation in the epithelial cells of the insect gut, since it alone is able to insert itself into lipid bilayers. The second domain, of approximately 200 amino acids, consists of three β-sheets and is responsible for binding to the "receptor" glycoprotein(s) on the gut surface, thus determining the specificity of action of the toxin, since binding to the gut surface appears to be necessary for effective pore formation to take place. Protein engineering experiments have shown that "swapping" domain II between different toxins also exchanges the specificity of insecticidal action of the toxins. Domain III, of approximately 150 amino acids, is again predominantly composed of β-sheets, folded in a "β-sandwich," and does not have a clearly defined functional role; it may be concerned with stabilising the structure of the entire molecule, but may also play a role in determining specificity or pore formation.

Attempts to express Bt toxin genes containing complete protoxin coding sequences in plants have been uniformly unsuccessful; protoxin expression levels obtained were undetectable or very low at best (of the order of .0001% (ng/mg) of total protein), which was too low to show any insecticidal effects (Barton et al. 1987; Vaeck et al. 1987). It was thus necessary to alter the expressed protein sequence and to express truncated toxin genes that only encoded the N-terminal region of the protein containing the active toxin. Expression levels of the active toxin molecules were one to two orders of magnitude higher in transgenic plants, up to 0.01% of total protein, and this level of expression was sufficient to show that transgenic Bt-toxin expressing plants showed enhanced resistance to insect pests. In the initial experiments, transformed tobacco plants were produced expressing various Cry1A toxins, which significantly decreased survival of larvae of tobacco hornworm (*Manduca sexta*) feeding on them (Barton et al. 1987; Vaeck et al. 1987). Similarly, transformed tomato also expressing Cry1A was protected from feeding damage by larvae of two major lepidopteran crop pests, *Helicoverpa armigera* and *Heliothis zea* (Fischhoff et al. 1987).

8.2.1.2 Changes to Gene Sequence

The levels of Bt toxin expressed in transgenic plants using constitutive promoters such as the Cauliflower Mosaic Virus (CaMV) 35S promoter were still two orders of magnitude lower than those obtained for other foreign proteins. It was apparent that higher levels of expression should be possible, and would be desirable in order to improve the protection against insect pests afforded by Bt transgenes. Engineering Bt toxin genes to improve expression levels has been a tour de force for molecular biology, achieved at the cost of many man-years of research to identify and remove the causes of poor expression (Perlak et al. 1991; van Aarsen et al. 1995). Two major factors were identified that resulted in poor expression: first, the codon usage of the bacterial gene was markedly different to typical plant genes, due to the bacterial genome having a high A+T content, whereas the plant genome has a high G+C content, leading to inefficient translation of the mRNA; second, the high A+T content of the bacterial genes was resulting in truncated transcripts (mRNAs), which were either unstable or could not produce functional protein, due to the presence of sequences that functioned as polyadenylation addition signals and intron processing signals in the plant. Genes encoding Bt toxins were thus reconstructed by a combination of mutagenesis and oligonucleotide synthesis to produce synthetic genes, which encoded the same proteins but which had codon usages typical for plant genomes, and which had all aberrant processing signals removed. Expression levels of Bt toxins from these synthetic genes was increased by nearly two orders of magnitude (to up to 0.3% of total protein (Perlak et al. 1991)) when expressed in transgenic plants. At this level of expression, the protection afforded by expression of Bt toxins approaches that achievable with chemical pesticides, with mortality of susceptible insect species approaching 100% over a time scale of days when exposed to transgenic plants (Wilson et al. 1992).

The synthetic Bt toxin genes have formed the basis of all the gene constructs that have been, and are being, used for the production of insect-resistant plants intended for commercial agriculture. A variety of constitutive, wound-induced and tissue-specific promoters are being used, which have been optimised for different host plants and different target pests (e.g., Koziel et al. 1993; Jansens et al. 1995); several specific cases are considered below. An alternative approach, which has as yet not been exploited commercially, has been to use a developing technology based on homologous recombination to target the Bt gene to the chloroplast genome instead of the nuclear genome. This strategy avoids the necessity to modify the toxin gene, since the chloroplast genome is bacterial in nature; thus, an unmodified Cry1A protoxin gene was integrated into the genome of tobacco chloroplasts, resulting in expression levels of protoxin protein of 3 to 5% of total protein in plants regenerated from the transformation (McBride et al. 1995).

8.2.1.3 Examples of Insect-Resistant Transgenic Plants Expressing Bt Toxins

Three commercial transgenic crops have been introduced that contain Bt toxin encoding genes for insect control: cotton, maize (corn), and potato. In two cases, cotton and potato, the impetus to deploy transgenic crops has been the development of almost complete resistance to acceptable insecticides in their major insect pests due to overreliance on insecticide usage (Roush 1997); in cotton the major pests are lepidopteran larvae of the bollworm species *Pectinophera gossypiella*, *Heliothis virescens*, and *Helicoverpa armigera*, whereas in potato the major pest is the coleopteran Colorado potato beetle, *Leptinotarsa decemlineata*. In the third case, that of maize, a major target pest is the lepidopteran European corn borer (*Ostrinia nubilalis*), where the larvae tunnel inside the stalks of the plants and are inaccessible to conventional insecticide sprays.

In transgenic cotton and corn, modified *cry1Ab* genes have been used to attempt to control the lepidopteran pests. With cotton, both laboratory (Perlak et al. 1990) and field trials (Wilson et al. 1992) gave high levels of control, not only of bollworms, but also in the field trial of beet armyworm and cotton leaf perforator. Transgenic corn containing a maize-optimised gene construct also gave excellent control of corn borer when tested in the field (Koziel et al. 1993; Carozzi and Koziel 1997). In the case of potato, not only have plants been engineered to express a modified *cry3A* gene to protect them against Colorado potato beetle (Perlak et al. 1993), but a *cry1Ab* gene construct has also been used to protect the tubers against damage by potato tuber moth larvae when in storage (Jansens et al. 1995).

Many other crops, including cereals, root crops, leafy vegetables, forage crops, and trees are now also being engineered to express Bt toxins (Schuler et al. 1998). Special mention may be made of rice (Fujimoto et al. 1993), where an international project, partly funded by the Rockefeller Foundation and coordinated through the International Rice Research Institute, is engineering *cry1Ab* and *cry1Ac* genes into rice to combat stem borers of several species (Wünn et al. 1996; Bennett et al. 1997). It is intended that these rice varieties will be freely available as a basis for breeding programmes in rice growing areas in the developing world.

8.3 INSECT-RESISTANT TRANSGENIC PLANTS EXPRESSING INHIBITORS OF INSECT DIGESTIVE ENZYMES

Whereas the strategy of employing genes encoding Bt toxins to produce insect-resistant transgenic plants has its origins in established practices with conventional insecticides, where an exogenous compound is used to protect the host plant, a number of other strategies for protecting crops from insect pests take as their starting point the endogenous resistance shown by plants to most insect predators. Although agricultural losses may obscure the fact, most plants survive attack by most potential insect predators, and as a result of selection pressure extending back at least 250 million years, have evolved many strategies of endogenous resistance (Ehrlich and Raven 1964). As well as physical defences, and ecological strategies such as dispersal and growth habits, plants make extensive use of biochemical defences, based primarily on a rich and varied secondary metabolism (Harborne 1988), but also on the use of defensive proteins. Genes encoding endogenous plant defensive proteins were thus obvious candidates for enhancing the resistance of crops to insect pests.

Interfering with digestion, and thus affecting the nutritional status of the insect, is a strategy widely employed by plants to defend themselves against pests. A major factor in inhibition of digestion is the presence of protein inhibitors of digestive enzymes (both proteinases and amylases) in plant tissues. These proteins interact with digestive enzymes, binding tightly to the active site and preventing access of the normal substrates (Garcia-Olmedo et al. 1987). In the case of proteinase inhibitors, binding is accompanied by hydrolysis of a target peptide bond in the inhibitor, which determines its specificity toward a particular type of protease. The enzyme inhibitor complex is both thermodynamically and kinetically very stable (some proteinase-proteinase inhibitor complexes have half lives of the order of weeks), and thus stoichiometric inhibition of the enzyme is achieved. The inhibition of digestive enzymes not only has a direct effect on the insect's nutritional status, but is also thought to lead to secondary effects where oversynthesis of digestive enzymes occurs as a feedback mechanism in an attempt to utilise ingested food (Figure 8.2). If the insect cannot overcome the inhibition of digestion, death by starvation occurs.

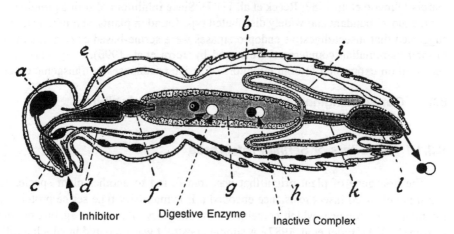

Inhibitor   Digestive Enzyme   Inactive Complex

Figure 8.2 Mechanism of antimetabolic action of digestive enzyme inhibitors. The insect consumes material containing the inhibitor, which passes down the gut to the midgut region (g), where digestive enzymes are secreted by the cells lining the gut. The inhibitor combines with the digestive enzyme to form a stable complex, inactivating the enzyme. Antimetabolic effects are exerted through direct suppression of digestion, leading to starvation of nutrients, and by effects on enzyme synthesis and recycling. In the presence of proteinase inhibitors, enzyme recycling will be less efficient because proteolysis is suppressed, leading to the loss of amino acids, which would normally be recovered from digestive enzymes; in addition, enzyme synthesis may be up-regulated to attempt to overcome inhibition of digestion, leading to further shortcomings in recycling of amino acids used for gut protein synthesis.

Evidence for a role of inhibitors of digestive enzymes in plant defence is provided by consideration of the sites of synthesis and accumulation of these proteins. They are normally accumulated in storage tissues, both in seeds and vegetative storage tissues such as potato tubers, and can reach concentrations as high as 2% of total protein. Since plant survival is dependent on protection of storage tissues against predators, this pattern of accumulation supports the defensive role; there is little evidence that inhibitors accumulated in these tissues function as a storage reserve by being broken down on germination or sprouting. Direct evidence for a defensive role of protein inhibitors of digestive enzymes is shown by the induced synthesis of serine proteinase inhibitors that occurs when many plant species are wounded (Ryan 1984), which can be caused by insect feeding, or mimicked by mechanical damage. The wound response in plants has been extensively investigated over recent years, and involves a variety of changes in the physiological state of the tissue, both locally, and in some cases systemically (Farmer and Ryan 1992); many different proteins increase in amount in plant tissues on wounding, but proteinase inhibitor synthesis remains a major feature of this response. Recent evidence suggests that insect feeding may lead to more rapid accumulation of inhibitors than simple wounding, providing further evidence of their defensive role against insect predators (Korth and Dixon 1997). Inhibitors of proteinases and amylases form extensive families of proteins in plants, and have been the object of much study in recent years; for a general review of the properties of these proteins, several reviews may be consulted (Ryan 1981; Garcia-Olmedo et al. 1987; Reeck et al. 1997). Since inhibitors of serine proteinases are the most abundant and widely distributed type found in plants, and initial studies suggested that insect digestive endoproteinases were serine-based enzymes, similar to their mammalian counterparts (reviewed by Terra et al. 1996), early work concentrated on expressing foreign inhibitors of serine proteinases in transgenic crops.

8.3.1 Genetic Engineering of Plants to Express Inhibitors of Digestive Proteinases

8.3.1.1 Inhibitors of Serine Proteinases

The first gene of plant origin that was transferred to another plant species to result in enhanced insect resistance encoded a Bowman-Birk type serine proteinase inhibitor from cowpea, which contained two inhibitory sites active against bovine trypsin (CpTI) (Hilder et al. 1987). A simple construct was prepared in which a full-length coding sequence derived from a cDNA clone was placed under the control of the constitutively expressed Cauliflower Mosaic Virus (CaMV) 35S promoter. Transgenic tobacco plants were produced by a standard *Agrobacterium tumefaciens*–mediated transformation protocol using a binary vector system. Transformants were screened for CpTI expression, which showed that many of the resulting plants expressed CpTI at levels greater than 0.1% of total soluble protein; subsequent experience has shown that this is generally the case for expression of genes of plant origin encoding defensive proteins in transgenic plants, in contrast to the very low levels of expression observed for unmodified toxin genes of bacterial origin. Plants expressing CpTI at the highest levels (approximately 1% of total soluble protein) were clonally propagated and used for insect bioassays against larvae of the tobacco budworm (*Heliothis virescens*). With these clonal plants, and subsequent generations derived from their self-set seed, the CpTI expressing plants showed reduced damage (by up to approximately 50%) compared to the control plants, and reduced insect survival and biomass (again, by as much as approximately 50%). The antimetabolic effects of CpTI expressed in transgenic tobacco have also been observed with other lepidopteran pests including *H. zea, Spodoptera littoralis,* and *Manduca sexta.* Subsequent trials carried out in California showed that expression of CpTI in tobacco afforded significant protection in the field against *H. zea* (Hoffman et al. 1991). Following on from the study using tobacco as a model system, the gene encoding CpTI has also been expressed in a range of different crops. For example, constitutive expression of CpTI in rice conferred significantly enhanced levels of resistance toward two species of rice stem borer (*Sesamia inferens* and *Chilo suppressalis*) in the field (Xu et al. 1996).

Other serine protease inhibitor encoding genes have also been tested as protective agents for crops. For example, the tomato inhibitor II gene (which encodes a trypsin inhibitor with some chymotrypsin inhibitory activity), when expressed in tobacco, was also shown to confer insect resistance (Johnson et al. 1989) when expressed constitutively using the CaMV 35S promoter, but, interestingly, not when expressed from a wound-inducible promoter. The bioassays showed that the decrease in larval weight in insects reared on transgenic plants was roughly proportional to the level of PI-ll being expressed. Several of the transgenic plants were shown to contain inhibitor levels over 200 µg/g tissue; these levels are within the range that is routinely induced by wounding leaves of either tomato or potato plants (Graham et al. 1986). However, tobacco plants expressing tomato inhibitor I at similar levels had no deleterious effects upon larval development, showing the specificity of interactions between inhibitors and insect species. McManus et al. (1994) obtained similar results with potato PI-II, when expressed in tobacco, against the noctuid lepidopteran *Chrysodeixis eriosoma* (green looper). The wound-inducible potato PIs (PI-I and PI-II) have now been constitutively expressed in a range of crops where they have been shown to confer resistance; for example, as with CpTI, expression of PI-II in rice conferred significant levels of protection in the field toward rice stem borers (Duan et al. 1996).

Although the range of serine proteinase inhibitors used, and the range of crops transformed, is ever increasing, as shown in Table 8.2, the commercial viability of this strategy has yet to be proven. In contrast to the situation where *Bt* toxins are expressed in transgenic plants at adequate levels, expression of serine proteinase inhibitors rarely results in high levels of mortality in the insect pest, and the levels of protection achieved, although often better than 50% in terms of reduction in plant damage, and decrease in insect biomass, do not reach the benchmark of >95% normally considered necessary for commercial viability. Several laboratories are actively addressing ways in which to increase the efficacy of serine proteinase inhibitors as protective agents against insect pests. However, a greater understanding of the mechanism of action of proteinase inhibitors in insects, both at the biochemical and molecular levels, will be necessary before the technology can be fully exploited. For example, it has become apparent in recent years that some insects exposed to

Table 8.2   Proteinase Inhibitors Expressed in Transgenic Plants for Insect Resistance. Only those examples published as research papers in refereed journals are given. Key to insect species: (C) = coleopteran, (H) = homopteran, (L) = lepidopteran.

Proteinase inhibitor	Transformed plant	Reference	Target pest
CpTI (cowpea trypsin inhibitor)	Tobacco	(Hilder et al. 1987)	*Heliothis virescens* (L)
	Rice	(Xu et al. 1996)	*Chilo suppressalis, Sesamia inferens* (L)
	Potato	(Gatehouse et al. 1997)	*Lacanobia oleracea* (L)
	Strawberry	(Graham et al. 1997)	*Otiorhynchus sulcatus* (C)
PI-II (potato proteinase inhibitor II)	Tobacco	(Johnson et al. 1989; McManus et al. 1994; Jongsma et al. 1995)	*Manduca sexta, Chrysodeixis eriosoma, Spodoptera exigua* (L)
	Rice	(Duan et al. 1996)	*Chilo suppressalis, Sesamia inferens* (L)
NaPI (*Nicotiana alata* multi-functional inhibitor)	Tobacco	(Heath et al. 1997)	*Helicoverpa punctigera* (L)
SKTI (soya bean Kunitz trypsin inhibitor)	Potato	(Gatehouse et al. 1999)	*Lacanobia oleracea* (L)
OC-I (rice oryzacystatin)	Poplar	(Leplé et al. 1995)	*Chrysomela tremulae* (C)
Insect serpins (*Manduca sexta* haemolymph proteinase inhibitors)	Tobacco	(Thomas et al. 1995)	*Bemisia tabaci* (H)

dietary proteinase inhibitors are able to adapt to overcome any antimetabolic effects (Broadway 1995; Jongsma et al. 1995), by drawing on inherent resources of preexisting families of proteinase encoding genes (Bown et al. 1997), and switching proteinase expression to favour enzymes that are insensitive to inhibition.

8.3.1.2 Inhibitors of Cysteine Proteinases

Whereas insects from the orders Lepidoptera, Diptera, Orthoptera, and Hymenoptera have been shown to employ proteinases based on serine as the catalytically active residue as their major digestive endoproteinases, similar to digestive proteinases in higher animals, many coleopteran species have been shown to employ digestive proteinases based on cysteine as the catalytically active residue (Houseman and Downe 1980; Gatehouse et al. 1985; Murdock et al. 1987); reviewed by Terra et al. (1996). These proteinases are not inhibited by typical plant protein inhibitors of serine proteinases. However, cysteine proteinases are used by plants for protein mobilisation, and by many animals for intracellular lysozomal protein digestion, and protein inhibitors of cysteine proteinases (cystatins) are widely distributed throughout all living organisms to regulate these endogenous proteinases, even if they are usually present in small amount (Garcia-Olmedo et al. 1987). By analogy to the use of serine proteinase inhibitor genes to control insect pests using serine digestive proteinases, genes encoding cysteine proteinase inhibitors have been suggested for use in transgenic plants for control of coleopteran insects. These inhibitors are effective *in vitro*, where inhibition of digestive proteinases of various coleopteran pests by cysteine proteinase inhibitors has been reported by a number of studies

(Liang et al. 1991; Michaud et al. 1993). They also have been shown to have dele-terious effects against coleopteran species when incorporated into artificial diets (Chen et al. 1991; Orr et al. 1994; Edmonds et al. 1996). However, as yet there are few published reports describing their insecticidal effects *in planta*. One example is the use of the gene encoding a rice cysteine proteinase inhibitor, oryzacystatin, which has been expressed constitutively in transgenic poplar trees, conferring resistance toward the coleopteran pest *Chrysomela tremulae* (Leplé et al. 1995).

8.3.1.3 Genetic Engineering of Plants to Express Inhibitors of Digestive Amylases

Nutrition in phytophagous insects is normally limited by the availability of nitrogen from amino acids rather than carbon skeletons from starch, and thus inhib-itors of starch digestion would not be expected to be as potent in their antimetabolic effects as proteinase inhibitors. Nevertheless, inhibitors of α-amylases from both higher animals and insects are widespread in plants, and are accumulated in similar tissues as proteinase inhibitors (although, as far as is known, they are not involved in responses to wounding).

Insecticidal effects of amylase inhibitors against lepidopteran pest species have not proved easy to demonstrate, and it is unlikely that these proteins play a major role in plant resistance to Lepidoptera. However, they have significant insecticidal activity toward phytophagous coleopterans, particularly pests of stored seeds. For example, α-amylase inhibitors purified from wheat and *Phaseolus vulgaris* have been shown to be insecticidal to coleopteran species that do not normally feed on these species when tested in artificial diet (Gatehouse et al. 1986; Ishimoto and Kitamura 1988).

The α-amylase inhibitor of *Phaseolus vulgaris* is encoded by a gene designated LLP (Moreno and Chrispeels 1989); it is in fact homologous to the seed lectin (q.v.) in *P. vulgaris* and represents an interesting example of evolution of protein function based on a common sequence. A chimeric gene, consisting of the coding sequence of the lectin gene that encodes LLP, and the 5' and 3' flanking sequences of the gene that encodes a lectin subunit, PHα-2, has been constructed and expressed in tobacco (Altabella and Chrispeels 1990). The promoter in this construct is seed-specific, and thus the transgene product should only be accumulated in seeds. Seeds from these transgenic plants expressed the bean α-amylase inhibitor, and contained inhibitory activity against both porcine pancreatic α-amylase and the α-amylase present in the midgut of mealworm, *Tenebrio molitor*. Although suitable insect bioassays could not be carried out with tobacco seeds, the inhibitory activity of the transgene product against insect α-amylase led to the suggestion that introduction of the bean amylase inhibitor gene into other leguminous plants may be a strategy to protect the seeds from seed-eating larvae of coleoptera. This suggestion was verified in a series of elegant experiments (Shade et al. 1994; Schroeder et al. 1995) in which transgenic garden peas were produced using a construct similar to that described above. Trans-formation was by an improved *Agrobacterium tumefaciens* vector system. Seeds of these plants contained significant levels of bean amylase inhibitor (up to approxi-mately 4% of total protein) and were highly resistant to attack by larvae of the coleopteran storage pests (bruchid beetles) *Bruchus pisorum* and *Callosobruchus maculatus,* with levels of mortality up to 100% being achieved at the highest expression levels (reviewed by Chrispeels 1997). Unfortunately, this inhibitor is unlikely to be useful against lepidopteran pests, as it is inactive at the alkaline pH of the lepidopteran gut. More recently, expression of the *P. vulgaris* α-amylase inhibitor in Adzuki bean has also been shown to confer resistance to larvae of bruchid beetles (Ishimoto et al. 1996), and it seems likely that this strategy will be generally applicable in protecting starchy grain legumes against bruchid pests.

8.4   INSECT-RESISTANT TRANSGENIC PLANTS EXPRESSING LECTINS

Lectins form a large and diverse group of proteins that are found throughout the range of living organisms but are identified by a common property of specific binding to carbohydrate residues, either as free sugars or more commonly, as part of oligo- or polysaccharides. They are distinguished from enzymes by having no action on the carbohydrate other than binding to it. Most lectin molecules contain multiple binding sites and thus can cross-link oligo- or polysaccharides. They are also called agglutinins for this reason, since the cross-linking of carbohydrate side chains on cell surface glycoproteins leads to the formation of aggregates of cells and a visual agglutination of cells such as red blood cells.

Plants were the first known source of lectins and accumulate lectins in many storage tissues; seeds are an abundant source, but other storage tissues such as bulbs, or bark, also contain lectins. They can be accumulated at levels up to 1%, or even higher, of total protein. The distribution of lectins is universal, but amounts accumulated vary widely; viable null mutants for some seed lectins are known (e.g., in pea). The role of lectins in plants has been a source of much speculation, and it has become evident that these proteins fill more than one role. At a fundamental level, they are involved in cell-cell interactions, and in legumes are known to be involved in the interaction between the plant and the symbiotic nitrogen-fixing bacterium *Rhizogenes* spp. However, the levels of accumulation of lectins in storage tissues is far in excess of that required for any role in cell interactions (at least two orders of magnitude greater in seeds than in roots in pea), and roles as storage proteins, or as an aid to packing storage proteins together, have been proposed. Based on the known toxicity of some lectins toward higher animals, more recent work has given emphasis to the possibility that these proteins, like enzyme inhibitors, are also part of the defensive mechanism of plants against insects and other pests and pathogens (Chrispeels and Raikhel 1991; Peumans and Van Damme 1995).

A role for lectins as defensive proteins in plants against insect predators was first proposed by Janzen et al. (1976), who suggested that the common bean (*P. vulgaris*) lectin was responsible for the resistance of these seeds to attack by coleopteran storage pests. Although subsequent work has shown that the major factor in causing resistance in this example was an α-amylase inhibitor (Huesing et al. 1991) (see above), the insecticidal properties of plant lectins have been demonstrated in numerous other studies where purified proteins were fed to insects in artificial diet bioassays. For example, 17 commercially available plant lectins were screened for insecticidal activity against the storage pest *Callsobruchus maculatus* (a bruchid beetle) (Murdock et al. 1990). Five lectins were found to cause a significant delay in larval development at dietary levels of 0.2% and 1% (w/w; approximately 1–5% of total protein). Czalpa and Lang (1990) took a similar approach when they screened a range of lectins for activity against the coleopteran species Southern corn rootworm (*Diabrotica undecimpunctata*), a major economic pest of corn, and the lepidopteran, European corn borer. In general, toxicity of lectins toward lepidopteran larvae has proved more difficult to demonstrate (Shukle and Murdock 1983), but artificial diet bioassays have shown that snowdrop lectin (GNA) significantly decreased growth and retarded development when fed to larvae of tomato moth (*Lacanobia oleracea*) at 2% of total protein, although little effect on survival was observed (Fitches et al. 1997).

Lectins are currently receiving most interest as insecticidal agents against homopteran plant pests. This important group of pests includes aphids, leafhoppers, and planthoppers, and which routinely feed by phloem abstraction. They contain little or no proteolytic activity in their guts, and thus are not in general susceptible to proteinase inhibitors, nor are Bt toxins with specificity toward homopterans known at present. An artificial diet bioassay system was used to test a series of lectins against the rice brown planthopper (*Nilaparvata lugens*), an important pest of rice in southeast Asia, and certain lectins were found to decrease insect survival significantly (Powell et al. 1993). The two most effective proteins tested were the lectins from snowdrop (GNA; mannose-specific) and wheat germ (WGA; GlcNAc-specific), each of which gave approximately 80% corrected mortality at a concentration of 0.1% w/v in the diet. The LC_{50} value for GNA against brown planthopper was found to be 0.02%, or approximately 6 μM (Powell et al. 1995). GNA was also found to be toxic to another sucking pest of rice, the rice green leafhopper, *Nephotettix cinciteps*. Habibi et al. (1993) carried out similar bio-assays in order to identify lectins that might be suitable in the control of the potato leafhopper (*Empoasca fabae*); those found to be effective were from jackfruit, pea, lentil, horse gram, common bean, and wheat germ (WGA). The lectin from *Canavalia ensiformis* (Con A) was also shown to be a potent toxin of the pea aphid *Acyrthosiphon pisum*, having a significant effect upon both survival and growth (Rahbé and Febvay 1993). Chitin-binding lectins from wheat germ (WGA), stinging nettle and *Brassica* spp. were also reported to cause high levels of mortality to the cabbage aphid *Brevicornye brassicae* when incorporated into artificial diet (Cole 1994). Subsequent experiments have shown that the snowdrop lectin (GNA) is also inhibitory to aphid development in both the peach-potato aphid *Myzus persicae*, and the potato aphid *Aulacorthum solanum*, although effects on survival were small (Down et al. 1996; Sauvion et al. 1996). GNA also significantly reduced female fecundity in mature aphids. This effect would be significant in preventing the build-up of an insect population.

Despite attempts to demonstrate a correlation, the specificity of binding to carbohydrate residues for a given lectin is not necessarily a good indicator of its potential insecticidal properties (Harper et al. 1995), and thus it is still necessary to test each lectin against a target pest on a case by case basis. Since the mechanism(s) by which some lectins are toxic to higher animals are not yet fully elucidated, it is

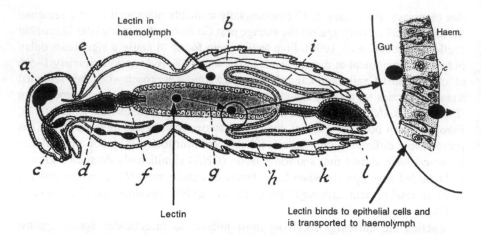

Figure 8.3 Mechanism of toxicity of lectins toward insects. The insect ingests material containing the lectin, which passes into the midgut (g). The lectin then binds to specific receptor glycoproteins (determined by the carbohydrate binding specificity of the lectin) on the surface of the epithelial cells lining the gut. This leads to potential disruption of the gut, or effects on gut metabolism, as lectins interact with receptors for growth factors. Lectin also passes across these cells, and appears in the haemolymph, leading to the possibility of systemic effects. Lectin binding can also be observed in the Malpighian tubules (i), and may occur to chemoreceptors in the mouth parts (c), leading to antifeedant effects.

perhaps not surprising that mechanisms of lectin toxicity to insects are largely unexplored. Binding of lectins to cells lining the gut in insects has been demonstrated in a number of species, but it is also clear that binding is not in itself sufficient to produce toxicity (Gatehouse et al. 1984; Harper et al. 1995; Powell et al. 1998). Lectins may also bind to the peritrophic membrane, as opposed to the epithelium, in the midgut region of insects (Eisemann et al. 1994), and it has been suggested that this may cause physical blockage of the normally porous membrane and interfere with nutrient uptake. Binding occurs to specific glycopolypeptides, as shown by separation of brush border membrane proteins by electrophoresis, followed by blotting techniques using labeled lectins (Harper et al. 1995; Powell et al. 1998), but it is not clear whether the specificity of binding can be related to toxicity, as the "receptors" have not been characterised. The observation that lectins can be transported across the gut wall and can be detected in the insect haemolymph (Fitches and Gatehouse 1998) suggests that systemic effects may also be important in lectin toxicity (Figure 8.3).

8.4.1 Transgenic Plants Expressing Foreign Lectins

A gene encoding the pea lectin (P-Lec) has been expressed in transgenic tobacco plants using a simple construct where expression of a complete lectin coding sequence was driven by the constitutive CaMV 35S promoter. Plants expressing the pea lectin at up to 1% of total protein were then tested in bioassay for enhanced levels of resistance to larvae of the lepidopteran pest *Heliothis virescens* (tobacco budworm). Larval biomass was significantly reduced on the transgenic plants, compared with that on control plants, and leaf damage, as determined by computer-aided image analysis, was also reduced (Boulter et al. 1990). Transgenic tobacco plants expressing both the cowpea trypsin inhibitor (CpTI) and P-Lec were obtained by crossing plants derived from the two primary transformed lines, and screened for enhanced resistance to *H. virescens*. The insecticidal effects of the two genes were additive, with insect biomass on plants expressing both transgenes reduced by nearly 90% compared to control plants, whereas plants expressing either CpTI or P-Lec alone only reduced biomass by approximately 50%. Leaf damage was also the least on the double-expressing plants. This experiment not only showed that the products of lectin genes could enhance resistance to insect attack in transgenic plants, but also demonstrated that additive protective effects could be obtained from different plant-derived insect-resistant genes.

A gene encoding the snowdrop lectin (GNA) has also been engineered into transgenic plants; a cDNA clone described by Van Damme et al. (1991) was used in constructs similar to those described for P-Lec above. Constructs containing a complete coding sequence for the preproprotein gave rise to levels of GNA up to 1.5% of total protein in leaf tissue of transgenic potato (Gatehouse et al. 1997). The functional integrity of GNA expressed in the transgenic plants was demonstrated by haemagglutination assay. GNA-expressing potatoes were significantly protected against attack by larvae of the tomato moth, *Lacanobia oleracea*; although reduction of larval survival was less than 25% in the transgenic plants when compared to controls, highly significant reductions in larval biomass (>50%) and in leaf damage to the plants (>70%) were observed in laboratory assays (Gatehouse et al. 1997). Similar results were obtained in a large-scale bioassay in the glass house (Fitches et al. 1997). The GNA-expressing transgenic potato plants also reduced the fecundity of the aphids *Myzus persicae* and *Aulacorthum solani*; production of parthenogenetic offspring of the latter was reduced by up to 80% in laboratory assays, and its population build-up was significantly reduced in a glass house experiment (Down et al. 1996; Gatehouse et al. 1996). Reduction in fecundity of the cereal aphid *Sitobion avenae* has also been observed on transgenic wheat plants expressing GNA (Stoger et al. 1999).

GNA has also been expressed in transgenic tobacco and rice plants in a phloem-specific manner, using a gene construct containing the GNA coding sequence driven by the promoter from the rice sucrose synthase gene *RSs1*. The *RSs1* promoter directs expression of a *gus* reporter gene in the phloem tissue of leaves, stems, petioles, and roots of transgenic tobacco plants with no detectable expression in other tissues (Shi et al. 1994), and can thus be used to produce constructs where expression of insecticidal genes is optimised for protection of transgenic plants against phloem-feeding homopteran insect pests. Transgenic rice containing an *RSs1–GNA* construct has been shown to accumulate GNA in vascular and epidermal tissue (Sudhakar et al. 1998), and decreases survival of rice brown planthoppers, which are exposed to the plants by up to 60% (Rao et al. 1998). Expression of GNA from a constitutive promoter (maize ubiquitin) gave similar results. In these assays, a feeding avoidance effect was observed in the transgenic plants expressing GNA, in agreement with earlier assays that had shown that GNA in artificial diet deterred feeding by brown planthoppers (Powell et al. 1995). This effect could be significant in reducing damage in the field.

As with proteinase inhibitors, the levels of protection against pests observed in transgenic plants expressing foreign lectins are generally not high enough to be considered commercially viable. However, the absence of genes with proven high insecticidal activity against homopteran pests may well mean that transgenic crops with partial resistance may still find acceptance in agriculture, especially if the resistance produced by the lectin transgenes proves to be additive to the existing endogenous partial resistance genes, or other transgenes also conferring partial resistance.

8.5 OTHER STRATEGIES FOR PRODUCING INSECT-RESISTANT TRANSGENIC PLANTS

8.5.1 Hydrolytic Enzymes

Some evidence suggests that certain hydrolytic enzymes, especially chitinases, may have potential for control of insect pests when expressed in transgenic plants. While the effects of plant chitinases on insects seem marginal at best (e.g., slight deleterious effects of bean chitinase on aphids when fed in artificial diet (Down 1998), transgenic tobacco plants expressing a chitinase from tobacco hornworm were protected from feeding by larvae of tobacco budworm, with decreases in larval survival and biomass (Kramer et al. 1997). Proteinases such as trypsin are also significantly toxic to aphids (Rahbé and Febvay 1993; Purcell et al. 1994), but it is not clear how a strategy to express these in transgenic plants could be executed.

8.5.2 Oxidative Enzymes

The involvement of the oxidative enzymes peroxidase and polyphenol oxidase in defensive reactions of plants against insect pests has been suggested by a number of authors (reviewed by Dowd and Lagrimi 1997; Felton 1996). Polyphenol oxidase, in particular, seems to be involved in defensive responses since its expression is induced by wounding (Constable et al. 1995). In both cases the enzymes have the potential to oxidise components present in plant tissues into potentially toxic compounds; polyphenol oxidase can significantly reduce the protein quality of an insect diet and thus reduce the growth of larvae feeding on that diet (Felton et al. 1992). However, direct evidence for toxicity of either enzyme to insects is lacking, and such transgenic plant trials as have been carried out have not given any clear evidence for protection by overexpression of either enzyme.

8.5.3 Lipid Oxidases

Two enzymes that oxidise lipids, lipoxygenase and cholesterol oxidase, have been put forward as potential insecticidal proteins for expression in transgenic plants. Lipoxygenase oxidises long-chain unsaturated fatty acids to their peroxides and is responsible for the development of rancidity in legume flours. Expression of lipoxygenase genes is induced by wounding and pathogen attack in many plant species (Felton et al. 1994), and this enzyme has come to be considered a part of the endogenous plant defensive system, although it is still not clear exactly what role it plays. Lipoxygenase has been shown to be toxic to lepidopteran larvae when fed in artificial diet (Shukle and Murdock 1983; Mohri et al. 1990). It is also toxic to the homopteran rice brown planthopper (Powell et al. 1993). However, transgenic plants overexpressing this enzyme do not appear to have been used in insect bioassays; it is not clear whether overexpression of lipoxygenase will have an adverse effect on the plant as well as its predator.

In contrast to the situation with lipoxygenase, cholesterol oxidase was identified as a larval insecticidal protein toward cotton boll weevil on the basis of a mass screen of culture isolates from microorganisms (Purcell et al. 1993). Since insects are unable to carry out sterol metabolism, and rely on either ingested sterols or symbiotic bacteria and/or yeasts for sterol supply, affecting the metabolism of these essential membrane components could be predicted to have an adverse effect on the insect. Purified cholesterol oxidase from a *Streptomyces* species was found to be toxic to boll weevil larvae at levels of approximately 20 µg/ml in artificial diet (reviewed by Purcell 1997). Experiments to express its encoding gene in transgenic plants are under way.

8.5.4 Manipulation of Secondary Metabolism

Since the majority of examples of endogenous plant resistance to insect pests involve secondary metabolites as the most significant factors in determining host resistance, the reliance on defensive proteins noted in plant genetic engineering for insect resistance may seem misplaced. However, present genetic engineering technology does not permit extensive manipulations of metabolic pathways, and many of the genes encoding enzymes involved in secondary metabolism remain uncharacterised. Nevertheless, genetic manipulation of secondary metabolism remains a goal for plant genetic engineers, and its exploitation to increase insect resistance in transgenic crops is inevitable. For reviews of possible candidate pathways that could be exploited for this purpose, the reader should consult articles by Hallahan et al. (1992) and Chilton (1997).

8.6 MANAGING PEST RESISTANCE TO TRANSGENIC PLANTS

When transgenic crops were in the development stage, the widespread adoption of insect-resistant transgenic plants in commercial agriculture was seen as posing a significant problem: the appearance of resistance to the products of insecticidal transgenes in the pest population, as had already been seen when synthetic pesticides became widely adopted (Chrispeels and Savada 1994). The imminent release of Bt toxin–expressing crops resulted in a number of studies to devise strategies to minimise the risk of pest populations with resistance to Bt toxins developing. Pest resistance to Bt toxins was known to develop in the laboratory within as few as four to five generations (Stone et al. 1989), and also occurred in the field, where Bt-resistant Diamond Backed Moth (*Plutella* sp.) had been observed as a result of repeated spraying with toxin formulations (Tabashnik et al. 1990). Further, while most examples of Bt toxin resistance in pests were specific to a particular toxin, so that resistance to Cry1Ac, for example, did not give resistance to Cry2, examples of broad-range resistance had also been observed (Gould et al. 1992).

Fortunately, experience with managing resistance to pesticides was relevant to the situation with transgenic crops, and a number of ecological simulation studies have been carried out to deduce recommended practices for deploying Bt-expressing crops (reviewed by Roush 1997). The main recommendations to emerge from these studies are:

1. To minimise the build-up of resistance in the pest population, the effectiveness of the toxin must be as high as possible, so that only individual insects homozygous for the resistance gene are able to survive, and all heterozygotes are killed.
2. To minimise the number of individual insects that are produced homozygous for the toxin resistance gene, refuges containing nontransgenic plants to sustain a population of susceptible insects (i.e., homozygous negative for the toxin resistance gene) must be planted adjacent to the transgenic crop.
3. The deployment of plants that express more than one toxin, while not absolutely necessary, will further delay the appearance of resistance genes in the pest population.

Recommendation (1) has been the main driving force behind efforts to increase expression levels of Bt toxins in transgenic plants (see above). Commercial deployment of transgenic Bt-expressing crops has so far followed recommendations (1) and (2), but not (3), whereas the IRRI programme for transgenic rice has emphasised (3) also, possibly because it is seen that the refuge strategy may be difficult to enforce in less developed agricultural systems.

8.7   INSECT-RESISTANT TRANSGENIC PLANTS IN IPM STRATEGIES; POTENTIAL EFFECTS ON BENEFICIAL INSECTS

If transgenic insect-resistant plants are to play a useful role in decreasing pesticide usage, it is apparent that they must be compatible with the other components of integrated pest management (IPM). Indeed, the recommended practices for deploying transgenic crops are all based on IPM. As discussed above, this is partly to decrease the potential for pests developing resistance to the transgene products; however, it also recognises that crops are attacked by a whole range of pests, and that genes that give protection against certain pests may not be effective against others. In the absence of an IPM strategy to manage pests on transgenic crops, the farmer will continue to use a broad-spectrum pesticide, which at least has the virtue of killing both the expected and unexpected pest species. Biological control, the use of natural enemies of pest species to keep the pest population at low levels, plays an important role in IPM.

Ideally, genes expressed in transgenic plants for control of pest species should at the same time produce no directly deleterious effects on predators or parasitoids, which may play an important role in biological control. Inevitably, removal of the host or prey for beneficial insects will be deleterious, but any toxic effect of the product of the transgene will decrease the viability of biological control. It may also indicate that widespread use of transgenic plants may have undesirable ecological consequences, such as global reductions in beneficial insect populations, with obvious regulatory implications.

The high level of specificity shown by Bt toxins suggests that the encoding genes are unlikely to cause deleterious effects on predators when expressed in transgenic plants. The high level of effectiveness of the Bt toxin genes toward targeted pests has led most commentators to consider that biological control is not a necessary part of IPM strategies for Bt-expressing transgenics (e.g., Roush 1997), with strategies to manage pest resistance to the toxins taking the main priority. Decreased pesticide usage is assumed, in any case, to increase survival of beneficial insects. However, the specificity of Bt toxin genes may pose problems in dealing with secondary pests, which can still cause significant yield losses if pesticides are not applied. The limited assays that have been published suggest that Bt toxins do indeed have low, or no, toxicity toward beneficial insects. For example, plants expressing Bt toxin were used as hosts for aphids (toward which the toxin had no protective effect), and were shown to have no deleterious effects on ladybirds, which were fed on the aphids raised on the transgenic plants (Dogan et al. 1996). Other studies found no deleterious effects on beneficial insects in transgenic cotton (Wilson et al. 1992), or potatoes (Riddick and Barbosa 1998), or corn (Pilcher et al. 1997). On the other hand, some toxicity of a Bt toxin toward another beneficial predator insect, the lacewing, was observed (Hilbeck et al. 1998), and the use of biological control of leuipdoptera by parasitoids is not fully compatible with the use of Bt toxins, due to the early and rapid mortality of the host (Blumberg et al. 1997). The lack of interest in biological control shown by large-scale commercial agriculture in the developed world largely accounts for the absence of more extensive data on Bt toxicity toward beneficial insects.

In the case of transgenes whose products do not cause complete or almost complete mortality in the target pest, the situation is very different. The control of pests produced by expression of these genes in transgenic plants could be considerably enhanced if it were combined with biological control of the pest. Much interest at present is therefore being placed on the effects of transgenes, including GNA and PIs, on predators and ladybirds. As mentioned above, expression of GNA in potato plants results in significant levels of resistance to certain aphid species upon which ladybirds prey. Recent studies have shown that when adult 2-spot ladybirds (*Adalia bipunctata*) were fed on aphids (*Myzus persicae*) colonising transgenic potato plants expressing GNA, ladybird fecundity, egg viability, and adult longevity were adversley affected, although no acute toxicity effects were observed. Although direct effects cannot be ruled out, at least some of the deleterious effects on ladybirds may well have been due to the nutritional quality of the food source being affected, i.e., the aphids being less fit (Birch et al. 1998). More recently, neonate larvae have been reared to adulthood on either GNA feed or control fed *M. persicae* using an artificial diet. In these studies it was possible to deliver known amounts of GNA to the developing ladybird larvae and to demonstrate accumulation of the transgene within the aphid. Under these conditions, GNA failed to show any deleterious effects upon ladybird larval survival or development, except in a very small, but statistically significant, delay in development time taken from 3rd to 4th instar (Down et al. 1999). However, it was also noted that ladybird larvae consumed more GNA-fed aphids compared to those fed control aphids since the former aphids were significantly smaller.

The effects of GNA on the ability of the gregarious ectoparasitoid wasp *Eulophus pennicornis* to parasitise lepidopteran larvae has also been investigated recently. In these studies larvae of the tomato moth, *Lacanobia oleracae*, were selected as the target pest (host) species since transgenic potato plants expressing GNA were previously shown to be significantly resistant to attack by this species (Gatehouse et al. 1997). Tomato moth larvae were reared either on an artificial diet containing GNA at 2% of the total protein, or on GNA expressing transgenic potato plants. Results from these studies showed that success of the wasp to parasitise the host was not affected by the presence of GNA, and that the resulting progeny of these wasps showed no significant differences in size, longevity, and egg load when compared to those that had developed on hosts fed the respective control diets (Bell et al. 1999). Interestingly, in the few instances when the presence of GNA did have a significant effect on parasitoid performance, it was always a beneficial effect, presumably due to the host (pest) being compromised. Similar experiments to those carried out on *E. pennicornis* with GNA are in the process of being set up to test the effects of CpTI on parasitoid performance using both transgenic plants and artificial diet bioassays.

Thus the data available to date for both CpTI and GNA would suggest that expression of genes encoding these insecticidal proteins are not likely in themselves to be acutely toxic to either the predator or parasitoid which have been tested to date. Trials to test the effects of CpTI on ladybird performance are at present in the process of being carried out. Previous work has shown that while potato plants expressing CpTI did significantly affect tomato moth, it had no effect upon the aphid species tested (Gatehouse et al. 1997). Furthermore, recent characterisation studies on 2-spot ladybird have revealed that the major digestive proteases present are cysteine and not serine proteases and confirmed that CpTI was not able to inhibit this enzyme activity *in vitro* (Walker et al. 1998). This information would suggest that CpTI is unlikely to have a deleterious effect upon ladybirds.

Pollination is another factor that must be considered in terms of possible effects of transgene products on beneficial insects. Possible effects of a number of transgene products, including Bt toxins, proteinase inhibitors and lectins, on bees have been studied in artificial diet. Results have suggested low or no toxicity for Bt toxins, with variable effects for plant-derived proteins, some having no deleterious effects, others showing similar effects to those seen with pest species (PicardNizou et al. 1997). However, these experiments do not reflect the situation *in plenta*, since the levels of the proteins employed in feeding assays are orders of magnitude higher than expression in pollen. It would also be possible to block expression of the transgene product in pollen by using antisense technology, if this was considered desirable.

8.8 POSSIBLE EFFECTS OF INSECT-RESISTANT TRANSGENIC PLANTS ON HIGHER ANIMALS

From the human consumer's point of view, the most important consideration for transgenic crops containing insecticidal proteins produced by the introduced trans-genes is that the product is safe to eat. This has proved a contentious and difficult issue (Käppell and Auberson 1998).

For Bt toxins, evaluations by the Environmental Protection Agency and the Food and Drug Administration in the U.S. have concluded that Bt-expressing potatoes (reviewed by Feldman and Stone 1997) and maize (reviewed by Carozzi and Kolziel 1997) pose no threats to human and animal health and nutrition. While this information is in the public domain, comparatively little has been published through normal scientific, peer-reviewed journals, which has led to certain groups concluding that some kind of "cover-up" has been executed. This is unfortunate, as there is no evidence that Bt toxins are anything other than a very safe product, which certainly do not have the undesirable effects (potential or actual) of the pesticides they are designed to replace.

In the case of plant-derived genes, there is an extensive literature on the effects of plant defensive proteins on higher animals, which has shown that some proteinase inhibitors and lectins do have anti-nutritional effects. However, other inhibitors and lectins have shown no deleterious effects on higher animals in feeding trials, which has allowed the selection of CpTI and GNA as products for expression in plants to confer resistance to insect pests, since neither show growth depressive effects in rat feeding trials (Pusztai et al. 1992; Pusztai et al. 1996). However, the fact that some proteinase inhibitors and lectins can be shown to be deleterious has affected public perception of the use of these genes.

It is an interesting reflection that if the kind of regulation of transgenic crops that is being requested by some pressure groups — namely, that no product containing a compound that can be shown to have adverse effects on human nutrition or health under any circumstances were applied to existing foodstuffs — most foods, including all fresh fruit and vegetables, would be withdrawn from sale. Hopefully, greater familiarity with genetically modified crops will change public perception of a technology that is viewed by some as inherently undesirable on ethical grounds, or even because it "takes mankind into realms that belong to God and to God alone" (Windsor 1998), but in fact offers great benefits to both humanity and the environment.

ACKNOWLEDGMENTS

The authors would like to thank The Scottish Office, The European Commission, BBSRC, MAFF, and the Rockefeller Foundation for financial support, and the many colleagues who have contributed to this review.

REFERENCES

Altabella, T. and M.J. Chrispeels. Tobacco plants transformed with the bean ai gene express an inhibitor of α-amylase in their seeds. *Plant Physiol.* 93, 805–810, 1990.

Barton, K., H. Whiteley and N.-S. Yang. Bacillus thuringiensis δ-endotoxin in transgenic Nicotiana tabacum provides resistance to lepidopteran insects. *Plant Physiol.* 85, 1103–1109, 1987.

Bell, H.A., E. Fitches, R.E. Down, G.C. Marris, J.P. Edwards, J.A. Gatehouse and A.M.R. Gatehouse. The effect of snowdrop lectin (GNA) delivered via artifical diet and transgenic plants on *Eulophus pennicornis* (Hymenoptera: Eulophidae), a parasitoid of the tomato moth *Lacanobia oleracea* (Lepidoptera: Noctuidae). *J. Insect Physiol.* In press 1999.

Bennett, J., M.B. Cohen, S.K. Katiyar, B. Ghareyazie and G.S. Khush, Enhancing insect resistance in rice through biotechnology, in *Advances in Insect Control: The Role of Transgenic Plants*, Carozzi, N. and M.G. Koziel, Eds., Taylor and Francis, London, 75–93, 1997.

Birch, A.N.E., I.E. Geoghegan, M.E.N. Majerus, J.W. McNichol, C. Hackett, A.M.R. Gatehouse and J.A. Gatehouse. Ecological impact on predatory 2-spot ladybirds of transgenic potaotes expressing snowdrop lectin for aphid resistance. *Molecular Breeding.* 5, 75–83, 1998.

Blumberg, D., A. Navon, S. Keren, S. Goldenberg and S.M. Ferkovich. Interactions among *Helicoverpa armigera* (Lepidoptera: Noctuidae), its larval endoparasitoid *Microplitis croceipes* (Hymenoptera: Braconidae), and *Bacillus thuringiensis*. *J. Econ. Entomol.* 90, 1181–1186, 1997.

Boulter, D., G.A. Edwards, A.M.R. Gatehouse, J.A. Gatehouse and V.A. Hilder. Additive protective effects of incorporating two different higher plant derived insect resistance genes in transgenic tobacco plants. *Crop Protection* 9, 351–354, 1990.

Bown, D.P., H.S. Wilkinson and J.A. Gatehouse. Differentially regulated inhibitor-sensitive and insensitive protease genes from the phytophagous insect pest, *Helicoverpa armigera*, are members of complex multigene families. *Insect Biochem. Mol. Biol.* 27, 625–638, 1997.

Broadway, R.M. Are insects resistant to plant proteinase inhibitors? *J. Insect Physiol.* 41, 107–116, 1995.

Carozzi, N. and M.G. Koziel, Transgenic maize expressing a Bacillus thuringiensis insecticidal protein for control of European corn borer, in *Advances in Insect Control: The Role of Transgenic Plants*, Carozzi, N. and M.G. Koziel, Eds., Taylor and Francis, London, 63–74, 1997.

Chen, M., B. Johnson, L. Wen, S. Muthukrishnan, K. Kramer, T. Morgan and G. Reeck. Rice cystatin: bacterial expression, purification, cysteine proteinase inhibitory activity and insect growth suppressing activity of a truncated form of the protein. *Protein Expr. Purif.* 3, 41–49, 1991.

Chilton, S. Genetic engineering of plant secondary metabolism for insect protection, in *Advances in Insect Control: The Role of Transgenic Plants*, Carozzi, N. and M.G. Koziel, Eds., Taylor and Francis, London, 237–269, 1997.

Choma, C.T., W.K. Surewicz, P.R. Carey, M. Pozsgay, T. Raynor and H. Kaplan. Unusual proteolysis of the protoxin and toxin of *Bacillus thuringiensis* — structural implications. *Eur. J. Biochem.* 189, 523–527, 1990.

Chrispeels, M.J., Transfer of bruchid resistance from the common bean to other starchy grain legumes by genetic engineering with the α-amylase inhibitor gene, in *Advances in Insect Control: The Role of Transgenic Plants*, Carozzi, N. and M.G. Koziel, Eds., Taylor and Francis, London, 139–156, 1997.

Chrispeels, M.J. and N.V. Raikhel. Lectins, lectin genes, and their role in plant defence. *Plant Cell* 3, 1–9, 1991.

Chrispeels, M.J. and D.E. Savada, *Plants, Genes and Agriculture*. Jones and Bartlett, Boston, 478, 1994.

Cole, R.A. Isolation of a chitin-binding lectin, with insecticidal activity in chemically-defined synthetic diets, from two wild brassica species with resistance to cabbage aphid *Brevicornye brassicae. Entomol. Exp. Appl.* 72, 181–187, 1994.

Constable, C.P., D.R. Bergey and C.A. Ryan. Systemin induces synthesis of wound-inducible tomato leaf polyphenol oxidase via the octadecanoid signalling pathway. *Proc. Natl. Acad. Sci. USA.* 92, 407–411, 1995.

Czalpa, T.H. and B.A. Lang. Effect of plant lectins on the larval development of European corn borer (Lepidoptera: Pyralidae) and southern corn rootworm (Coleoptera: Chrysomelidae). *J. Econ. Entomol.* 83, 2480–2485, 1990.

Dogan, E.B., R.E. Berry, G.L. Reed and P.A. Rossignol. Biological parameters of convergent lady beetle (Coleoptera: Coccinellidae) feeding on aphids (Homiptera: Aphididae) on transgenic potatoes. *J. Econ. Entomol.* 89, 1105–1108, 1996.

Dowd, P.F. and L.M. Lagrimini, The role of peroxidase in host insect defenses, in *Advances in Insect Control: The Role of Transgenic Plants*, Carozzi, N. and M.G. Koziel, Eds., Taylor and Francis, London, 195–223, 1997.

Down, R.E., *Use of endogenous plant defensive proteins to confer resistance to aphids in crop plants.* Ph.D. thesis, Department of Biological Sciences, University of Durham, 1998, 200.

Down, R.E., A.M.R. Gatehouse, W.D.O. Hamilton and J.A. Gatehouse. Snowdrop lectin inhibits development and fecundity of the glasshouse potato aphid (*Aulacorthum solani*) when administered *in vitro* and via transgenic plants, both in laboratory and glasshouse trials. *J. Insect Physiol.* 42, 1035–1045, 1996.

Down, R.E., L. Ford, S.D. Woodhouse, R.J.M. Raemaekers, B. Leitch, J.A. Gatehouse and A.M.R. Gatehouse. Snowdrop lectin (GNA) has no acute toxic effects on a beneficial insect predator, the 2-spot ladybird (*Adalia bipunctata* L.). *J. Insect Physiol.* In press, 1999.

Duan, X., X. Li, Q. Xue, M. Abo-El-Saad, D. Xu and R. Wu. Transgenic rice plants harboring an introduced potato proteinase inhibitor II gene are insect resistant. *Nat. Biotechnol.* 14, 494–496, 1996.

Duck, N. and S. Evola, Use of transgenes to increase host plant resistance to insects: opportunities and challenges, in *Advances in Insect Control: The Role of Transgenic Plants*, Carozzi, N. and M.G. Koziel, Eds., Taylor and Francis, London, 1–20, 1997.

Edmonds, H.S., L.N. Gatehouse, V.A. Hilder and J.A. Gatehouse. The antimetabolic effects of oryzacystatin on larvae of the Southern Corn Rootworm (Diabrotica undecimpunctata howardi): use of a bacterial expression system for oryzacystatin. *Entomol. Exp. Appl.* 78, 83–94, 1996.

Ehrlich, P.R. and P.H. Raven. Butterflies and plants: a study in coevolution. *Evolution* 18, 586–608, 1964.

Eisemann, C.H., R.A. Donaldson, R.D. Pearson, L.C. Cadagon, T. Vuocolo and R.L. Tellam. Larvicidal activity of lectins on Lucilla cuprina: mechanism of action. *Entomol. Exp. Appl.* 72, 1–11, 1994.

Farmer, E.E. and C.A. Ryan. Octadecanoid precursors of jasmonic acid activate the synthesis of wound-inducible proteinase inhibitors. *Plant Cell* 4, 129–134, 1992.

Feldman, J. and T. Stone. The development of a comprehensive resistance management plan for potatoes expressing the Cry3A endotoxin, in *Advances in Insect Control: The Role of Transgenic Plants*, Carozzi, N. and M.G. Koziel, Eds., Taylor and Francis, London, 49–61, 1997.

Felton, G.W. Nutritive quality of plant protein: sources of variation and insect herbivore responses. *Arch. Insect Biochem. Biophysiol.* 32, 107–130, 1996.

Felton, G.W., K. Donato, R.J. Del Vecchio and S.S. Duffet. Impact of oxidised plant phenolics on the nutritional quality of dietary protein to a noctuid herbivore, *Spodoptera exigua.* *J. Insect Physiol.* 38, 277–285, 1992.

Felton, G.W., C.B. Summers and A.J. Mueller. Oxidative responses in soybean foliage to herbivory by bean leaf beetle and three-cornered alfalfa hopper. *J. Chem. Ecol.* 20, 639–650, 1994.

Fischhoff, D.A., K.S. Bowdish, F.J. Perlak, P.G. Marrone, S.M. McCormick, J.G. Niedermeyer, D.A. Dean, K. Kusano-Kretzmer, E.J. Meyer, D.E. Rochester, S.G. Rogers and R.T. Fraley. Insect tolerant transgenic tomato plants. *Bio/Technology* 5, 807–813, 1987.

Fitches, E. and J.A. Gatehouse. A comparison of the short and long term effects of insecticidal lectins on the activities of soluble and brush border enzymes of tomato moth larvae (*Lacanobia oleracea*). *J. Insect Physiol.* 44, 1213–1224, 1998.

Fitches, E., A.M.R. Gatehouse and J.A. Gatehouse. Effects of snowdrop lectin (GNA) delivered via artificial diet and transgenic plants on the development of tomato moth (*Lacanobia oleracea*) larvae in laboratory and glasshouse trials. *J. Insect Physiol.* In press, 1997.

Fujimoto, H., K. Itoh, M. Yamamoto, J. Kyozuka and K. Shimamoto. Insect resistant rice generated by introduction of a modified δ-endotoxin gene of *Bacillus thuringiensis.* *Bio/Technology* 11, 1151–1155, 1993.

Garcia-Olmedo, F., G. Salcedo, R. Sanchez-Monge, L. Gomez, J. Royo and P. Carbonero. Plant proteinaceous inhibitors of proteinases and α-amylases. *Oxford Surv. Plant Mol. Cell. Biol.* 4, 275–334, 1987.

Gatehouse, A.M.R., K.J. Butler, K.A. Fenton and J.A. Gatehouse. Presence and partial characterisation of a major proteolytic enzyme in the larval gut of *Callosobruchus maculatus.* *Entomol. Exp. Appl.* 39, 279–286, 1985.

Gatehouse, A.M.R., G.M. Davison, C.A. Newell, A. Merryweather, W.D.O. Hamilton, E.P.J. Burgess, R.J.C. Gilbert and J.A. Gatehouse. Transgenic potato plants with enhanced resistance to the tomato moth *Lacanobia oleracea*; growthroom trials. *Mol. Breed.* 3, 49–63, 1997.

Gatehouse, A.M.R., F.M. Dewey, J. Dove, K.A. Fenton and A. Pusztai. Effect of seed lectin from *Phaseolus vulgaris* on the larvae of *Callosobruchus maculatus*; mechanism of toxicity. *J. Sci. Food Agric.* 35, 373–380, 1984.

Gatehouse, A.M.R., R.E. Down, K.S. Powell, N. Sauvion, Y. Rahbé, C.A. Newell, A. Merryweather and J.A. Gatehouse. Effects of GNA-expressing transgenic potato plants on peach-potato aphid, *Myzus persicae. Entomol. Exp. Appl.* 79, 295–307, 1996.

Gatehouse, A.M.R., K.A. Fenton, I. Jepson and D.J. Pavey. The effects of α-amylase inhibitors on insect storage pests: inhibition of α-amylase *in vitro* and effects on development *in vivo. J. Sci. Food Agric.* 55, 63–74, 1986.

Gatehouse, A.M.R., E. Norton, G.M. Davison, S.M. Babbé, C.A. Newell and J.A. Gatehouse. Digestive proteolytic activity in larvae of tomato moth, *Lacanobia oleracea*; effects of plant protease inhibitors *in vitro* and *in vivo. J. Insect Physiol.*, 45, 545–558, 1999.

Gould, F., A. Martinez-Ramirez, A. Anderson, J. Ferré, F.J. Silva and W. Moar. Broad-spectrum resistance to *Bacillus thuringiensis* toxins in *Heliothis virescens. Proc. Natl. Acad. Sci. USA.* 89, 7986–7990, 1992.

Graham, J., S.C. Gordon and R.J. McNicol. The effect of the CpTI gene in strawberry against attack by vine weevil (*Otiorhynchus sulcatus* F. Coleoptera: Curculionidae). *Ann. Appl. Biol.* 131, 133–139, 1997.

Graham, J.S., G. Hall and C.A. Ryan. Regulation of synthesis of proteinase inhibitors 1 and 11 mRNAs in leaves of wounded tomato plants. *Planta* 169, 399–405, 1986.

Grochulski, P., L. Masson, S. Borisova, M. Pusztai-Carey, J.-L. Schwartz, R. Brousseau and M. Cygler. Bacillus thuringiensis CryIA(a) insecticidal toxin; crystal structure and channel formation. *J. Mol. Biol.* 254, 447–464, 1995.

Habibi, J., E.A. Backus and T.H. Czalpa. Plant lectins affect survival of the potato leafhopper. *J. Econ. Entomol.* 86, 945–951, 1993.

Hallahan, D.L., J.A. Pickett, L.J. Wadhams, R.M. Wallsgrove and C. Woodcock, Potential of secondary metabolites in the genetic engineering of crops for resistance, in *Plant Genetic Manipulation for Crop Protection*, Gatehouse, A.M.R., V.A. Hilder and D. Boulter, Eds., CAB International, Wallingford, Oxon., U.K., 215–248, 1992.

Harborne, J.B., *Introduction to Ecological Biochemistry.* Academic Press, London, 1988, 356.

Harper, S.M., R.W. Crenshaw, M.A. Mullins and L.S. Privalle. Lectin binding to insect brush border membranes. *J. Econ. Entomol.* 88, 1197–1202, 1995.

Heath, R., G. McDonald, J.T. Christeller, M. Lee, K. Bateman, J. West, R. van Heeswijck and M.A. Anderson. Proteinase inhibitors from *Nicotiana alata* enhance plant resistance to insect pests. *J. Insect Physiol.*, 1997.

Hilbeck, A., M. Baumgartner, P.M. Fried and F. Bigler. Effects of transgenic Bacillus thuringiensis corn-fed prey on mortality and development time of immature Chrysoperla carnea (Neuroptera: Chrysopidae). *Environ. Entomol.* 27, 480–487, 1998.

Hilder, V.A., A.M.R. Gatehouse, S.E. Sherman, R.F. Barker and D. Boulter. A novel mechanism for insect resistance engineered into tobacco. *Nature* 330, 160–163, 1987.

Hoffman, M.P., F.G. Zalom, J.M. Smilanick, L.D. Malyj, J. Kiser, L.T. Wilson, V.A. Hilder and W.M. Barnes. Field evaluation of transgenic tobacco containing genes encoding *Bacillus thuringiensis* δ-endotoxin or cowpea trypsin inhibitor: Efficacy against *Helicoverpa zea* (Lepiodptera: Noctuidae). *J. Econ. Entomol.* 85, 2516–2522, 1991.

Höfte, H., H. De Greve, J. Seurinck, S. Jansens, J. Mahillon, C. Ampe, J. Vanderkerckhove, H. VanderBruggen, M. Van Montagu, M. Zabeau and M. Vaeck. Structural and functional analysis of a cloned delta endotoxin of Bacillus thuringiensis Berliner 1715. *Eur. J. Biochem.* 161, 273–280, 1986.

Höfte, H. and H.R. Whiteley. Insecticidal crystal proteins of *Bacillus thuringiensis. Microbiol. Rev.* 53, 1989.

Houseman, J.G. and A.E.R. Downe. Endoproteinase activity in the posterior midgut of *Rhodnius prolixus* Stal (Hemiptera: Reduviidae). *Insect Biochem.* 10, 363–366, 1980.

Huesing, J.E., R.E. Shade, M.J. Chrispeels and L.L. Murdock. α-Amylase inhibitor, not phytohemagglutinin, explains resistance of common bean seeds to cowpea weevil. *Plant Physiol.* 96, 993–996, 1991.

Ishimoto, M. and K. Kitamura. Identification of the growth inhibitor onazuki bean weevil in kidney bean (*Phaseolus vulgaris* L.). *Jpn. J. Breed.* 38, 1988.

Ishimoto, M., J. Sato, M.J. Chrispeels and K. Kitamura. Bruchid resistance of transgenic azuki bean expressing seed α-amylase inhibitor in the common bean. *Entomol. Exp. Appl.* 79, 309–315, 1996.

Jansens, S., M. Cornelissen, R. De Clercq, A. Reynaerts and M. Peferoen. *Phthorimaea operculella* (Lepidoptera: Gelechiidae) resistance in potato by expression of the *Bacillus thuringiensis* CryIA(b) insecticidal crystal protein. *J. Econ. Entomol.* 88, 1469–1476, 1995.

Janzen, D.H., H.B. Juster and I.E. Liener. Insecticidal action of the phytohemagglutinin in black beans on a bruchid beetle. *Science* 192, 795–796, 1976.

Johnson, R., J. Narvaez, G. An and C. Ryan. Expression of proteinase inhibitors I and II in transgenic tobacco plants: Effects on natural defense against *Manduca sexta* larvae. *Proc. Natl. Acad. Sci. USA.* 86, 9871–9875, 1989.

Jongsma, M.A., P.L. Bakker, J. Peters, D. Bosch and W.J. Stiekma. Adaptations of *Spodoptera exigua* larvae to plant proteinase inhibitors by induction of gut proteinase activity insensitive to inhibition. *Proc. Natl. Acad. Sci. USA.* 92, 8041–8045, 1995.

Käppell, O. and L. Auberson. How safe is safe enough in plant genetic engineering? *Trends Plant Sci.* 3, 276–281, 1998.

Knowles, B.H. and J.A.T. Dow. The crystal δ-endotoxins of Bacillus thuringiensis: models for their mechanism of action on the insect gut. *BioEssays* 15, 469–476, 1993.

Korth, K.L. and R.A. Dixon. Evidence for chewing insect-specific molecular events distinct from a general wound response in leaves. *Plant Physiol.* 115, 1299–1305, 1997.

Koziel, M.G., G.L. Beland, C. Bowman, N.B. Carozzi, R. Crenshaw, L. Crossland, J. Dawson, N. Desai, M. Hill, S. Kadwell, K. Launis, K. Lewis, D. Maddox, K. McPherson, M.R. Meghji, E. Merlin, R. Rhodes, G.W. Warren, M. Wright and S.V. Evola. Field performance of elite transgenic maize plants expressing an insecticidal protein derived from *Bacillus thuringiensis. Bio/Technology* 11, 194–200, 1993.

Koziel, M.G., N.B. Carozzi, T.C. Currier, G.W. Warren and S.V. Evola. The insecticidal crystal proteins of *Bacillus thuringiensis*: past, present and future uses. *Biotechnol. Genet. Engin. Rev.* 11, 171–228, 1993.

Kramer, K.J., S. Muthukrishnan, L. Johnson and F. White, Chitinases for insect control, in *Advances in Insect Control: The Role of Transgenic Plants*, Carozzi, N. and M.G. Koziel, Eds., Taylor and Francis, London, 185–193, 1997.

Leplé, J.C., M. Bonadé-Bottino, S. Augustin, G. Pilate, V.D. Le Tân, A. Delplanque, D. Cornu and L. Jouanin. Toxicity to *Chrysomela tremulae* (Coleoptera: Chrysomelidae) of transgenic poplars expressing a cysteine proteinase inhibitor. *Mol. Breed.* 1, 1995.

Li, J., J. Carroll and D.J. Ellar. Crystal structure of insecticidal δ-endotoxin at 2.5Å resolution. *Nature* 353, 815–823, 1991.

Liang, C., G. Brookhart, G. Feng, G. Reeck and K. Kramer. Inhibition of digestive proteinases of stored grain coleoptera by oryzacystatin, a cysteine proteinase inhibitor from rice seed. *FEBS Letts.* 278, 139–142, 1991.

Liang, Y., S. Patel and D.H. Dean. Irreversible binding of *Bacillus thuringiensis* δ-endotoxin to *Lymantria dispar* brush border membrane vesicles is directly correlated to toxicity. *J. Biol. Chem.* 270, 24719–24727, 1995.

Manthavan, S., P.M. Sudha and S.M. Pechimuthu. Effect of *Bacillus thuringiensis* on the midgut cells of *Bombyx mori*: a histopathological and histochemical study. *J. Invert. Pathol.* 53, 217–227, 1989.

McBride, K.E., Z. Svab, D.J. Schaaf, P.S. Hogan, D.M. Stalker and P. Maliga. Amplification of a chimeric *Bacillus* gene in chloroplasts leads to an extraordinary level of an insecticidal protein in tobacco. *Bio/Technology* 13, 362–365, 1995.

McManus, M.T., D.W.R. White and P.G. McGregor. Accumulation of a chymotrypsin inhibitor in transgenic tobacco can affect the growth of insect pests. *Transgenic Res.* 3, 50–58, 1994.

Michaud, D., B. Nguyen-Quoc and S. Yelle. Selective inhibition of Colorado potato beetle cathepsin H by oryzacystatins I and II. *FEBS Letts.* 331, 1993.

Mohri, S., Y. Endo, K. Matsuda, K. Kitamura and K. Fujimoto. Physiological effects of soybean seed lipoxygesases on insects. *Agric. Biol. Chem.* 54, 2265–2270, 1990.

Moreno, J. and M.J. Chrispeels. A lectin gene encodes the α-amylase inhibitor of common bean. *Proc. Natl. Acad. Sci. USA.* 86, 7885–7889, 1989.

Murdock, L., G. Brookhart, P. Dunn, D. Foard, S. Kelley, L. Kitch, R. Shade, R. Shukle and J. Wolfson. Cysteine digestive proteinases in coleoptera. *Comp. Biochem. Physiol.* 87B, 783–787, 1987.

Murdock, L.L., J.E. Huesing, S.S. Nielsen, R.C. Pratt and R.E. Shade. Biological effects of plant lectins on the cowpea weevil. *Phytochemistry* 29, 85–89, 1990.

Orr, G., J. Strickland and T. Walsh. Inhibition of *Diabrotica* larval growth by multicystatin from potato tubers. *J. Insect Physiol.* 40, 893–900, 1994.

Peferoen, M., Insect control with transgenic plants expressing *Bacillus thuringiensis* crystal proteins, in *Advances in Insect Control: The Role of Transgenic Plants*, Carozzi, N. and M.G. Koziel, Eds., Taylor and Francis, London, 1997.

Perlak, F.J., R.W. Deaton, T.A. Armstrong, R.L. Fuchs, S.S. Sims, J.T. Greenplate and D.A. Fischhoff. Insect resistant cotton plants. *Bio/Technology* 8, 939–943, 1990.

Perlak, F.J., R.L. Fuchs, D.A. Dean, S.L. McPherson and D.A. Fischhoff. Modification of the coding sequence enhances plant expression of insect control protein genes. *Proc. Natl. Acad. Sci. USA.* 88, 3324–3328, 1991.

Perlak, F.J., T.B. Stone, Y.M. Muskopf, L.J. Peterson, G.B. Parker, S.A. McPherson, J. Wyman, S. Love, G. Reed, D. Biever and D.A. Fischhoff. Genetically improved potatoes: protection from damage by Colorado potato beetles. *Plant Mol. Biol.* 22, 1993.

Peumans, W.J. and E.J.M. Van Damme. Lectins as plant defense proteins. *Plant Physiol.* 109, 347–352, 1995.

PicardNizou, A.L., R. Grison, L. Olsen, C. Pioche, G. Arnold and M.H. PhamDelegue. Impact of proteins used in plant genetic engineering: toxicity and behavioural study in the honeybee. *J. Econ. Entomol.* 90, 1710–1716, 1997.

Pilcher, C.D., J.J. Obrycki, M.E. Rice and L.C. Lewis. Preimaginal development, survival and field abundance of insect predators on transgenic *Bacillus thuringiensis* corn. *Environ. Entomol.* 26, 446–454, 1997.

Powell, K.S., A.M.R. Gatehouse, V.A. Hilder and J.A. Gatehouse. Antifeedant effects of plant lectins and enzymes on the adult stage of the rice brown planthopper, *Nilaparvata lugens*. *Entomol. Exp. Appl.* 75, 51–59, 1995.

Powell, K.S., A.M.R. Gatehouse, V.A. Hilder and J.A. Gatehouse. Antimetabolic effects of plant lectins and fungal enzymes on the nymphal stages of two important rice pests, *Nilaparvata lugens* and *Nephotettix cinciteps*. *Entomol. Exp. Appl.* 66, 119–126, 1993.

Powell, K.S., A.M.R. Gatehouse, V.A. Hilder, E.J.M. van Damme, W.J. Peumans, J. Boonjawat, K. Horsham and J.A. Gatehouse. Different antimetabolic effects of related lectins toward nymphal stages of *Nilaparvata lugens*. *Entomol. Exp. Appl.* 75, 61–65, 1995.

Powell, K.S., J. Spence, M. Bharathi, J.A. Gatehouse and A.M.R. Gatehouse. Immunohistochemical and developmental studies to elucidate the mechanism of action of the snowdrop lectin on the rice brown planthopper, *Nilaparvata lugens* (Stal). *J. Insect Physiol.* 44, 529–539, 1998.

Purcell, J.P., Cholesterol oxidase for the control of boll weevil, in *Advances in Insect Control: the Role of Transgenic Plants*, Carozzi, N. and M.G. Koziel, Eds., Taylor and Francis, London, 95–108, 1997.

Purcell, J.P., J.P. Greenplate, N.B. Duck, R.D. Sammons and R.J. Stonard. Insecticidal activity of proteinases. *FASEB J.* 8, A1372, 1994.

Purcell, J.P., J.T. Greenplate, M.G. Jennings, J.S. Ryerse, J.C. Pershing, S.R. Sims, M.J. Prinsen, D.R. Corbin, M. Tran, R.D. Sammons and R.J. Stonard. Cholesterol oxidase: a potent insecticidal protein active against boll weevil larvae. *Biochem. Biophys. Res. Commun.* 146, 1406–1413, 1993.

Pusztai, A., G. Grant, J.C. Stewart, S. Bardocz, S.W.B. Ewen and A.M.R. Gatehouse. Nutritional evaluation of the trypsin inhibitor from cowpea. *Br. J.Nutr.* 68, 783–791, 1992.

Pusztai, A., J. Koninkx, H. Hendriks, W. Kok, S. Hulscher, E.J.M. VanDamme, W.J. Peumans, G. Grant and S. Bardocz. Effect of the insecticidal *Galanthus nivalis* agglutinin on metabolism and the activities of brush border enzymes in the rat small intestine. *J. Nutr. Biochem.* 7, 677–682, 1996.

Rahbé, Y. and G. Febvay. Protein toxicity to aphids: an *in vitro* test on *Acyrthosiphon pisum*. *Entomol. Exp. Appl.* 67, 149–160, 1993.

Rao, K.V., K.S. Rathore, T.K. Hodges, X. Fu, E. Stoger, D. Sudhakar, S. Williams, P. Christou, M. Bharathi, D.P. Bown, K.S. Powell, J. Spence, A.M.R. Gatehouse and J.A. Gatehouse. Expression of snowdrop lectin (GNA) in the phloem of transgenic rice plants confers resistance to rice brown planthopper. *Plant J.* 15, 469–477, 1998.

Reeck, G.R., K.J. Kramer, J.E. Baker, M.R. Kanost, J.A. Fabrick and C.A. Behnke, Proteinase inhibitors and resistance of transgenic plants to insects, in *Advances in Insect Control: the Role of Transgenic Plants*, Carozzi, N. and M.G. Koziel, Eds., Taylor and Francis, London, 157–183, 1997.

Riddick, E.W. and P. Barbosa. Impact of Cry3α-intoxicated *Leptinotarsa decemlineata* (Coleoptera: Chrysomelidae) and pollen on consumption, development and fecundity of *Coleomegilla maculata* (Coleoptera: Coccinellidae). *Ann. Entomol. Soc. America.* 91, 303–307, 1998.

Roush, R., Managing resistance to transgenic crops, in *Advances in Insect Control: The Role of Transgenic Plants*, Carozzi, N. and M.G. Koziel, Eds., Taylor and Francis, London, 271–294, 1997.

Ryan, C.A., Defense responses of plants, in *Plant Gene Research: Genes Involved in Microbe-Plant Interactions*, Verma, D.P.S. and T. Hohn, Eds., Springer-Verlag, New York, 375–386, 1984.

Ryan, C.A., Proteinase inhibitors, in *The Biochemistry of Plants*, Marcus, A., Ed., Academic Press, New York, 351–370, 1981.

Sauvion, N., Y. Rahbé, W.J. Peumans, E. van Damme, J.A. Gatehouse and A.M.R. Gatehouse. Effects of GNA and other mannose-binding lectins on development and fecundity of the peach-potato aphid. *Entomol. Exp. Appl.* 79, 285–293, 1996.

Schroeder, H.E., S. Gollasch, A. Moore, L.M. Tabe, S. Craig, D.C. Hardie, M.J. Chrispeels, D. Spencer and T.J.V. Higgins. Bean α-amylase inhibitor confers resistance to the pea weevil (*Bruchus pisorum*) in transgenic peas (*Pisum sativum* L.). *Plant Physiol.* 107, 1233–1239, 1995.

Schuler, T.H., G.M. Poppy, B.R. Kerry and I. Denholm. Insect-resistant transgenic plants. *Trends Biotechnol.* 16, 168–175, 1998.

Shade, R.E., H.E. Schroeder, J.J. Pueto, L.M. Tabe, L.L. Murdock, T.J.V. Higgins and M.J. Chrispeels. Transgenic pea seeds expressing the alpha amylase inhibitor of the common bean are resistant to bruchid beetles. *Bio/Technology* 12, 793–796, 1994.

Shi, Y., M.B. Wang, V.A. Hilder, A.M.R. Gatehouse, D. Boulter and J.A. Gatehouse. Use of the rice sucrose synthase-1 promoter to direct phloem-specific expression of b-glucuronidase and snowdrop lectin genes in transgenic plants. *J. Exp. Bot.* 45, 623–631, 1994.

Shukle, R.H. and L.L. Murdock. Lipoxygenase, trypsin inhibitor and lectin from soybeans: effects on larval growth of *Manduca sexta* (Lepidoptera: Sphingidae). *Environ. Entomol.* 12, 787–791, 1983.

Stoger, E., S. Williams, P. Christou, R.E. Down and J.A. Gatehouse. Expression of the insecticidal lectin from snowdrop (*Galanthus nivalis* agglutinin; GNA) in transgenic wheat plants: effects on predation by the grain aphid *Sitobion avenuae*. *Molecular Breed.* 5, 65–73, 1999.

Stone, T.B., S.R. Sims and P.G. Marrone. Selection of tobacco budworm for resistance to a genetically engineered Pseudomonas fluorescens containing the δ-endotoxin of Bacillus thuringiensis subsp. kurstaki. *J. Invert. Pathol.* 53, 228–234, 1989.

Sudhakar, D., X. Fu, E. Stoger, S. Williams, J. Spence, D.P. Bown, M. Bharathi, J.A. Gatehouse and P. Christou. Expression and immunolocalization of the snowdrop lectin, GNA in transgenic rice plants. *Transgenic Res.* 7, 371–378, 1998.

Tabashnik, B.E., N.L. Cushing, N. Finson and M.W. Johnson. Field development of resistance to *Bacillus thuringiensis* in diamondback moth (Lepidoptera: Plutellidae). *J. Econ. Entomol.* 83, 1671–1676, 1990.

Terra, W.R., C. Ferreira, B.P. Jordao and R.J. Dillon, Digestive enzymes, in *Biology of the Insect Midgut*, Lehane, M.J. and P.F. Billingsley, Eds., Chapman and Hall, London, 153–194, 1996.

Thomas, J.C., D.G. Adams, V.D. Keppenne, C.C. Wasmann, J.K. Brown, M.R. Kanost and H.J. Bohnert. *Manduca sexta* encoded protease inhibitors expressed in *Nicotiana tabacum* provide protection against insects. *Plant Physiol. Biochem.* 33, 611–614, 1995.

Vaeck, M., A. Reynaerts, H. Höfte, S. Jansens, M. De Beuckeleer, C. Dean, M. Zabeau, M. Van Montagu and J. Leemans. Transgenic plants protected from insect attack. *Nature* 327, 33–37, 1987.

van Aarsen, R., P. Soetart, M. Stam, J. Dockx, V. Gosselé, J. Seurinck, A. Reynaerts and M. Cornelissen. CryIA(b) transcript formation in tobacco is inefficient. *Plant Mol. Biol.* 28, 513–524, 1995.

Van Damme, E.J.M., N. DeClerq, F. Claessens, K. Hemschoote, B. Peeters and W. Peumans. Molecular cloning and characterization of multiple isoforms of the snowdrop (*Galanthus nivalis* L.) lectin. *Planta.* 186, 35–43, 1991.

Walker, A.J., L. Ford, M.E.N. Majerus, I.E. Geoghegan, A.N.E. Birch, J.A. Gatehouse and A.M.R. Gatehouse. Characterisation of the proteolytic activity in the larval midgut of two-spot ladybird (*Adalia bipunctata* L.) and its sensitivity to proteinase inhibitors. *Insect Biochem. Mol. Biol.* 28, 173–180, 1998.

Wilson, F.D., H.M. Flint, W.R. Deaton, D.A. Fischhoff, F.J. Perlak, T.A. Armstrong, R.L. Fuchs, S.A. Berberich, N.J. Parks and B.A. Stapp. Resistance of cotton lines containing a *Bacillus thuringiensis* toxin to pink bollworm (lepidoptera: Gelechiidae) and other insects. *J. Econ. Entomol.* 84, 1516–1521, 1992.

Windsor, C. (HRH the Prince of Wales). Seeds of disaster. *The Daily Telegraph.* London, p. 16, 8th June 1998.

Wünn, J., A. Kloti, P.K. Burckhardt, G.C. Ghosh Biswas, K. Launis, V.A. Iglesias and I. Potrykus. Transgenic indica rice breeding line IR58 expressing a synthetic cryIA(b) gene from Bacillus thuringiensis provides effective insect pest control. *Bio/Technology* 14, 171–176, 1996.

Xu, D.P., Q.Z. Xue, D. McElroy, Y. Mawal, V.A. Hilder and R. Wu. Constitutive expression of a cowpea trypsin-inhibitor gene, CpTI, intransgenic rice plants confers resistance to 2 major rice insect pests. *Mol. Breed.* 2, pp. 167, 1996.

CHAPTER 9

Genetic Engineering of
Biocontrol Agents for Insects

Robert L. Harrison and Bryony C. Bonning

CONTENTS

1-56670-479-0/00/$0.00+$.50

9.1   INTRODUCTION

In order to reduce environmental contamination from chemical pesticides, renewed emphasis has been placed on the development of effective biocontrol agents for management of insect and mite pests. The Environmental Protection Agency Office of Pesticide Programs, for example, has implemented an accelerated review process for biological products. Among the advantages of bioinsecticides are their safety to nontarget organisms and the environment, lack of mammalian toxicity, and absence of toxic residues. Some of the limitations of bioinsecticides, such as slow action, restricted host range, and limited persistence in the field, can be addressed by various strategies involving genetic manipulation. In addition to providing a means for improving a given insecticidal product, genetic engineering can also provide increased understanding of the biology and pathogenicity of the organism.

In this chapter, we outline recent progress that has been made in enhancing the efficacy of various natural enemies of insects and mites by using recombinant DNA technology and speculate on future strategies for optimization of these agents by genetic engineering. We use the broader definition of biocontrol (National Academy of Sciences USA, 1987) to include the use of toxins derived from *Bacillus* spp. that act more in the manner of classical chemical insecticides than biological control agents that suppress pest populations (Van Driesche and Bellows, 1996). Development of transgenic plants with toxins derived from insect pathogens is discussed in Chapter 8 of this volume and will not be described here.

9.2   GENETIC ENGINEERING OF PREDATORS AND PARASITOIDS

9.2.1   Introduction

Here we briefly summarize developments relating to the use of recombinant DNA methods to create genetically improved strains of arthropod natural enemies. Although we are a long way from the deployment of transgenic arthropods as part of pest control programs, there has been progress toward the development of techniques for the introduction and stable expression of genes in arthropods. Excellent comprehensive reviews of arthropod transformation and its application to biological control are available (Ashburner et al., 1998; Heilmann et al., 1994; Hoy, 1994; O'Brochta and Atkinson, 1996; O'Brochta and Atkinson, 1997).

Genetic improvement of arthropod natural enemies to enhance their capacity to control pests has been achieved previously by artificial selection (Beckendorf and Hoy, 1985; Johnson and Tabashnik, 1994). An integrated pest management program featuring a predatory mite strain selected for insecticide resistance has been successfully implemented (Headley and Hoy, 1987). However, the development of recombinant DNA techniques has made it possible or at least conceivable to transfer genes specifying beneficial traits directly to arthropods (Ashburner et al., 1998; Heilmann et al., 1994; Hoy, 1994). The use of genetic engineering methods for the improvement of beneficial arthropods has two advantages over artificial selection:

(1) The goal of genetic improvement can be achieved rapidly, without the generations of rearing required for classical selection protocols; (2) Rather than selecting solely from the available gene pool of the arthropod natural enemy, any gene from any species can be used, in principle, for genetic improvement.

Genetic engineering of arthropod natural enemies to improve their effectiveness as biological control agents requires the identification of beneficial traits, the cloning of genes that influence such traits, and the development of techniques for introducing these genes into the natural enemy species in such a way that they are appropriately expressed and stably transmitted to progeny. Heilmann et al. (1994) identified three beneficial traits that can be conferred to natural enemies: (1) pesticide and disease resistance; (2) cold hardiness; (3) sex ratio alteration for species where only one sex attacks the pest. These traits are monogenic (controlled by one gene) or are likely to be influenced strongly by a single gene. Complex, polygenic traits such as host-finding ability and host preference are not well-understood at the molecular level and hence are less amenable to modification. A list of potentially useful pesticide resistance genes for transformation of parasites and predators is provided by Hoy (1994).

Improving natural enemies by genetic engineering requires vectors for the stable and heritable introduction and expression of foreign genes in arthropods. Progress toward the routine transformation of arthropods other than drosophilid flies has been reported with transposable elements, and "paratransformation" has been achieved by engineering symbiotic bacteria (Beard et al., 1992; Coates et al., 1998; Durvasula et al., 1997; Handler et al., 1998; Jasinskiene et al., 1998; Loukeris et al., 1995). Pantropic retrovirus vectors also show much potential as vectors for the transformation of a wide variety of arthropods (Jordan et al., 1998; Matsubara et al., 1996). However, with the exceptions described below, these efforts have focused on the transformation of disease vectors (such as mosquitoes) or on the Medfly, *Ceratitis capitata*.

9.2.2 Predatory Mites and Parasitic Wasps

Arthropod transformation techniques currently being developed often involve microinjection of embryos. Embryo microinjection will likely require the development of new methods for egg collection and processing for each new species and is not feasible for some species, such as parasitoids and viviparous insects. An alternative gene delivery method, maternal microinjection, was developed to transform the western predatory mite, *Metaseiulus occidentalis* (Presnail and Hoy, 1992). This species is a member of the Family Phytoseiidae, a group of mites that are mass-reared for the control of spider mites and that are often wiped out by pesticides sprayed to control other pest species. Hence, transforming mites of this family with genes conferring pesticide resistance may enhance their pest control capacity. Efforts to transform *M. occidentalis* by microinjection of eggs resulted in substantial mortality and little success (Presnail and Hoy, 1992). To bypass the difficulties involved in injecting eggs of this species, gravid females were injected (Presnail and Hoy, 1992). *M. occidentalis* eggs are developed one at a time and are visible through the cuticle. These properties allowed for the injection of DNA directly into or close to the maturing eggs within the female mite.

Maternal microinjection of mites proved to be easier than microinjection of mite eggs. For the first reported maternal microinjection experiment, gravid female mites were injected with a plasmid that contained the *lacZ* gene (that encodes β-galactosidase) under control of the *Drosophila melanogaster hsp70* promoter (Presnail and Hoy, 1992). Seven of the 48 lines established from eggs laid by injected females produced G1 larvae that expressed *lacZ* upon induction with heat shock. Two of these lines contained plasmid sequences after six generations. However, these lines did not persist long enough to provide the amount of DNA required to detect chromosomal integration of the plasmid. A study on the fate of plasmid DNA injected into *M. occidentalis* and another phytoseiid mite, *Amblyseius finlandicus*, revealed that plasmid sequences could be detected by PCR at a high frequency in multiple eggs laid by microinjected females. This result suggested that the injected DNA persisted in some form for an extended period and was capable of either entering or forming a stable association with the eggs (Presnail and Hoy, 1994).

To determine whether the foregoing method could be used to produce mites with foreign sequences integrated in their chromosomes, four more transformed lines were produced and maintained in culture for more than 150 generations (Presnail et al., 1997). After 100 generations, mites from all lines still contained plasmid sequences. Reverse transcription and PCR analysis of RNA from the transformed lines indicated that the *lacZ* sequence was being transcribed in all four lines, but no β-galactosidase expression was detected. When genomic DNA from the transformed lines was analyzed by restriction digests and hybridized with labeled plasmid DNA, only one line produced results consistent with integration of the plasmid into genomic DNA. Crosses of males of this line with wild-type females produced daughters that contained the plasmid sequences. When these females (which were heterozygous for the plasmid DNA) were mated with wild-type males, approximately half of the progeny contained plasmid sequences. These results are consistent with the association of the microinjected DNA with the nuclear genome and with Mendelian inheritance.

Maternal microinjection was also attempted with a braconid parasitoid wasp, *Cardiochiles diaphaniae* (Presnail and Hoy, 1996). This species was imported into the United States to control lepidopteran pests of the genus *Diaphania*. Because the cuticle of this wasp is pigmented and opaque, it was necessary to carry out dissection and preliminary injections to determine where to insert the microinjection pipette in order to deliver DNA into the ovaries. Wasps were microinjected with a plasmid containing the paraoxon resistance (parathion hydrolase) gene and selected for transgenic G_1 progeny with parathion. This procedure yielded one survivor with DNA that displayed a hybridization pattern with a plasmid probe that was consistent with chromosomal integration of plasmid sequences.

Field releases of transformed *M. occidentalis* at a Florida research plot were carried out in the spring and fall of 1996. The goal of these releases was to determine the persistence of the plasmid-derived *lacZ* gene in the predatory mites under field conditions, the capacity of transgenic mites to control spider mites, and the ability of transgenic mites to disperse from the release site (USDA-APHIS Regulation of Transgenic Arthropods web site). For the initial trial (in April 1996), spider mites and transgenic predatory mites were released. Both predatory and spider mite populations declined rapidly due to heavy rains and low temperatures, forcing the termination of the experiment. By this time, few of the predatory mites contained the plasmid sequences, indicating that the sequence was unstable in the field. This was unexpected, because the plasmid sequence had persisted in the transgenic mites for more than 150 generations under laboratory conditions, and no differences in several fitness parameters were detected between transgenic and wild-type mites (Li and Hoy, 1996). The plasmid sequence was also rapidly lost during a second trial carried out in October 1996 with mites from six additional transgenic lines. Little to no dispersal of the transgenic mites was observed. *M. occidentalis* is not adapted to the climate in Florida, and field trials in the western United States, which is the native habitat of this species, may have yielded different results.

9.2.3 Future Prospects

A variety of approaches hold promise for transformation of predators and parasitoids of targeted insect pest species. These include the use of transposable elements for introduction of foreign DNA into the genome, and engineering of the bacterium *Wolbachia*, to drive foreign genes through populations of arthropod natural enemies.

Among the techniques under investigation for arthropod transformation is the use of transposable elements to deliver foreign DNA into the genomes of target species. Transposable elements are independently mobile segments of DNA that can move from one site in the genome to another. The widespread distribution of the *mariner* and *hAT* transposable elements implies that these elements may be active in a wide variety of species, including beneficial arthropods (O'Brochta and Atkinson, 1996; Robertson, 1995). *Mariner* elements have been discovered in the genome of the green lacewing, *Chrysoperla plorabunda*, a member of the Family Chrysopidae, several members of which are reared commercially for biological control (Robertson et al., 1992). The transposable elements *hermes, mariner*, and *piggyBac* have been shown to undergo transposition in embryos and cell lines of a number of dipteran and lepidopteran species, suggesting that the mobility of these elements is not limited by a stringent requirement for host-specific factors. Hence, there is much potential for these or related elements to be developed as germ-line transformation vectors for beneficial arthropods.

With the use of transposons to integrate genes into the germ-lines of arthropods, there is the prospect of spreading a gene encoding a beneficial trait through a natural enemy population (Kidwell and Ribeiro, 1992). This property would obviate the need for repeated mass releases of transgenic arthropods. The P element isolated from the fruit fly *D. melanogaster* has been demonstrated to spread rapidly in both natural and laboratory populations (Anxolabéhère et al., 1988; Good et al., 1989), and a marker gene borne by a P element vector rapidly increased in frequency when introduced into bottle and cage populations (Carareto et al., 1997). However, the ecological risks of replacing a native population with a genetically engineered population may proscribe against this approach (Hoy, 1995; Tiedje et al., 1989). For example, the numerous instances of apparent horizontal transposition of elements between different taxa raises the possibility that a gene for a beneficial trait (such as pesticide resistance) may be transmitted to a pest species, making the pest more difficult to control.

Bacterial and viral genomes are significantly easier to manipulate than eukaryotic genomes, and efforts have been made to develop bacteria and viruses as vehicles for stable, heritable foreign gene expression. Bacterial symbionts of the hematophagous bug, *Rhodnius prolixus*, and the tsetse flies, *Glossina* spp., have been successfully cultured outside their hosts and transformed with plasmids (Beard et al., 1992; Beard et al., 1993). Genetically engineered bacterial symbionts of *R. prolixus* effectively reduced the quantity of *Trypanosoma cruzi* (the causative agent of American trypanosomiasis or Chagas' disease) carried by this vector (Durvasula et al., 1997). However, symbionts are most frequently found in insects that feed on a restricted diet (such as blood, plant sap, or cellulose) and are rare among predatory and parasitoid species (Dasch et al., 1984), whereas bacteria of the genus *Wolbachia* are relatively widespread. *Wolbachia* spp. are found in several insect orders and are estimated to occur in up to 15% of insect species. *Wolbachia* spp. have also been discovered in mites, crustaceans, and a nematode (Werren and O'Neill, 1997).

Wolbachia spp. are transmitted in egg cytoplasm and infect gonadal tissues, causing a number of reproductive distortions that provide a selective advantage to the bacteria (Werren, 1997). The most common reproductive alteration exhibited by *Wolbachia*-infected insects is cytoplasmic incompatibility (CI). CI results when uninfected females mate with *Wolbachia*-infected males. The progeny of this cross do not develop due to failure of karyogamy in the fertilized eggs. Infected females, however, can produce offspring with either infected or uninfected males. Hence, *Wolbachia* confers a tremendous reproductive advantage to infected females that allows the bacteria to rapidly spread through a population. A *Wolbachia* strain of *Drosophila simulans* was shown to spread through *D. simulans* populations in California at 100 km/year (Turelli and Hoffmann, 1991). Consequently, there is much interest in using CI-inducing *Wolbachia* to drive foreign genes through a population (Sinkins et al., 1997). Another *Wolbachia*-associated reproductive distortion is parthenogenesis induction (PI: Werren, 1997). Infection with PI *Wolbachia* is associated with thelytoky, the exclusive production of female progeny, in some parasitoid wasps. Because only the females parasitize hosts, there is an interest in using PI *Wolbachia* to confer thelytoky to parasitoid wasps currently used in biological control programs (Sinkins et al., 1997). Thelytoky is expected to eliminate the waste of resources used to rear males, both in a mass rearing facility and in the field (Stouthamer, 1993). Also, thelytokous parasitoids do not need to spend time mating in order to reproduce, which is expected to be an advantage in low-density wasp populations where females may have difficulty locating males (Stouthamer, 1993).

Although *Wolbachia* could potentially serve as a vector for the stable and heritable expression of foreign genes in beneficial arthropods, it has been reported to reside primarily in the gonadal tissue of its host, where foreign gene expression may not have the desired effect. However, there is unpublished evidence that *Wolbachia* is present in other tissues (Pettigrew and O'Neill, 1997). A more serious problem with *Wolbachia* as a vector for natural enemies is the occurrence of horizontal transmission of *Wolbachia* between species (Werren, 1997), including an instance of apparent transmission between a parasitoid and its host (Werren et al., 1995).

9.3 GENETIC ENGINEERING OF INSECT PATHOGENS

9.3.1 Introduction

The ability to enhance the insecticidal properties of each group of entomopathogens is related to the degree of understanding of the biology of pathogenicity and the ease of genetic manipulation. Most progress has been made toward optimization of baculovirus insecticides at the genetic level. The sequences of several baculovirus genomes are now known (Ahrens et al., 1997; Ayres et al., 1994; Gomi et al., 1999) and analysis of this information has greatly facilitated understanding of the biology of baculovirus–host interaction. *Bacillus thuringiensis* has also received considerable attention because of the insecticidal toxins produced by different strains. Because these toxins are active after ingestion by a variety of insect pests, they have been widely exploited for production of transgenic plants and optimization of the bacterium itself as a microbial insecticide.

Genetic engineering of the entomopathogenic nematodes and fungi is in its infancy: researchers working on optimization of entomopathogenic nematodes at the genetic level have been able to exploit information gained from the relatively closely related *Caenorhabditis elegans*, which has been extensively studied as a model organism for developmental processes. Despite a potentially wide array of pesticidal proteins produced by entomopathogenic fungi, fungal genes have played little part in agricultural biotechnology to date. The complexities of the interaction between entomopathogenic fungi and their insect hosts are gradually being unraveled, and some elegant research has resulted in the first genetically engineered entomopathogenic fungus with enhanced properties for control of lepidopteran larvae. For all insect pathogens, isolation and mutagenesis of putative virulence genes will facilitate understanding of pathogenicity and provide the groundwork for optimization for use as biocontrol agents using recombinant DNA technology.

9.3.2 Fungi

The broad host range of some entomopathogenic fungi, such as *Beauveria bassiana* and *Metarhizium anisopliae*, is an attractive characteristic for insect pest control. One major limitation of the use of entomopathogenic fungi, however, is that the effectiveness of control is dependent on the relative humidity after application of spores. Certain aspects of the insecticidal efficacy of entomopathogenic fungi such as production, stability, and application, have been optimized by nongenetic means, but to date there is only one example of a genetically optimized entomopathogenic fungus. Optimization of entomopathogenic fungi by genetic engineering is in its infancy because of a limited knowledge of the molecular and biochemical bases for fungal pathogenesis. Identification of genes involved in pathogenesis is necessary for improvement of insecticidal efficacy by genetic manipulation, and there are likely to be several broad classes of genes involved with pathogenicity at the different stages of fungus–host interaction (St. Leger, 1993).

Molecular research on the approximately 700 species of entomopathogenic fungi is limited in part because good cloning systems are available for only a few species of deuteromycete entomopathogens (Goettel et al., 1989). These fungi are relatively easy to culture and have relatively simple interactions with the host insect. The molecular and biochemical bases of pathogenicity of *M. anisopliae,* which causes green muscardine disease, have been particularly well studied (St. Leger, 1995; 1993). Various genes relating to formation of the appressorium (a specialized structure involved in penetration of the insect cuticle by the fungus), virulence, and nutritional stress have been cloned from *M. anisopliae.*

Research on fungal pathogenicity has focused on penetration of the host cuticle by the fungus, which involves the combined action of physical force and multiple enzymes. Recent evidence suggests that pH acts in conjunction with induction by the host cuticle to trigger secretion of fungal molecules that modify the cuticle (St. Leger et al., 1998). The enzymes secreted by the fungus serve the dual function of facilitating penetration through the cuticle and providing nutrients for fungal growth, and include subtilisin- and trypsin-like proteases, metalloproteases, and exopeptidases (Joshi et al., 1997; St. Leger, 1995). The specificity of the proteases confers some degree of host specificity to the fungus.

M. anisopliae normally takes from 5 to 10 days to kill a host insect. In order to enhance insecticidal efficacy, additional copies of the *Pr1* gene, which encodes a subtilisin-like protease involved in host cuticle penetration, were engineered into the genome of *M. anisopliae* (St. Leger et al., 1996). A constitutive promoter, *gpd* from *Aspergillus nidulans,* was used to drive expression. Four transformants were selected and each contained threee to six copies of the *Pr1* gene. While Pr1 is normally only expressed during growth on the cuticle, Pr1 was produced by the recombinant fungus in the hemocoel of lepidopteran larvae infected with the recombinant fungi. Infection with the recombinant strains resulted in partial hydrolysis of hemolymph proteins and extensive melanization of the larvae. Pr1 was found to activate trypsins, which in turn activated the phenoloxidase system involved in initiation of the melanization process and tyrosine metabolism in general. Larvae infected with the recombinant strains died from these effects 25% sooner than larvae infected with wild-type *M. anisopliae* and feeding damage was reduced by 40%. Moreover, the melanized cadavers were poor substrates for fungal sporulation, which would restrict dissemination of the recombinant fungus, thereby limiting the environmental impact and necessitating repeat sales of the control agent. Although it would be possible to incorporate other insecticidal genes into *M. anisopliae,* use of a homologous gene such as *Pr1* may be more acceptable form a regulatory viewpoint (St. Leger et al., 1996).

For regulation of recombinant entomopathogenic fungi, special consideration would be paid to the reproductive potential and nature of the genetic enhancement. A sexual cycle has not been identified for *M. anisopliae,* but genetic recombination through the parasexual cycle has been demonstrated under laboratory conditions. While the significance of this to the field is unknown, the prospect of using recombinant fungi for insect pest control highlights the need for further research. Genetic exchange between strains could be monitored readily by use of isolate-specific molecular markers (e.g., Random Amplification of Polymorphic DNA bands), which could also be used to monitor the fate of engineered strains in the field (Clarkson, 1996).

The host specificity of entomopathogenic fungi is determined in part by the specificity of proteases produced on contact with the host cuticle. One goal of future research in this area would be to tailor a fungus for control of targeted pest species by manipulation of protease genes. Site-directed mutagenesis of enzymes involved in pathogenesis could also be used to elucidate the catalytic mechanisms of the enzymes with the potential for enhancement through genetic modification.

9.3.3 Nematodes

Entomopathogenic nematodes in the families Steinernematidae and Heterorhabditidae (Order Rhabditida) are obligate parasites and infect more than 250 species of insects. The species susceptible to nematode infection include numerous soil-dwelling insect larvae such as white grubs and chafers for which control measures are restricted because of the environmental persistence, toxicity, or leaching problems associated with chemical insecticides in the soil. The entomopathogenic nematodes are highly virulent, safe for nontarget organisms, amenable to genetic selection, and are exempt from registration in the United States. All of these factors make entomopathogenic nematodes attractive for insect pest control, and they are currently used in a variety of niche markets (Gaugler and Hashmi, 1996; Kaya and Gaugler, 1993). One of the major limitations of entomopathogenic nematodes for pest control, however, is their susceptibility to environmental stress, temperature extremes, solar radiation, and desiccation, and the potential of genetic engineering to enhance these traits is under investigation.

Entomopathogenic nematodes have a symbiotic association with bacteria in the genus *Xenorhabdus* (Family: Enterobacteriaceae). The symbiosis is specific in that each nematode species carries its own unique species of bacterium. The infective, free-living, nonfeeding forms of the nematode carry the bacteria in specialized intestinal vesicles (Bird and Akhurst, 1983). When the nematode locates a suitable host insect, it enters the host usually through natural orifices, and releases its bacterial symbionts into the hemocoel. The bacteria grow in the hemocoel and kill the host insect within 48 hr. The bacteria produce extracellular enzymes that break down proteins and lipids in the insect carcass, thereby providing nutrients for bacterial and nematode growth. The bacteria also suppress secondary infection of the host by producing antibiotic substances, provide an optimal environment for nematode reproduction, and serve as food for the nematode. The nematode provides the bacteria with protection and transport between hosts. Upon depletion of the host resources, the nematodes molt into third instar nonfeeding larvae, acquire the bacterial symbionts, and emerge from the cadaver in search of new hosts. Genetic engineering of the symbiotic Enterobacteriaceae is discussed in the section entitled *Enterobacteroaceae*.

There is natural variation among nematode strains in virulence, desiccation tolerance, host-finding ability, and activity at low temperatures, and hence classical selection can be used to optimize these parameters (Gaugler and Hashmi, 1996).

Ethylmethanesulfonate-induced mutagenesis has also been carried out on *Heterorhabditis bacteriophora*, but multiple loci can be affected using this technique, which complicates analysis of the genetic basis for a mutant phenotype (Rahimi et al., 1993). Moreover, genetic heterogeneity may be lost on colonization of nematodes, and the traits associated with virulence and the genetic basis for those traits are poorly understood.

Steinernematidae and Heterorhabditidae are in the same superfamily as *Caenorhabditis elegans*, which is a model organism for molecular genetics. Conserved genes in *C. elegans* are likely to be homologous in the Steinernematidae and Heterorhabditidae (Fodor et al., 1990). Efforts to engineer entomopathogenic nematodes have relied heavily on knowledge and techniques for manipulation of the *C. elegans* genome. This research has focused particularly on enhancing the environmental stability of the nematode *H. bacteriophora* strain HP88 with respect to heat tolerance. The ease of culture, short generation time, and ability to establish pure lines from hermaphroditic individuals facilitate genetic studies of the Heterorhabditidae.

H. bacteriophora has been transformed by microinjection of plasmid vectors used for *C. elegans* transformation, containing a *C. elegans* Hsp16/*E.coli lacZ* fusion transgene under the control of the *hsp16* promoter, and the *C. elegans rol6* gene for the "roller" phenotype (Hashmi et al., 1995a). The *rol-6* is a dominant allele of a collagen-producing locus that causes *C. elegans* to roll over. The DNAs were injected into the nematode hermaphrodite gonad based on established techniques for *C. elegans*. Half of the progeny were assayed and 6 to 7% were transformed, as shown by β-galactosidase activity in the muscle and hypodermis after exposure to heat shock. The frequency of transformation declined after three or four generations, indicating that the introduced cDNAs were not incorporated into the genome. The DNA was maintained in the nematode as a concatamer of the introduced sequences. The roller phenotype was not observed in *H. bacteriophora*, although the gene was present in some progeny (Hashmi et al., 1995a). It appears that the rol6 collagen does not interact appropriately with the basement membrane to cause the roller phenotype in *H. bacteriophora* as it does in *C. elegans* (Gaugler and Hashmi, 1996).

Using the *hsp16-lacZ* fusion construct, Hashmi et al. (1995b) developed and tested a novel apparatus for nematode transformation. Silicon fabrication technology was used to construct arrays of sharp pyramidal microprobes on a silicon wafer. DNA was added on to the array, and a concentrated suspension of *H. bacteriophora* pipetted on top of the microprobes. Nematodes were injected by the microprobes. By using this labor-saving technique, 8% of the nematodes were heritably transformed as shown by expression of β-galactosidase.

H. bacteriophora was then engineered to express the green fluorescent protein (GFP) from the jellyfish *Aequorea victoria* to investigate the potential of GFP as a marker enzyme for genetic studies (Hashmi et al., 1997). Because Steinernematidae and Heterorhabditidae autofluoresce from accumulation of lipofucsin (Davis et al., 1982), the *C. elegans* mec-4 promoter, which is only active in the tail region away from sites of autofluorescence, was used to drive GFP expression. The plasmid DNA was injected into the gonads of adult hermaphroditic nematodes with four to six eggs. Two percent of the progeny were transformed and GFP fluorescence was detected and easily distinguishable from autofluorescence (Hashmi et al., 1997).

H. *bacteriophora* was engineered to express *C. elegans* Hsp70A to enhance tolerance of high temperatures (Hashmi et al., 1998). Five to ten copies of the *Hsp70A* gene were incorporated into the nematodes, and 90% of the transformed nematodes survived exposure to 40°C compared to only 2 to 3% of the wild-type nematodes. The transformed nematodes exhibited normal growth and development and no changes were detected in fitness, other than improved thermotolerance (Gaugler et al., 1997; Hashmi et al., 1998). Although the transgene was inherited extrachromosomally, the transformation was stable for 15 generations.

Genetic engineering may result in changes in fitness, persistence, or host range of the genetically modified organism, and risk assessment studies are required to determine the indirect effects of transgenes. Gaugler et al. (1997) conducted field trials with the transformed *H. bacteriophora* carrying the *C. elegans Hsp70A* gene in 1996, with the objectives of comparing the field persistence of wild-type and transgenic nematodes, and testing the regulatory procedures involved with field trials of transgenic nematodes (Gaugler et al., 1997). There was no apparent variation in field persistence of the transgenic nematodes, although temperatures in the field did not reach the level at which thermotolerance would have been conferred by the *Hsp70A* gene. There was no apparent effect on infection rates of oriental beetles, *Exomala orientalis*, that were introduced into the field plot, demonstrating that the thermotolerance trait is unrelated to infection parameters such as host finding and recognition, attachment, and penetration.

Engineering of entomopathogenic nematodes could potentially be used to enhance pathogenicity, improve environmental tolerance, and alter host range once the genetic bases for these traits have been elucidated (Poinar, 1991). In addition, genes that confer resistance to insecticides and fungicides could be incorporated for protective purposes. It is possible that the Tc1-Tc6 transposons in *C. elegans* (Zwaal et al., 1993), or transposons of Steinernematidae and Heterorhabditidae themselves could be used as insertion vectors to increase the efficiency and stability of transformation of entomopathogenic nematodes. Interruption of specific genes by transposons would facilitate elucidation of gene function. However, as for the other insect biocontrol agents, recombinant DNA technology will not provide a panacea. Very little is known of the basic genetics of entomopathogenic nematodes and while the use of *C. elegans* as a model will facilitate progress in this area, research focused on pathogenicity traits of the Steinernematidae and Heterorhabditidae themselves will be required.

9.3.4 Bacteria

There are numerous entomopathogenic bacteria, but few have been engineered for enhanced insecticidal properties for use in biocontrol. Relative to other biocontrol agents, the bacteria are particularly amenable to genetic manipulation via the plasmids that they carry and because of their ease of culture. Most attention has been given to improving the insecticidal properties of *Bacillus thuringiensis* (Bt) and Bt toxins, and to engineering different bacteria to provide alternative delivery systems for the insecticidal Bt toxins. Bt toxins are particularly attractive for insect pest control because of their specificity and because they are environmentally benign. Research has also been conducted on genetic improvement of Enterobacteriaceae, which are harbored in the guts of entomopathogenic nematodes and released into the hemocoel of the host insect (discussed above), and on *B. sphaericus* for mosquito control.

Bacteria provide an outstanding resource for discovery of novel insecticidal toxins for transgenic plants and other biotechnological applications. Newly discovered toxin groups such as the vegetative insecticidal proteins (VIPs) from *B. cereus* (Estruch et al., 1996; Yu et al., 1997), and the Pht toxins from *Photorhabdus luminescens* (Bowen et al., 1998) will be used to enhance the insecticidal properties of transgenic plants and/or pathogens, according to the site of action of each toxin.

9.3.4.1 Enterobacteriaceae

In addition to the genetic modification of entomopathogenic nematodes described above, the potential for engineering the bacteria associated with these nematodes has been examined. The bacteria in the genus *Xenorhabdus* exist in two distinct forms, phase 1 and phase 2. Phase 1 bacteria are present in infective nematodes and provide optimum conditions for nematode reproduction, while phase 2 bacteria appear on culture of phase 1 bacteria. The bioluminescence of *X. luminescens* has been studied at the molecular level in part to understand the mechanism of conversion from phase 1 to phase 2 (Frackman et al., 1990; Frackman and Nealson, 1990). In order to study genes involved in the virulence of *Xenorhabdus* species, a transformation system was developed (Xu et al., 1989), and avirulent mutants of *X. nematophilus* were produced by insertion of the transposon Tn5 (Hurlbert, 1994; Xu et al., 1991). However, because there is no morphological trait associated with loss of virulence, mutants had to be screened for virulence by inoculation into larvae of the wax moth, *Galleria mellonella* (Hurlbert, 1994). Five avirulent mutants were identified, all of which differed in their biochemical, mutational, and colonial characteristics. Of particular note was a protein of 32.5 kD that was missing from three of the avirulent mutants. The ultimate goal of this research is to identify the genes involved with virulence and study their role in pathogenesis. The use of Tn5 mutagenesis for induction of mutations in *X. nematophilus* will facilitate identification of other genes involved in production of antibiotics for example, and ultimately will enhance our understanding of the bacteria–nematode symbiotic relationship. Ultimately, it may be possible to engineer the Enterobacteriaceae with nematode specificity genes, and insecticidal genes.

9.3.4.2 Bacillus thuringiensis

Bacillus thuringiensis (Bt) is a gram-positive, spore-forming bacterium. Bt is effective for control of agricultural and household pests as well as vectors of animal and human disease. Bt products comprise 95% of the biopesticide market, and most of these products are based on a single natural isolate, Bt subspecies *kurstaki* HD-1, which was isolated 30 years ago (Dulmage, 1970). Management of resistance to specific Bt toxins is a critical issue for Bt-based products. Bt produces several insecticidal crystal proteins (ICPs) during the late phase of bacterial growth, and the ICPs accumulate in the cytoplasm as crystalline inclusions. The ICPs can account for up to 30% of the dry weight of a sporulated culture. The ICPs are classified as α-, β- and γ-exotoxins, and δ-endotoxin, and of these the β-exotoxin and δ-endotoxin have been used in agriculture. The most extensively studied toxin is the δ-endotoxin, which is active against larvae within Lepidoptera, Diptera, and Coleoptera.

The genes encoding the ICPs of Bt, which are referred to as *cry* genes (Höfte and Whiteley, 1989), are generally located on large, nonessential extrachromosomal plasmids (>30 MDa). A single strain can harbor from 2 to 12 *cry* genes. The sequences of more than 50 *cry* genes have been published to date. The genes have been grouped roughly according to specificity and sequence similarity, into cryI to cryVI (Rajamohan and Dean, 1996). The ICPs are processed in the gut of the insect host by solubilization of the crystal protein and proteolytic processing to produce the active form. The toxin then binds to a receptor on the microvillar brush border membranes that line the midgut of susceptible insects (Pietrantonio and Gill, 1996). The gut and mouthparts of the insect become paralyzed and the gut epithelium disintegrates. Different ICPs can be active against different target insects.

Chimeric genes can be used to determine regions conferring species specificity, and fusion of the proteins CryIAb and CryIAc results in a hybrid that has the toxicity of both individual components (Honee et al., 1990). The crystal structures of CryIAa, CryIIIA, and CryIIIB2 have been studied (Borisova et al., 1994; Cody et al., 1992; Li et al., 1991), and partial functions have been assigned to different regions based on biochemical and genetic studies (Thompson et al., 1995). Ultimately modification of ICPs through recombinant DNA technology, in addition to discovery of toxins with novel modes of action, should facilitate effective management of insect resistance to Bt toxins through integrated pest management.

The pathogenicity and host range of Bt has been improved by using both recombinant and nonrecombinant means. For example, the plasmid complement of a Bt strain can be altered without use of recombinant DNA methods: ICP genes can be transferred between strains by a conjugation-like process. A strain with generally good insecticidal activity is identified, and a variant that has lost plasmids with ICP genes or has ICP genes with relatively poor activity is used for incorporation of plasmids with genes encoding highly insecticidal ICPs. Because different ICPs are effective against different target hosts, manipulation and combination of ICP genes can radically affect both the virulence and host range of the bacterium.

New strains of Bt can also be constructed with new combinations of ICP genes by genetic means. A Bt-based cloning vector system has been developed that allows selection of recombinants with ICP genes in *E. coli* or in Bt (Carlton, 1996). This system allows production of Bt strains containing only Bt DNA, which is advantageous in view of acceptance by regulatory agencies. A Bt transposon system facilitates the cloning and selection of strains with new ICP combinations (Baum, 1994).

Genetic engineering has been used to provide alternative delivery systems for the Bt toxins (Koziel et al., 1998). In addition to spraying Bt for crop protection against insect pests, Bt toxins can be engineered directly into plants (described in Chapter 8) or delivered via plant epiphytes or plant endophytes, for example via root-colonizing or xylem-limited bacteria to target pests feeding at specific locations. These alternative strategies enhance the field persistence of the toxin, which is limited for spore–crystal formulations.

Two limitations of Bt are the relatively poor persistence under field conditions, which results in the need for repeated spray application, and the dissemination of large numbers of spores, which is perceived to be a problem in some countries in which products with viable Bt spores are not authorized. Two approaches have been taken to address these two factors concurrently, both of which involve engineering a bacterium that is not normally pathogenic to insects. In the first case, ICP genes were cloned into a nonpathogenic strain of *Pseudomonas fluorescens* (Carlton, 1996). The bacteria were killed, resulting in encapsulated ICPs that had enhanced residual properties in the field and no Bt spores. The EPA approved small-scale field trials of this product in 1985, making it the first recombinant Bt product to be approved for outdoor testing. In the second case, a *spoOA* mutant strain of *B. subtilis* was engineered to express CryIIIA, which is active against Coleoptera (Lereclus et al., 1995). The gene *spoOA* is involved in initiation of sporulation, and disruption of this gene prevents sporulation. The result of this procedure is that the mutant strain can be cultured continuously, no spores are produced, and the insecticidal toxin is encapsulated within a cell and partially protected. Both of these novel formulations provide environmentally safe and stabilized Bt-based biopesticides.

Bt δ-endotoxins have been engineered into the chromosomes of *Pseudomonas fluorescens*, which is a ubiquitous nonpathogenic soil bacterium that also colonizes corn roots (Watrud et al., 1985; Obukowicz et al., 1986). Engineering this bacterium may be useful for increasing the range of habitats that can be exploited for Bt-based insect pest control. This strategy provided some toxicity to the black cutworm *Agrotis ipsilon*, but was not effective against a corn rootworm in the genus *Diabrotica*, which is a much more serious pest (Watrud et al., 1985). The gene encoding Cry1Ac has also been engineered into the endophytic, xylem-limited bacterium *Clavibacter xyli* (Tomasino et al., 1995). The engineered bacterium was introduced into corn seedlings by wound or seed inoculation with high efficiency, and tunneling damage caused by the European corn borer *Ostrinia nubilalis* was significantly reduced by 55 to 65% (Tomasino et al., 1995). However, there was no significant increase in the yield of grain, which could have been inhibited by physical stress on the plant caused by the bacteria. Endophytic microbes provide an alternative approach for potentially enhanced delivery of toxins to leaf and stem feeding Lepidoptera, but optimization in terms of colonization efficiency, attenuation of the detrimental effects on the host plant, and increased production of insecticidal proteins is required.

For Bt, genetic engineering can be used to increase the activity of a particular strain against targeted pest species, extend the host range, increase field stability, and provide alternative constructs to facilitate management of resistance.

9.3.4.3 Bacillus sphaericus

Initial interest in *Bt israelensis* for control of aquatic Diptera such as mosquito and blackfly larvae subsided because of toxin instability in the field. Attention has now turned to *B. sphaericus* for mosquito control because of its greater persistence (Rajamohan and Dean, 1996). Mosquito larvicidal genes from *B. sphaericus* have been cloned from total cellular DNA and the insecticidal genes in this species do not seem to be associated with extrachromosomal plasmids. A binary toxin of 51 and

42 kD has been identified from highly larvicidal strains, in addition to a weakly active 100 kD toxin. When expressed in a protease-deficient strain of *B. sphaericus*, it was found that the 100 kD toxin had enhanced insecticidal efficacy. This result suggests that the low activity of this toxin may be the result of poor expression, or proteolytic degradation during sporulation (Thanabalu and Porter, 1995). The use of protease deficient bacteria could enhance the delivery of toxins to their targeted pests. In *B. sphaericus*, the toxin inclusion bodies are held adjacent to the spores by a membrane envelope. The mechanism of toxin action is incompletely understood, but *B. sphaericus* will have the potential for genetic modification to enhance the efficacy against targeted aquatic Diptera.

9.3.5 Protozoa

The majority of protozoans with potential for insect pest control are in the phylum Microspora. While analysis of ribosomal RNA has been used to determine phylogenetic and evolutionary relationships (Baker et al., 1998; Vossbrinck et al., 1987), the basic knowledge and tools required for genetic engineering of this pathogen group are lacking (Heckmann, 1996). New techniques are required for *in vitro* culture, formulation, and application of entomopathogenic protozoa. The protozoa, and microsporidia in particular, characteristically have low pathogenicity and hence are most likely to be used in combination with other insect control agents. In theory, genetic engineering could be applied to the microsporidia to enhance host range and pathogenicity. In practice, the fact that the microsporidia infect all vertebrate classes in addition to many invertebrates, and that insect and mammalian pathogenic microsporidia are closely related taxonomically, raises concerns over the safety of genetic manipulation (Kurtti and Munderloh, 1987). A clearer understanding of host range determinants is required before genetically engineered protozoa could be considered for insect pest control purposes.

9.3.6 Viruses

There are a variety of proteins and peptides such as insect neurotoxins that act within the hemocoel that are not active against an insect pest by ingestion or by topical application. Insect viruses such as baculoviruses provide a delivery system for these toxins into the hemocoel of the insect, where they have access to the site of action. Such toxins provide a second line of offense against the host insect in addition to the virus itself. The only insect viruses that have been engineered for enhanced insecticidal performance are the baculoviruses. Genetic engineering has been applied to enhancement of baculovirus insecticides since the late 1980s and has resulted in products that now approach the efficacy of chemical insecticides.

9.3.6.1 Construction of Recombinant Baculoviruses

Methods for producing recombinant baculoviruses bearing genetic alterations (insertions of foreign genes or alterations/deletions of native genes) are described in a number of baculovirus expression vector technique manuals (King and Possee,

1992; O'Reilly et al., 1992; Richardson, 1995; Summers and Smith, 1987) and have been reviewed recently (Jarvis, 1997). The most common approach for the construction of recombinant baculoviruses involves the cotransfection of an insect cell line that supports viral replication with naked viral genomic DNA and a plasmid transfer vector. The transfer vector consists of a viral genomic restriction fragment containing the desired alteration. Homologous recombination between the parental virus genome and the transfer vector on either side of the alteration (a double crossover event) results in the incorporation of the alteration into the viral genome (Figure 9.1). Because naked baculovirus DNA is infectious when introduced into a permissive cell line, the transfection leads to the production of progeny virus, some of which are recombinant viruses carrying the desired alteration. Clonal isolates of the recombinant virus are identified and isolated in a plaque assay. Methods for the direct cloning of genes into the *Autographa californica* multicapsid nucleopolyhedrovirus (AcMNPV) genome (Lu and Miller, 1996a; Luckow et al., 1993) and for using a PCR product for the insertion of foreign genes (Gritsun et al., 1997) have also been developed, but it remains to be seen if these methods will supplant the use of plasmid transfer vectors.

Baculovirus expression vectors initially used the very strong polyhedrin (*polh*) promoter to drive foreign gene expression. The *polh* gene is dispensable for growth in cell culture. Consequently, most transfer vectors are designed to integrate genes into the *polh* locus. For recombinant viruses to be tested for pesticidal activity, transfer vectors are available that have an intact *polh* gene so that the recombinant viruses will produce polyhedra for use in bioassays (Bishop, 1992; Wang et al., 1991). Foreign genes in these transfer vectors are positioned upstream of the *polh* gene and next to a viral promoter that will drive expression. Transfer vectors for the deletion/interruption of other viral genes (O'Reilly and Miller, 1991; Zuidema et al., 1989) and the insertion of foreign genes in other regions of the AcMNPV genome (Vlak et al., 1990) have also been developed. Either the transfer vector or the parental virus must carry a marker gene that will allow for the identification of plaques formed by recombinant progeny viruses. The *lacZ* gene is a popular choice for this purpose, as it allows blue/white color selection. The presence of *lacZ* in an AcMNPV parental virus also allows the virus to be linearized with the restriction enzyme *Bsu36* I, which has a site in *lacZ* but not in the AcMNPV genome. Linearization of parental virus DNA has been shown to dramatically increase the proportion of recombinant viruses arising from transfection (Kitts and Possee, 1993).

Methods for scaling up recombinant clones and analyzing viral DNA for the correct incorporation of the desired alteration have been described (O'Reilly et al., 1992). Polyhedra for bioassays are generally isolated from larvae to eliminate polyhedra formation mutants that often accumulate during serial passage of baculoviruses in insect cell lines (Fraser, 1987; Kumar and Miller, 1987; Slavicek et al., 1995).

9.3.6.2 *Recombinant Baculovirus Insecticides*

Approximately 20 recombinant baculoviruses have been successfully engineered for improved insecticidal efficacy in terms of reduced survival time and feeding damage caused by infected larvae. The requirements for insecticidal peptides or

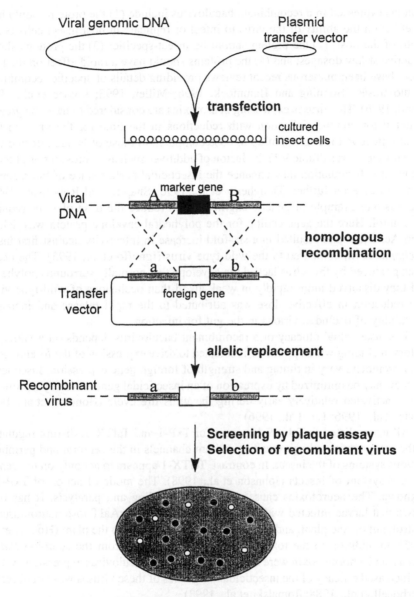

Figure 9.1   Construction of recombinant baculoviruses. The insertion of alterations (foreign genes or deleted/interrupted viral genes) into viral genomic DNA occurs by homologous recombination between the viral genomic DNA and a transfer vector containing a viral restriction fragment with the foreign or altered gene. A double crossover event in the areas flanking the alteration (a) and (b) results in the replacement of the native sequence in the viral genome with the altered sequence from the transfer vector. The recombination event takes place in insect tissue culture cells into which the baculovirus and transfer vector DNAs have been introduced by transfection. Progeny virus from the transfected cells are screened by plaque assay, and plaques exhibiting the desired phenotype are picked for further purification and study.

proteins expressed in a recombinant baculovirus include (1) the protein should have no effect on the ability of the virus to infect or replicate in host insect cells before death of the host; (2) the protein should be insect-specific; (3) the proteins should be active at low dosages; and (4) the proteins should have a rapid effect on the host. There have been numerous recent reviews providing details of specific recombinant baculoviruses (Bonning and Hammock, 1996; Miller, 1995; Possee et al., 1997; Wood, 1996). The viruses expressing neurotoxins are considered to have the greatest potential for commercialization with reductions in the time for the virus to take effect (effective time: ET_{50}) of up to 60% compared to those of larvae infected with the wild-type virus (Table 9.1). Inclusion of additives such as fluorescent brighteners in the virus formulation may enhance the insecticidal performance of the recombinant viruses even further (Dougherty et al., 1996; Shapiro and Robertson, 1992). There is one example of genetic engineering to reduce the dose of virus required for control. Here the gene coding for the polyhedral envelope protein was deleted from AcMNPV and resulted in a six-fold increase in infectivity against first instar, *Trichoplusia ni*, compared to the wild-type virus (Ignoffo et al., 1995). The occlusions produced by this virus lacked the envelope that normally surrounds polyhedra, and they dissolved more rapidly in weak alkali than occlusions of wild-type virus. The reduction in effective dose was attributed to the rapid release and increased availability of occluded virions in the gut for infection.

The insecticidal efficacy of a recombinant baculovirus depends on a variety of factors, including which promoter is selected to drive expression of the foreign gene. Viral promoters vary in timing and strength of foreign gene expression. Insecticidal efficacy may be optimized by expression of an insecticidal gene driven by a promoter that is activated relatively early during the virus life cycle (Bonning et al., 1994; Jarvis et al., 1996; Lu et al., 1996).

All toxins listed in Table 9.1, except for TxP-1 and TalTX-1, disrupt regulation of the presynaptic, voltage-sensitive sodium channels in the cerebral and peripheral nervous systems of the insect. In contrast, TalTX-1 appears to act only on the central nervous system of insects (Johnson et al., 1998). The mode of action of TxP-1 is unknown. The neurotoxins cause cessation of feeding and paralysis. It has been shown that larvae infected with the recombinant virus AcAaIT lose neuromuscular control, fall off the plant, and are unable to climb back on to the plant (Hoover et al., 1995). In addition to the toxins listed, a toxin derived from the scorpion *Buthus eupeus*, and a hornet toxin were also expressed in the baculovirus expression system, but increased efficacy of the insecticidal properties of these viruses was not observed (Carbonell et al., 1988; Tomalski et al., 1993).

The neurotoxins that target the sodium channel act at multiple sites, some being excitatory in action and others inhibitory. Co-injection of insect-specific neurotoxins showed that toxins that act at different sites on the sodium channel can act synergistically, and hence coexpression of toxins by the same baculovirus may further improve the insecticidal efficacy of the virus (Herrmann et al., 1995). To date, there is one report of a recombinant baculovirus expressing multiple toxins: this virus, vMAg4Sat2, coexpresses the spider toxin μ-Aga-IV and the sea anemone toxin AsII (Prikhod'ko et al., 1998). In this case, a slight reduction in the feeding damage caused by larvae infected with the recombinant virus was seen compared to larvae

Table 9.1 Recombinant Baculoviruses Expressing Neurotoxic Peptides

Origin of toxin	Toxin	Virus name	Efficacy of virus		Lepidopteran host tested	Reference
			Reduction in ET_{50}	Reduction in feeding		
			Mite			
Pyemotes tritici straw itch mite	TxP-1[a]	vSp-Tox34 vp6.9tox34 HzEGTDA –26tox34	40% 60% 40%		*Trichoplusia ni* *T. ni*, *Spodoptera frugiperda* *Helicopverpa zea*	(Tomalski & Miller, 1991/92) (Lu et al., 1996) (Popham et al., 1997)
			Scorpion			
Androctonus australis North African (Algerian) scorpion	AaIT	AcST-3 AcAaIT BmAaIT Ro6.9AaIT	10–38% 35%	55–62%	*T. ni* *Heliothis virescens* *Bombyx mori*[c] *Ostrinia nubilalis*	(Cory et al., 1994; Hoover et al., 1995; Maeda et al., 1991; McCutchen et al., 1991; Stewart et al., 1991; Bonning & Harrison[d]
Leiurus quinquestriatus hebraeus Israeli yellow scorpion	LqhIT1 LqhIT2 LqhIT3 Lqhα IT[b]	AcLIT1.p10 AcLIT2.pol BmLqhIT2 AcLqhIT3 AcLα22	24% 32% 20% 35%		*Helicoverpa armigera* *H. armigera* *B. mori*[c] *H. virescens* *H. armigera*	(Gershburg et al., 1998) (Gershburg et al., 1998) Imai, Aly & Maeda[d] (Herrmann et al., 1995) (Chejanovsky et al., 1995)

Table 9.1 (continued) Recombinant Baculoviruses Expressing Neurotoxic Peptides

Origin of toxin	Toxin	Virus name	Efficacy of virus		Lepidopteran host tested	Reference
			Reduction in ET$_{50}$	Reduction in feeding		
Sea anemone						
Anemonia sulcata	AsII	vSAt2p+	38%	48%	*T. ni*	(Prikhod'ko et al., 1998; Prikhod'ko et al., 1996)
Stichadactyla helianthus	ShI	vSSh1p+	36%		*T. ni*	(Prikhod'ko et al., 1996)
Spider						
Agelenopsis sulcata American funnel web spider	μ-Aga-IV	vMAg4p+	37%	50%	*Spodoptera frugiperda* *T. ni*	(Prikhod'ko et al., 1998; Prikhod'ko et al., 1996)
Tegenaria agrestis	TalTX-1	vTalTX-1	18–33%	16–39%	*T. ni* *Spodoptera exigua* *H. virescens*	(Hughes et al., 1997)
Diguetia canities primitive weaving spider	DTX9.2	vAcDTX9.2	9–24%	30–40%	*T. ni* *S. exigua* *H. virescens*	(Hughes et al., 1997)
Toxin combination	μ-Aga-IV + AsII	vMAg4Sat2	43%	63%	*T. ni* *S. frugiperda*	(Prikhod'ko et al., 1998)

[a] Mechanism of action currently unknown.
[b] Nonselective neurotoxin. Weak mammalian toxicity.
[c] Larvae infected by injection.
[d] Unpublished data.

infected with recombinant viruses expressing the individual toxins (Prikhod'ko et al., 1998). It has also been reported that application of low levels of a pyrethroid insecticide (which targets the sodium channels) or a carbamate insecticide (which targets acetylcholinesterase) were found to act synergistically with the virus AcAaIT (McCutchen et al., 1997). Application of a pyrethroid did not enhance the efficacy of viruses expressing the toxins μ-Aga-IV, ShI, or AsII, however (Popham et al., 1998), which may result from the different bioassay methodology used in this study, or from differences in the target sites of the respective toxins.

Toxins derived from *Bacillus thuringiensis* have been expressed in the baculovirus expression system, which provides a useful supply of toxin for physiological studies (Martens et al., 1990, 1995; Merryweather et al., 1990; Pang et al., 1992). However, because the target of the Bt toxins are receptors in the gut of the insect, toxin contamination of virus preparations caused feeding inhibition (Merryweather et al., 1990), and intracellular production of the toxins by the virus is not be expected to enhance the insecticidal efficacy of the virus.

Various risk-assessment studies have been conducted to assess the impact of neurotoxins such as AaIT on nontarget organisms such as predators and parasites of the target pest (Heinz et al., 1995; McCutchen et al., 1996; McNitt et al., 1995; Treacy, 1997; Treacy et al., 1991). The development time and size of the parasitoid *Microplitis croceipes* developing in *Heliothis virescens* was reduced by infection of the host with AcAaIT (McCutchen et al., 1996). No deleterious impact of the toxin or recombinant viruses on generalist predators or the honey bee were found, however. Because recombinant baculoviruses kill the host insect more rapidly, fewer progeny viruses are produced in the infected insect compared to those infected with the wild-type virus. In addition, cadavers of larvae killed by the recombinant virus do not lyse, which, in turn, impairs efficient virus dispersal in the environment. The fact that larvae killed by some of the baculovirus-expressed neurotoxins fall off the plant may be more desirable for consumers and regulatory agencies, but the effects of toxin-laden cadavers on ground-scavenging beetles for example should be investigated. The continued study of baculovirus ecology to determine the effects of genetic manipulation on such parameters as population growth, density, and dispersal, will be extremely important for predicting risks associated with recombinant baculovirus insecticides (Cory et al., 1997; Fuxa et al., 1998).

Table 9.2 lists seven physiological agents (six proteins and one antisense sequence), which, when expressed by recombinant baculoviruses, have deleterious physiological effects resulting in enhanced insecticidal performance of the virus and reduced survival time (ST_{50}) of infected larvae. In addition, deletion of the ecdysteroid glucosyl transferase (egt) gene from the virus has been shown to reduce the survival time of infected insects (O'Reilly, 1995; O'Reilly, 1997: see O'Reilly, 1995 for review of the action of egt). For some of these agents, such as the PBAN cDNA, which expressed five different peptides (Ma et al., 1998), the mechanism of insecticidal activity remains to be determined.

In addition to the physiological effectors that improved the efficacy of the recombinant baculoviruses, a number of proteins expressed either had no effect or had a deleterious impact on the virulence of the virus. Baculovirus expression of juvenile hormone esterase (nonmodified: Bonning et al., 1992; Eldridge et al., 1992),

Table 9.2

Optimization of Baculovirus Insecticides by Expression of Physiological Effectors

Agent	Virus	Target	Physiological effect	Efficacy/reduction in ST_{50}	Lepidopteran host tested	Reference
Diuretic hormone	BmDH5	Malpighian tubules	Disrupt water balance	20%	B. mori [a]	(Maeda, 1989)
Pheromone biosynthesis activating neuropeptide (PBAN)	AcBX-PBAN	Pheromone biosynthesis	Unknown (5 peptides expressed)	19–16%	T. ni	(Ma et al., 1998)
Juvenile hormone esterase (modified)	AcJHE-SG	Unknown	Disruption of molt or contraction paralysis	30% / 66% reduction in feeding damage	T. ni / H. virescens	(Bonning et al., 1995)
	AcJHE-KK	Juvenile hormone	Disruption of lysosome targeting in pericardial cells	50% reduction in feeding damage	H. virescens	(Bonning et al., 1997)
Chitinase	vAcMNPV.chi	Chitin		22–23%	S. frugiperda [a]	(Gopalakrishnan et al., 1995)
Maize URF13	BV13T BV13.3940	Mitochondria	Proteins bind to cell membranes	40%	T. ni [a]	(Korth and Levings, 1993)
Antisense c-myc	MycAS	Cell regulation		15–20% feeding inhibition	S. frugiperda [a]	(Lee et al., 1997)
egt deletion	vEGTDEL	Unclear — deletion of virus egt gene removes virus inactivation of ecdysteroids	Degeneration of Malpighian tubules	20%	S. frugiperda	(Flipsen et al., 1995a; Flipsen et al., 1995b; O'Reilly and Miller, 1991)

[a] Larvae injected with virus

and eclosion hormone (Eldridge et al., 1991) had no effect on the efficacy of the recombinant baculoviruses compared to the wild-type counterparts (although these recombinant viruses do provide a convenient supply of recombinant protein for physiological studies). In contrast, expression of prothoracicotropic hormone increased the LC_{50} of the baculovirus 150-fold (O'Reilly, 1995), and expression of trehalase inhibited virus replication in the insect and increased the LT_{50} of the recombinant virus relative to the wild-type virus (Sato et al., 1997). These observations on the deleterious impact of recombinant proteins provide potential leads for examination of factors involved in determining virulence. The failure of some of these physiological effectors to disrupt insect homeostasis may result from feedback mechanisms operating in the host. Greater understanding of the regulation of insect physiological processes will facilitate future optimization of viruses expressing such proteins.

Although recombinant baculoviruses expressing insect-specific neurotoxins are considered to have the greatest potential for insect pest control, baculovirus expression of a physiological effector derived from the insect itself may be more acceptable from the regulatory viewpoint. With increasing concern over genetically modified organisms and foods, particularly in Europe, this issue may increase in importance in the future.

9.3.6.3 Progress Toward Commercialization of Recombinant Baculovirus Insecticides

Many aspects pertaining to commercialization of recombinant baculovirus insecticides also pertain to commercialization of their wild-type counterparts. A comprehensive review has been published recently (Black et al., 1997) with details of production, formulation, and application that apply to both wild-type and recombinant viruses.

The commercialization of recombinant baculovirus insecticides has been pursued by American Cyanamid and DuPont (Black et al., 1997). Both companies have conducted U.S. EPA-approved, open field trials with baculoviruses expressing insect-specific neurotoxins in the United States. For small-scale field trials, the EPA requires comprehensive information, including potential effects on nontarget organisms, persistence of the engineered virus, host range, and potential for recombination (Black et al., 1997). American Cyanamid conducted field trials in 1993 and 1994 with the virus vEGTDEL, which has the *egt* gene deleted, and found that its insecticidal efficacy was comparable to that of two Bt products. In 1995 and 1996, American Cyanamid field tested a recombinant baculovirus expressing AaIT, with the *egt* gene deleted, and found improved performance relative to vEGTDEL (Black et al., 1997). In 1996 DuPont tested a recombinant baculovirus expressing the toxin LqhIT2 (DuPont, 1996), although the results of these trials have not been published.

Recombinant baculovirus insecticides may soon be on the market given the apparent success of small-scale field trials and the relative safety of these agents to nontarget organisms. With increasing interest in finding environmentally benign insect pest control agents, recombinant baculovirus insecticides are likely to find application for control of lepidopteran pests on cotton and vegetables, for example.

9.3.6.4   Future Directions in Genetic Engineering of Baculoviruses

Numerous insect-specific neurotoxins have been isolated from a variety of different venoms and used as probes for the study of insect neurophysiology (Adams, 1994). In addition, there is an ongoing effort within commercial, government, and academic labs to isolate toxins from a variety of creatures that use venoms to immobilize their prey, that specifically target the insect nervous system (Krapcho et al., 1995; Nakagawa et al., 1997, 1998). With increased understanding of toxin structure, specificity, and mechanism of action, it will in theory be possible to engineer improved insect-specific toxins. It is likely that newly discovered or genetically modified toxins with even greater potency will further improve the insecticidal efficacy of recombinant baculoviruses. Further research into the physiological effects of insect neuropeptides may also provide leads for enhancing baculovirus efficacy.

Almost all efforts to improve the performance of baculoviruses as control agents by recombinant DNA methods have focused on reducing the time required for baculoviruses to kill or incapacitate its host by insertion of genes for neurotoxins or physiological effectors. Although these efforts have been successful and have also resulted in reduced feeding, there are other limitations to the pesticidal efficacy of baculoviruses that have not been addressed. The host range of baculoviruses is one such limitation (Miller and Lu, 1997). A beneficial trait for a baculovirus to be used as a pesticide is the ability to infect and kill all or most of the pests attacking a crop. However, most baculoviruses have a narrow host range consisting of one or a few species. AcMNPV and related viruses have a much broader host range, but the efficiency of infection and replication of these viruses varies widely in different hosts, with only a few species being highly permissive for virus infection and replication. The effectiveness of baculoviruses as control agents is also limited by "developmental resistance," which is the increase in the dose required to kill larvae as they age (Engelhard and Volkman, 1995). The level of resistance in the last larval instars can be particularly high (Kirkpatrick et al., 1998). It would be advantageous, therefore, to engineer baculoviruses with expanded host ranges and reduced lethal dose.

To date, there is one report of an engineered virus with reduced lethal dose, which is discussed in Section 9.3.6.2 (Ignoffo et al., 1995). In a number of instances, the insertion or alteration of a single viral gene has been found to influence host range. Alterations in the AcMNPV *p143* gene, which encodes a DNA helicase homologue, allowed AcMNPV to infect cell lines and larvae of *Bombyx mori*, a species that is normally completely resistant to AcMNPV infection (Argaud et al., 1998; Kamita and Maeda, 1997). The *hrf-1* (host range factor 1) gene of *Lymantria dispar* MNPV (LdMNPV), when inserted into the AcMNPV genome, enabled AcMNPV to infect *L. dispar* cell lines and larvae (Chen et al., 1998). The AcMNPV *hcf-1* (host cell-specific factor 1) gene is required for efficient replication in *Trichoplusia ni* cell lines but not in a *Spodoptera frugiperda* cell line (Lu and Miller, 1996b). This gene was necessary for efficient infection of *T. ni* larvae when injected into the hemocoel, and *hcf-1* mutation resulted in a 20 to 30% increase in lethal time by oral infection. The AcMNPV *lef-7* (late expression factor 7) is required for efficient infection of a *S. frugiperda* and a *Spodoptera exigua* cell line, but not a *T. ni* cell line (Chen and

Thiem, 1997). Finally, the baculovirus anti-apoptotic gene *p35* allows AcMNPV to replicate in a *S. frugiperda* cell line and in larvae, but is not required for replication in *T. ni* cells and larvae (Clem et al., 1991; Clem and Miller, 1993; Clem et al., 1994).

Many of the genes described above appear to function by blocking cellular defense mechanisms triggered by infection (Miller and Lu, 1997). There is also a report indicating that the host immune response can define the host range of a virus (Washburn et al., 1996). AcMNPV can infect *Helicoverpa zea*, but this species is only semipermissive for AcMNPV replication, and high doses are required to kill larvae (Vail and Vail, 1987). Dissection of AcMNPV-infected *H. zea* larvae revealed that infected cells and tissues were encapsulated by hemocytes (Washburn et al., 1996). Engineering AcMNPV with immunosuppressant genes may allow for a more efficient systemic infection of *H. zea* and other pests.

There are two potential physical barriers to baculovirus infection within insect hosts that could influence lethal dose. One of these barriers is the peritrophic matrix, an extracellular layer of protein and chitin that separates food in the midgut lumen from the midgut epithelium (Lehane, 1997). *T. ni* granulovirus (TnGV) encodes a zinc metalloprotease called enhancin, which digests a mucin-like protein in the peritrophic matrix of *T. ni* larvae (Lepore et al., 1996; Wang and Granados, 1997). Enhancin augments the mortality resulting from AcMNPV infections when this enzyme is mixed with AcMNPV polyhedra (Wang and Granados, 1997; 1998), as does chitinase (Shapiro et al., 1987). These observations indicate that the peritrophic membrane limits the degree of primary infection in the midgut. Engineering viruses to express peritrophic matrix-degrading enzymes on the surface of occluded virions (Hong et al., 1997) or within the matrix of occlusions is therefore anticipated to reduce the dose of virus required to kill the targeted insect.

Another potential barrier to spread of infection within the insect is the basal lamina. Basal laminae are extracellular sheets of proteins that surround the tissues of all animals, providing structural support, a filtration function, and a surface for cell attachment, migration, and differentiation (Rohrbach and Timpl, 1993). A study of the lepidopteran *Calpodes ethlius* revealed that the basal lamina of this insect did not allow passage of gold particles larger than 15 nm (Reddy and Locke, 1990). It is unlikely that baculovirus virions, which measure 40 to 60 nm wide by 250 to 300 nm long (Adams and McClintock, 1991) can freely diffuse through a barrier with a 15 nm size exclusion limit. Substantial accumulations of virions in the extracellular spaces between basal laminae and the plasma membranes of midgut and fat body cells have been observed with infections of TnGV (Hess and Falcon, 1987), suggesting that the TnGV virions could not freely diffuse through the midgut basal lamina. Although there is some controversy over the effect that basal lamina has on systemic infection (Federici, 1997), these observations strongly suggest that the movement and systemic infection by budded virus is constrained by the basal lamina barrier. Although baculoviruses seem to bypass the basal lamina by traveling through the epidermal cells of tracheae (Engelhard et al., 1994), a recombinant virus engineered with enzymes that degrade basal lamina components might be able to establish a widespread infection more rapidly within the host insect.

Most research on genetic manipulation of insect viruses has been conducted on baculoviruses and particularly on the prototype AcMNPV. However, in order to target different lepidopteran species, the same genetic engineering techniques are now being applied to a variety of other baculoviruses, such as *Helicoverpa zea* SNPV (Popham et al., 1997) and *Rachiplusia ou* MNPV (=*Anagrapha falcifera* MNPV; Harrison and Bonning, 1999). It is also likely that a variety of other insect virus groups such as entomopoxviruses and the *Drosophila* C virus will be engineered in the future for insect pest control purposes. Different insect viruses will provide alternative delivery systems for insect-specific neurotoxins to insect pests that are not susceptible to infection by baculoviruses. Hence new types of insect pests can be targeted by these engineered viruses.

9.4   CONCLUDING REMARKS

For the predators and parasitoids of insect pests, some progress has been made toward developing the techniques necessary for transformation to introduce beneficial traits into the genome of these organisms. Questions concerning the ultimate use of this technology relate to the ecological impact of transformed organisms, and the potential spread of transgenes such as pesticide resistance genes to other species, including the targeted pest. Of the insect pathogens, genetically modified bacteria have reached the commercial sector, and recombinant baculovirus insecticides are likely to be marketed within the foreseeable future. Even though significant progress has been made, a considerable amount remains to be learned of baculovirus biology and the functions of many baculovirus genes remain to be elucidated. The engineering of entomopathogenic nematodes and fungi is in its infancy, slowed in part by the greater genetic complexity of these organisms. One of the most limiting characteristics of entomopathogenic fungi with respect to insect pest control is the requirement for humid conditions for germination. The genetic basis for this trait is unknown but provides a potential target for genetic manipulation, to broaden the range of conditions under which germination will occur. For all of the insect pathogens, enhanced understanding of determinants of pathogenicity, host range, and insecticidal properties will pave the way for enhancement as insecticidal agents by recombinant DNA technology. Progress toward understanding the molecular determinants of baculovirus host range, for example (Miller and Lu, 1997), may ultimately enable researchers to manipulate the host range of the virus to suit the requirements of the producer. The limited host range of pathogens such as baculoviruses is desirable in terms of risk assessment considerations, but not practical for control of multiple pest species on a single crop.

Genetic engineering provides a valuable tool for enhancement of some of the properties of insect natural enemies, but does not provide a panacea. It is unlikely that all desirable traits can be adjusted by genetic means, and fundamental research on behavioral and ecological aspects of natural enemy biology remains extremely important. There is wide variation in insecticidal efficacy between natural isolates of the same organism, and the importance of classical selection for natural isolates with optimal traits cannot be overemphasized. Strains selected for optimal virulence should provide the basis for improvement by genetic modification.

In the future we are likely to see expansion of the effort to optimize insect natural enemies through genetic manipulation, particularly to different groups of entomopathogenic viruses and bacteria. Ecological considerations may make it less likely that we will see field deployment of transgenic arthropods and nematodes in the near future, while the insect viruses show the best potential for actual and routine use in the field. Whether or not transgenic nematodes, fungi, and arthropod natural enemies are developed will depend to a large extent on the size of the potential market for the product. Whether or not any of the genetically engineered biocontrol agents will be marketed will also depend on production costs and availability of alternative technologies for control of the targeted pest species.

ACKNOWLEDGMENTS

We thank Anthony Boughton, Drs. Elliot Krafsur, Les Lewis, and John Obrycki for critical reading of this chapter. This project was funded in part by grants from the Illinois-Missouri Biotechnology Alliance (96-3), USDA (9702936) and a NATO collaborative research grant (CRG 951318). Journal Paper No. J -18047 of the Iowa Agriculture and Home Economics Experiment Station, Ames, Iowa, Project Nos. 3301 and 3483, and supported by Hatch Act and State of Iowa funds.

REFERENCES

Adams, J. R. and McClintock, J. T., Baculoviridae. Nuclear polyhedrosis viruses, Part 1: Nuclear polyhedrosis viruses of insects, in *Atlas of Invertebrate Viruses,* Adams, J. R. and Bonami, J. R., Eds., CRC Press, Boca Raton, FL, 1991, 87.

Adams, M. E., Venom toxins reveal ion channel diversity, in *Natural and Engineered Pest Management Agents,* Hedin, P. A., Menn, J. J., Hollingworth, R. M., Eds., American Chemical Society, Washington D.C., 1994, 249.

Ahrens, C. H., Russell, R. L. Q., and Rohrmann, G. F., The sequence of the *Orgyia pseudotsugata* multinucleocapsid nuclear polyhedrosis virus genome, *Virology,* 229, 381, 1997.

Anxolabéhère, D., Kidwell, M. G., and Periquet, G., Molecular characteristics of diverse populations are consistent with the hypothesis of a recent invasion of *Drosophila melanogaster* by mobile P elements, *Mol. Biol. Evol.,* 5, 252, 1988.

Argaud, O., Croizier, L., López-Ferber, M., and Croizier, G., Two key mutations in the host-range specificity domain of the *p143* gene of *Autographa californica* nucleopolyhedrovirus are required to kill *Bombyx mori* larvae, *J. Gen. Virol.,* 79, 931, 1998.

Ashburner, M., Hoy, M. A., and Peloquin, J. J., Prospects for the genetic transformation of arthropods, *Insect Molec. Biol.,* 7, 201, 1998.

Ayres, M. D., Howard, S. C., Kuzio, J., Lopez-Ferber, M., and Possee, R. D., The complete sequence of *Autographa californica* nuclear polyhedrosis virus, *Virology,* 202, 586, 1994.

Baker, M. D., Vossbrinck, C. R., Becnel, J. J., and Andreadis, T. G., Phylogeny of *Amblyospora* (Microsporida: Amblyosporidae) and related genera based on small subunit ribosomal DNA data: A possible explanation of host parasite cospeciation, *J. Invert. Pathol.,* 71, 199, 1998.

Baum, J., Tn5401: A new class II transposable element from *Bacillus thuringiensis,* *J. Bacteriol.,* 176, 2835, 1994.

Beard, C. B., Mason, P. W., Aksoy, S., Tesh, R. B., and Richards, F. F., Transformation of an insect symbiont and expression of a foreign gene in the Chaga's disease vector *Rhodnius prolixus, Am. J. Trop. Med. Hyg.,* 46, 195, 1992.

Beard, C. B., O'Neill, S. L., Mason, P., Mandelco, L., Woese, C. R., Tesh, R. B., and Richards, F. F., Genetic transformation and phylogeny of bacterial symbionts from tsetse, *Insect Molec. Biol.,* 1, 123, 1993.

Beckendorf, S. K. and Hoy, M. A., Genetic improvement of arthropod natural enemies through selection, hybridization or genetic engineering techniques, in *Biological Control in Agricultural IPM Systems,* Hoy, M. A. and Herzog, D. C., Eds., Academic Press, Orlando, FL, 1985, 167.

Bird, A. F. and Akhurst, R. J., The nature of the intestinal vesicle of the family Steinernematidae, *Int. J. Parasitol.,* 13, 599, 1983.

Bishop, D. H. L., Baculovirus expression vectors, *Semin. Virol.,* 3, 253, 1992.

Black, B. C., Brennan, L. A., Dierks, P. M., and Gard, I. E., Commercialization of baculoviral insecticides, in *The Baculoviruses,* Miller, L. K., Ed., Plenum Press, New York, 1997, 341.

Bonning, B. C. and Hammock, B. D., Development of recombinant baculoviruses for insect control, *Annu. Rev. Entomol.,* 41, 129, 1996.

Bonning, B. C., Hirst, M., Possee, R. D., and Hammock, B. D., Further development of a recombinant baculovirus insecticide expressing the enzyme juvenile hormone esterse from *Heliothis virescens, Insect Biochem. Molec. Biol.,* 22, 453, 1992.

Bonning, B. C., Hoover, K., Booth, T. F., Duffey, S., and Hammock, B. D., Development of a recombinant baculovirus expressing a modified juvenile hormone esterase with potential for insect control, *Arch. Insect Biochem. Physiol.,* 30, 177, 1995.

Bonning, B. C., Roelvink, P. W., Vlak, J. M., Possee, R. D., and Hammock, B. D., Superior expression of juvenile hormone esterase and beta-galactosidase from the basic protein promoter of *Autographa californica* nuclear polyhedrosis virus compared to the p10 protein and polyhedrin promoters, *J. Gen. Virology,* 75, 1551, 1994.

Bonning, B. C., Ward, V. K., van Meer, M. v., Booth, T. F., and Hammock, B. D., Disruption of lysosomal targeting is associated with insecticidal potency of juvenile hormone esterase, *Proc. Natl. Acad. Sci. USA,* 94, 6007, 1997.

Borisova, S., Grochulski, P., van Faassen, H., Pusztai-Carey, M., Masson, L., and Cygler, M., Crystallization and preliminary X-ray diffraction studies of the Lepidopteran-specific insecticidal crystal protein CryIA(a), *J. Mol. Biol.,* 243, 530, 1994.

Bowen, D., Rocheleau, T. A., Blackburn, M., Andreev, O., Golubeva, E., Bhartia, R., and ffrench-Constant, R. H., Insecticidal toxins from the bacterium *Photorhabdus luminescens, Science,* 280, 2129, 1998.

Carareto, C. M. A., Kim, W., Wojciechowski, M. F., O'Grady, P., Prokchorova, A. V., Silva, J. C., and Kidwell, M. G., Testing transposable elements as genetic drive mechanisms using *Drosophila* P element constructs as a model system, *Genetica,* 101, 13, 1997.

Carbonell, L. F., Hodge, M. R., Tomalski, M. D., and Miller, L. K., Synthesis of a gene coding for an insect-specific scorpion neurotoxin and attempts to express it using baculovirus vectors, *Gene,* 73, 409, 1988.

Carlton, B. C., Development and commercialization of new and improved biopesticides, *Ann. New York Acad. Sci.,* 792, 154, 1996.

Chejanovsky, N., Zilberberg, N., Rivkin, H., Zlotkin, E., and Gurevitz, M., Functional expression of an alpha anti-insect scorpion neurotoxin in insect cells and lepidopterous larvae, *FEBS Lett.,* 376, 181, 1995.

Chen, C.-J., Quentin, M. E., Brennan, L. A., Kukel, C., and Theim, S. M., *Lymantria dispar* nucleopolyhedrovirus *hrf-1* expands the larval host range of *Autographa californica* nucleopolyhedrovirus, *J. Virol.,* 72, 2526, 1998.

Chen, C.-J. and Thiem, S. M., Differential infectivity of two *Autographa californica* nucleopolyhedrovirus mutants on three permissive cell lines is the result of *lef-7* deletion, *Virology,* 227, 88, 1997.

Clarkson, J. M., Molecular biology of fungi for the control of insects., in *Molecular Biology of the Biological Control of Pests and Diseases of Plants,* Gunasekaran, M. and Weber, D. J., Eds., CRC Press, Boca Raton, FL, 1996, 123.

Clem, R. J., Fechheimer, M., and Miller, L. K., Prevention of apoptosis by a baculovirus gene during infection of insect cells, *Science,* 254, 1388, 1991.

Clem, R. J. and Miller, L. K., Apoptosis reduces both the *in vitro* replication and the *in vivo* infectivity of a baculovirus, *J. Virol.,* 67, 3730, 1993.

Clem, R. J., Robson, M., and Miller, L. K., Influence of infection route on the infectivity of baculovirus mutants lacking the apoptosis-inhibitory gene *p35* and the adjacent gene *p94,* *J. Virol.,* 68, 6759, 1994.

Coates, C. J., Jasinskiene, N., Miyashiro, L., and James, A. A., *Mariner* transposition and transformation of the yellow fever mosquito, *Aedes aegypti, Proc. Natl. Acad. Sci. USA,* 95, 3748, 1998.

Cody, V., Luft, J., Jensen, E., Pangborn, W., and English, L., Purification and crystallization of insecticidal delta-endotoxin CryIIIB2 from *Bacillus thuringiensis, Proteins: Struct. Funct. Genet.,* 14, 324, 1992.

Cory, J. S., Hails, R. S., and Sait, S. M., Baculovirus ecology, in *The Baculoviruses.,* Miller, L. K., Ed., Plenum Press, New York, 1997, 301.

Cory, J. S., Hirst, M. L., Williams, T., Hails, R. S., Goulson, D., Green, B. M., and Carty, T., Field trial of a genetically improved baculovirus insecticide, *Nature (London),* 370, 138, 1994.

Dasch, G. A., Weiss, E., and Chang, K.-P., Endosymbionts of insects, in *Bergey's Manual of Systematic Bacteriology,* Krieg, N. R., Ed., Williams and Wilkins, Baltimore, 1984, 811.

Davis, B. O., Anderson, G. L., and Dusenbury, D. B., Total luminescence spectroscopy of fluorescence changes during aging in *Caenorhabditis elegans, Biochemistry,* 21, 4089, 1982.

Dougherty, E. M., Guthrie, K. P., and Shapiro, M., Optical brighteners provide baculovirus activity enhancement and UV radiation protection, *Biol. Control,* 7, 71, 1996.

Dulmage, H., Insecticidal activity of HD-1: A new isolate of *Bacillus thuringiensis* var. *alesti,* *J. Invertebr. Pathol.,* 15, 232, 1970.

DuPont, Notification to conduct small-scale field testing of a genetically altered baculovirus, *EPA,* No. 352-NMP-4, 1996.

Durvasula, R. V., Gumbs, A., Panackal, A., Kruglov, O., Aksoy, S., Merrifield, R. B., and Richards, F. F., Prevention of insect-borne disease: An approach using transgenic symbiotic bacteria, *Proc. Natl. Acad. Sci. USA,* 94, 3274, 1997.

Eldridge, R., Horodyski, F. M., Morton, D. B., O'Reilly, D. R., Truman, J. W., Riddiford, L. M., and Miller, L. K., Expression of an eclosion hormone gene in insect cells using baculovirus vectors, *Insect Biochem.,* 21, 341, 1991.

Eldridge, R., O'Reilly, D. R., Hammock, B. D., and Miller, L. K., Insecticidal properties of genetically engineered baculoviruses expressing an insect juvenile hormone esterase gene, *Appl. Environ. Microbiol.,* 58, 1583, 1992.

Engelhard, E. K., Kam-Morgan, L. N. W., Washburn, J. O., and Volkman, L. E., The insect tracheal system: A conduit for the systemic spread of *Autographa californica* M nuclear polyhedrosis virus, *Proc. Natl. Acad. Sci. USA,* 91, 3224, 1994.

Engelhard, E. K., and Volkman, L. E., Developmental resistance in fourth instar *Trichoplusia ni* orally inoculated with *Autographa californica* M nuclear polyhedrosis virus, *Virology,* 209, 384, 1995.

Estruch, J. J., Warren, G. W., Mullins, M. A., Nye, G. J., Craig, J. A., and Koziel, M. G., Vip3A, a novel *Bacillus thuringiensis* vegetative insecticidal protein with a wide spectrum of activities against lepidopteran insects, *Proc. Natl. Acad. Sci. USA*, 93, 5389, 1996.

Federici, B. A., Baculovirus pathogenesis, in *The Baculoviruses*, Miller, L. K., Ed., Plenum Press, New York, 1997, 33.

Flipsen, J. T. M., Mans, R. M. W., and Vlak, J. M., Erratum: Deletion of the baculovirus ecdysteroid UDP-glucosyltransferase gene induces early degeneration of Malpighian tubules, *J. Virol.*, 69, 7380, 1995a.

Flipsen, J. T. M., Mars, R. M. W., Kleefsman, A. W. F., Knebel-Morsdorf, D., and Vlak, J. M., Deletion of the baculovirus UDP-glucosyltransferase gene induces early degeneration of Malpighian tubules in infected insects, *J. Virol.*, 69, 4592, 1995b.

Fodor, A., Vecseri, G., and Farkas, T., *Caenorhabditis elegans* as a model for the study of entomopathogenic nematodes, in *Entomopathogenic Nematodes in Biological Control*, Gaugler, R. and Kaya, H. K., Eds., CRC Press, Boca Raton, FL, 1990, 249.

Frackman, S., Anhalt, M., and Nealson, K. H., Cloning, organization, and expression of the bioluminescence genes of *Xenorhabdus luminescens*, *J. Bacteriol.*, 172, 5767, 1990.

Frackman, S. and Nealson, K. H., The molecular genetics of *Xenorhabdus*, in *Entomopathogenic Nematodes in Biological Control*, Gaugler, R. and Kaya H. K., Eds., CRC Press, Boca Raton, FL, 1990, 285.

Fraser, M. J., FP mutation of nuclear polyhedrosis viruses: A novel system for the study of transposon-mediated mutagenesis, in *Biotechnology in Invertebrate Pathology and Cell Culture*, Maramorosch, K., Ed., Academic Press, San Diego, 1987, 265.

Fuxa, J. A., Fuxa, J. R., and Richter, A. R., Host-insect survival time and disintegration in relation to population density and dispersion of recombinant and wild-type nucleopolyhedroviruses, *Biol. Control*, 12, 143, 1998.

Gaugler, R. and Hashmi, S., Genetic engineering of an insect parasite, *Genetic Engineering: Principles and Methods*, 18, 135, 1996.

Gaugler, R., Wilson, M., and Shearer, P., Field release and environmental fate of a transgenic entomopathogenic nematode, *Biol. Control*, 9, 75, 1997.

Gershburg, E., Stockholm, D., and Chejanovsky, N., Baculovirus-mediated expression of a scorpion depressant toxin improves the insecticidal efficacy achieved with excitatory toxins, *FEBS Lett.*, 422, 132, 1998.

Goettel, M. S., St. Leger, R. J., Bhairi, S., Jung, M. K., Oakley, B. R., and Staples, R. C., Transformation of the entomopathogenic fungus *Metarhizium anisopliae* using the *Aspergillus nidulans benA3* gene, *Curr. Genet.*, 17, 129, 1989.

Gomi, S. , Majima, K., and Maeda, S., Sequence analysis of the genome of *Bombyx mori* nucleopolyhedrovirus, *J. Gen. Virol.* 80, 1323, 1999.

Good, A. G., Meister, G., Brock, H., and Grigliatti, T. A., Rapid spread of transposable P elements in experimental populations of *Drosophila melanogaster*, *Genetics*, 122, 387, 1989.

Gopalakrishnan, B., Muthukrishnan, S., and Kramer, K. J., Baculovirus-mediated expression of a *Manduca sexta* chitinase gene: Properties of the recombinant protein, *Insect Biochem. Molec. Biol.*, 25, 255, 1995.

Gritsun, T. S., Mikhailov, M. V., Roy, P., and Gould, E. A., A new, rapid and simple procedure for direct cloning of PCR products into baculoviruses, *Nucleic Acids Res.*, 25, 1864, 1997.

Handler, A. M., McCombs, S. D., Fraser, M. J., and Saul, S. H., The lepidopteran transposon vector, *piggyBac*, mediates germ-line transformation in the Mediterranean fruit fly, 95, 7520, 1998.

Harrison, R.L. and Bonning, B.C., The nucleopolyhedroviruses of *Rachiplusiaou* and *Anagrapha falcifera* are isolates of the same virus, *J. Gen. Virol.* in press, 1999.

Hashmi, S., Hashmi, G., and Gaugler, R., Genetic transformation of an entomopathogenic nematode by microinjection, *J. Invertebr. Pathol.*, 66, 293, 1995a.

Hashmi, S., Hashmi, G., Glazer, I., and Gaugler, R., Thermal response of *Heterorhabditis bacteriophora* transformed with the *Caenorhabditis elegans hsp70* encoding gene, *J. Exper. Zool.*, 281, 164, 1998.

Hashmi, S., Hatab, M. A. A., and Gaugler, R. R., GFP: green fluorescent protein a versatile gene marker for entomopathogenic nematodes, *Fundam. Appl. Nematol.*, 20, 323, 1997.

Hashmi, S., Ling, P., Hashmi, G., Reed, M., Gaugler, R., and Trimmer, W., Genetic transformation of nematodes using arrays of micromechanical piercing structures, *BioTechniques*, 19, 766, 1995b.

Headley, J. C. and Hoy, M. A., Benefit/cost analysis of an integrated mite management program for almonds, *J. Econ. Entomol.*, 80, 555, 1987.

Heckmann, R. A., Molecular biology of Protozoa for biological control of harmful insects, in *Molecular Biology of the Biological Control of Pests and Diseases of Plants*, Gunasekaran, M. and Weber, D. J., Eds., CRC Press, Boca Raton, FL, 1996, 137.

Heilmann, L. J., DeVault, J. D., Leopold, R. L., and Narang, S. K., Improvement of natural enemies for biological control: A genetic engineering approach, in *Applications of Genetics to Arthropods of Biological Control Significance*, Narang, S. K., Bartlett, A. C., Faust, R. M., Eds., CRC Press, Boca Raton, FL, 1994, 167.

Heinz, K. M., McCutchen, B. F., Herrman, R., Parrella, M. P., and Hammock, B. D., Direct effects of recombinant nuclear polyhedrosis viruses on selected nontarget organisms, *J. Econ. Entomol.*, 88, 259, 1995.

Herrmann, R., Moskowitz, H., Zlotkin, E., and Hammock, B. D., Positive cooperativity among insecticidal scorpion neurotoxins, *Toxicon*, 33, 1099, 1995.

Hess, R. T. and Falcon, L. A., Temporal events in the invasion of the codling moth, *Cydia pomonella*, by a granulosis virus: An electron microscopic study, *J. Invertebr. Pathol.*, 50, 85, 1987.

Höfte, H. and Whiteley, H.R., Insecticidal crystal proteins of *Bacillus thuringiensis*, *Microbiol. Rev.* 53, 242, 1989.

Honee, G., Vriezen, W., and Visser, B., Translation fusion product of two different insecticidal crystal protein genes of *Bacillus thuringiensis* exhibits an enlarged insecticidal spectrum, *Appli. Environ. Microbiol.*, 56, 823, 1990.

Hong, T., Summers, M. D., and Braunagel, S. C., N-terminal sequences from *Autographa californica* nuclear polyhedrosis virus envelope proteins ODV-E66 and ODV-E25 are sufficient to direct reporter proteins to the nuclear envelope, intranuclear microvesicles, and the envelope of occlusion derived virus, *Proc. Natl. Acad. Sci. USA*, 94, 4050, 1997.

Hoover, K., Schultz, C. M., Lane, S. S., Bonning, B. C., Duffey, S. S., and Hammock, B. D., Reduction in damage to cotton plants by a recombinant baculovirus that causes moribund larvae of *Heliothis virescens* to fall off the plant, *Biol. Control*, 5, 419, 1995.

Hoy, M. A., Impact of risk analysis on pest-management programs employing transgenic arthropods, *Parasitol. Today*, 11, 229, 1995.

Hoy, M. A., Transgenic pest and beneficial arthropods for pest management programs, in *Insect Molecular Genetics: An Introduction to Principles and Applications*, Hoy, M. A. Ed., Academic Press, San Diego, 1994, 431.

Hughes, P. R., Wood, H. A., Breen, J. P., Simpson, S. F., Duggan, A. J., and Dybas, J. A., Enhanced bioactivity of recombinant baculoviruses expressing insect-specific spider toxins in lepidopteran crop pests, *J. Invertebr. Pathol.*, 69, 112, 1997.

Hurlbert, R. E., Investigations into the pathogenic mechanisms of the bacterium-nematode complex, *ASM News*, 60, 473, 1994.

Ignoffo, C. M., Garcia, C., Zuidema, D., and Vlak, J. M., Relative *in vivo* activity and simulated sunlight-uv stability of inclusion bodies of a wild-type and an engineered polyhedral envelope-negative isolate of the nucleopolyhedrovirus of *Autographa californica, J. Invertebr. Pathol.,* 66, 212, 1995.

Jarvis, D.L., Baculovirus expression vectors, in *The Baculoviruses,* Miller, L. K., Ed., Plenum Press, New York, 1997, 389.

Jarvis, D. L., Reilly, L. M., Hoover, K., Schultz, C., Hammock, B. D., and Guarino, L. A., Construction and characterization of immediate early baculovirus pesticides, *Biol. Control,* 7, 228, 1996.

Jasinskiene, N., Coates, C. J., Benedict, M. Q., Cornel, A. J., Rafferty, C. S., James, A. A., and Collins, F. H., Stable transformation of the yellow fever mosquito, *Aedes aegypti,* with the *Hermes* element from the housefly, *Proc. Natl. Acad. Sci. USA,* 95, 3743, 1998.

Johnson, M. W., and Tabashnik, B. E., Laboratory selection for pesticide resistance in natural enemies, in *Applications of Genetics to Arthropods of Biological Control Significance,* Narang, S. K., Bartlett, A. C., Faust, R. M., Eds., CRC Press, Boca Raton, FL, 1994, 91.

Johnson, J. H., Bloomquist, J. R., Krapcho, K. J., Kral, R. M., Troato, R., Eppler, K. G., and Morgan, T. K., Novel insecticidal peptides from *Tegenaria agrestis* spider venom may have a direct effect on the insect central nervous system, *Arch. Insect Biochem. Physiol.,* 38, 19, 1998.

Jordan, T. V., Shike, H., Boulo, V., Cedeno, V., Fang, Q., Davis, B. S., Jacobs-Lorena, J., Pantropic retroviral vectors mediate somatic cell transformation and expression of foreign genes in dipteran insects, *Insect Molec. Biol.,* 7, 215, 1998.

Joshi, L., St. Leger, R. J., and Roberts, D. W., Isolation of a cDNA encoding a novel subtilisin-like protease (Pr1B) from the entomopathogenic fungus, *Metarhizium anisopliae* using differential display-RT-PCR, *Gene,* 197, 1, 1997.

Kamita, S. G. and Maeda, S., Sequencing of the putative DNA helicase-encoding gene of the *Bombyx mori* nuclear polyhedrosis virus and fine-mapping of a region involved in host range expansion, *Gene,* 190, 173, 1997.

Kaya, H. K. and Gaugler, R., Entomopathogenic nematodes, *Annu. Rev. Entomol.,* 38, 181, 1993.

Kidwell, M. G. and Ribeiro, J. M. C., Can transposable elements be used to drive disease refractoriness genes into vector populations? *Parasitol. Today,* 8, 325, 1992.

King, L. A. and Possee, R. D., *The Baculovirus Expression System,* Chapman & Hall, London, 1992.

Kirkpatrick, B. A., Washburn, J. O., and Volkman, L. E., AcMNPV pathogenesis and developmental resistance in fifth instar *Heliothis virescens, J. Invertebr. Pathol.,* 72, 63, 1998.

Kitts, P. A. and Possee, R. D., A method for producing recombinant baculovirus expression vectors at high frequency, *BioTechniques,* 14, 810, 1993.

Korth, K. L. and Levings, C. S. L., Baculovirus expression of the maize mitochondrial protein URF13 confers insecticidal activity in cell cultures and larvae, *Proc. Natl. Acad. Sci. USA,* 90, 3388, 1993.

Koziel, M.G., Carozzi, N.B., Currier, T.C., Warren, G.W., and Evola, S.V., The insecticidal crystal proteins of *Bacillus thuringiensis*: past, present and future uses, *Biotechnol. Genet. Eng. Rev.* 11, 171, 1993.

Krapcho, K. J., Kral Jr., R. M., and Morgan, T. K., Characterization and cloning of insecticidal peptides from the primitive weaving spider *Diguetia canities, Insect Biochem. Molec. Biol.,* 25, 991, 1995.

Kumar, S. and Miller, L. K., Effects of serial passage of *Autographa californica* nuclear polyhedrosis virus in cell culture, *Virus Res.,* 7, 335, 1987.

Kurtti, T. J. and Munderloh, U. G., Biotechnological application of invertebrate cell culture to the development of microsporidian insecticides, in *Biotechnology in Invertebrate Pathology and Cell Culture*, Maramorosch, K., Ed., Academic Press, San Diego, 1987, 327.

Lee, S.-Y., Qu, X., Chen, W., Poloumienko, A., MacAfee, N., Morin, B., Lucarotti, C., Insecticidal activity of a recombinant baculovirus containing an antisense *c-myc* fragment, *J. Gen. Virology*, 78, 273, 1997.

Lehane, M. J., Peritrophic matrix structure and function, *Annu. Rev. Entomol.*, 42, 525, 1997.

Lepore, L. S., Roelvink, P. R., and Granados, R. R., Enhancin, the granulosis virus protein that facilitates nucleopolyhedrovirus (NPV) infections, is a metalloprotease, *J. Invertebr. Pathol.*, 68, 131, 1996.

Lereclus, D., Agaisse, H., Gominet, M., and Chaufaux, J., Overproduction of encapsulated insecticidal crystal proteins in a *Bacillus thuringiensis spoOA* mutant, *Biotechnology*, 13, 67, 1995.

Li, J., Carroll, J., and Ellar, D., Crystal structure of insecticidal delta-endotoxin from *Bacillus thuringiensis* at 2.5 A resolution, *Nature*, 353, 815, 1991.

Li, J., and Hoy, M. A., Adaptability and efficacy of transgenic and wild-type *Metaseiulus occidentalis* (Acari: Phytoseiidae) compared as part of a risk assessment, *Exper. Appl. Acarol.*, 20, 563, 1996.

Loukeris, T. G., Livadaras, I., Arca, B., Zabalou, S., and Savakis, C., Gene transfer into the Medfly, *Ceratitis capitata*, using a *Drosophila hydei* transposable element, *Science*, 270, 2002, 1995.

Lu, A. and Miller, L. K., Generation of recombinant baculoviruses by direct cloning, *Bio-Techniques*, 21, 63, 1996a.

Lu, A. and Miller, L. K., Species-specific effects of the *hcf-1* gene on baculovirus virulence, *J. Virol.*, 70, 5123, 1996b.

Lu, A., Seshagiri, S., and Miller, L. K., Signal sequence and promoter effects on the efficacy of toxin-expressing baculoviruses as biopesticides, *Biol. Control*, 7, 320, 1996.

Luckow, V. A., Lee, S.C., Barry, G.F., and Olins, P.O., Efficient generation of infectious recombinant baculoviruses by site-specific transposon-mediated insertion of foreign genes into a baculovirus genome propagated in *Escherichia coli*, *J. Virol.* 67, 4566, 1993.

Ma, P. W. K., Davis, T. R., Wood, H. A., Knipple, D. C., and Roelofs, W. L., Baculovirus expression of an insect gene that encodes multiple neuropeptides, *Insect Biochem. Molec. Biol.*, 28, 239, 1998.

Maeda, S., Increased insecticidal effect by a recombinant baculovirus carrying a synthetic diuretic hormone gene, *Biochem. Biophys. Res. Commun.*, 165, 1177, 1989.

Maeda, S., Volrath, S. L., Hanzlik, T. N., Harper, S. A., Maddox, D. W., Hammock, B. D., and Fowler, E., Insecticidal effects of an insect-specific neurotoxin expressed by a recombinant baculovirus, *Virology*, 184, 777, 1991.

Martens, J. W. M., Honee, G., Zuidema, D., van Lent, J. W. M., Visser, B., and Vlak, J. M., Insecticidal activity of a bacterial crystal protein expressed by a recombinant baculovirus in insect cells, *Appl. Environ. Microbiol.*, 56, 2764, 1990.

Martens, J. W. M., Knoester, M., and Vlak, J. M., Characterization of baculovirus insecticides expressing tailored *Bacillus thuringiensis* CryIA(b) crystal proteins, *J. Invertebr. Pathol.*, 66, 249, 1995.

Matsubara, T., Beeman, R. W., Shike, H., Besansky, N. J., Mukabayire, O., Higgs, S., and James, A. A., Pantropic retroviral vectors integrate and express in cells of the malaria mosquito, *Anopheles gambiae*, *Proc. Natl. Acad. Sci USA*, 93, 6181, 1996.

McCutchen, B. F., Choudary, P. V., Crenshaw, R., Maddox, D., Kamita, S. G., Palekar, N., and Volrath, S., Development of a recombinant baculovirus expressing an insect-specific neurotoxin: Potential for pest control, *Bio/Technol.*, 9, 848, 1991.

McCutchen, B. F., Herrmann, R., and Hammock, B. D., Effects of recombinant baculoviruses on a nontarget endoparasitoid of *Heliothis virescens*, *Biol. Control.*, 6, 45, 1996.

McCutchen, B. F., Hoover, K., Preisler, H. K., Betana, M. D., Herrmann, R., Robertson, J. L., and Hammock, B. D., Interactions of recombinant and wild-type baculoviruses with classical insecticides and pyrethroid-resistant tobacco budworm (Lepidoptera: Noctuidae), *J. Econ. Entomol.*, 90, 1170, 1997.

McNitt, L., Espielie, K. E., and Miller, L. K., Assessing the safety of toxin-producing baculovirus biopesticides to a nontarget predator, the social wasp *Polistes metricus* Say, *Biol. Control*, 5, 267, 1995.

Merryweather, A. T., Weyer, U., Harris, M. P. G., Hirst, M., Booth, T., and Possee, R. D., Construction of genetically engineered baculovirus insecticides containing the *Bacillus thuringiensis* subsp. kurstaki HD-73 delta endotoxin, *J. Gen. Virology*, 71, 1535, 1990.

Miller, L. K., Genetically engineered insect virus pesticides: Present and future, *J. Invertebr. Pathol.*, 65, 211, 1995.

Miller, L. K. and Lu, A., The molecular basis of baculovirus host range, in *The Baculoviruses*, Miller, L. K., Ed., Plenum Press, New York, 1997, 217.

Nakagawa, Y., Lee, Y. M., Lehmberg, E., Herrmann, R., Herrmann, R., Moskowitz, H., and Jones, A. D., Anti-insect toxin 5 (AaIT5) from *Androctonus australis*, *Eur. J. Biochem.*, 246, 496, 1997.

Nakagawa, Y., Sadilek, M., Lehmberg, E., Herrmann, R., Herrmann, R., Moskowitz, H., and Lee, T. M., Rapid purification and molecular modeling of AaIT peptides from venom of *Androctonus australis*, *Arch. Insect Biochem. Physiol.*, 38, 53, 1998.

National Academy of Sciences USA, *Report of the Research Briefing Panel on Biological Control in Managed Ecosystems*. National Academy Press, Washington D.C., 1987.

O'Brochta, D. A., and Atkinson, P. W., Recent developments in transgenic insect technology, *Parasitol. Today*, 13, 99, 1997.

O'Brochta, D. A. and Atkinson, P. W., Transposable elements and gene transformation in non-drosophilid insects, *Insect Biochem. Molec. Biol.*, 26, 739, 1996.

Obukowicz, M. G., Perlak, F. J., Kusano-Kretzmer, K., Mayer, E. J., and Watrud, L. S., Integration of the delta-endotoxin gene of *Bacillus thuringiensis* into the chromosome of root-colonizing strains of pseudomonads using Tn5, *Gene*, 45, 327, 1986.

O'Reilly, D. R., Auxiliary genes of baculoviruses, in *The Baculoviruses*, Miller, L. K., Ed., Plenum Press, New York, 1997, 267.

O'Reilly, D. R., Baculovirus-encoded ecdysteroid UDP-glucosyltransferases, *Insect Biochem. Molec. Biol.*, 25, 541, 1995.

O'Reilly, D. R. and Miller, L. K., Improvement of a baculovirus pesticide by deletion of the EGT gene, *Bio/Technol.*, 9, 1086, 1991.

O'Reilly, D. R., Miller, L. K., and Luckow, V. A. *Baculovirus Expression Vectors: A Laboratory Manual*, Freeman, New York, 1992.

Pang, Y., Frutos, R., and Federici, B. A., Synthesis and toxicity of full-length and truncated bacterial CryIVD mosquitocidal proteins expressed in lepidopteran cells using a baculovirus vector, *J. Gen. Virology*, 73, 89, 1992.

Pettigrew, M. M. and O'Neill, S. L., Control of vector-borne disease by genetic manipulation of insect populations: Technological requirements and research priorities, *Aust. J. Entomol.*, 36, 309, 1997.

Pietrantonio, P. V. and Gill, S. S., *Bacillus thuringiensis* endotoxins: action on the insect midgut, in *Biology of the Insect Midgut,* Lehane, M. J. and Billingsley, P. F., Eds., Chapman & Hall, London, 1996, 345.

Poinar, G. O., Genetic engineering of nematodes for pest control, in *Biotechnology for Biological Control of Pests and Vectors,* Maramorosch, K., Ed., CRC Press, Boca Raton, FL, 1991, 77.

Popham, H. J. R., Li, Y., and Miller, L. K., Genetic improvement of *Helicoverpa zea* nuclear polyhedrosis virus as a biopesticide, *Biol. Control,* 10, 83, 1997.

Popham, H. J. R., Prikhod'ko, G. G., and Miller, L. K., Effect of deltamethrin treatment on lepidopteran larvae infected with baculoviruses expressing insect-specific toxins mu-Aga-IV, As II, or Sh 1, *Biol. Control,* 12, 79, 1998.

Possee, R. D., Barnett, A. L., Hawtin, R. E., and King, L. A., Engineered baculoviruses for pest control, *Pestic. Sci.,* 51, 462, 1997.

Presnail, J. K. and Hoy, M. A., Maternal microinjection of the endoparasitoid *Cardiochiles diaphaniae* (Hymenoptera: Braconidae), *Ann. Entomol. Soc. Am.,* 89, 576, 1996.

Presnail, J. K. and Hoy, M. A., Stable genetic transformation of a beneficial arthropod, *Metaseiulus occidentalis* (Acari: Phytoseiidae), by a microinjection technique, *Proc. Natl. Acad. Sci. USA,* 89, 7732, 1992.

Presnail, J. K. and Hoy, M. A., Transmission of injected DNA sequences to multiple eggs of *Metaseiulus occidentalis* and *Amblyseius finlandicus* (Acari: Phytoseiidae) following maternal microinjection, *Exp. Appl. Acarol.,* 18, 319, 1994.

Presnail, J. K., Jeyaprakash, A., Li, J., and Hoy, M. A., Genetic analysis of four lines of *Metaseiulus occidentalis* (Acari: Phytoseiidae) transformed by maternal microinjection, *Ann. Entomol. Soc. Am.,* 90, 237, 1997.

Prikhod'ko, G. G., Popham, H. J. R., Felcetto, T. J., Ostlind, D. A., Warren, V. A., Smith, M. M., and Garsky, V. M., Effects of simultaneous expression of two sodium channel toxin genes on the properties of baculoviruses as biopesticides, *Biol. Control,* 12, 66, 1998.

Prikhod'ko, G. G., Robson, M., and Miller, L. K., Properties of three baculovirus-expressing genes that encode insect-selective toxins: mu-Aga-IV, As II, and Sh I, *Biol. Control,* 7, 236, 1996.

Rahimi, F. R., McGuire, T. R., and Gaugler, R., Morphological mutant in the entomopathogenic nematode, *Heterorhabditis bacteriophora.*, *J. Hered.,* 84, 475, 1993.

Rajamohan, F., and Dean, D. H., Molecular biology of bacteria for the biological control of insects., in *Molecular Biology of the Biological Control of Pests and Diseases of Plants,* Gunasekaran, M. and Weber, D. J., Eds., CRC Press, Boca Raton, FL, 1996, 105.

Reddy, J. T. and Locke, M., The size limited penetration of gold particles through insect basal laminae, *J. Insect Physiol.,* 36, 397, 1990.

Richardson, C. D. *Baculovirus Expression Protocols,* Vol. 39, Humana Press, Totowa, NJ, 1995.

Robertson, H. M., The *Tc1-mariner* superfamily of transposons in animals, *J. Insect Physiol.,* 41, 99, 1995.

Robertson, H. M., Lampe, D. J., and MacLeod, E. G., A *mariner* transposable element from a lacewing, *Nucleic Acids Res.,* 20, 6409, 1992.

Rohrbach, D. H. and Timpl, R., Eds. *Molecular and Cellular Aspects of Basement Membranes,* Academic Press, New York, 1993.

St. Leger, R., Biology and mechanisms of insect-cuticle invasion by deuteromycete fungal pathogens, in *Parasites and Pathogens of Insects,* Beckage, N.E., Thompson, S.N., Federici, B.A. Eds., Academic Press, New York, 1993, 211.

St. Leger, R. J., The role of cuticle-degrading proteases in fungal pathogenesis of insects, *Can. J. Bot. (Suppl. 1)*, 73, S1119, 1995.

St. Leger, R. J., Joshi, L., and Roberts, D., Ambient pH is a major determinant in the expression of cuticle-degrading enzymes and hydrophobin by *Metarhizium anisopliae*, *Appl. Environ. Microbiol.*, 64, 709, 1998.

St. Leger, R. J., Joshi, L., Bidochka, M. J., and Roberts, D. W., Construction of an improved mycoinsecticide overexpressing a toxic protease, *Proc. Natl. Acad. Sci. USA*, 93, 6349, 1996.

Sato, K., Komoto, M., Sato, T., Enei, H., Kobayashi, M., and Yaginuma, T., Baculovirus-mediated expression of a gene for trehalase of the mealworm beetle, *Tenebrio molitor*, in insect cells, Sf-9, and larvae of the cabbage armyworm, *Mamestra brassicae*, *Insect Biochem. Molec. Biol.*, 27, 1007, 1997.

Shapiro, M., Preisler, H. K., and Robertson, J. L., Enhancement of baculovirus activity on gypsy moth (Lepidoptera: Lymantriidae) by chitinase, *J. Econ. Entomol.*, 80, 1113, 1987.

Shapiro, M., and Robertson, J. L., Enhancement of gypsy moth (Lepidoptera: Lymantriidae) baculovirus activity by optical brighteners, *J. Econ. Entomol.*, 85, 1120, 1992.

Sinkins, S. P., Curtis, C. F., and O'Neill, S. L., The potential application of inherited symbiont systems to pest control, in *Influential Passengers: Inherited Microorganisms and Arthropod Reproduction*, O'Neill, S. L., Hoffmann, A. A., Werren, J. H., Eds., Oxford University Press, New York, 1997, 155.

Slavicek, J. M., Hayes-Plazolles, N., and Kelly, M. E., Rapid formation of few polyhedra mutants of the *Lymantria dispar* multinucleocapsid nuclear polyhedrosis virus during serial passage in cell culture, *Biol. Control*, 5, 251, 1995.

Stewart, L. M. D., Hirst, M., Lopez-Ferber, M., Merryweather, A. T., Cayley, P. J., and Possee, R. D., Construction of an improved baculovirus insecticide containing an insect-specific toxin gene, *Nature*, 352, 85, 1991.

Stouthamer, R., The use of sexual versus asexual wasps in biological control, *Entomophaga*, 38, 3, 1993.

Summers, M. D. and Smith, G. E. *A Manual of Methods for Baculovirus Vectors and Insect Cell Culture Procedures*, Texas Agr. Exp. Stn. Bull. Vol. 1555, 1987.

Thanabalu, T. and Porter, A. G., Efficient expression of a 100-kilodalton mosquitocidal toxin in protease-deficient recombinant *Bacillus sphaericus*, *Appl. Environ. Microbiol.*, 61, 4031, 1995.

Thompson, M. A., Schnepf, H. E., and Feitelson, J. S., Structure, function and engineering of *Bacillus thuringiensis* toxins, *Genetic Engineering, Principles and Methods*, 17, 99, 1995.

Tiedje, J. M., Colwell, R. K., Grossman, Y. L., Hodson, R. E., Lenski, R. E., Mack, R. N., and Regal, P. J., The planned introduction of genetically engineered organisms: Ecological considerations and recommendations, *Ecology*, 70, 298, 1989.

Tomalski, M. D. and Miller, L. K., Expression of a paralytic neurotoxin gene to improve insect baculoviruses as biopesticides, *Bio/Technol.*, 10, 545, 1992.

Tomalski, M. D. and Miller, L. K., Insect paralysis by baculovirus-mediated expression of a mite neurotoxin gene, *Nature*, 352, 82, 1991.

Tomalski, M. D., King, T. P., and Miller, L. K., Expression of hornet genes encoding venom allergen antigen 5 in insects, *Arch. Insect Biochem. Physiol.*, 22, 303, 1993.

Tomasino, S. F., Leister, R. T., Dimock, M. B., Beach, R. M., and Kelly, J. L., Field performance of *Clavibacter xyli* subsp. *cynodontis* expressing the insecticidal protein gene *cryIA(c)* of *Bacillus thuringiensis* against European corn borer in field corn, *Biol. Control*, 5, 442, 1995.

Treacy, M., Efficacy and nontarget arthropod safety of an AaIT gene-inserted baculovirus: results from field and laboratory studies conducted 1995-1996, in *Biopesticides and Transgenic Plants. New Technologies to Improve Efficacy, Safety and Profitability*, IBC Library Series, 1997.

Treacy, M. F., All, J. N., and Kukel, C. F., Invertebrate selectivity of a recombinant baculovirus: Case study on AaHIT gene-inserted *Autographa californica* nuclear polyhedrosis virus, in *New Developments in Entomology*, Bondari, K., Ed., Research Signpost, London, 1997, 57.

Turelli, M, and Hoffmann, A. A., Rapid spread of an inherited incompatibility factor in California *Drosophila*, *Nature*, 353, 440, 1991.

USDA-APHIS Regulation of Transgenic Arthropods web site: http://www.aphis.usda.gov/bbep/bp/arthropod/permits/9532602r/32602rrp.html

Vail, P. V. and Vail, S. S., Comparative replication of *Autographa californica* nuclear polyhedrosis virus in tissues of *Heliothis* spp. (Lepidoptera: Noctuidae), *Ann. Entomol. Soc. Am.*, 80, 734, 1987.

Van Driesche, R. G. and Bellows, T. S. *Biological Control*, Chapman & Hall, New York, 1996.

Vlak, J. M., Schouten, A., Usmany, M., Belsham, G. J., Klinge-Roode, E. C., Maule, A. J., and van Lent, J. W. M., Expression of cauliflower mosaic virus gene I using a baculovirus vector based on the p10 gene and a novel selection method, *Virology*, 179, 312, 1990.

Vossbrinck, C. R., Maddox, J. V., Friedman, S., Debrunner-Vossbrinck, B. A., and Woese, C. R., Ribosomal RNA sequence suggests microsporidia are extremely ancient eukaryotes, *Nature*, 326, 411, 1987.

Wang, P. and Granados, R. R., An intestinal mucin is the target substrate for a baculovirus enhancin, *Proc. Natl. Acad. Sci. USA*, 94, 6977, 1997.

Wang, P. and Granados, R. R., Observations on the presence of the peritrophic membrane in larval *Trichoplusia ni* and its role in limiting baculovirus infection, *J. Invertebr. Pathol.*, 72, 57, 1998.

Wang, X., Ooi, B. G., and Miller, L. K., Baculovirus vectors for multiple gene expression and for occluded virus production, *Gene*, 100, 131, 1991.

Washburn, J. O., Kirkpatrick, B. A., and Volkman, L. E., Insect protection against viruses, *Nature*, 383, 767, 1996.

Watrud, L.S., Peralk, F.J., Tran, M., Kusano, K., Mayer, E.J., Miller-Wideman, M.A., Obukowicz, M.G., Nelson, D.R., Kreitinger, J.P., and Kaufman, R.J. Cloning of the *Bacillus thuringiensis* subsp. *kurstaki* delta-endotoxin gene into *Pseudomonas fluorescens*: Molecular biology and ecology of engineered microbial pesticide, in Engineered Organisms in the Environment: Scientific Issues, Halvorson, H.O., Pramer, D., Rogul, M., Eds., Amer. Soc. Microbiol., Washington, D.C. 1985, 40.

Werren, J. H., Biology of *Wolbachia*, *Annu. Rev. Entomol.*, 42, 587, 1997.

Werren, J. H. and O'Neill, S. L., The evolution of heritable symbionts, in *Influential Passengers: Inherited Microorganisms and Arthropod Reproduction*, O'Neill, S. L., Hoffmann, A. A., Werren, J. H., Eds., Oxford University Press, New York, 1997, 1.

Werren, J. H., Zhang, W., and Guo, L. R., Evolution and phylogeny of *Wolbachia*: reproductive parasites of arthropods, *Proc. Royal Soc. London Ser. B*, 251, 55, 1995.

Wood, H. A., Genetically enhanced baculovirus insecticides, in *Molecular Biology of the Biological Control of Pests and Plant Diseases*, Gunasekaran, M. and Weber, D. J., Eds., CRC Press, Boca Raton, FL, 1996, 91.

Xu, J., Lohrke, S., Hurlbert, I. M., and Hurlbert, R. E., Transformation of *Xenorhabdus nematophilus*, *Appl. Environ. Microbiol.*, 55, 806, 1989.

Xu, J., Olson, M. E., Kahn, M. L., and Hurlbert, R. E., Characterization of Tn5-induced mutants of *Xenorhabdus nematophilus* ATCC 19061, *Appl. Enrivon. Microbiol.,* 57, 1173, 1991.

Yu, C. G., Mullins, M. A., Warren, G. W., Koziel, M. G., and Estruch, J. J., The *Bacillus thuringiensis* vegetative insecticidal protein Vip3A lyses midgut epithelium cells of susceptible insects, *Appl. Environ. Microbiol.,* 63, 532, 1997.

Zuidema, D., Klinge-Roode, E. C., van Lent, T. W. M., and Vlak, J. M., Construction and analysis of an *Autographa californica* nuclear polyhedrosis mutant lacking the polyhedral envelope, *Virology,* 173, 98, 1989.

Zwaal, R. R., Broeks, A., Van-Meurs, J., Groenen, J. T., and Plasterk, R. H., Target-selected gene inactivation in *Caenorhabditis elegans* by using a frozen transposon insertion mutant bank, *Proc. Natl. Acad. Sci. USA,* 90, 7431, 1993.

CHAPTER 10

Environmental Impact
of Biotechnology

Robert G. Shatters, Jr.

CONTENTS

1-56670-479-0/00/$0.00+$.50
© 2000 by CRC Press LLC

10.1   INTRODUCTION

Before discussing the effect of biotechnology on the environment it is important to set the boundaries of the discussion, that is, to define biotechnology. In its broadest sense and as defined by the U.S. Congress, biotechnology includes any technique that uses living organisms (or parts of organisms) to make or modify products, to improve plants or animals, or to develop microorganisms for specific uses (Office of Technology Assessment, 1993). However, for the purposes of this chapter, biotechnology will be limited to using recombinant DNA techniques to develop genetically engineered plants (GEPs) for the purpose of pest insect control. A genetically engineered, or transgenic, plant is defined as one that has had foreign genetic material purposefully introduced and stably incorporated into the plant genome through means other than those that naturally occur in the environment. This new genetic material becomes an integral part of the plant genetic material and is therefore inherited in subsequent generations in a fashion consistent with the rest of the genome complement within which the new DNA is inserted. Therefore, the new genetic material can be transferred through pollen (assuming that the DNA was integrated within the nuclear genome and not plastids) and ovules. The power of this technique is that virtually any genetic material, whether it comes from other plants, animals, bacteria, or viruses, or even completely synthetic genetic material, can be added to an organism's genome. The environmental concerns that arise from this are based on the inability to precisely predict what effect this greatly increased fluidity of genetic material among living organisms will have.

Although transgenic plants are unique, since combinations of genetic material within a plant can be generated that presumably would never have occurred before in nature, it is important to note that nature over the course of evolution, and standard breeding practices being used for hundreds of years have also created unique genetic combinations. This occurs in nature when natural mutations create novel sequences that produce altered gene products with unique capabilities. Using standard breeding techniques, humans have taken advantage of genetic diversity by selectively crossing related plants each with desirable characteristics, and then carried the progeny of these crosses all over the earth to grow in close relationship with plants native to the new areas. In some cases the introduced crop plants can exchange genetic material with native plants in the new areas if the two species are related. Therefore, even though the individual plants have evolved for many thousands of years in isolation from each other, humans bring the new genetic material back together, creating new combinations. This has resulted in a successful agricultural industry that is providing for the food needs of the world. The novelty of genetic engineering is not about the general ability to recombine genetic material in producing new crop plants. It is the scope of this combinatorial ability that is greatly increased. This is the single point that makes genetic engineering an extremely powerful tool that could aid in greatly improving agricultural productivity, but it is this single point that also is at the center of the controversy over the safety of this new biological tool.

Perhaps the most controversial issue with the use of biotechnology for crop improvement is the potential for disruption of, or damage to, the environment. It is important to understand that this controversy is based on theoretical risk, since there have been no instances of GEPs causing environmental damage. Despite the lack of examples of how genetic engineering of plants could cause environmental problems, it is pertinent to discuss this issue in a theoretical sense, since once genetically engineered plants (GEPs) are released it is difficult, or impossible, to reverse the effects of interactions between these plants and the environment.

What is the source of the potential risk to the environment? As previously stated the risk arises from the greatly expanded ability to create new combinations of genetic information, i.e., the ability to freely introduce limited amounts of genetic material into a specific plant species, and the inability to precisely predict how the newly developed GEP will perform in the environment or how the introduced genetic material will behave in the new genetic background. Areas of concern include the direct impact of GEPs in cultivated fields and natural ecosystems, the transfer (escape) of the introduced genetic material to related plants through sexual reproduction, the transfer of the genetic material to nonrelated organisms (horizontal gene transfer), and finally any changes in agricultural management practices to support the growth of GEPs that have a negative impact on the environment.

Specific questions with respect to environmental impact of genetically engineering plants designed to be resistant to insects include: (1) Could the elimination of natural pests' ability to control the proliferation of genetically engineered crop plant create a weed problem? (2) If the introduced genetic material for insect resistance is transferred through standard sexual transmission to weedy plants closely related to the genetically engineered crop plant, could the weed become more noxious? (3) If the introduced genetic material encodes a protein toxic to insects, could it have adverse effects on nontarget beneficial insects? (4) Could there be a detrimental effect of products of introduced genes on a broad range of fauna — noninsect wildlife that ingests the genetically altered plants? (5) Will overuse of a specific biological control strategy through the development of transgenic plants stimulate the rate of insect tolerance to these biological control mechanisms, rendering the mechanism ineffective in alternative nontransgenic plant strategies utilizing the same biocontrol strategy?

The impact of specific biotechnology approaches using GEPs to reduce insect pest problems will be discussed with respect to each of these concerns.

To date, only a single genetic engineering approach for insect control, development of transgenic plants expressing the *Bacillus thuringiensis* δ-endotoxins (bt-toxins), has been released commercially. However, it would be a disservice to limit the discussion to the use of bt-toxins, as much as discussions in the early part of this century about the future impact of automobile transportation on our society and environment would have been ineffective if it had been limited to the development and use of the Model A Ford. Instead, this chapter presents a review of the biotechnological approaches being developed for insect control in agriculture (both short-term and long-term projects), and a discussion of the potential impact of these methods on the environment. This chapter is written with the view that the question should not be: Should biotechnology be used to improve agricultural crops? Instead, the question should be: What is the appropriate use of biotechnology to support environmentally friendly agricultural practices?

10.2 REVIEW OF BIOTECHNOLOGICAL APPROACHES TO PEST INSECT CONTROL

10.2.1 Separating the Method from the Concept

Biotechnology is a method to produce a plant with altered characteristics. Environmental impact is not relatable to the techniques being used to insert foreign DNA, but is relatable to the type of foreign DNA being inserted and the species that it is being inserted into. There is a great diversity of crop plants and of the types of genetic material that could be inserted into a plant, and as a result, environmental impact of each individual genetic engineering strategy will have to be assessed independently. For example, inserting a gene encoding resistance to only a specific insect pest in a plant that has no native, weedy, or potentially weedy relative to which the insect resistance gene could be transferred would have much less potential for creating an environmental problem than inserting genetic material encoding a product that is toxic to a broad range of insect and other animals, including mammals, into a plant that readily exchanges genetic material with closely related native and weedy species. However, the range of problems that could arise as a result of the release of GEPs can be categorized, and the impact of each strategy can be assessed by relating it to each of the potential problems. To address the concerns related to plants genetically engineered for pest insect resistance we must first understand what strategies show promise in insect control.

10.2.2 Current GEP Strategies for Insect Control

10.2.2.1 Bacillus thuringiensis δ-Endotoxins

Although biotechnological control strategies are covered elsewhere in this book, a brief review of the technologies being addressed in this chapter is in order. Current technology limits the types of novel compounds that can be produced in plants as a result of genetic engineering. Although it is theoretically possible to introduce many genes, which encode different proteins with different functions, technology only allows one to several individual genes to be inserted, and there are only a small number of genes that are currently well characterized that produce compounds that reduce insect feeding damage.

Unquestionably, the most well known and most successful biotechnological approach toward improving plant insect resistance has been the use of a gene encoding an insect toxin protein isolated from the bacterium, *Bacillus thuringiensis*. This microbe has been used as a biopesticide for more than 30 years (Feitelson et al., 1992) due to the insecticidal activity of a class of proteins termed δ-endotoxins that the bacterium produces during sporulation. Numerous strains of *B. thuringiensis* have been isolated that produce related toxin proteins with different insect specificities. Toxins are known that control Lepidopteran, Dipteran, and Coleopteran insects (Höfte and Whiteley, 1989; Lereclus et al., 1992). These toxins have a very limited range of insects upon which they act, and are harmless to mammals, proving therefore to be an environmentally sound method for insect control. Numerous field trials have been performed with genetically engineered plants expressing the bt-toxins since 1986, and commercial bt-toxin expressing cotton has been available since 1996. Continued analysis of toxins produced by Bacillus bacteria resulted in a recent finding of a new class of insect toxins called vegetative insecticidal proteins (VIPs) (Warren et al., 1994). These proteins have activity against insects with tolerance to the δ-endotoxins, thereby increasing the possible uses of *B. thuringiensis* produced insect toxins as a biotechnological tool for developing insect resistant crops.

10.2.2.2 Lectins

Simple gene products produced in a diverse array of organisms have also been shown to function in controlling insect damage to plants (Hilder et al., 1990). One group, lectin and lectin-like proteins, are carbohydrate binding molecules that are produced by many organisms and are especially abundant in seeds and storage tissues of plants (Etzler, 1986). It has been suggested that a major role for these molecules is in plant defense against insects (Chrispeels and Raikhel, 1991). The toxicity of these molecules to susceptible insects is thought to occur as a result of binding to receptors on the surface of the midgut epithelial cells. This apparently inhibits nutrient uptake and facilitates the absorption of potentially harmful substances (Gatehouse et al., 1984, 1989, and 1992). Insects that are harmful to crop plants include those that feed directly on the plant structures (i.e., leafs, stems, roots, etc.) as well as the sap-sucking insect. Since the sap-sucking insects only feed on the phloem exudates, biotechnological approaches aimed at controlling these insects require that the insect deterrent compound is present in the phloem translocation stream. A lectin from the snowdrop plant (*Galanthus nivalis*) was shown to be the first protein to have a toxicity effect on a sap-sucking insects when expressed in transgenic plants (Hilder et al., 1995). The protein was introduced in the phloem exudate by placing the gene encoding this lectin under the control of a promoter (a switch that activates the transcription of the gene, resulting in the production of the corresponding protein) that functioned specifically in the phloem cells. A lectin from pea (*Pisum sativum*) seeds was also shown to cause increased mortality of tobacco budworm larvea (*Heliothis virescens*) when the gene encoding this protein was expressed in transgenic tobacco (Boulter et al., 1990). One concern with the use of lectins is that they have relatively high mammalian toxicities and therefore are not suitable if expressed in edible parts of food crops. These proteins are also strong allergens in humans, which further complicates the ability to use them in transgenic plant approaches.

10.2.2.3 Protease and Amylase Inhibitors

Protease inhibitors represent another group of single gene products that have insecticidal/antimetabolic activity in insects and have been proven to reduce insect damage to transgenic plants expressing these proteins (Hilder, 1987; Johnson et al., 1989). Although the mechanism of action is not completely understood, the anti-insect activity appears to be the result of more complicated interactions than just inhibition of digestive enzymes (for review: Gatehouse et al., 1992). These molecules display a wide range of activity, being effective against Lepidopteran, Orthopteran, and Coleopteran insects (Höfte and Whiteley, 1989). Alpha-amylase inhibitors are another class of enzyme inhibitor isolated from plants and shown to have insecticidal/antimetabolic activities. Transgenic pea expressing an alpha-amylase inhibitor at 1.2% of total protein displayed increased resistance to both cowpea weevil and Azuki bean weevils (Shade et al., 1994).

The single gene enzyme inhibitors have to be expressed at high levels, typically with greater than 0.1% (w/w) and often around 1.0% of total protein to be effective, with the exception of the bt-toxins, which are active at 10^{-7}%. Another single gene product with insecticidal activity and greater specific activity than the enzyme inhibitors or lectins is cholesterol oxidase. The mode of action of cholesterol oxidase also involves the perturbation of midgut cells, thus inhibiting nutrient uptake (Purcell et al., 1993). This enzyme has strong insecticidal activity against the boll weevil larvae (*Anthonomus grandis grandis* Boheman) at concentrations of 2×10^{-3}% (w/w). Therefore there is precedence for simple gene products other than the bt-toxins to have strong insecticidal activity at relatively low concentrations.

Major active components in certain arachnid and scorpion venoms are known to be proteins with potential use in the biotechnology arena. Genes encoding toxin proteins from a scorpion (*Androctonus australis*) have been cloned and shown to produce toxins active against insects when expressed in baculovirus insecticide systems (baculovirus is a virus that specifically infects insect cells) (Hoover et al., 1995). However, the utility of these proteins in genetically engineered crops is still in question since they have mammalian toxicities and they are often broken down rapidly when taken up through the digestive system. Perhaps future engineering of this class of proteins can be used to develop new insect toxins with greater activity and less mammalian toxicity.

10.2.3 Future Strategies

The previously described approaches to increasing insect resistance in crop plants are ones that have already been shown to function in either field trials or laboratory tests. These represent the first generation of plant biotechnology. As technology advances, it can be assumed that future protocols will involve the use of even more single gene products as they become available and the genetic engineering of more complex metabolic processes that will require the insertion of multiple genes in a metabolic pathway. Current limitations to this approach include (1) the lack of knowledge about the enzymes in these metabolic pathways; (2) the high number of genes required to introduce a novel metabolic pathway; and (3) the lack of understanding of the pleotrophic nature of perturbations of existing metabolic pathways. An example of complex metabolic pathways involved in pest insect control is the production of insect hormone analogs in plants. It has been known for quite some time that plants can produce organic compounds that affect insect growth and development (Whittaker and Feeny, 1971; Beck and Reese, 1976), and insect hormone analogs have been found in numerous plant species (Bergamasco and Horn, 1983). It is speculated that these function in protecting the plant from insect damage. Synthesis of these complex molecules requires numerous enzymatic reactions, and each enzyme is synthesized by one or several genes. Therefore, engineering a plant to synthesize a single insect hormone analog may require the introduction of at least several genes. However, plants typically produce numerous secondary metabolites that are the precursors to the active hormone structures, so depending on the plant and the hormone structure of interest, many of the synthetic steps may already be present. Future research is needed to understand the precursors in the insect hormone biosynthetic pathway that are already present in plants and characterization and isolation of the genes that encode the enzymes necessary for the desired metabolic pathway.

Other complex organic molecules that may provide insect resistance include a number of host defense response chemicals and antifeedant molecules. Plants are known to have inducible defense systems where antimicrobial or anti-insecticidal compounds are synthesized in response to infection or feeding damage. It may therefore be theoretically possible to move genes encoding an effective insect control mechanism from one plant species to another that does not have this capability. However, to date there are no reports of genetic engineering being used to success-fully modify the synthesis of these types of molecules resulting in greater protection from insect damage. Antifeedant molecules that prevent insects from feeding on specific tissues have been identified from some plants. Future characterization of the genes involved in the synthesis of these compounds may also allow genetic engineering strategies to be employed to develop desirable crop plants that produce these molecules.

Finally, as plant development becomes better understood, opportunities may arise to use genetic engineering to alter plant morphology or structure to limit insect feeding on desirable crops. Compatibility between insect feeding structures/behavior and plant design plays a role in host–pest recognition and could be exploited as a way of preventing feeding on the plants. Examples include increased lignification of epidermis, or changes in epidermal hairs or trichome structures that increase insect resistance. The ability of plant sap-sucking insects to extract nutrients from a crop plant may be inhibited by changing aspects of the plant's vascular structure. Increased lignification of the cell walls of this specific tissue could make them less penetrable by the insect's piercing mouth parts, or plants with vascular bundles deeper within the stem tissue could carry on nutrient transport in cells that cannot be accessed by the insects. Also, it is known that when plants are damaged by insect feeding, certain plants can release volatile molecules that function as attractants to insects that feed on or are parasitic on the plant pest insect (Dicke et al., 1990; Turlings et al., 1990; Takabayashi and Dicke, 1996; McCall et al., 1993, 1994; Loughrin et al., 1995). Engineering this ability into desirable crop plants that may not be able to attract the desired protective insects may also improve crop perfor-mance. The ultimate goal is to expand our ability to control pest damage that is limiting crop productivity, while at the same time reducing our need for environ-mentally damaging chemicals and agricultural practices. The purpose of evaluating the environmental impact of these biotechnological approaches is to assure that we do not trade the use of some environmentally damaging practices (the use of dan-gerous pesticides) for another equally or more damaging practice.

Table 10.1 Categorized Environmental Concerns Associated with Crops Genetically Engineered for Insect Resistance

- Direct impact of genetically engineered crop on the environment
  - The GEP becomes a weed.
  - Environmental contamination with genetically engineered product produced in GEP.
  - Toxicity to wildlife (including beneficial insects).
- Increased fluidity of genetic material
  - Transfer of genetic material to nonweedy relatives of the GEP (creating new weeds).
  - Transfer of genetic material to weedy relatives of the GEP (creating worse weeds).
  - Transfer of genetic material to unrelated microorganisms.
- Changes in management practices as a result of the use of genetically engineered crops
  - Reduced reliance on sustainable agricultural practices.

10.3 EVALUATION OF THEORETICAL NEGATIVE ENVIRONMENTAL IMPACT FROM RELEASE OF GEPS

As our knowledge of the interaction of plants and insects increases and the capabilities of biotechnology are expanded, it is clear that a great number of approaches utilizing biotechnology will offer improvements in our need to control crop pest insects. The great diversity of potential approaches is a signal that some will be great ideas and some will not, and appropriately some will help develop more environmentally friendly agricultural practices while others will not. A priori evaluation of the proposed approach is therefore crucial to offer insight about what the potential impact on the environment could be. As stated in the introduction, potential environmental problems related to the release of GEPs are related to the increased combinatorial ability of genetic information. These concerns can be divided into three main categories, and these categories can again be divided into specific concerns (Table 10.1). Each will be discussed with respect to the creation of GEP as an insect control strategy.

10.3.1 The Direct Impact of Genetically Modified Plants on the Environment

10.3.1.1 Creating a Weed

The question here is how well can we expect to predict the behavior of the GEP? For example, one of the most common arguments is that improving the fitness of a crop plant could create a significant weed problem in agricultural fields or an invasive plant in natural ecosystems. In the context of this chapter, the argument would be that increasing resistance to a group of insect pests could cause the plant to become more aggressive as a weed. Rapeseed genetically engineered for insect resistance was shown to have a better reproductive chance than nontransgenic rapeseed in experiments imposing strong herbivorous insect selective pressure (Stewart et al., 1997). However, it was not shown that this resulted in greater weediness of the plant in native conditions. To understand weediness, a number of characteristics have been identified that make a plant a weed (Table 10.2), and typically weeds have all but a

Table 10.2 Weediness Characteristics^a

- Successful plant establishment occurs over a broad range of environmental conditions.
- Controls internal to the seed permit discontinuous germination (throughout the year) .
- Seeds are long lived.
- Continuous seed production.
- Self-fertilizing, or if cross-pollinated, it is done so by wind or unspecialized insects.
- High seed production under optimal conditions (some seed production even diverse environments).
- Efficient seed dispersal both short- and long-range.
- Rapid growth (life cycle).
- Perennials have efficient vegetative reproduction or regeneration from fragments.
- Perennials are not easily uprooted.
- Growth characteristics (rosette, thick matte growth) or biochemical basis (allelopathy) allow plant to be highly competitive for resources

^a Compiled from information in Baker, 1967 and 1974

couple of these characteristics (Baker 1967, 1974). Each of these characteristics is controlled by at least one gene and most likely by a group of genes. Crop plants have only five to six of these characteristics, indicating that a single gene inserted into a GEP cannot confer weediness on a crop plant. Furthermore, because genetic engineering is a precise process where the genetic material being introduced into a plant is well characterized, inferences about how this genetic material affects each of the weediness characteristics can be made. For example, it is safe to infer that a gene encoding an insect toxic protein only in the roots of a plant will not affect the seed dissemination mechanism.

Even if insect damage was the only limiting factor that prevents a commercial crop plant from becoming a severe weed pest, incorporation of genetic material that confers resistance to the insect, allowing the plant to become weedy, would be a problem whether the resistance to the insect were incorporated by either standard breeding techniques or genetic engineering. Therefore, this is a concern about improving insect resistance of a crop in general and is not a concern limited strictly to genetically engineered plants. Standard breeding practices performed by humans for hundreds of years have resulted in increased insect pest resistance of populations of many crops. However, there has never been a report where release of new insect resistance varieties from standard breeding programs has been the factor causing the cultivar to become a devastating weed problem. It is highly unlikely then that a crop plant genetically engineered for insect resistance would become a significantly greater weed problem than the parent plant from which it was derived.

10.3.1.2 Environmental Contamination with the Genetically Engineered Product

Because GEPs can continuously produce the products of the introduced genetic material, there is a concern that the gene products could contaminate the environment. Plants produce hundreds of molecules in their cells that remain within the cell or are transported out of the cell. These products can therefore be released into the environment, either by secretion from living cells or release of cellular contents when the cells die. Therefore, products produced as a result of genetically engineering a plant could leak into the environment. However, biologically produced molecules typically have very short half-lives in the environment due to breakdown by soil microbes, and as a result these substances do not accumulate in the soil or contaminate groundwater. It is therefore extremely unlikely that harmful effects to the environment would result from release of insecticidal proteins or other molecules produced in genetically engineered plants. This may be a concern if a plant is engineered to produce novel synthetic compounds not previously produced in nature, and that are not readily biodegradable. However, there are currently no indications that such compounds would be expressed in GEPs for insect control. If such a control strategy was developed, experiments should be required to determine residual half-life of the products in the environment.

10.3.1.3 Impact on Wildlife and Beneficial Insects

Crop plants become an integral part of the environment and can be a food source or home to many beneficial or nontarget insects or wildlife. As natural habitats shrink, beneficial insects and wildlife have become more and more dependent on agricultural land for food and shelter. Although wildlife can often avoid being directly sprayed with chemical pesticides, interaction with residues left on the plants is a certainty; however, residual chemicals remain for only a limited amount of time after application. Alternatively, plants genetically engineered to produce insecticidal proteins are in the field continuously; therefore the impact of exposure to wildlife is an important consideration. Although the products of GEPs are continuously present within the plant, exposure to the insect-controlling compounds would be limited to insects that ingest the plant material. The potential for harmful effects to the wildlife would be limited to those organisms that ingest the plant material or those that feed on the insects that ingest the plant material.

If GEPs express proteins toxic to beneficial insects and/or wildlife, certain precautions can be taken to minimize the direct uptake of the toxins by the beneficial organisms. It is currently possible to have the genes encoding the insect control proteins expressed only in the cells that are targeted by the pest insect as food. For example, promoters can be used that turn the gene on only in leaf and stem tissues and not floral parts or roots. It will also be possible to place the expression in specific tissues under developmental control, being turned on in specific tissues only at certain periods during plant development. If pest insects are only a problem in young leaves, it is conceivable to have the genes responsible for insect resistance turned on only in young leaves and turned off as the leaves age. An example of the benefit of this capability in GEPs is the toxicity of an insecticidal trypsin protease inhibitor to honey bees (Malone et al., 1995) and the finding that GEPs expressing insect toxins are toxic to bees (Crabb, 1997). Toxicity of insecticidal protein expressing GEPs to bees occurs when the toxin is expressed in the pollen. However, using tissue-specific promoters, expression of insecticidal compounds in the pollen and nectar can be prevented. It is also possible to use a promoter that is stress induced. Specific genes are turned on in plants in response to damage such as insect feeding. Using promoters isolated from these genes would limit the production of the insect toxins to those times that the plant is experiencing insect attack. Harm to beneficial insects should be considered a serious concern and, when appropriate, promoters should be used that prevent pollen/nectar expression and limit beneficial insect contact to the toxins.

Exposure of insects and other animals that feed on plant pest insects to the GEP-produced insect toxins is not as easily addressed. If the plant pest feeds on a plant expressing a novel insect toxin, and a predatory insect subsequently feeds on this plant pest, it is likely that the predatory and beneficial insect will also be exposed to the insect-controlling substance. However, since these insect control molecules are derived from biological synthesis, they will most likely be rapidly biodegraded. There should be no residual build-up along the food-chain, as has been proven to occur with many chemical pesticides resulting in severe harm to many species of animals. This is, however, an important enough concern that the potential for the passage of the insect control compounds along the food chain should be considered for each novel product that is synthesized in GEPs. This concern will become increasingly important as the technology of GEP production improves, allowing different types of insect control molecules to be expressed.

10.3.2 Environmental Risk Associated with Fluidity of Genetic Material Within and Between Species

Another area of environmental risk is the potential "escape" of genetic material from the GEP to other organisms including weedy relatives and completely unrelated microorganisms (Keeler and Turner, 1991; Rabould and Gray, 1993, Kerlan et al., 1993; Darmency, 1994). It is well understood that, in nature, genetic material flows between crop plants and their weedy or native relatives. It has even been shown that genes introduced into rapeseed by genetic engineering techniques can combine with genetic material from related weedy species by interspecific hybridization (Kerlan et al., 1993; Darmency, 1994, Chevre et al., 1996). Exchange of genetic material between direct-seeded rice and wild rice has also been shown to occur naturally (Aswidinnoor et al., 1995). Wild rice is a significant weed problem in direct-seeded rice. It is therefore conceivable that genetic material introduced into certain crop plants will find its way to recombine with genetic information from weedy relatives, and we should therefore consider this fact before introducing new genetic information into desired crops.

The impact of the escape of transgenes on the environment will depend on the genes and plants in question. For example, the transfer of herbicide resistance genes from a genetically engineered crop to a weedy relative could cause a problem in agriculture but would probably have no impact in the natural ecosystem balance. This is because herbicides are not used to maintain an ecological balance in nature. However, transfer of an insect resistance gene from a crop to a weedy relative could in theory influence the ecological balance. Insect resistance is a constantly evolving phenomenon in natural populations of plants; however, increasing the gene flow among living things could greatly increase the rate at which resistance develops. Of course for this to change the ecological balance, insect damage to the weedy/native crop would have to be a major limiting factor preventing the plant's spread in the natural ecosystem. The points to raise here are the same ones raised when discussing concerns over the possibility that the genetically engineered plant directly becomes a weed problem. Additionally, the interspecific hybrid between the crop and its related weed would need to be fertile, or vegetative propagation would have to be an important mechanism of dispersal. In most instances, interspecific hybrids are sterile, and if not sterile, at least they are typically less well adapted for survival as a natural population. The probability is small that all of the factors are present that are necessary to create an aggressive weed. It is more likely that a noticeable problem would result from transfer of insect resistance genes to a related plant that is already a weed, and thereby strengthening its weedy characteristics. However, this possibility should be considered prior to the release of crops, whether they are produced by genetic engineering or classical breeding.

If we compare the aims of plant genetic engineering to those of classical breeding, we see that using classical breeding, insect resistance is commonly bred into important varieties of crops from insect-resistant germplasm collected from wild populations all over the world. Despite the thousands of years of plant breeding and expansion of the typical range of plant growth, there has not been a catastrophic combination of genetic material created that caused significant environmental problems. The environment apparently has a buffering capacity to deal with a certain amount of genetic fluidity among organisms, without the occurrence of major ecological upheavals. Of course, the question is: Will the increase in genetic fluidity resulting from commercialization of GEPs surpass this buffering capacity? It is unlikely that using genetic engineering to improve traits, such as insect resistance, that also were improved historically using standard plant breeding practices will challenge this capacity.

If escape of a specific gene being used in development of GEPs is deemed possible due to proximity of related native plants, and its escape is potentially worrisome, certain biotechnological manipulations can be done to greatly reduce the risk of transfer of foreign genetic material from a crop to its weedy or natural relatives. The main avenue for escape of genetic material from one species is by pollen transmittance. Pollen has the capability of escaping the cultivated area and traveling to populations of species that are closely enough related to allow interspecific hybrids to form. However, it is now possible to add the new genetic material to a GEP in a manner that prevents this DNA from being transmitted paternally by the pollen (Daniell et al., 1998). This is done by targeting foreign DNA to the plastidic DNA complement of the recipient plant. Plastidic inheritance offers the advantage of high levels of expression of the introduced gene along with strict maternal inheritance in some crop plants. Since the plastidic genome is not transferred via pollen during fertilization, the only way the foreign DNA can recombine with other genetic material in interspecific hybridizations is through the pollenation of the GEP by the weedy or native relative. Since these plants are harvested, the probability of the escape of the transgene material is greatly reduced. This ability to limit the flow of genetic material is not possible when only classical breeding schemes are used in crop development.

A very controversial area of concern over the effect of biotechnology on the environment involves the lateral transfer of DNA between unrelated organisms

(Rissler and Melon, 1993). For example, the uptake of eukaryotic DNA by micro-organisms. It is theoretically possible that DNA from decaying plant material could be taken up by soil bacteria. Microorganisms can take up DNA from their environment, and it is conceivable that DNA could survive bound to soil particles long enough for the microbes to have access to it; however, this has never been documented in nature. The problem of using this as a caution against the release of GEP is that it is a circular argument. If genetic material is indeed that free-flowing among organisms, the genetic engineering of a plant with an insect resistance gene from another organism should not be so unique to the environment and therefore less a digression from what naturally occurs. In fact, in a report from a World Bank Panel on Transgenic Crops (Kendall et al., 1997), it is considered highly unlikely that this type of gene flow from genetically engineered plants to soil microorganisms would occur. Even if the transfer of genetic material from eukaryotes to prokaryotes was considered a concern, there are biotechnological approaches that would reduce this risk even more substantially. Single genes from eukaryotes often are not contiguous stretches of DNA. Partial coding regions of the gene are often separated by long sequences of DNA that is not used to produce the gene products. Eukaryotic RNA transcripts synthesized from such genes must be spliced to remove the intervening sequences before the RNA can be translated into a functional protein. Prokaryotic organisms do not have this splicing capability. Inserting artificial intervening sequences into genes being introduced into plants by genetic engineering would therefore prevent the gene from functioning if it escaped into a bacterial genetic background.

10.3.3 Changes in Crop Management Practices Resulting from Use of GEPs

When attempting to determine the impact of any technology on the environment, an unrealistic approach is to compare the impact of the technology to a static environment that has no interactions and does not change. Using this scenario, all technological advances are destined to fail. In reality, agriculture has greatly impacted our environment and will continue to do so, whether or not biotechnology becomes a major part of our agricultural industry. It is therefore more appropriate to compare the impact of biotechnology with that of the most recent historical agricultural practices or other new emerging techniques that could be employed in the place of using biotechnological tools to improve agricultural productivity. The most obvious comparison would be the continued reliance on pesticide sprays as an alternative to the use of genetic engineering to increase crop pest resistance. Since it is the negative impact of this technology on the environment that has stimulated the need for alternative approaches, it is difficult to see how properly designed biotechnological strategies for insect control would be anything but more beneficial to the environment.

Without the use of biotechnology to develop more insect-resistant crops, it could be assumed that testing of new plant introductions would be a viable alternative (as it currently is a common practice). Using this approach, wild relatives to existing crops would be screened for greater insect resistance. If found, standard breeding programs would be used to introgress the desired insect resistance trait into a suitable cultivated variety. Much larger portions of genetic material from the introduced wild relative, collected from anywhere in the world, would be combined with the genetic material of the cultivated crop than would occur using genetic engineering. The complexity of this genetic combination would be far greater than that produced by engineering precise genes into a plant, resulting in less of an ability to predict all the potential characteristics that would be bestowed on the newly developed crop and how these would interact in the environment. Therefore, although genetic engineering allows a greater diversity of possible genetic combinations, the potential impact on the environment should be more predictable.

If genetic engineering were not available, testing of new alternative crops from other regions of the world would also be an alternative for certain insect sensitive crops. This process of plant introduction provides the least ability to predict how the crop will behave in the new environment and therefore the greatest environmental risk. Historically, this approach has provided great improvements in our agriculture (e.g., corn, rice, soybeans, alfalfa, sugarcane) but has also resulted in the introduction of some of the greatest weed problems (e.g., tumbleweed, melileuca, kudzu). In an ideal world, we would know what species introductions would provide great benefit with minimal impact on the environment before the range of potentially important plants is tested in the environment. In reality, we have to weigh the cost and the benefit of allowing relatively open testing of introduced plants as potential new crops. Although the need to improve the relationship between the environment and our agricultural practices is clear, the benefit of the use of our major crop plants has been worth the associated problems.

If used properly, genetic engineering provides an avenue for improving the productivity and utility of crops we currently are using, which simplifies the ability to predict the impact of specific GEPs on our environment. For example, improving the performance of currently grown crops under stress conditions (i.e., improving insect resistance), or changing the biochemical makeup of the harvested product (i.e., changing the oil composition of the seed). Improving stress tolerance has the obvious advantage of improving production efficiency, thereby providing a higher profit margin and a greater incentive to grow the crop. Using biotechnology to alter chemical composition of the crop would increase the market demand for the product of a single species. This would allow the growers to fill the demand for numerous chemical commodities by growing a single species, albeit different genetically engineered varieties of the single species. This is currently being done with rapeseed, where genetic engineering has been used to change the type of oil produced in the seed to fit commodity niches (Knutzon et al., 1992; Topfer and Martini, 1998). As demands for unique biochemical feedstocks develop, using genetically engineered versions of existing crop species will prevent the need to expand the range of new crops about which there is no information on how they will interact with the new environment in which they would grow. The advantage to the environment results because historical information from the hundreds of years that many of the current crops have been grown in their existing regions can be used to understand how the plants will interact with their environment when they are changed subtly by only one or a few genes. This would allow a much more educated and fact supported way of deciding what types of improvements would be best for the industry and the environment.

10.4 BIOTECHNOLOGY AS A COMPONENT OF
ENVIRONMENTALLY FRIENDLY AGRICULTURE

When weighing the potential environmental problems associated with biotechnology in agriculture, it is arguable that they are at least equal to those concerns
over the existing agricultural practices, and depending on the biotechnological
approach being used, biotechnology may provide solutions for improving the environmental impact of agriculture. As long as society demands a year-round supply
of a great diversity of relatively low-cost foods, intensive agriculture is the only
alternative currently available. Although food products derived from strict organic
farming practices are filling a growing niche in the marketplace, there is no indication
that these practices could be incorporated into the mainstream production system
while assuring the year-round availability and low cost enjoyed from our current
agricultural practices. The goal is therefore to incorporate environmentally friendly
sustainable agricultural practices that will fit this profile. The added benefit of biotechnological approaches to crop improvement is that they can augment new sustainable agricultural methods. It is hoped that biocontrol strategies involving beneficial
insects and microorganisms that attack pest insects will provide sustainable control
practices that work in harmony with GEPs. If chemical pesticide or fungicide sprays
are used on a crop, the adverse effect on beneficial insects or microorganisms will
preclude the success of this type of biocontrol. When we monoculture anything, as
we do in many agricultural systems, we are upsetting the complex interactions nature
has setup, so it may be unrealistic to expect natural control mechanisms to work
completely. The best we can hope for are mechanisms that work in harmony with
natural control, and use of GEPs can be one of these mechanisms.

Plants genetically engineered for resistance to specific insects or fungi can work
in harmony with the biocontrol strategies, since the GEP produces substances that
typically only affect the organisms that are attaching the plant. An important argument is that insect toxins in GEPs may also harm beneficial insects that feed on
plant pest insects. Although they do not take up the plant-produced toxin directly,
they would be exposed by feeding on the plant pests that have ingested this compound. If this were true, then the use of insect resistant GEPs expressing a toxin
would prevent the use of certain types of biocontrol strategies. However, in field
trials with GEPs expressing the bt-toxin the efficacy of this protein in transgenic
plants was greater in the field than in the greenhouse. This finding was used to
propose that a synergy between bt-toxin expression in plants and natural predator
and parasite control exists; however, this has yet to be proven.

It has been stated that humans are trying to recreate nature in their view with
many of the ways in which we are interacting with our environment and with the
way in which we are manipulating plants and animals for our needs. This is presented
as an unacceptable practice and it is argued that it should stop. This is more a
sociopolitical point and not really related to the scientific discussion about the impact
of biotechnology on the environment; however, it needs to be addressed in the context
of setting the boundaries for this scientific discussion. Since the first human walked
on this planet we have left our footprint, or impact, forever changing the direction
of evolution of the earth's environment. This is a fact that is realized for all organisms

from the largest dinosaurs to all plants and microorganisms. The tremendous resources required to support our current population worldwide guarantees that we will continue to impact nature. As we continue the struggle to provide for the worldwide human population it is therefore important to take control of how we impact our environment. It is important not to recreate nature in our view, but to understand and, in a rational manner, minimize the negative impact we have on our environment.

The way in which we try to control our impact on the environment is also in dispute. Some argue that we need to greatly reduce our reliance on technology to save the environment. Conversely and based on fact, some of the worst environmental damage was done when developed countries were developing, as is currently being done in Third World countries. It is often reported that the only way to reduce this damage is by helping the countries improve their standard of living by increasing their technological base and providing them with more modern alternatives to old practices that are not in harmony with the environment. From a policy standpoint this makes sense, since a population is more willing to exert political pressure in support of environmental policies when they feel secure with their food and economic situations. It is clear that embracing sustainable and environmentally friendly technologies, new and old, is the key to a better and more environmentally friendly agricultural system worldwide.

As long as our society makes the sociopolitical decision to support a high standard of living based on a variety of readily available, inexpensive foods, continuously advancing medical technologies, industrialization, and convenient local and global transportation, the human impact on the environment will be tremendous. Plant genetic engineering is one tool that can be used as only one of a group of integrated tools to help reduce the negative aspects of the impact of agriculture on natural ecosystems and at the same time allow the optimization of crop productivity under a variety of environmental conditions. If used properly, biotechnology can be a component of integrated pest management schemes that minimize environmental damage, especially with respect to pest insect control. The management practices associated with the first attempts at commercializing genetically engineered crops expressing the bt-toxins support this view.

Commercialization of GEPs expressing the bt-toxins is an example of using a strategy that is already available as an environmental biopesticide. As a result, proponents of the biopesticide use of *B. thuringiensis* are worried that widespread use of bt-toxin-expressing GEPs may increase the rate at which insects develop resistance to the toxins. This would have a negative impact on the environment because growers who had previously relied on *B. thuringiensis* biopesticide applications would now have to resort to chemical pesticides or suffer tremendous losses. To reduce this risk, the U.S. EPA has put restrictions on the sale of cotton containing bt-toxins to ensure that every U.S. farm has some fields planted with varieties that do not produce the bt-toxin proteins (Kendall et al., 1997). This source of non-engineered cotton provides a refuge for a population of pest insects to produce a full life-cycle not under the selective pressure of the bt-toxin. Therefore the gene pool of pest insects that arose from the refuge plots has developed without selective pressure that could amplify the frequency of the occurrence of resistance in the insect germplasm pool. Mixing of this population with the very few that may have made it to maturity by developing on the bt-toxin-expressing plants dilutes the amplification of the bt-toxin resistance trait, thereby greatly slowing the development of resistance within the insect populations.

This refugia technique has the added benefit of providing a haven for the development and maintenance of beneficial or other natural populations of neutral insects in the refuge plots of the crop. Since the refuge area, or refugia, is purposefully not treated with pesticides, complete life cycle of many insects will be allowed to occur. Also, a pool of beneficial insects can be maintained in this refugia that may function synergistically with the genetically engineered insect control strategy.

The maximum benefit for the growth of genetically engineered crops will often come to the grower only if the grower is willing to maintain close management of the crop. The bt-toxin of these GEPs is only a component of an integrated pest management approach to crop production. Fields need to be continually monitored for the presence of pests. The advantage of using the GEP is that pest outbreaks should be less severe and, when observed, should be controlled by regional application of the necessary pesticide. Since the need for broad application of a pesticide is reduced, other biocontrol strategies will have a greater chance of working as part of the integrated pest management approach. It is therefore clear that increased management of the crop is a crucial component if the bt-toxin-expressing plants are going to provide an advantage to the grower. This is especially true since reduced pesticide applications are necessary to offset the higher price paid for the GEP seeds. One concern over the use of the bt-toxin-expressing plants is that if the crop is not properly managed, not only will the benefit to the grower not be realized, but the development of insect resistance to the bt-toxin may be accelerated. This acceleration being the result of exposure of insects to suboptimal levels of the toxin that allows the insects to complete a full life-cycle. Lack of appropriate management practices may be a more pressing concern if the GEPs become available in Third World countries. If these countries do not have the infrastructure to educate the growers on the proper management practices, the benefits of the GEP may not be realized and, in the case of bt-toxin-expressing GEPs, resistance to the bt-toxin may be accelerated. Control of the use of GEPs should therefore be limited to areas where proper management can be assured and monitored.

An interesting result of the commercialization of GEPs is the involvement of the companies in setting guidelines and management practices. One of the main arguments against the release of GEPs has been that the large corporations producing these plants were doing so with only profit as a motive and little concern over the environmental impact. As it turns out, even if this is the case, the corporate view on profit potential is requiring them to understand the environmental impact prior to release of a GEP. Profits are only returned from the high cost of developing GEPs if the crop is successful for a long period of time. Therefore, it is crucial to the financial planning for companies to understand potential problems resulting from the release of the crop. Rapid development of insect resistance to the bt-toxin would limit the usefulness of the GEPs expressing this toxin. If it becomes an ineffective tool for insect control, the growers will refuse to pay the high price for the seed, and a market for the bt-toxin-expressing crops will be lost.

It is said that the road to hell is paved with good intentions. In practice, without a strong motivational force, good intentions are often left as just that. Economics is perhaps one of the strongest driving forces in our society as a whole and it is this motivational force that pushes the grower of GEPs to manage their crop to get the return out of the premium seed prices. It is also economics that drives the companies to understand what management practices are necessary to maintain the efficacy of their GEP-based control strategy. This is the driving force for the development of strong education programs to accompany the release of these products, and the requirement that the growers agree to the management terms designed to reduce the development of insect resistance.

Caution needs to be taken to avoid the distribution of bt-toxin-expressing GEPs outside of the range in which they were tested for reasons other than lack of management capabilities. Directly distributing seed of GEPs to diverse geographical regions without first testing the expression level of the toxin under the new growth conditions, and the efficacy of the plants on the insect pests indigenous to the new area, could result in ineffective control. For example, it would be a mistake to assume that bt-toxin-expressing cotton designed and tested to improve resistance to the boll weevil in the southern U.S. would provide adequate resistance against major cotton pests in other parts of the world where cotton is a major crop. Inability to obtain a high level of control of the pest insect due either to low level of bt-toxin expression in the plant or a higher level of tolerance to the toxin in the insect pest could stimulate a rapid rise of resistance in the insect population, again preventing the future use of the bt-toxins in any type of control strategy. This would cause growers to resort to alternatives such as pesticide applications that would be more harmful to the environment.

10.5 SUMMARY

It is estimated that 1.5 billion ha of land are utilized in agriculture worldwide (Kendall et al., 1997). Additional arable land used in agriculture production could increase 25% to a total of approximately 2 billion ha. However, during this time the population will increase 100% in a quarter century or so; therefore the farmland per capita will continue to decrease rapidly. Simple arithmetic can be used to show that productivity on a per ha basis will have to increase on a global scale to maintain the per capita demand for food. This can be accomplished in part by reducing the losses in productivity due to stresses imposed during the growth of the crop. One of the major stresses on crop yield is insect pest load. A major area of agricultural concern is therefore the development of environmental friendly strategies for insect control.

These needs for reducing losses in crop yields are occurring at a time when there is a need and strong social pressure to reduce the negative impact that humans have on the environment. The initial response may be to think that demanding greater productivity from the land while reducing the adverse effect we are having on our environment are divergent or contradictory goals. However, research is beginning to show that there may be alternative agricultural management and production methods for maintaining high yields without the use of environmentally damaging cultivation techniques and reducing the need for toxic pesticides. Plant genetic engineering is one of the many tools that if used properly will allow the manipulation of biological systems to reduce pest damage to crops and thereby increase crop yields. With respect to insect control, biotechnology provides a means of improving crop yields often with a greatly reduced need for use of environmentally damaging chemicals. Cost savings are realized due to the need for less pesticides and because fossil fuel use is reduced if growers do not have to drive tractors through the fields as frequently to spray the crop. Based on the growing needs for food and the increasing demands to develop more environmentally friendly agricultural practices, it will be important to pursue biotechnological strategies for pest insect control.

Prior to implementation of a plant genetic engineering strategy for insect control, it will be important to review the specific strategy considering the risk related to performance of the GEP in the environment, the escape of genetic material into other plant species, and the related changes in agricultural management that will be necessary to support the GEP-based strategy. Once this information is obtained, a rational decision can be made whether or not to pursue the commercialization of the concept. Is there an incentive for industry to conduct this type of review? Experience with bt-toxin-expressing plants indicates that with respect to insect control, industry does have a strong economic incentive to assure that the GEP will perform for a long period of time, in a fashion that will not result in the loss of efficacy of the strategy due to adverse effects on the environment. Biotechnological approaches to insect control therefore offer much promise in developing agricultural practices that are in closer harmony with the environment than we have had in the past, indicating that the environmental impact of biotechnology should be a favorable one.

REFERENCES

Aswidinnoor, H., R.J. Nelson, and J.P. Gustafson. Genome-Specific Repetitive DNA Probes Detect Introgression of *Oryza minuta* Genome into Cultivated Rice, *Oryza sativa. Asia Pacific Journal of Molecular Biology and Biotechnology* 3(3), 215–223, 1995.

Baker, H.G. Characteristics and Modes of Origin of Weeds, in *The Genetics of Colonizing Species,* Baker, H.G., and Stebbins, G.L., Eds., Academic Press, New York and London, 1967, 147–172.

Baker, H.G. The Evolution of Weeds. *Annual Review of Ecology and Systematics* 5, 1–24, 1974.

Beck, S.D., and J.C. Reese. Insect-Plant Interactions: Nutrition and Metabolism, in *Biochemical Interactions Between Plants and Insects. Recent Advances in Phytochemistry,* Vol. 10, Wallace, J.W., and Mansell, R.L., Eds., Plenum Press, New York, London, 1976, 41–92.

Bergamasco, R., and D.H.S. Horn. Distribution and Role of Insect Hormones in Plants, in *Endocrinology of Insects,* Downer, R.G.H., and Laufer, H., Eds., Alan R. Liss, New York, 1983, 627–654.

Boulter, D., A.M.R. Gatehouse, and V. Hilder. Engineering Enhanced Natural Resistance to Insect Pests — a Case Study. *UCLA Symposia on Molecular and Cellular Biology* 129, 267–273, 1990.

Chevre, A.M., F. Eber, A. Baranger, M.C. Kerlan, P. Barretn, G. Festoc, P. Vallee, and M. Renard. Interspecific Gene Flow as a Component of Risk Assessment for Transgenic Brassicas. *Acta Horticulturae* 407, 69–179, 1996.

Chrispeels, M.J., and N.V. Raikhel. Lectins, Lectin Genes, and their Role in Plant Defense. *Plant Cell* 3, 1–9, 1991.

Crabb, C. Sting in the Tale for Bees. *New Scientist* 155, 14, 1997.

Daniell, H., R. Datta, S. Varma, S. Gray, and S.B. Lee. Containment of Herbicide Resistance Through Genetic Engineering of the Chloroplast Genome. *Nature Biotechnology* 16(4), 345–348, 1998.

Darmency, H. The Impact of Hybrids Between Genetically Modified Crop Plants and Their Related Species: Introgression and Weediness. *Molecular Ecology* 3, 37, 1994.

Dicke J., T.A. van Beek, M.A. Posthumus, N. Ben Dom, H. Van Bokehoven, and A.E. De Groot. Isolation and Identification of Volatile Kairomone That Affects Acarine Predator-Prey Interactions. Involvement of Host Plant in its Production. *Journal of Chemical Ecology* 16, 381–396, 1990.

Etzler, M.E. Distribution and Function of Plant Lectins, in *The Lectins*, Liener, I.E., Sharon, N., and Goldstein, I.J., Eds., Academic Press, San Diego, 1986, 371–435.

Feitelson, J.S., J. Payne, and L. Kim. *Bacillus thuringiensis*: Insects and Beyond. *Biotechnology* 10, 271–275, 1992.

Gatehouse, A., V. Hilder, and D. Boulter. *Plant Genetic Manipulation for Crop Protection*, C.A.B. International, Wallingford, Oxon, U.K., 1992, p. 266.

Gatehouse, A.M.R., F.M. Dewey, J. Dove, K.A. Fenton, and A. Pusztai. Effects of Seed Lectins from *Phaseolus vulgaris* on the Development of Larvae of *Callosobruchus maculatus*; Mechanism of Toxicity. *Journal of the Science of Food and Agriculture*, 35, 373–380, 1984.

Gatehouse, A.M.R., V.A. Hilder, K.S. Powell, M. Wang, G.M. Davison, L.N. Gatehouse, R.E. Down, H.S. Edmonds, D. Boulter, C.A. Newell, A. Merryweather, W.D.O. Hamilton, and J.A. Gatehouse. Insect-Resistant Transgenic Plants: Choosing the Gene to do the Job. *Biochemical Society Transcripts* 22, 944–949, 1994.

Gatehouse, A.M.R., S.J. Shackley, K.A. Fenton, J. Bryden, and A. Pusztai. Mechanism of Seed Lectin Tolerance by a Major Insect Storage Pest of *Phaseolus vulgaris*, *Acanthoscelides obtectus*. *Journal of the Science of Food and Agriculture* 47(3), 269–280, 1989.

Hilder, V.A., A.M. Gatehouse, and D. Boulter, Genetic engineering of crops for insect resistance using genes of plant origin, in *Genetic Engineering of Crop Plants* Lycett, G.W. and Grierson, D. Eds., Butterworth, London, 1990, pp. 51–56.

Hilder, V.A., A.M. Gatehouse, S. E. Sheerman, R. F. Barker, and D. Boulter. A Novel Mechanism of Insect Resistance Engineered into Tobacco. *Nature* 330, 160–163, 1987.

Hilder, V.A., K.S. Powell, A.M.R. Gatehouse, J.A. Gatehouse, L.N. Gatehouse, Y. Shi, W.D.O. Hamilton, A. Merryweather, C. Newell, J.C. Timans, W.J. Peumans, E.J.M. Van Damme, and D. Boulter. Expression of Snowdrop Lectin in Transgenic Tobacco Plants Results in Added Protection Against Aphids. *Transgenic Research* 4, 18–25, 1995.

Höfte, H., and H.R. Whiteley. Insecticidal Crystal Proteins of *Bacillus thuringiensis*. *Microbiological Reviews* 53, 242–255, 1989.

Hoover, K., C.M. Schultz, S.S. Lane, B.C. Bonning, S.S. Duffy, B.F. McCutchen, and B.D. Hammond. Reduction in Damage to Cotton Plants by a Recombinant Baculovirus that Knocks Moribund Larvae of *Heliothis virescens* off the Plant. *Biological Control: Theory and Applications* 5(3), 419–426, 1995.

Johnson, R., J. Narsvaez, G. An, and C. Ryan. Expression of Proteinase Inhibitors I and II in Transgenic Tobacco Plants: Effects on Natural Defense Against *Manduca sexta* Larvae. *Proceedings of the National Academy of Science USA* 86, 9871–9875, 1989.

Keeler, K.H., and C.E. Turner. Management of Transgenic Plants in the Environment, in *Risk Assessment in Genetic Engineering,* Levin, M., and Strauss, H., Eds., McGraw-Hill, New York, 1991, pp. 189–218.

Kendall, H.W., R. Beachy, T. Eisner, F. Gould, R. Herdt, P.H. Raven, J.S. Schell, and M.S. Swaminathan. Bioengineering of Crops: Report of the World Bank Panel on Transgenic Crops, in Environmentally and Socially Sustainable Development Studies and Monographs, Series 23, The International Bank for Reconstruction and Development/The World Bank. 1818 H Street, N.W. Washington D.C., 1997.

Kerlan, M.C., A.M. Chevre, and F. Eber. Interspecific Hybrids Between a Transgenic Rapeseed (*Brassica napus*) and Related Species: Cytogenetical Characterization and Detection of the Transgene. *Genome* 36, 1099, 1993.

Knutzon, D.S., G.A. Thompson, S.E. Radke, W.B. Johnson, V.C. Knauf, and J.C. Kridle, Modification of Brassica Seed Oil by Antisense Expression of Stearoyl-Acyl Carrier Protein Desaturase Gene, *Proceedings of the National Academy of Sciences USA* 89(7), 2624–2628, 1992.

Lereclus, D., M. Vallade, J. Chaufaux, O. Arantes, and S. Rambaud. Expansion of Insecticidal Host Range of *Bacillus thuringiensis* by *in vivo* Genetic Recombination. *Bio/technology* 10 (4), 418–421, 1992.

Loughrin, J.H., A. Manukian, R.R. Heath, T.C.J. Turlings, and J.H. Tumlinson. Volatiles Emitted by Different Cotton Varieties Damaged by Feeding Beet Armyworm Larvae. *Journal of Chemical Ecology* 21, 1217–1227, 1995.

Malone, L.A., H.A. Giacon, E.P.J. Burgess, J.Z. Maxwell, J.T. Christeller, and W. A. Laing. Toxicity of Trypsin Endopeptidase Inhibitor to Honey Bees (Hymenoptera: Apidae). *Apiculture and Social Insects* 88(1), 46–50, 1995.

McCall, P.J., T.C.J. Turling, W.J. Lewis, and J.H. Tumlinson. Role of Plant Volatiles in Host Location by the Specialist Parasitoid *Microplitis croceipes* Cresson (Braconidae: Hymenoptera). *Journal of Insect Behavior* 6, 625–639, 1993.

McCall, P.J., T.C.J. Turling, J.H. Loughrin, A.T. Proveaux, and J.H. Tumlinson. Herbivore-Induced Volatile Emissions from Cotton (*Gossypium hirsutum* L.) Seedlings. *Journal of Chemical Ecology* 20, 3039–3050, 1994.

Office of Technology Assessment. Biotechnology on a Global Economy, U.S. Senate Committee Report, 1993.

Purcell, J.P., J.T. Greenplate, M.G. Jennings, J.S. Ryerse, J.C. Pershing, S.R. Sims, M.J. Prinsen, D.R. Corbin, M. Tran, R.D. Sammons, and R.J. Stonard. Cholesterol Oxidase: a Potent Insecticidal Potein Active Against Boll Weevil Larvae. *Biochemical Biophysical Research Communications* 196, 1406–1413, 1993.

Raybould, A.F., and A.J. Gray. Genetically Modified Crops and Hybridization with Wild Relatives: a UK Perspective. *Journal of Applied Ecology* 30, 199, 1993.

Rissler, J., and M. Mellon. Perils Amidst the Promise, Ecological Risks of Transgenic Crops in a Global Market. Union of Concerned Scientists, Washington, D.C., December 1993.

Shade, R.E., H.E. Schroeder, J.J. Pueyo, L.M. Tabe, L.L. Murdock, T.J.V. Higgins, and M.J. Chrispeels. Transgenic Pea Seeds Expressing the Alpha-Amylase Inhibitor of the Common bean are Resistant to Bruchid Beetles. *Bio/technology* 12, 793–796, 1994.

Stewart, Jr., C.N., J.N. All, P.L. Raymer, S. Ramachandran. Increased Fitness of Transgenic Insecticidal Rapeseed Under Insect Selection Pressure. *Molecular Evolution* 6(8), 773–779, 1997.

Takabayashi, J., and M. Dicke. Plant-Carnivore Mutualism Through Herbivore-Induced Carnivore Attractants. *Trends in Plant Science* 1, 109–113, 1996.

Topfer, R., and N. Martini. Engineering of Crop Plants for Industrial Traits, in *Agricultural Biotechnology,* Altman A., Ed., Marcel Dekker, New York, 1998, pp. 161–181.

Turlings, T.C.J., J.H. Tumlinson, and W.J. Lewis. Exploitation of Herbivore-Induced Plant Odors by Host Seeking Parasitic Wasps. *Science* 250, 1251–1253, 1990.

Warren, G.W., M.G. Koziel, M.A. Mullins, G.J. Nye, N.M. Desai, B. Carr, and K.N. Kostichka. Novel Pesticidal Proteins and Strains. Patent Application No. PCT WO 94/21795, 1994.

Whittaker, R.H., and P. Feeny. Allelochemics: Chemical Interactions Between Species. *Science* 171, 757–770, 1971.

Regulation

CHAPTER 11

Regulatory Aspects of Biological Control Agents and Products Derived by Biotechnology

J. Thomas McClintock, Nikolai A. M. van Beek, John L. Kough,
Michael L. Mendelsohn, and Phillip O. Hutton

CONTENTS

11.1 INTRODUCTION

The Federal Insecticide, Fungicide, and Rodenticide Act (FIFRA) authorizes the U.S. Environmental Protection Agency (EPA) to regulate pesticides to ensure that their use in commerce does not cause unreasonable adverse effects to humans and the environment. Registrants of pesticides are responsible for submitting specific data to the EPA, which is subsequently reviewed by Agency scientists to assess their effects on human health and the environment. Once a pesticide is registered by the Agency, it may be sold and distributed in the United States and used as specified on the approved label.

A pesticide or active pesticidal ingredient is defined as "any substance (or group of structurally similar substances if specified by the Agency) intended to prevent, destroy, repel, or mitigate any pest, or that functions as a plant regulator, desiccant, or defoliant, and any nitrogen stabilizer " (FIFRA Section 2). Products that are intended to exclude pests only by providing a physical barrier against access are not considered pesticides. Exemptions for pesticides of a character not requiring FIFRA regulation are outlined in Part 40 of the Code of Federal Regulations (CFR) 152.25 and are discussed elsewhere (McClintock, 1995).

The EPA recognizes two broad classes of pesticides: conventional chemical pesticides and biological pesticides. Chemical pesticides includes such synthetic compounds as carbamates, organophosphate esters, and pyrethroids. Biological pesticides or biopesticides can be divided into three categories: biochemical pesticides, which are naturally occurring and have a nontoxic mode of action and contain a pheromone, a hormone, or certain insect or plant growth regulator as the active pesticidal ingredient; microbial pesticides, which contain a bacterium, virus, fungus, protozoa, or alga as the active pesticidal ingredient; and more recently, plant-pesticides, which are certain pesticidal substances expressed in transgenic plants to confer resistance to a plant pest.

Since most biological pesticides display a more narrow host range than chemical pesticides, natural predators and beneficial insect species are less at risk. However, the extremely narrow host range typical of some biological pesticides may be considered disadvantageous not only from an agronomic pest control viewpoint, but also from a commercial perspective. Also, some biological pesticides are less stable than conventional chemical pesticides, so that shelf-life is reduced and storage and handling may be an additional cost. Other disadvantages include the fact that, when compared to chemical pesticides, some biopesticides work slowly on the targeted pests, may be more rapidly degraded in the environment, and may require careful monitoring for correct application. Plant pesticides, however, offer comparable and often superior pest control compared to conventional chemical pesticides.

New pest management methods being developed focus on biological approaches, including the use of biotechnology to alter either the genome of the plant or the microbial pesticide active ingredient. Biotechnology, as defined here, refers to those methods or techniques that use living organisms or substances from such organisms that make or modify a product(s), to improve plants or to develop microorganisms for specific uses. The new tools of molecular biology, with the capability of effecting genetic changes that are precise and rapid, can help significantly in the development of new pest control strategies for agricultural crops.

Biotechnology can be used to develop more efficacious or potent microbial pesticides, to improve the physiological tolerance of biological control agents to stresses encountered in nature, and to expand host range. The tools of biotechnology can be used to improve the delivery of the active pesticidal ingredient of the biological pesticide to the target. One example includes the various application methods used to deliver the insect toxin produced by the bacterium *Bacillus thuringiensis* (commonly referred to as B.t.), to the insect pest. *Bacillus thuringiensis* toxin genes have also been incorporated into the genomes of a variety of crops such as cotton, corn, and potatoes, and when the toxins are expressed, the crop is protected against susceptible herbivorous insect pests. Another example is the genetic engineering of baculoviruses, which have been altered to express the scorpion-toxin gene to accelerate their lethal effects on lepidopteran larvae susceptible to this family of insect viruses.

The purpose of this chapter is to discuss the current registration process of naturally occurring and genetically altered biopesticides by the Office of Pesticides Program (OPP) at the EPA. This review will discuss the data and information appropriate for the evaluation of human health and environmental risks associated with the widespread use and distribution of biopesticides, and the existing mechanisms and incentives that encourage the development and registration of these pesticides. In addition, case studies will be presented to demonstrate the mechanisms of the regulatory process for biopesticidal products derived from biotechnology.

11.2 OVERVIEW OF THE REGULATORY PROCESS

Registration actions for all biological pesticide products are handled in the Biopesticides and Pollution Prevention Division (BPPD) of OPP (http://www.epa.gov/oppbppd1/biopesticides). BPPD is a multidisciplinary division with science reviewers, regulatory and pollution prevention staff working together to streamline the registration and reregistration of biopesticides while encouraging their development and use. Registration of biopesticide products, whether naturally occurring or derived from the use of biotechnology, generally involves a presubmission conference, data development, application preparation and submission, followed by an Agency screen of the application, data review and decision regarding the registration.

11.2.1 Presubmission Conference

Although not mandatory, a presubmission conference with the appropriate Registration Action Leader (RAL) of BPPD is recommended before developing the required human health and safety data and preparing the application. The presubmission conference is important if the applicant is trying to determine whether the pesticidal product is a conventional chemical pesticide or a biochemical pesticide, contains a new active or inert ingredient, provides a new use of a currently registered pesticide product, or represents a "me-too" analog of an existing, registered product. The presubmission conference also provides the registrant an opportunity to develop a proposed data set with input from Agency scientists that will address the perceived risks associated with an active pesticidal ingredient.

During the past few years, there has been renewed interest in the use of biochemical pesticides as effective pest control agents. Several pheromone products have been marketed primarily because of the development of resistance to conventional chemical pesticides in the target pest(s) and adverse environmental effects caused by these conventional pesticides. This renewed interest is reflected in the number of requests made by registrants for classification of their active pesticidal ingredient as biochemical pesticides. If a registrant believes that their product meets the criteria for classification as a biochemical pesticide, the Agency can be requested to make such a determination. The advantage of having an active ingredient classified as a biochemical pesticide vs. a conventional chemical pesticide resides in the potential for reduced data requirements for the former group. The Agency recommends that the registrant consult with BPPD on the format and amount of information needed to justify a biochemical pesticide classification (for further information see McClintock et al., 1994).

11.2.2 Data Development

The generic and product-specific data requirements for biochemical and microbial pesticides are specified in 40 CFR, Parts 158.690 and 158.740, respectively. This information specifies the types of studies and data the Agency requires in order to make regulatory judgments about the risks and benefits of various kinds of pesticide products and to determine whether to approve an experimental-use permit (EUP) or registration application. These data and information address concerns relating to the identity, composition, potential adverse effects, and environmental fate of the biopesticide. The data requirements, or studies to be completed to support an EUP or registration of a biopesticide, are determined based on the proposed use pattern. A complete description of all data requirements and study protocols for microbial and biochemical pesticides is available in advisory documents (collectively referred to as Pesticide Assessment Guidelines) through the National Technical Information Service, U.S. Department of Commerce, Springfield, VA. These documents are also available on the Internet at the EPA's homepage under guidelines in the Office of Prevention, Pesticides and Toxic Substances (OPPTS) section (http://www.epa.gov/opptsfrs/home/guidelin.htm.). It should be noted that specific guidelines have not been developed for genetically modified plants expressing pesticidal traits, termed plant-pesticides, by EPA.

11.2.3 Application Preparation and Screening Process

Any person seeking to obtain a registration for a new pesticide product must submit an application for registration that contains information on the applicant, the authorized agent, if appropriate, various forms, and a listing of the data submitted with the application along with a brief description of the results of the studies (40 CFR 152.50). Each application must be formatted correctly as described in Pesticide Registration Notice 86-5, and any information claimed as confidential business information must be properly identified. If the product is intended for food or feed crop use, or if the intended use of the pesticide may be expected to result, directly or indirectly, in pesticide residues in or on food or feed, the applicant must submit a statement indicating whether such residues are covered by a tolerance or an exemption from a tolerance regulation issued under Section 408 of the Federal Food, Drug and Cosmetic Act (FFDCA) as amended by the Food Quality Protection Act (FQPA) in 1996. If such residues have not been authorized, the application must also be accompanied by a petition for the establishment of appropriate tolerances or exemptions in accordance with Part 180 of 40 CFR. Tolerance petitions or an exemption from the requirement of a tolerance are required for an EUP if the treated crop will enter commerce. A tolerance petition must also accompany a registration involving a new active ingredient or an application involving a change in the food or feed use pattern of a currently registered pesticide. A summary of the human health risk endpoints, as outlined in FQPA, should accompany tolerance petitions in both paper and electronic formats.

The Food Quality Protection Act amendments to the FFDCA have changed some of the requirements for determining a pesticide food tolerance, including what information must be submitted with a tolerance petition. Among the changes was the specification of nine points to be covered for every tolerance determination, including the applicant's original tolerance petition. The nine points include the following areas of information: (1) the validity, completeness, and reliability of the available data from studies of the pesticide chemical; (2) the nature of any toxic effects shown to be caused by the pesticide; (3) available information concerning the relationship of the observed toxic effects to human risk; (4) information concerning the dietary consumption patterns of consumers, including major identifiable subgroups of consumers; (5) available information concerning the cumulative effects of pesticides and other substances having a common mechanism of toxicity; (6) available information concerning the aggregate exposure to the pesticide chemical and related substances, including dietary exposure and other non-occupational exposures; (7) available information concerning the variability of sensitivities of major identifiable subgroups of consumers, including infants and children; (8) an examination of any possible endocrine effects related to the pesticide; and (9) safety factors that are generally recognized as appropriate for the use of animal experimentation data. Once the data are reviewed and a determination for a food tolerance is made, a publication of the scientific findings to justify the tolerance must also include a final accounting of the nine FQPA points for that pesticide.

Upon receipt of an EUP or registration application, the Agency examines the information for administrative completeness. This screening is referred to as Front-End Processing (FEP). If data are contained in the submission, it is screened for compliance with Pesticide Registration Notice 86-5 (the standard formatting procedures required when submitting data to the Agency to support a pesticide registration). Within 45 days of receipt, the Agency must notify the applicant in writing with information on the completeness of the application. If complete, the application is forwarded to BPPD for further processing and scientific review. If the application is incomplete or insufficient, the Agency informs the applicant of the identified deficiencies. After the deficiencies are addressed the applicant can submit a revised application. Applications deemed complete, but which have studies that do not pass

PR Notice 86-5, are forwarded to BPPD. BPPD then notifies the applicant of the formatting deficiencies.

11.3 DATA REQUIREMENTS FOR MICROBIAL PESTICIDES

The information and/or data required by the EPA for an EUP or for registration of microbial pesticides includes a thorough taxonomic characterization of the active microbial ingredient, as well as a description of the manufacturing process, including quality control procedures used to minimize the presence of contaminating organisms. Newly prepared batches or lots of manufactured microbial pesticides are required to be screened for the presence of likely contaminants, including human pathogens. In addition, the potential pathogenicity and toxicity of the microbe are determined by testing the active microbial ingredient together with fermentation medium in laboratory animals and nontarget organisms. Guidelines for each of the subject matters discussed below are available in Subdivision M (U.S. EPA, 1989) or on the Internet at the EPA's homepage under the OPPTS section (http://www.epa. gov/opptsfrs/home/guidelin.htm.).

11.3.1 Product Identity/Analysis

The product identity/analysis requirement for a microbial pesticide includes submission of detailed information on the identity and characterization of the active and inert ingredients, a description of the manufacturing process, including any unintentional ingredients formed, and if appropriate, specification of the analytical method used. The product analysis requirement should include data and/or information on the taxonomic classification of the microbe, including results of identification methods such as biochemical and morphological tests, serotype, composition, and strain of the microorganism, and the unique nature and composition of the active microbial ingredient.

For microorganisms genetically altered to enhance their pesticidal activity, characterization should include information as described in 40 CFR 172.48. This section delineates the data necessary for a notification to the Agency prior to field release of a genetically modified microbial pesticide to determine if an EUP is required. These data include, but are not limited to, identification of the donor and recipient organisms, information on the inserted gene sequence(s) to be expressed, and, if appropriate, regulatory regions or sequences to be inserted into the recipient microorganism, as well as information on the level of expression of the inserted gene or gene sequences. This information should also include a description of the phenotypic traits gained or lost and the genetic stability of the altered genetic region.

There are certain microorganisms that are not readily amenable to adequate characterization from standard taxonomic procedures, either because of inadequate or nonsustainable culture systems, growth only in association with a particular host organism, or the system of taxonomy used is based on morphological characteristics of which the microorganism has few to no unique structures. Therefore, because historical experience often is lacking on adverse effects that might occur when humans are exposed to high numbers of such environmentally isolated microorganisms, the Agency requires a battery of acute pathogenicity/toxicity studies in surrogate laboratory animals.

11.3.2 Description of Manufacturing Process

While the taxonomic data and the acute mammalian toxicity studies provide information useful in assessing toxicity of protein components of the active microbial ingredient, it is information on the manufacturing process that addresses the likelihood of adverse effects from the presence of contaminating microorganisms. Particular attention is given to the measures that pesticide manufacturers use to minimize the potential for growth of contaminating organisms. A description of the quality control/quality assurance procedures used to ensure a uniform or standardized product should include: (1) proper maintenance of stock and "seed" cultures used to begin the fermentation of a microbial agent as well as analyses for biological purity; (2) a description of sterilization procedures of growth media and of fermentation vessels; (3) monitoring of appropriate physical parameters conditions during fermentation ($e.g., O_2, CO_2$, pH); and (4) analysis of lots for quality assurance/quality control when fermentation is completed. The Agency requests that the pesticide manufacturer present this information as it provides a framework for a discussion on the likelihood of toxic or sensitizing materials arising from growth of contaminating microorganisms in the pesticide product. If the standardization technique(s) includes a bioassay against a target pest for product acceptance, these methods should be described. EPA is particularly interested in the QA/QC procedures that control or remove ingredients that may be toxic or sensitizing to humans and other nontarget organisms.

If the production method can support growth of human or animal pathogens each batch of a microbial pesticide should be analyzed for the presence of pathogens (e.g., *Shigella*, *Salmonella*, and *Vibrio* or an indicator organism) and for unexpected toxins (via injection into laboratory animals). The applications also should state proposed methodologies for detecting these pathogens, and/or their elimination from the production batch if contaminated batches are not discarded.

For *B. thuringiensis* fermentation batches, each lot is tested "...by subcutaneous injection of at least 1 million spores into each of five laboratory test mice." The test results should show "...no evidence of infection or injury in the test animals when observed for 7 days following injection" (40 CFR 180.1011). In addition each "master seed lot" is screened for the isolate's ability to produce β-exotoxin or, if appropriate, production batches are periodically examined for the presence of β-exotoxin to ensure that manufacturing procedures eliminate the exotoxin from the final product. Other specific issues or data related to the registration and reregistration of *B. thuringiensis* is discussed in the Reregistration Eligibility Document for *Bacillus thuringiensis* (U.S. EPA, 1998a).

11.3.3 Toxicity Testing of Microbial Pesticides in Laboratory Animals

Table 11.1   Mammalian Toxicology Data Requirements for Microbial Pesticides (40 CFR 158.740)

Acute toxicity studies	Guideline Reference No.*
Tier I Studies	
Acute oral toxicity/pathogenicity (rat)	152-10 (885.3050)
Acute dermal toxicity (rat/mouse)	152-11 (885.3100)
Acute pulmonary toxicity/pathogenicity (rat/mouse)	152-12 (885.3150)
Acute injection toxicity/pathogenicity (rat/mouse)	152-13 (885.3200)
Primary eye irritation (rabbit)	152-14
Reporting of hypersensitivity incidents	152-15 (885.3400)
Cell culture tests with viral pest control agents	152-16 (885.3500)
Tier II Studies	
Acute toxicity	152-20 (885.3550)
Subchronic toxicity/pathogenicity studies	152-21 (885.3600)
Tier III Studies	
Reproductive and fertility effects	152-30 (885.3650)
Oncogenicity study	152-31
Immunodeficiency studies	152-32
Primate infectivity/pathogenicity study	152-33

* Revised guideline numbers are listed in parentheses

The data and information obtained from the product characterization can be used to establish the mammalian toxicology data necessary to determine the risks associated with human and domestic animal exposure. The current mammalian toxicology data requirements are structured in a tiered testing system so as to provide focus only on those studies considered necessary for an adequate human health risk evaluation (Table 11.1). Studies that are usually required in Tier I for registration of a microbial pesticide for use on a terrestrial food crop include acute infectivity/toxicity tests with the technical grade active ingredient (oral, pulmonary, and injection exposures) and mammalian cell culture studies for pesticides containing an insect virus as the active pesticidal ingredient. In addition, tests on the toxicity and irritation of the formulated end-use product are required. After dosing, test animals are evaluated by recording mortality, body weight gain, and making cageside observations for clinical signs of toxicity. Test animals are also assessed by performing a gross necropsy and evaluating the pattern of clearance of the microorganism from the animals. For the latter endpoint, the microorganism is periodically enumerated from appropriate organs, tissues, and body fluids of test animals to verify the lack of pathogenicity/infectivity or persistence in mammals and to document normal immunological processing of the microbial inoculum. These studies would also be required at the experimental-use permit (EUP) stage, if the treated food crop is not to be destroyed and a food tolerance is not in place.

The information from these acute toxicity/pathogenicity studies allows an assessment for the potential of the microorganism to be pathogenic to, or toxic to, mammals. In most cases, lack of adverse effects allows for the reasonable conclusion that the protein components of the microorganism and fermentation residues are not toxic to mammals. However, if toxicity is observed in the test animals — in the absence of signs of pathogenicity — then the toxic components in the test material are to be identified, and, to the extent practical, isolated to determine an LD_{50} value. Further testing in laboratory animals with the toxic components usually will be required to provide an estimate of the amount of material needed to elicit toxic or lethal effects.

The potential toxicity of proteins and other components in the growth or fermentation medium can be evaluated by including the growth/fermentation materials in the dosing material for the acute oral, pulmonary, or injection studies. It is important to enumerate the number of microbial units (e.g., colony-forming units, plaque-forming units) in the dosing material. It may be inappropriate to include significant amounts of fermentation ingredients when dosing rodents via the intravenous route, since lethality from nonspecific toxicity may occur. For example, particulates in the fermentation material may result in mechanical blockage of capillaries. On some occasions nonspecific toxicity may result from reaction to injection of significant amounts of foreign protein into the bloodstream. Also, it should be expected that intravenous injection of large numbers of Gram-negative bacteria would cause rapid mortality due to the shock reaction to the lipopolysaccharide (endotoxin) component of bacterial cell wall material.

Hypersensitivity (i.e., dermal sensitization) studies are generally not required for registration of microbial pesticide products, since injection induction and challenge with microbial pesticides that include proteinaceous components into the commonly used laboratory animal (i.e., guinea pig) would be expected to yield a positive response. Conversely, topical induction and challenge with the active microbial ingredient would most likely lead to a negative response. This, coupled with the historical experience with fermentation products have allowed for the conclusion that reporting of observed allergic responses to microbial pesticides during manufacture and use should be adequate to address the potential risk for dermal sensitization. However, registrants must submit to EPA any information/data on incidents of hypersensitivity, including immediate-type and delayed-type reactions of humans or domestic animals that occur during the production or testing of the technical grade of the active ingredient, the manufacturing-use product, or the end-use product. For incident reporting, refer to the requirements in connection with Section 6(a) (2) of FIFRA and 40 CFR Part 159.

Cell culture tests are required to support the registration of products whose active ingredient is a virus (e.g., baculovirus). These studies provide information on the ability of these viral agents to infect, replicate in, transform, or cause toxicity in, mammalian cell lines. Using the most infectious form or preparation of the virus that gives optimal infection in a susceptible insect cell culture or insect (if a cell line is not available), human or mammalian cell lines are challenged and observed daily for the appearance of cytopathic or cytotoxic effects as well as the ability of the virus to infect or replicate in the host cell. Cytopathic effects include such endpoints as morphological or biochemical changes, and include but are not limited to, cell growth, attachment, morphology, nucleus size and shape, and cellular processes such as macromolecular synthesis. Toxicity evaluation focuses on the ability of the virus to inflict injury or damage to host cells where infection by, and/or replication of the virus are not necessarily required. Toxicity can also be the ability of non-viral components of a preparation to inflict injury or damage to the host cell(s). These non-viral components should be minimal in the cell culture tests, since these tests require inoculation with the most infectious form of the virus, usually a purified extract of the expected product.

Prior to viral challenge, the inoculum should be titered by the most sensitive assay available. When a plaque assay for the virus is available, a minimum of five plaque-forming units (PFUs)/mammalian cell is required. If a plaque assay is unavailable, seven times the LD_{50} unit from the permissive insect host system per mammalian cell can be used as a dose. For each series of tests, the viral inoculum should be tested in the permissive cell line or host organism as a positive control and for direct reference to the data obtained from the vertebrate cell lines. Current protocols for these studies are found in the Cell Culture protocol of the Toxicity Test Guidelines for the Office of Prevention, Pesticides and Toxic Substances at OPPTS 885.3500 or Subdivision M of the Pesticide Testing Guidelines: Microbial and Biochemical Pest Control Agents (U.S. EPA, 1989). These guidelines can also be found on the Internet at the EPA's homepage under the OPPTS section for microbial guidelines at OPPTS 885.3500 (http://www.epa.gov/opptsfrs/home/guidelin.htm.). Additional information and procedures describing assays of insect virus for toxic effects in mammalian cells are described elsewhere (Hartig et al., 1989).

11.3.4 Nontarget Organism Data Requirements

The data and information required to assess hazards to nontarget organisms are derived from tests to determine pesticidal effects on birds, mammals, fish, terrestrial and aquatic invertebrates, and plants. These tests include short-term acute, subacute, reproduction, simulated and/or actual field studies arranged in a tier system that progresses from the basic laboratory tests to the applied field tests (Table 11.2). The test species are those expected to be exposed and can include indicator species such as bobwhite quail, mallard duck, sunfish, rainbow trout, *Daphnia*, honeybee, nontarget insects and nontarget plants. For genetically altered microorganisms, information on the toxicity of the pesticidal substance produced, or modified as a result of the genetic insertion, would be required as well as the fate and effect of the inserted genetic material and the resulting recombinant to nontarget organisms and the environment.

In the acute toxicity/pathogenicity tests currently required, avian wildlife are exposed through the oral and sometimes the respiratory tract. The avian oral toxicity/pathogenicity study provides data on any direct toxic effects to avian wildlife following oral exposure to the naturally occurring or genetically modified microorganism or any toxins that may be produced during fermentation. This test would also provide data on pathogenic effects due to the microbial agent following oral exposure. The avian respiratory pathogenicity test provides information on the pathogenic effects of the active microbial ingredient on birds following exposure due to spray drift or aerosolation. The duration of both the avian oral and respiratory studies

Table 11.2   Nontarget Organism Data Requirements (40 CFR 158.740)

	Guideline Reference No.*
Tier I Studies	
Avian oral toxicity/pathogenicity (Bobwhite quail/mallard duck)	154-16 (885.4050)
Avian respiratory pathogenicity (Bobwhite quail/mallard duck)	154-17 (885.4100)
Wild mammal toxicity/pathogenicity	154-18 (885.4150)
Freshwater fish testing (rainbow trout)	154-19 (885.4200)
Freshwater aquatic invertebrate	154-20 (885.4240)
Estuarine and marine animal test	154-21 (885.4280)
Nontarget plant studies	154-22 (885.4300)
Nontarget insect testing	154-23 (885.4340)
Honeybee testing	154-24 (885.4380)
Tier II Studies	
Terrestrial environmental testing	155-18
Freshwater environmental testing	155-19
Marine or estuarine environmental expression	155-20
Tier III Studies	
Terrestrial wildlife and aquatic organism testing	154-25
Chronic avian pathogenicity and reproduction test	154-26 (885.4600)
Aquatic invertebrate range testing	154-27 (885.4650)
Fish life cycle studies	154-28 (885.4700)
Aquatic ecosystem test	154-29 (885.4750)
Nontarget plant studies	154-31
Tier IV Studies	
Simulated and actual field tests (Birds and mammals)	154-33
Simulated and actual field tests (Aquatic organisms)	154-34
Simulated and actual field tests (Insect predators and parasites)	154-35
Simulated and actual field tests (Insect pollinators)	154-36

* Revised Guideline Numbers are listed in parentheses

should be at least 30 days to permit time for incubation, infection, and manifestation of pathogenic effects in the test organism. In the instances where pathogenesis is suspected, attempts should be made to isolate the causative organism to determine if it is the active microbial pesticide ingredient.

In both the avian oral toxicity/pathogenicity and respiratory pathogenicity tests, the test animals are evaluated by noting mortality, changes in behavior, pathogenic or toxic effects, gross necropsy, and histopathological examination, including culture and isolation of the causal microbe from exposure sites, tissues, or other organs showing anatomical or physiological abnormalities. In cases where cell or tissue preferences are known or suspected, those tissues should be examined whether or not gross anatomical or physiological changes are observed. If no toxic or pathogenic effects are observed after exposure via oral and respiratory routes, no further testing in birds is required. If effects are observed, Tier II environmental expressions tests would be required.

Data on wild mammal toxicity/pathogenicity are required on a case-by-case basis when data indicate that there is considerable variation in the sensitivity of different mammalian species to the effects of a microbial-based pesticide or where wild mammals would be expected to be exposed to a high dose under normal use. However, the toxicity/pathogenicity data in laboratory rodents submitted to evaluate hazards to humans are normally adequate to indicate potential hazards to wild mammals. Usually if no toxicity/pathogenicity effects are observed in these tests, no further testing of wild mammals would be required.

Important considerations in aquatic studies is the need to keep the microbial pesticide test substance in suspension and to measure the actual concentration of the test substance in the water column. The actual measured and the nominal concentration are usually different. The challenge of keeping as much material in suspension as possible is more difficult for *Daphnia* than for fish. For microbial pesticides applied in terrestrial use patterns, where direct aquatic exposure is not anticipated, one freshwater fish and one freshwater aquatic invertebrate should be tested to assess toxicity and pathogenicity. These tests should be conducted as 30-day (for fish) or 21-day (for aquatic invertebrate) static renewal bioassays where the microbial inoculum is administered as a suspension in water, and also in the diet for fish in the form of diseased host insects or treated feed. If mortality is observed during the course of the study, the cause of death (e.g., toxicity, pathogenicity) should be determined, if possible, and reisolation of the microorganism from the test organism's tissues should be attempted. Individual test animals should be removed periodically, if necessary, throughout the test period and at test termination for examination to assess pathogenicity.

Assessment of potential risk to nontarget insects from the use of naturally occurring and/or recombinant microorganisms that are insect pathogens is also an environmental concern and is evaluated by an examination of the published scientific literature and toxicity/pathogenicity studies. For recombinant microbes, several issues need to be considered prior to field trials and widespread commercial use. Such issues include modification of host range, stability and persistence of the microbial construct in the environment that could increase its potential for uncontrolled spread, and the potential for genetic exchange of the foreign insecticidal gene with other naturally occurring microbes. Similar concerns exist for assessing the potential hazards to nontarget plants for other microbial products that are potential plant pathogens.

In spite of the factors cited above, the nontarget organism tier testing scheme is adequate to address many of these issues and concerns. The tier-testing scheme is based on a fairly extensive first tier that assesses toxicity and pathogenicity of the microbe to the honeybee and to three species of predaceous and parasitic insects. Selection of the predator/parasitic species should be representative of groups that will be exposed under the condition of proposed use and, if possible, that have some relationship to the target pest. Tier I testing also includes toxicity/pathogenicity testing with *Daphnia*, and, if available and appropriate, an aquatic insect species depending on use pattern. Data derived from the Tier I tests are used in conjunction with available information on use patterns, specificity of host range, fate, and other factors, to assess potential for adverse effects. If the results indicate no adverse effects, no further testing is required. By contrast, if toxicity or pathogenic effects are observed, Tier II testing, environmental expression, would be required. It should also be noted that the best routes of exposure in the Tier I tests will depend on the developmental stage and location of the nontarget insect.

The data and information obtained from the nontarget organism and environmental expression tests described above allow the Agency to assess potential hazards from microbial pesticide exposure. However, in some cases data waivers may be appropriate for nontarget testing requirements. Where nontarget fish, plant, insect, or bird exposure to the microbe in question can be documented in the scientific literature, and there are no reports of pathogenicity, a waiver may be entertained by the Agency for pathogenicity testing. Consequently, nontarget pathogenicity testing may not be necessary if the microbial pesticide's natural environmental distribution includes the habitat of the nontarget organism species normally tested in Tier I pathogenicity studies, and the microbial pesticide has never been found in association with nontarget organism infectivity and disease. Waivers may also be justified if there is a reasonable argument that the nontarget organisms for that particular environmental niche will not be exposed to the microbial pesticide. Toxicity, in addition to pathogenicity, must be addressed due to the impact that the fermentation and post-fermentation processing may have on the production and elimination or concentration of microbial toxins and metabolites.

11.4   DATA REQUIREMENTS FOR BIOCHEMICAL PESTICIDES

Biochemical pesticides are distinguished from conventional chemical pesticides by their natural occurrence and nontoxic or indirect mode of action *to the target pest*. Often biochemical pesticides also display a narrow range of target species and are effective at low application rates. Due to the unique characteristics of biochemical pesticides, OPP recognized that appropriate and, in some instances, reduced data requirements were justified to adequately evaluate the safety of these pest control agents. Part 158.690 of 40 CFR specifies the kind of data and information appropriate for the evaluation of human health and environmental risks associated with the widespread use and distribution of biochemical pesticides. The fundamental information necessary to evaluate such products for such risks include product analysis information and data on the toxicity of the active ingredient to laboratory mammals and other nontarget organisms. The key information is summarized below; however, for a complete description of study protocols for biochemical pesticides refer to the Pesticide Assessment Guidelines, Subdivision M: Guidelines for Testing Biorational Pesticides (U.S. EPA, 1983) or EPA's homepage under the appropriate section of the OPPTS guidelines (http://www.epa.gov/opptsfrs/home/guidelin.htm).

11.4.1 Classification of Active Ingredients as Biochemical Pesticides

Active pesticidal ingredients isolated from a natural source and demonstrated to be nontoxic to the target pest would be classified as a biochemical pesticide. Insect pheromones, certain plant growth regulators such as auxins, gibberellins, and cytokinins, and common food sources or components, such as garlic and cinnamon, are also by definition biochemical pesticides. However, some plant-extracted materials, although of "natural" origin, are not necessarily pesticidal by a nontoxic or indirect mode of action. For example, pyrethrins mitigate target pests via a toxic mechanism of action. Control of a pest by simple suffocation (e.g., by vegetable oil) also would be considered equivalent to a nontoxic mechanism of activity. Antibiotics from microorganisms, if used as pesticides, would not be considered biochemical pesticides, because by definition, these substances act via a toxic mode of action to the target pest.

Although natural occurrence is a criterion for classification as a biochemical pesticide, a number of active biochemical ingredients have been chemically synthesized. If synthesized, the active ingredient must be structurally similar to, and functionally identical to, a naturally occurring counterpart. For example, the active ingredient indole-3-butyric acid is classified as a biochemical pesticide, since the synthetic plant growth regulator is a structural analog of indole acetic acid (auxin) and also mimics the function of the natural plant hormone. In some instances, the synthesis of a biochemical pesticide or a structural analog, rather than isolation from naturally occurring material, may be preferred because sufficient quantities of the material can be generated economically and in a more highly purified form (e.g., an insect pheromone) or may yield products with increased efficacy and longevity in the environment (e.g., modified forms of the neem seed extract, azadirachtin).

The precise mode of action of a pesticidal active ingredient against a target pest may not be readily apparent, and consequently the determination of a nontoxic mode of activity cannot be precisely elucidated. In these cases, the best available scientific information and knowledge are applied to make the most appropriate decision on the candidate material. It is possible to conclude that a pesticidal substance is best classified as a biochemical pesticide, even though the precise mode of action against the target pest is not known. The pesticidal active ingredients that have thus far been classified as biochemical pesticides by the EPA are listed in Table 11.3.

11.4.2 General Guidance for Classification

If an active pesticidal ingredient meets the criteria for classification as a biochemical pesticide, the registrant can request that the Agency make such a determination to facilitate the review and processing within the Agency. A formal request, containing the basic information that supports the claim of natural occurrence and nontoxic mode of action to the target pest, can be submitted to the Biochemical Pesticides Branch of BPPD. The final decision for or against classification as a biochemical pesticide is then conveyed back to the petitioner through BPPD. If warranted, a registrant can contact BPPD directly for preliminary guidance on classification issues.

Table 11.3   Biochemical Active Ingredients

Chemical Name	Target Pest(s)
I. Pheromones	
Dodecenyl acetates, aldehydes, alcohols, and isomers	Grape berry moth, western pine shoot borer, codling moth, oriental fruit moth
Isomers of trimethyl dodecatriene	Tetranychid mite, aphids
Hexadecanyl acetates, aldehydes, alcohols, and isomers	Pink bollworm, artichoke plume moth
(R,Z)-5-(1-Decenyl)dihydro-2-(3)-furanone	Japanese beetle
Octadecadienyl acetates	Peachtree borer
Periplanone B	American cockroach
Tridecenyl acetates, aldehydes, and isomers	Tomato pinworm, tobacco budworm, cotton bollworm
Tetradecenyl acetate and alcohols	Grape berry moth, tufted apple bud moth
(Z)-9-Tricosene	Housefly
(E)-5-Decenol	Peach twig borer
(E)-5-Decenyl acetate	
Grandlure	Cotton boll weevil
Musculure	Housefly
Cis-7,8-epoxy-2-methyloctadecane (Disparlure)	Gypsy moth
II. Plant growth regulators	
N-6-Benzyladenine	Various ornamental plants and food crops
Natural plant extracts containing gibberellins, zeatins, IAA	
Cytokinin (6-Furfural(amino)purine)	
Ethylene	
Gibberellins and salts	
Indole-3-butyric acid	
Kinetin	
Pelargonic acid	
Aminoethoxyvinylglycine	Ornamentals, apples
1,4-Dimethylnaphthalene	Potato sprout inhibitor
Acetic acid	Herbicide
III. Floral lures/attractants/repellents	
Capsaicin	Insects, dog, bird
Castor oil	Moles
Cedarwood oil	Fleas and moths
Cinnamaldehyde	Corn rootworm, spotted cucumber beetle
Cinnamyl alcohol	Corn rootworm, spotted cucumber beetle
Dried blood	Rabbits, dogs
Eucalyptus oil	
Eugenol (2-methyl-4-(2-propenyl) phenol)	Japanese beetle
Garlic	Birds
Indole	Corn rootworm, spotted cucumber beetle
Lemongrass oil	Moths
Meat meal	Deer, rabbits, racoons, birds
Methyl anthranilate	Birds
4-Methoxybenzenethanol	Corn rootworm
4-Methyl cinnamaldehyde	Corn rootworm, spotted cucumber beetle
4-Methyl phenethyl alcohol	Corn rootworm, spotted cucumber beetle

Table 11.3 (continued)   Biochemical Active Ingredients

Chemical Name	Target Pest(s)
1-Octene-3-ol (octenol)	Mosquitoes, biting flies
Oil of citronella	Mosquitoes, ticks
Oil of geranium (geraniol)	Japanese beetle
3-Phenyl propanol	Corn rootworm, spotted cucumber beetle
Putrescent whole egg solids	Big game animals
Red pepper	Deer, rabbits, raccoons, birds
1,2,4 trimethyoxybenzene	Corn rootworm, spotted cucumber beetle

IV. Natural insect regulators

Azadirachtin	Insects
Dihydroazadirachtin	Insects
Trimethyl-dodecadienoates	
Hydroprene	Roaches
Kinoprene	Whiteflies, aphids, scales, gnats
Methoprene	Mosquitoes, hornflies

V. Fungicides

Clarified Hydropholic Extract of Neem Oil	Fungi, insects
Neem Oil	Fungi
Potassium Bicarbonate	Fungi
Sodium Bicarbonate	Fungi

VI. Other

Calcium sulfate	Fleas
Cellulose gum (sodium carboxymethylcellulose)	Insects, mites
Vegetable (soybean) oil	Insects, mites
Lactic acid	Antimicrobial

The kind of information and data essential for classification of active ingredients as biochemical pesticides include documentation by citation to, and submission of, references from the published literature that support the natural occurrence of the substance as well as indications that the active ingredient acts by a nontoxic mode of action to the target. If the active ingredient is chemically synthesized, the molecular structure of the substance and its structural relationship to a naturally occurring substance should be submitted along with a brief description of the manufacturing process. If the active ingredient is extracted as a mixture of substances from biological material(s), a description of the "manufacturing process" should include the nature of the source substance(s) to be extracted, extraction materials, any subsequent purification process and materials used, and a characterization of the extracted substance(s) using appropriate analytical methods.

Table 11.4　Pheromone Regulatory Relief Action Plan: A Historical Perspective

Action	Status
1. Exemption from requirement of a tolerance for inert materials in polymeric matrix dispensers	Final Rule published December 8, 1993. 58 FR 64493
2. Exempt from requirement of tolerance pheromones in polymeric matrix dispensers	Final Rule published March 30, 1994. 59 FR 14757
3. Raise EUP limit to 250 acres for pheromones in polymeric matrix dispensers	FR Notice published January 26, 1994. 59 FR 3681
4. Raise EUP limit to 250 acres for testing of nonfood use broadcast pheromones	FR Notice published July 7, 1994. 59 FR 34182
5. Raise EUP acreage limit to 250 acres for straight-chained lepidopteran pheromones (sprayables)	FR Notice published August 30, 1995 60 FR 168
6. Tolerance exemption for straight-chained lepidopteran pheromones (sprayables)	FR Notice published September 13, 1995
7. Exemption from requirement of a tolerance for inert polymers in sprayable formulations (beads)	FR Notice published February 21, 1996. 61FR6550.

11.4.3　Classes/Uses of Biochemical Pesticides Exempted from Regulation Under FIFRA

Under specified conditions of use the Agency has determined that arthropod pheromones have been exempted from all provisions of FIFRA. For a summary of the pheromone regulatory relief action plan see Table11.4. As stated in 40 CFR 152.25 (b), Subpart B (July 1, 1991), pheromones, and identical or "substantially similar" compounds, produced by arthropods and used only in traps, are exempt from regulation as long as the substance traps individuals of the same arthropod species and achieves pest control solely by removal of the target pests from the environment via attraction to the trap. The pheromone trap also cannot result in increased levels of pheromones or identical compounds over a significant portion of the treated area. For the purposes of this exemption "substantially similar" means that "...the only differences between the molecular structures are between the stereochemical isomer ratios of the... two compounds..." [40 CFR 152.25 (b) (2)]. The EPA, however, may determine that certain synthetic substances used in traps may possess many characteristics of a pheromone and thus meet the criterion of a "substantially similar" compound. Finally, products considered as "foods" that attract pests but do not contain active pesticidal ingredients also are exempt from regulation by EPA under FIFRA.

11.4.4　Product Identity/Analysis Data Requirements

The product identity/analysis data for biochemical pesticides closely parallel those for conventional chemical pesticides. The specific guidelines are found in the EPA's homepage under the OPPTS product chemistry series for conventional chemical pesticides or in the Subdivision M Guidelines for Biorational Pesticides (U.S. EPA, 1983). Detailed information about how the active ingredient is produced and

Table 11.5   Mammalian Toxicology Data Requirements for
             Biochemical Pesticides (40 CFR 158.690)

Acute Toxicity Studies	Guideline Reference No.*
Tier I Studies	
Acute oral (rat)	81-1
Acute dermal (rat/mouse)	81-2
Acute inhalation (rat/mouse)	81-3
Primary eye irritation (rabbit)	81-4
Primary dermal irritation (rabbit/guinea pig)	81-5
Dermal sensitization	81-6
Hypersensitivity incidents	—
Genotoxicity studies	84-2
a. Ames assay	
b. Forward gene mutation assay	
c. *In vivo* cytogenetics assay	
Subchronic Studies	
Immunotoxicity (1 *spp.*)	(880.3550)
90-day feeding, dermal, inhalation (1 *spp.*)	82-1, 82-3, 82-4
Developmental toxicity (1 *spp.*)	83-3
Tier II Studies	
Immune response (rodent)	—
Tier III Studies	
Chronic exposure (rodent)	83-1
Oncogenicity (rodent)	83-2

* Revised Guideline Numbers are listed in parentheses

the quality assurance/quality control techniques used to ensure a uniform or standardized product are required for the manufacturing process description. Product identity/analysis information encompasses three general areas: (1) product identity and composition, (2) analysis and certified limits, and (3) physical and chemical characteristics. Data on product composition include both the active ingredient and any intentionally added inert materials. Each product to be registered must be analyzed for the upper and lower concentrations (certified limits) for both the active ingredient and any intentionally added inert substance. In addition to composition of the final or "end-use" product, the product characterization data includes a description of starting materials, production and formulation process, and a discussion of the possible formation of impurities.

Data on physical and chemical characteristics of the pesticidal active ingredient and end-use products include, when appropriate, information on their physical state, stability, pH, specific gravity, melting/boiling point, flammability, viscosity, vapor pressure, oxidizing and reducing potential, storage stability, and corrosiveness.

11.4.5 Mammalian Toxicology Data Requirements

The current mammalian toxicology data requirements are set forth in 40 CFR 158.690 and are listed in Table 11.5. Specific guidance on methods and procedures for conduct of these studies is described in Subdivision M of the Pesticide Testing Guidelines (U.S. EPA, 1983) or at the EPA's homepage under the appropriate section of the OPPTS guidelines. The toxicology data requirements are structured in a tiered testing system so as to provide focus only on those studies considered necessary for an adequate health risk evaluation. Studies that are usually required in Tier I for registration of a biochemical pesticide for use on a terrestrial food crop include acute toxicity tests (oral, dermal, and inhalation), a primary eye and a dermal irritation study, a battery of genotoxicity studies, an immunotoxicity study, a 90-day feeding study, a developmental toxicity study, and reporting of hypersensitivity incidents. Unless a food tolerance has been established, these studies would also be required at the EUP stage, if the treated food is not to be destroyed.

Specific conditions, qualifications, or exceptions to the designated tests are provided in Part 158.690 (a) (1). For example, the acute oral and dermal toxicity study would not be required if the test material is a gas, or is sufficiently volatile so as to render performance of a test impractical. If the test material is corrosive to skin, then the acute dermal toxicity study and the primary eye and dermal irritation studies would not be required and the product would have appropriate warnings or signal words for these exposures. A dermal sensitization study is required at registration if there is repeated contact with human skin under the conditions of use. Although no specific tests are required, all incidents of hypersensitivity must be reported to the Agency immediately following their occurrence. However, the requirement for allergenic incident reports and specific lack of sensitization with prior wide-scale human exposure could provide the basis for requesting a waiver for the dermal sensitization study.

Studies to determine genotoxicity/mutagenicity are required to support any food/nonfood use if significant human exposure may result or if the active pesticidal ingredient is structurally related to a known mutagen or belongs to a class of chemical compounds containing known mutagens. The genotoxicity battery of studies includes those currently found most useful for evaluating mutagenicity potential of chemical pesticides; namely, gene mutation studies (i.e., the *Salmonella typhimurium* reverse mutation assay (Ames assay)), a forward gene mutation assay with mammalian cells in culture, chromosomal damage assays (i.e., an *in vivo* cytogenetics assay), and other studies evaluating DNA repair or unscheduled DNA synthesis. Current protocols for these mutagenicity studies are found in the U.S. EPA's OPP Health Effects Testing Guidelines (40 CFR Part 158, Subpart F — Genetic Toxicity) or on the Internet at the EPA's homepage under the OPPTS section (http://www.epa.gov/opptsfrs/home/guidelin.htm.).

If repeated human exposure to the pesticide is expected to occur, subchronic studies (90-day feeding, dermal, and/or inhalation) may be required. As with the acute toxicity studies, there are specific conditions, qualifications, or exceptions to the designated subchronic test requirements as described in Part 158.690 (a) (1). For example, the 90-day feeding study is conditionally required for nonfood use, but required if the use of the product results in repeated human exposure by the oral route or the use requires a food tolerance determination. If repeated contact with skin occurs, a 90-day dermal study in the rat is required. Likewise, if there is repeated pulmonary exposure to the pesticide at concentrations that are likely to be toxic, as indicated from the acute inhalation study, a 90-day inhalation study would be required. Although not specifically indicated, the oral and dermal subchronic studies requirements should be significantly reduced if the nature of the test material renders performance of a test impractical (i.e., the material is a gas at room temperature).

Data addressing immunotoxicity are conditionally required to support the registration of a pesticidal product, but essentially becomes required when there is a requirement for any of the subchronic studies reflecting, again, significant human exposure situations. Protocols for the immunotoxicity study are available from the Agency and have been summarized elsewhere (Sjoblad, 1988). Briefly, the study employs either the rat or the mouse as the test animal and assays are performed after 30 days of dosing to evaluate effects of the test substance on humoral, specific cell-mediated, and nonspecific cell-mediated immunity. It should be noted that a developmental toxicity study (Tier I) is required for food use and conditionally required for nonfood use when the use of the product is expected to result in significant exposure to females. If significant adverse effects in the immunotoxicity studies are observed at the Tier I level, a Tier II study may be needed to provide an estimate of risk related to these positive toxicity endpoints.

To assess potential hazard resulting from prolonged and repeated exposure, a chronic exposure study (Tier III) would be required if the potential for adverse effects were found in any of the Tier I subchronic studies and the use pattern indicated significant human exposure. A carcinogenicity study, also in Tier III, is required if the active ingredient (or any metabolites, degradates, or impurities thereof) causes morphological effects indicative of neoplastic potential (i.e., hyperplasia) in the subchronic study test animals, or if carcinogenic potential is indicated in the mutagenicity and/or immunotoxicity studies.

11.4.6 Nontarget Organism Testing

As with nontarget organism testing for microbial pesticides, the purpose of testing is to develop data necessary to assess potential hazard of biochemical pesticides to terrestrial wildlife, aquatic animals, plants, and beneficial insects. The Agency bases the hazard evaluation of biochemical pesticides on tests similar to those required to support registration of conventional chemical pesticides. However, recognizing the nature and nontoxic mode of action of most biochemical pesticides, the Agency has structured the data requirements in a tier-testing scheme. The use of tiered data requirements allow regulatory decisions to be made with fewer tests than for conventional chemical pesticides and results in much lower costs to the registrant and less time for the registration process.

In general, biochemical pesticides control behavior, growth, and/or development of target organisms. Ideally, Tier I tests should be capable of detecting adverse effects resulting from the primary mode of action on the nontarget organisms.

Table 11.6 Nontarget Organism and Environmental Expression
Data Requirements for Biochemical Pesticides (40 CFR 158.690)

	Guideline Reference No.
Tier I Studies	
Avian acute oral toxicity (bobwhite quail/mallard duck)	154-6
Avian dietary toxicity (bobwhite quail/mallard duck)	154-7
Freshwater fish LC_{50} testing (Rainbow trout)	154-8
Freshwater aquatic invertebrate LC_{50} testing	154-9
Nontarget plant studies	154-10
Nontarget insect testing	154-11
Tier II Studies	
Volatility	155-4
Dispenser water leaching	155-5
Absorption-desorption	155-6
Octanol/water partition	155-7
U.V. Absorption	155-8
Hydrolysis	155-9
Aerobic soil metabolism	155-10
Aerobic aquatic metabolism	155-11
Soil photolysis	155-12
Aquatic photolysis	155-13
Tier III Studies	
Terrestrial wildlife testing	154-12
Aquatic animal testing	154-13
Nontarget plant studies	154-14
Nontarget insect testing	154-15

The following criteria are used to determine the need for further testing of biochemical pesticides beyond the first tier:

1. If signs of abnormal behavior are reported in Tier I tests at levels equal to or less than the maximum expected environmental concentrations; or
2. If detrimental growth, developmental, or reproductive effects can be expected, based on Tier I test data, available fate data, use pattern information, results of the mammalian toxicology testing, and the phylogenetic similarity between target pest and nontarget organism(s); or
3. If the maximum expected environmental concentration is equal to or greater than one fifth the LC_{50} values established in Tier I terrestrial wildlife studies, or equal to or greater than one tenth the LC_{50} or EC_{50} values in Tier I aquatic animal studies.

In addition, both Tier I and Tier II tests would be required if the pesticide is not volatile, is applied directly to water, and has proposed high use rates. Tier II testing involves environmental fate testing to estimate environmental concentrations of the biochemical pesticides after application. Tier III consists of further acute, subacute, and chronic laboratory testing on nontarget organisms, and Tier IV consists of applied field tests encompassing both nontarget organisms and environmental fate. The results of each tier of tests must be evaluated to determine if further testing is necessary. Representative test species are dosed at high rates that represent a maximum challenge situation to evaluate adverse effects. Normally, if the results of Tier I testing indicate significant toxicity in the test organism, further testing at a higher tier level is required. The data requirements, as found in 40 CFR 158.740, are outlined in Table 11.6.

11.5 PLANT-PESTICIDES

Since the early 1980s, the introduction and expression of foreign genes in plant cells has been possible through the use of *Agrobacterium*-mediated transformation and biolistic technology. Such transformation technology has been used to genetically engineer plants to express pesticidal substances. The most common examples of pesticidal traits to date involve transgenic plants engineered to provide protection from insect attack (*B. thuringiensis* delta-endotoxin) and resistance to viral infections (viral coat proteins). EPA has published a proposed regulatory system for genetically engineered plants with pesticidal traits (*Federal Register*, November 23, 1994) and has since registered several pest resistance traits expressed in plants. Currently, EPA has no final data requirements or testing guidelines for plant-pesticides and has been advising applicants on a case-by-case basis.

The proposed definition of plant-pesticides includes substances expressed in plants to impart pest resistance as well as the genetic material necessary for its production and expression. The active pesticide ingredient, known as a plant-pesticide, is both the expressed pesticidal substance(s) and the genetic material introduced to produce the substance. The appropriate focus for a determination of hazard and risk is on the expressed pesticidal substance and the potential for gene transfer to other plants leading to new exposures for that pesticidal substance. Other information, as described below, is needed to effectively evaluate potential risks associated with novel human and environmental exposure.

The Agency recognizes that there are many substances in plants that provide resistance to insect or microbial damage and that some plant-produced substances have even been involved in herbicidal activity against other plant species. Some plant-pesticidal substances may occur naturally in food crops currently being consumed, implying a background exposure to these substances. However, some pesticidal traits from microbes, animals, or even other plants when introduced into a new plant species, may represent a novel exposure and perhaps a new risk for human health or the environment. It should be noted that the Agency has proposed not to regulate the plant *per se* but rather the pesticidal substance produced in the plant and the novel exposure that plant, and possibly related species, may provide for the plant-pesticide substance.

The Agency has identified a regulatory system that specifically exempts those compounds that are least likely to present a risk to human health or the environment. The exemptions from FIFRA as proposed include plant-pesticidal substances that are derived from plant species sexually compatible with the plant in question; pesticidal traits that act primarily on the plant as physical barriers (e.g., waxes, hairs), toxin inactivators and receptors responsible for the hypersensitive response; and the coat proteins from plant pathogenic viruses. The Agency has also proposed to exempt from FFDCA requirements, those pesticidal traits derived from sexually compatible plants, coat proteins from plant pathogenic viruses, and nucleic acids associated with the plant-pesticide traits. Fundamental information or data needed for a risk assessment by the Agency is a thorough description of the source and nature of the inserted genes or gene segments and a description of the novel products (e.g., proteins) encoded for by the genetic material. Presuming that the encoded products have been characterized adequately, this information would allow for a reasonable prediction of toxicology issues and for the type of data essential to the evaluation of potential risks.

EPA has divided the pesticidal substances into two categories: proteinaceous pesticides and non-proteinaceous pesticides. This approach is based on the fact that proteins, whether characterized or not, are significant components of human diets and are usually susceptible to acid and enzymatic digestion to amino acids prior to assimilation. Presuming that the new proteinaceous products are adequately characterized, minimum human health concerns would exist unless (1) the proteins have been implicated in mammalian toxicity including food allergy; (2) dietary exposure of the protein, although never implicated in mammalian toxicity through other routes of exposure, has not been documented; or (3) "novel" proteins are created via modification of the primary structure of the natural protein pesticide. Non-proteinaceous pesticidal substances expressed in plants may be evaluated separately in a manner analogous to that for conventional chemical or biochemical pesticides, although none have been submitted to date.

Product characterization is critical for assessing potential risks resulting from exposure of humans to plants expressing novel pesticidal substances. Characterization embraces four basic areas: (1) identification of the donor organism(s) and the gene sequence(s) to be inserted into the recipient plant; (2) identification and description of the vector or delivery system used to move the gene into the recipient plant; (3) identification of the recipient organism, including information on the insertion of the gene sequence; and (4) data and information on the stability and level of expression of the inserted gene sequence. This information is critical for assessing potential risks to humans, domestic animals, and other nontarget organisms exposed to novel plant pesticides. Specific data/information that is helpful for a risk evaluation by the EPA have been previously described (McClintock et al., 1991).

The product characterization data/information can help establish the mammalian toxicology and ecotoxicology data necessary to determine the potential risks associated with human, domestic animal, and nontarget organism exposure to transgenic plant pesticide products. Key factors determining the extent of data requirements would include the nature of the pesticidal product (i.e., purported mode of action, proteinaceous or nonproteinaceous) and whether or not the use pattern will result in dietary and/or nondietary exposure. Since dietary consumption is presumed to be the predominant route of exposure for food and feed crops engineered to express pesticidal substances, the potential toxicity of these unique substances could be assessed by oral toxicity studies using the purified pesticidal substance. For most proteinaceous pesticidal substances an acute oral toxicity test may suffice. Longer term studies such as subchronic feeding may be needed for non-proteinaceous substances with no previous dietary exposure or proteins known to be enzyme inhibitors or impediments to the uptake of vitamins and nutrients. An *in vitro* digestibility assay is needed to provide information about the potential for a protein to survive digestion and potentially induce food allergy. For most proteinaceous plant-pesticides, the Agency foresees no reasonable scenario for significant dermal or pulmonary exposure to a pesticidal substance expressed within the vegetative cells of a plant and would probably not require specific tests to address these routes of exposure. However, if plants were engineered to produce volatile pesticide components, pulmonary exposure might be significant even without a food use and that inhalation exposure may need to be addressed.

Environmental fate (persistence and movement in the environment) and effects (toxicity) endpoints for transgenic plants are often quite different from those used for conventional chemical pesticides. Unlike chemical pesticides where spray drift and movement in groundwater are important, fate endpoints for plant-pesticides address the movement of the gene trait to other crops and/or noncrop plants (biological fate) and stability and movement of the pesticidal product in the environment (chemical fate). Toxicity endpoints address the ability of the pesticide to cause adverse effects to nontarget organisms. Such effects could occur following consumption of the transgenic plant containing the pesticidal product by nontarget organisms.

In general, environmental fate and effects endpoints include, but may not be limited to, (1) pesticidal substance persistence and gene movement in the environment, (2) potential for enhanced weediness, (3) unplanned production of the pesticidal product offsite, leading to exposure to a new group on nontarget organisms, (4) disruption of the ecosystem by the establishment of a new trait in wild relatives, and (5) the effects on nontarget organisms and fate in the environment. While many crop plants are highly domesticated and do not survive outside cultivation, the potential for plant-pesticides to spread to other species and replicate in the environment is an endpoint not applied to conventional pesticides. If outcrossing occurs whereby the novel trait is transferred to a related wild plant species, the pesticidal product may be produced in unintended areas leading to exposure to a new group of nontarget organisms or lead to enhanced weediness if the wild relative is already an agronomic pest. If the engineered trait is transmitted and is able to become established in wild relatives, the newly acquired trait may provide the wild relatives a competitive advantage within the natural plant community and disrupt the ecosystem. The ability of the gene and the expressed product/trait to be acquired and persist are endpoints to consider for plants with new pesticidal traits.

Finally, effects of the gene/trait on nontarget organisms and fate in the environment are addressed by data requirements similar to those required for the registration of other biological pesticides. The suggested studies include an acute avian oral study, an avian dietary study, an acute fish study, an acute freshwater aquatic invertebrate study, a honeybee study, and perhaps nontarget insect studies. Currently, a collembola and earthworm study are performed if crop residue exposure is expected. Also pertinent to an ecological risk assessment is information relating to host range of the pesticidal substance; an assessment of outcrossing potential of the plant carrying the plant-pesticide; an evaluation of the potential competitiveness of the novel trait in the plant community; and an assessment of the ability of the pesticidal substance to degrade or persist in the environment.

11.6 LABELING

Pesticide products subject to FIFRA must bear an EPA approved label. Such products include FIFRA Section 3 commercial products, FIFRA Section 5 experimental-use products, and FIFRA Section 24 (C) special local need products. Required label elements include such items as the EPA registration or experimental-use permit number, precautionary statements, first aid statements, directions for use, and storage and disposal statements.

Labels for plant-pesticides are dealt with differently than other pesticide products. An important feature of EPA's approach to plant-pesticides is that the Agency will not register the plant but rather the plant pesticide. Plant material for plant pesticides approved under an experimental-use permit or seed increase registration must have an EPA approved label. However, for full commercial use, plant-pesticides will not contain a FIFRA-type label accompanying the seed or propagative materials sold in commerce, but rather will contain information that will instruct the grower as to what cultural practices need to be modified when growing the plant with a plant-pesticide. The registered label may require that such information accompany the propagative materials at the time of sale, similar to the information that accompanies articles or seeds treated with conventional pesticides. In general, this informational material describes the plant-pesticide that is expressed, along with the pests controlled.

Informational statements may also indicate to growers that a plant expressing a specific toxin may not need to be sprayed for certain pests, that certain resistance management techniques (such as refugia) should be employed, or that limitations on sale or distribution exist.

11.7 A CASE STUDY: GENETICALLY MODIFIED BACULOVIRUS-BASED INSECTICIDES, AN INDUSTRY PERSPECTIVE

11.7.1 Historical Perspective

The U.S. EPA has registered several insecticides with wild-type baculoviruses as the active ingredient. The first baculovirus-based insecticide, registered in 1975, contained the single-embedded nucleopolyhedrovirus of *Helicoverpa zea* (HzSNPV) and was commercialized by Sandoz Crop Protection under the trade name Elcar®. Other examples include insecticides based on a variety of baculoviruses (e.g., the nucleopolyhedroviruses of *Lymantria dispar*, *Orgyia pseudotsugata*, *Spodoptera exigua*, *Autographa californica*, *Anagrapha falcifera*, and the granulovirus of *Cydia pomonella*.

A major disadvantage using baculoviruses as insecticides is that they are relatively slow in killing the target insect pest. Unlike most synthetic chemical insecticides that exert their action shortly after contact, baculoviruses have to be ingested, and unlike *Bacillus thuringiensis* (Bt), have to replicate in their host insect before death occurs. In the field, feeding cessation may take from 4 to 10 days after application depending on the virus, the host, and ambient temperatures. In the mid-to-late 1980s the concept of expediting the action of baculoviruses by expression of heterologous insecticidal proteins encoded by baculoviruses was tested. Several companies, often in collaboration with academic or government laboratories, started research programs in this area (e.g., Sandoz Crop Protection, American Cyanamid, FMC, biosys, Zeneca, and DuPont), but the first successes were obtained in academic institutions (e.g., Maeda, 1989; Stewart et al., 1991; Tomalski and Miller, 1991; McCutchen et al., 1991). A wide variety of insecticidal genes, including those encoding insect neurohormones, insect regulatory enzymes, arthropod- and sea anemone-derived toxins, Bt δ-endotoxins, the maize mitochondrial gene URF13, and the human c-myc antisense gene have been inserted into the baculovirus genome (for a recent review see van Beek and Hughes, 1998). Perhaps considered as most promising for the near future are insect-specific scorpion toxin genes, which in laboratory assays can reduce the time to cessation of feeding to half that of the wild-type, progenitor virus.

Field evaluation is a critical step in the assessment of the potential of baculovirus-based insecticides for commercialization. In particular, in the case of viruses engineered with the aim to increase efficacy, it was presumed but not known whether a faster speed of action demonstrated in laboratory assays would lead to improved crop protection in the field. Moreover, it was not known *a priori* whether this presumed improved efficacy would manifest itself in reduction of damage due to burrowing larvae (e.g., tobacco budworm or bollworm), leaf-feeding larvae (e.g., cabbage looper, beet armyworm, or diamondback moth) or both.

After pioneering work involving limited field releases of a series of genetically altered viruses in England (Bishop, 1989), the first small-scale field releases in the U.S. were carried out with baculoviruses containing marker genes or gene deletions (Wood et al., 1994). In 1994, American Cyanamid Corporation requested permission from EPA to carry out small-scale field tests with *A. californica* nucleopolyhedrovirus (AcMNPV) encoding the insect-specific toxin gene of the scorpion *Androctonus australis* Hector (vEGTDEL-AaIT; EPA No. 241-NMP-2). After publication of the notification in the *Federal Register*, and after evaluation of the data and comments received from the public, permission for the field test was granted by the EPA and tests were conducted in 1995. In 1995, DuPont Agricultural Products requested a presubmission meeting with the EPA initiating the process of obtaining permission for small-scale field testing of genetically modified baculoviruses, the subject of this review.

11.7.2 Introduction to Genetically Modified Baculoviruses

To appreciate the difference between genetically engineered baculoviruses and their wild-type parental or progenitor viruses it is necessary to start with a brief description of the biology and genetics of baculoviruses, and the methods by which recombinant viruses are generated. The family Baculoviridae is subdivided into two genera, nucleopolyhedrovirus (NPV) and granulovirus (GV; Murphy et al., 1995). Of special interest in the context of this review are the NPVs, since they are amenable to genetic engineering.

During the NPV infection cycle two distinct phenotypes are produced: the occluded and the nonoccluded or budded virus. The occluded phenotype is orally infectious to susceptible insect larvae and consists of crystal-like structures called polyhedra, occlusion bodies (OBs), or polyhedral inclusion bodies (PIBs), within which many virus particles are embedded. The virus particles consist of one (single-embedded) or more (multiple-embedded) rod-shaped nucleocapsids surrounded by a membrane. Each nucleocapsid contains a double-stranded, covalently closed, circular DNA genome between 110 and 160 kb in size. Infection of an insect larva takes place after ingestion of polyhedra, which dissolve due to the high pH in the larval midgut releasing virions that enter the midgut epithelial cells. After virus multiplication in the infected cell nuclei, virus particles bud through the cell membranes and invade other tissues in the insect. During the early stages of the infection cycle, progeny virus particles bud from the cell (budded virus), while in later stages they remain in the cell nucleus where they are embedded in polyhedra (occluded virus). This process typically continues until most organs and/or tissues have liquefied, whereupon the larval integument ruptures, releasing the occluded virus into the environment.

A well-developed *in vitro* system (methods to transfect insect cells with viral and plasmid DNA, agar-overlay plaque assay), together with other beneficial characteristics of NPVs (e.g., presence of nonessential genes whose expression is linked to changes in phenotype and the absence of morphological constraints to viral genome length) has allowed the development of the widely used baculovirus expression vector system (e.g., Smith et al., 1983; Luckow and Summers, 1988; for recent review see Jarvis, 1997) and subsequently a method by which to potentially improve the pesticidal properties of the virus (Carbonell et al., 1988; Miller, 1995). Especially in the last decade, rapid progress has been made toward the understanding of the infection cycle of baculovirus at the molecular level (for review see Blissard, 1996) and, to some extent, the function of approximately 50 baculovirus genes is known. It is important to note that the entire nucleotide sequence of AcMNPV (C6, a laboratory clone) has been published (Ayres et al., 1994).

During infection, gene expression is temporally regulated with early genes being transcribed by host RNA polymerase II before the onset of viral DNA replication (for review see Friesen, 1997) and late genes being transcribed by an α-amanitin-resistant RNA polymerase activity during or after DNA replication (for review see Lu et al., 1997). Viral DNA replication is a prerequisite for late and very late gene transcription (for review see Lu and Miller, 1997). Genes that are involved in baculovirus occlusion such as the polyhedrin and p10 genes, are transcribed very late during infection and at very high levels. Also, various facets of the interaction between virus and its host are being discovered and understood (Miller and Lu, 1997; Clem, 1997; O'Reilley, 1997).

The two genetically modified viruses discussed in this case study were derived from the C6 clone of AcMNPV (Pharmingen, San Diego, CA; Possee, 1986) and a clone of HzSNPV (Ignoffo, 1965). The modified viruses contained all the genes and sequences of the wild-type parental clone as well as the sequences necessary for the expression of the heterologous gene product, the insect-specific toxin IT2, from the yellow scorpion *Leiurus quinquestriatus* Hebreus (LqhIT2). The LqhIT2 gene was inserted into a specific region of the AcMNPV genome by homologous recombination between linearized AcMNPV-C6 DNA, in which the polyhedrin gene and an essential open reading frame were deleted, and a transfer plasmid. The transfer plasmid contained the toxin gene, the missing viral genes and AcMNPV sequences that facilitate homologous recombination with the viral DNA. The toxin gene consisted of a viral promoter upstream of the bombyxin signal sequence fused to the LqhIT2 coding sequence. The viral promoters were derived from AcMNPV and consisted of either the very late p10 gene promoter, or the weaker, early promoter-enhancer combination of homologous repeat 5 (hr5) and IE1 (Jarvis et al., 1990). Thus, cells of AcMNPV-Lq(IE1)- or AcMNPV-Lq(p10)-infected host larvae secrete LqhIT2 toxin early or late during the infection cycle, respectively. The LqhIT2 toxin is a 65 amino acid polypeptide, of the class of depressant insect-selective neurotoxins. It acts through suppression of insect sodium channel conductance, resulting in larval paralysis (Zlotkin et al., 1993).

HzSNPV-LqhIT2 was made by homologous recombination between HzSNPV DNA and a transfer plasmid containing the β-glucuronidase marker gene (GUS) in addition to the toxin and polyhedrin genes with associated viral flanking sequences necessary for recombination. Recombinant HzSNPV was selected and purified from virus plaques expressing GUS. The GUS marker gene was then deleted, resulting in a final HzSNPV recombinant expressing LqhIT2 under control of the AcMNPV hr5/IE1 promoter.

11.7.3 The Notification Process

Field release of genetically modified baculoviruses was considered on the basis of the following set of premises: (1) the historical and safe use of several widely used, registered wild-type baculoviruses, (2) the scientifically based and well documented fact that baculoviruses do not replicate in vertebrate/mammalian cells, (3) the selectivity and insect-specificity of the LqhIT2 toxin, and (4) the absence of oral toxicity in mammals caused by LqhIT2.

The presubmission meeting between the EPA and DuPont established that a notification was required because laboratory tests had shown that the insecticidal properties of the altered virus were enhanced by the modifications that were made to the wild-type progenitor virus. Furthermore, it was determined that the request for small-scale field testing should be submitted under 40 CFR Part 172, subpart C, "Notification for Certain Genetically Modified Microbial Pesticides for Field Testing." The submitted data package needed to address all sections of Subdivision M of the Pesticides Testing Guidelines: Microbial and Biochemical Pest Control Agents (U.S. Environmental Protection Agency, Office of Pesticides and Toxic Substances, 1989). The EPA also reserved the right to ensure that all additional conditions were met before granting the small-scale field release. Such conditions primarily concerned measures to limit the spread of the released virus as well as provisions for sanitation and mitigation.

11.7.4 The Data Package

The following is a discussion of the information and data submitted by DuPont Agricultural Products (1996) that highlights the issues specific for recombinant baculoviruses. It should be noted that the data requirements discussed below were specific for DuPont's application and may vary with other baculovirus-based pesticide submissions.

11.7.4.1 *Information on the Host Range, an Assessment of Infectivity, and Pathogenicity to Nontarget Organisms*

A large body of experimental results has been published with regard to the safety of baculoviruses to mammals following various routes of exposure (e.g., Ignoffo, 1975; for review see Groener, 1986). One study demonstrated a low level of infection following the inoculation of cultured Chinese hamster cells at a high multiplicity of infection with wild-type baculovirus (McIntosh and Shamy, 1980). However, this claim has been disputed by other researchers (Reiman and Miltenburger, 1983; Groener et al., 1984). Based on these published reports, specific experiments were designed by DuPont to demonstrate the safety of recombinant baculoviruses to mammals. The data package submitted to the EPA summarized the results of a full set of comparative mammalian toxicology studies carried out with AcMNPV-Lq(IE1) and wild-type AcMNPV. In these studies rats were exposed to virus using oral (100 million OBs), intravenous (10 million OBs), and inhalation (37 million OBs) routes. The results of these studies demonstrated that both the wild-type and genetically modified AcMNPV were not toxic, infective, or pathogenic to mammals. In addition, both AcMNPV-Lq(IE1) and AcMNPV-Lq(p10) were tested for their ability to infect or cause cytotoxic effects in various human cell lines established from intestine, lung, or liver tissue, without any indication of infection, viral replication, or significant cytotoxic differences.

A number of environmental and nontarget organism studies were also included in the data submission forwarded to the EPA. These involved studies with AcMNPV-Lq(IE1) on quail (administered orally and via the intraperitoneal route), trout (diet and aqueous), and grass shrimp (aqueous exposure). AcMNPV-Lq(IE1)–infected larvae were also fed to quail and trout. No adverse effects were observed in any of these studies.

Finally, claims that the insect host range was not affected by the expression of the toxin were supported by literature studies using a recombinant AcMNPV, carrying the insect-specific toxin gene AaIT from the scorpion *Androctonus australis*. In one study, the relative infectivity of AcST-3 and wild-type AcMNPV was assessed using 5 permissive, 26 less susceptible, and 16 nonpermissive lepidopteran species endemic to Great Britain (Possee et al., 1993). This study found only one species, the less susceptible nut tree tussock moth (*Colocasia coryli*), that seemed to respond with markedly higher mortality to AcST-3 in a nonreplicated assay. Two studies have addressed the effect of AcMNPV-AaIT on the nontarget beneficial predators, *Chrysoperda carnea* and *Orius insidiosus* (Heinz et al., 1995) and the parasite *Microplitis croceipes* (McCutchen et al., 1996). No adverse effects were detected with predators; however, as expected the parasitoid showed reduced size and shorter development times. In a study by McNitt and co-workers (1995) no adverse effects were noticed when the solitary predatory wasp *Polistes metricus* was fed prey infected with AcMNPV expressing Tox-34, a toxin gene of the predatory mite *Pyemotes tritici* (Tomalski and Miller, 1991).

The experiments conducted by DuPont with AcMNPV-Lq(IE1) and AcMNPV-Lq(p10) on two susceptible (*Heliothis virescens* and *Trichoplusia ni*) and two less susceptible (*Spodoptera exigua* and *S. frugiperda*) insect species showed no significant differences in infectivity, based on the overlap of the 95% confidence intervals of the LC_{50} values. However, in *Helicoverpa zea,* an extremely marginal host for AcMNPV, the infectivity of AcMNPV-Lq(p10) and AcMNPV-Lq(IE1) was at least one and two orders of magnitude higher than wild-type, respectively. This result was attributed to a newly discovered immune response of *H. zea* larvae against AcMNPV, which was believed to limit or stop the infection during its initial stages (Washburn et al., 1996). Thus, a possible explanation was that *H. zea* larvae infected with wild-type AcMNPV may be able to contain the infection and survive, whereas larvae inoculated with the same dose of the recombinant virus might die due to the production of sufficient toxin to compromise its own defense mechanism to the extent that it cannot prevent spread of the infection. Data obtained later suggested that the observed differences in infectivity were not primarily related to the defense mechanism but rather to the presence of a minor contamination of HzSNPV in the recombinant virus preparation (see submission of additional data, DuPont, 1997).

11.7.4.2 Survival and Ability of the Microbial Pesticide to Perpetuate in the Environment

The levels of active (wild-type) baculovirus in the field vary with place and time but are generally thought to be relatively low except for a short period immediately following an epizootic (Evans, 1986). There was no *a priori* reason to believe that insertion of the toxin gene would significantly affect the characteristics of the engineered virus, when compared to the parental strain, with respect to its susceptibility to environmental degradation. The submitted results of a comparative study of baculovirus yields in fourth-instar *H. virescens* larvae demonstrated that the yield reduction compared to the progenitor virus was 90% and 72% for AcMNPV-Lq(IE1) and AcMNPV-Lq(p10), respectively. This result is a direct consequence of the considerably smaller window for replication of genetically engineered virus and is expected to apply to other insect pest species in cotton such as *H. zea* and *S. exigua,* which also are killed much faster by the engineered virus. Consequently, the production of secondary inoculum of engineered virus in the field should be markedly lower than that of wild-type virus.

11.7.4.3 Relative Environmental Competitiveness Compared to the Parental Strain

Competitiveness and the possibility of displacement of wild-type viruses depend on the competitive fitness of the engineered virus compared to its wild-type progenitor virus. Baculovirus survival is a balance between production of progeny virus and decline of active virus in the field. There is no reason to believe that insertion of the toxin gene results in an effect on the physical and biochemical properties of AcMNPV occlusion bodies that determine the rate of inactivation. Thus, the production of progeny virus can be considered the major competitive factor affecting survivability in the environment. As discussed previously, production of recombinant virus progeny is significantly impaired relative to wild-type due to the shorter survival times of recombinant virus-infected larvae. Consequently, when both AcM-NPV types are present the competitive advantage belongs to wild-type baculoviruses. Evidence that the genetically altered virus is less competitive than wild-type has been supported by the preliminary results of a greenhouse study in which direct competition between wild-type and AcMNPV-AaIT was found to result in a swift extinction of the recombinant virus (Fuxa et al., 1997).

11.7.4.4 *Data on the Potential for Genetic Transfer and Gene Stability*

The petitioner acknowledges that evidence exists from the scientific literature that transfer of genetic material between baculoviruses, as well as between the host and baculovirus, occurs in the laboratory and in nature. In these instances, the probability of such an event resulting in expression of toxin in the new host is extremely low, but it must be considered a mathematical possibility. The potential risk is very low since the host is unlikely to survive the combination of baculovirus infection and toxin expression and if the insect somehow survived, it would be difficult to see the presence of the scorpion toxin gene as a benefit to the host. In order for genetic transfer to occur between the recombinant virus and another microorganism in the infected host, certain conditions must be present. First, the other microorganism must replicate simultaneously within the nucleus of the same cell infected by the virus. Second, the potential recipient DNA must be very similar to the recombinant DNA in order to facilitate genetic recombination. Due to these constraints it is likely that genetic transfer of the toxin gene would occur between two similar baculoviruses. Such a transfer would lead to an organism similar to AcMNPV-Lq in the sense that it had acquired a gene that diminished its competitive fitness.

Gene stability of LqhIT2 was also addressed with regard to the possibility of mutational activity leading to a modified LqhIT2 with mammalian toxicity. It was established that differences in amino acid sequence between LqhIT2 and scorpion toxins with activity against mammals (Zlotkin et al., 1991) were too great to fortuitously change LqhIT2 into a mammalian toxin. This would require that the three-dimensional structure and the charge distribution undergo several very specific changes.

11.7.4.5 Description of the Proposed Testing Program: Monitoring and Disposal

DuPont proposed to monitor the spread and persistence of the recombinant viruses at three release sites. Soil, leaf, and host insect samples were to be collected from all treatments at weekly intervals. These samples were monitored for amount and type of virus present until crop destruction and plow down. After this, soil samples were collected monthly until the next field season.

It is well established that baculoviruses are extremely vulnerable to degradation in the environment, especially when a UV-protecting formulation is absent. Consequently, the crop on the test plots was to be left standing in the field for at least 2 weeks prior to destruction in order to allow degradation of the virus by sunlight.

11.7.4.6 Contaminants

The formulated viral products were to be examined for the level of microbial contamination using aerobic, heterotrophic plate counts. The microbial load was not to exceed 1×10^8 CFU/g. Furthermore, in order to ensure that no human or animal pathogens were present, all microbial isolates were to be identified or verified using one or more of the standard identification methods.

11.7.5 Public Comments

Public comments from two sources were received by the EPA. The comments were in essence the same, as the second source wrote to support the comments of the first. They were limited to two points, namely host range and containment.

The comment noted that the higher infectivity of AcMNPV-Lq for *H. zea* may have indicated possible undocumented genetic alterations and possible changes to the host range of the genetically engineered virus. A general concern was the possibility that baculoviruses may cause nonsymptomatic sublethal infections in nontarget insects, which could normally go unnoticed in the case of wild-type viruses, but which could lead to mortality with genetically altered viruses. Subsequently, in the case of a sublethal infection sufficient toxin could be expressed to kill the larva.

The second comment can be summarized by stating that the submitter did not consider that containment was adequately supported by the data on larval yield presented.

11.7.6 EPA Ruling

Based on the internal review of the submitted package and the public comments, the EPA ruled that an experimental-use permit was not required to perform the small-scale field tests. The most important directions that were given concerned risk mitigation/containment and monitoring and are summarized below:

Risk Mitigation/Containment. (1) It was established that there would be a 10-foot minimum unplanted buffer zone surrounding the entire test area, and within the test area AcMNPV-Lq treatments shall be separated by plant-free alleyways to minimize spread. (2) Plants within the AcMNPV-Lq-treated plots and immediately adjoining the alleyways would be inspected for target and nontarget organisms showing signs of recombinant virus infection or intoxication (flaccid paralysis) at least twice a week, beginning one week posttreatment. Such insects would be collected and removed from the test site. (3) Spray drift would be minimized by using small-scale hand-held application equipment and the equipment cleaned with a 0.1% hypochlorite solution following use with AcMNPV-LqhIT2. (4) Upon completion of the test the recombinant-virus treated plants would remain standing for at least 2 weeks for desiccation and UV-irradiation, after which dried material would be removed and burned. (5) The treated area must be sprayed with wild-type AcMNPV before the 2-week desiccation period, or after plant debris removal, depending on the likely exposure for insect feeding. (6) After the final monitoring, the treated areas would be limed following standard agricultural practices to encourage polyhedrin dissolution and increasing virus degradation.

Monitoring. Upon completion of the monitoring study, the results would be presented to the Agency. In addition to the above stipulations, the submitter was informed that additional data regarding host range alterations, viral yield, and nontarget testing would need to be submitted before any future submissions would be considered. The Agency also suggested that the susceptibility of *H. zea* larvae to the toxin be investigated, as higher susceptibility might be the cause of the discrepancy between the results in *H. zea* and other larvae tested. The first two points are a direct consequence of the public comments regarding the changed LC_{50} of the viral recombinant in *H. zea* larvae and the lack of extensive data in support of containment. The last point concerned a repeat of the earthworm and honeybee studies that had been reported as "no-tests" due to high control mortality. The last requirement was later dropped by the EPA since such studies were not needed for small-scale field testing.

11.7.7 Submission of Additional Data

The additional information submitted to the EPA addressed the concerns raised in the public comments. With regard to potential changes in host range, new results generated by DuPont were reported and literature was cited that indicated that no significant differences in toxicity toward the various insects existed (Herrmann et al., 1995; Herrmann et al., 1990; Quistad and Skinner, 1994; Krapcho et al., 1995). The conclusion that there was no significant difference between the infectivity of AcMNPV-LqhIT2 and the wild-type progenitor virus was based on evidence demonstrating HzSNPV contamination in last year's AcMNPV-LqhIT2 inoculum. First, when the inocula for the recombinant and the wild-type virus were propagated in *T. ni* larvae (a non-host for HzSNPV) the resulting infectivity toward *H. zea* larvae was the same, whereas the infectivity of the previous year's inoculum (propagated in *H. virescens* larvae) was, as expected, again much higher than that of the wild-type and the *T. ni*–propagated recombinant virus. Moreover, large amounts of HzSNPV were found in larvae that had died in the bioassay from treatment with the previous year's inoculum in contrast to those treated with the new, *T. ni*–propagated recombinant and wild-type inocula. Finally, PCR analyses demonstrated that HzSNPV was present in the previous year's inoculum that had been propagated in *H. virescens*.

The results of an extensive recombinant virus yield study were submitted in support of containment of the genetically altered virus due to its inability to produce large numbers of progeny viral occlusion bodies. This study compared the occlusion body yields in *H. virescens*, *T. ni*, and *H. zea* larvae infected with AcMNPV-LqhIT2(ie1) or wild-type AcMNPV. Inoculation of test insects at three developmental stages (second, third, and fourth instar) were included in the comparison. Both *H. virescens* and *T. ni* were inoculated with a moderate ($5 \times LC_{50}$) and a high dose ($100 \times LC_{50}$), whereas *H. zea* was inoculated with a single (high) dose resulting in mortality of 75, 50, and 25%, for second, third, and fourth instar larvae, respectively. The results demonstrated that in all cases the genetically altered virus produced less progeny occlusion bodies than the wild-type virus under similar conditions. More specifically, the reduction in yield relative to wild-type AcMNPV in *H. virescens* larvae ranged from 5- to 100-fold, and, in *T. ni* and *H. zea* larvae, the reductions ranged from 7- to 100-, and from 3- to 300-fold, respectively. These results show that AcMNPV-LqhIT2 was severely impaired in its ability to produce large numbers of progeny viral occlusion bodies. A study on competition between AcMNPV-AaIT and wild-type virus in the greenhouse provided further support for the lack of competitiveness of the recombinant virus (Fuxa et al., 1997).

The results of the monitoring study were submitted as part of the additional data package. The results at the time of submission were preliminary, since the monitoring study was still in progress. Two different geographically separated agroecosystems were monitored, cotton in Louisiana and cabbage in Texas. Using bioassays, PCR, and restriction endonuclease analysis, the persistence and movement of the wild-type and the two genetically engineered viruses were monitored on the leaf surface and in soil.

The cotton trial was monitored for 16 days, covering four virus applications with soil samples collected monthly after crop destruction. In all cases, the amount of active virus on the leaf surface was ca. 1000-fold increased immediately following a treatment but dropped to almost prespray levels within 3 days. The virus load in the soil was moderate and did not fluctuate much during the period (approximately two- and three-fold increases for the recombinant and wild-type viruses, respectively, differences that were not statistically significant). The two sets of monthly samples also contained low levels of virus, which was almost exclusively wild-type virus.

The cabbage trial was monitored for 7 weeks (including 4 weeks after the last of the five applications) and AcMNPV-lqhIT2 (ie1) consisted of formulated and unformulated treatments. On the leaf surface similar rates of viral degradation were observed as in cotton with virus declining to prespray levels. In the soil, the increase in virus seemed most pronounced in the case of the wild-type (14-fold, significant), less with the unformulated recombinant (11-fold, not significant) and least with the formulated recombinant (2-fold, not significant). The amount of active virus had dropped in all three cases back to prespray levels at 2 weeks after the last treatment, but again increased to roughly double prespray level at 4 weeks. Thus in both trials there was a tendency for the wild-type virus to accumulate to higher levels in the soil, although this observation was not statistically significant. If this trend was biologically significant it can be attributed to the higher secondary inoculum output of the wild-type virus. The spread of the viruses was very limited, since most viruses were identified as corresponding to the treatment of the plot where the insect or soil was collected.

After review of the data and consideration of the public comments, the Agency decided that the additional data provided satisfactory explanations of the points raised in 1996 and consequently accepted the submission as presented for the 1997 field season for notification and review without additional data required.

11.7.8 The Second Year: AcMNPV- and HzSNPV-LqhIT2 Data Package

After obtaining permission for a small-scale release and successfully carrying out a limited number of field tests, another data package was assembled for a planned second field season. In this case the virus of interest was HzSNPV genetically engineered to express the same LqhIT2 toxin.

The differences between the first and second year of proposed field tests were in essence: (1) AcMNPV-LqhIT2(p10) was dropped from the testing program, (2) a new parental virus, HzSNPV-LqhIT2(ie1) was added to the program, (3) HzSNPV-LqhIT2(ie1) was generated via an intermediate virus containing the marker gene (GUS), and (4) two modified viruses were to be applied as a mixture. The second notification (DuPont Agricultural Products, 1997) contained many issues dealt with in the first notification and consequently no further attention to these issues was required. Since the toxin gene cassette inserted into HzSNPV was identical to that of AcMNPV-LqhIT2, there was no need to reestablish the safety and selectivity of the toxin. The parental virus, HzSNPV, was already registered by two different companies and had undergone extensive safety testing. In addition, DuPont's toxicity studies with AcMNPV-LqhIT2 had shown no effects caused by exposure of nontarget animals (quail, trout, shrimp) to the occlusion bodies via various exposure routes. As a result, the nontarget ecotoxicology studies were not repeated with HzSNPV-LqhIT2. The major difference from the prior submission were the data needed for verifying host range and effects on nontarget organisms, and to address the survival and competitiveness of the altered virus.

Data supporting the safety of HzSNPV to nontarget organisms consisted of a host range study with susceptible and nonsusceptible insect hosts. HzSNPV is a very selective baculovirus whose known hosts are limited to eight species of the genera *Helicoverpa* and *Heliothis* (Groener, 1986). Hosts are very susceptible to infection by HzSNPV and less susceptible hosts are not known. The activity of wild-type and recombinant HzSNPV in two susceptible species (*H. zea* and *H. virescens*) and three nonsusceptible species (*S. exigua*, *S. frugiperda*, and *T. ni*) were evaluated. The presence of possible contaminating viruses was avoided by using cell-culture-propagated viral occlusion bodies in these studies. No significant differences in infectivity between wild-type and recombinant virus was observed in any of these species.

While the safety of parental HzSNPV (Ignoffo, 1975) and the LqhIT2 toxin (DuPont Agricultural Products, 1996) to mammals had been firmly established, the safety of HzSNPV-LqhIT2 for mammals was addressed in a mammalian cell line study. Cultured human liver cells were exposed to 10^6 pfu/ml budded virus of recombinant HzSNPV alone or exposed to both recombinant AcMNPV and HzSNPV simultaneously. No signs of viral infection, replication, or differences in cytotoxicity between wild-type and recombinant viruses were found in these mammalian cell lines.

The second issue pertains to the competitiveness, survival, and ability of the microbial pesticide to increase in numbers (biomass) in the environment. The results of a comparative viral yield study involving the two major susceptible species in cotton (*H. zea* and *H. virescens*) were presented. Second instar larvae of both species were inoculated with HzSNPV-LqhIT2 or wild-type HzSNPV at low dose and high dose and fourth instars at high dose only. The exact doses, expressed as LC_{50} equivalents for third instar larvae, were $5 \times LC_{50}$ for the low and $100 \times LC_{50}$ for the high dose. Occlusion bodies yields from recombinant virus-infected *H. zea* and *H. virescens* ranged from 7.5 to 10.6% and from 13.6 to 31.8% of wild-type yields, respectively. Previous arguments applied here, where the combination of decreased recombinant OB yield coupled to an unchanged rate of degradation in the environment means that the likelihood of an increase in biomass for the genetically engineered virus is extremely low. The com,petitiveness of HzSNPV-LghIT2 in the environment compared to the parental strain would have been also negatively affected by the reduced virus yield.

11.7.9 Public Comments and EPA Ruling

No public comments were received. The EPA ruled that no experimental-use permit was required and approved the small-scale field testing. DuPont was also allowed to conduct future small-scale field tests with the described recombinant baculoviruses without notifying the Agency. Overspraying with wild-type viruses at the conclusion of a test was also no longer required if wild-type virus was used as a control treatment in the particular field test. Tests could not be conducted in areas adjacent to or containing potentially susceptible, endangered insect species.

11.7.10 Conclusion

The development of a recombinant baculovirus into a commercial product is a complex process of which pesticide registration is only one facet, albeit a critical one. However, before the registration process is initiated, it is necessary to demonstrate the field efficacy of the product and its market potential in a highly competitive insecticide business that is dominated by relatively inexpensive synthetic chemical insecticides. In the case of baculoviruses, this requires a development process involving the testing of multiple parental viruses, the evaluation of various genetic regulatory elements for toxin expression, and the evaluation of different formulations.

It is the responsibility of the Agency and the registrant that this development process be carried out with public input and in a careful and prudent way to ensure maximal safety and minimal risks to human health and the environment. Limited, small-scale, well-controlled field releases were undertaken only after the registrant demonstrated the safety of the baculovirus and of its foreign gene product (i.e., the insect-selective scorpion toxin) to human health and the environment. The recombinant baculoviruses were impaired in their ability to compete with their wild-type progenitors because of relatively low progeny virus production, thus providing some degree of biological containment. Also, the infectivity and host-selectivity of the recombinant viruses was compared to their wild-type progenitors in order to assess the risk of unforeseen effects on beneficial and nontarget insects or unintentional spread of the recombinant virus. But even after the most extensive laboratory testing, it is only by releasing the virus that the effects of the virus on the environment can be fully assessed to determine potential risks (Godfray, 1995). Therefore, the field tests were intensively monitored for persistence and spread of the recombinant virus construct. The data collected from the release sites over prolonged periods of time proved very valuable and supported the claim of no unusual persistence or spread, which was based initially on laboratory studies only.

While the testing programs conducted over the last 3 years have increased our knowledge considerably and have given no cause for concerns related to field release of recombinant baculoviruses, future tests with substantially different insecticidal genes will be subjected to the same scrutiny. The need for infectious and virulent viruses for specific target pest complexes dictates the continuation of virus development and discovery efforts. The viruses to be tested in small-scale tests in the immediate future may be derived from more virulent parental viruses, including virus isolates from outside the U.S., and may possess altered insect host ranges, possibly as a result of genetic engineering. After a number of seasons of small-scale field testing in cotton and vegetables by two companies (DuPont Agricultural Products and American Cyanamid), the next logical step in commercialization of recombinant baculovirus-based insecticides is the request to the EPA for permission to conduct larger scale field testing of candidate viruses.

11.8   A GENERIC CASE STUDY: *BACILLUS THURINGIENSIS* WITH ALTERED INSECTICIDAL TOXINS

One advantage of utilizing *B. thuringiensis* as a biological pest control agent is the possibility of exploiting the myriad of insecticidal toxins expressed by the various subspecies (e.g., *kurstaki, aizawai, tenebrionis,* and *israelensis*). Although several of these *B. thuringiensis* subspecies have been registered, many others have been described but have not been developed commercially. To better understand the process of registering a microbial pesticide that has been genetically altered to express a new insecticidal toxin, a hypothetical case study is presented here with a discussion of the data and information needed by the Agency to evaluate the human health and environmental risks associated with the use and wide-scale distribution of such products.

11.8.1   Product Characterization

Product characterization is intended to provide a description of the pesticide product that includes an accurate designation of the taxonomic position of the active microbial ingredient and a complete description of the manufacturing process. The manufacturing process includes a description of the quality control/quality assurance steps taken to maintain pure stock cultures for subsequent seed cultures needed for future fermentations and to ensure the final pesticide product is free of human pathogens. For strains altered to enhance pesticidal activity, a description of each genetic manipulation or alteration to produce the final construct must be provided. In the engineered *B. thuringiensis*–based products submitted to date, the specific steps used to generate the new product are considered confidential business information and cannot be disclosed. For most genetically engineered microbial pesticides, the stability and mobility of the introduced trait is important, especially if the novel trait presents a new exposure or alters the potential for movement of that trait into other microorganisms. However, the processes used to alter the *B. thuringiensis* strains seen to date generally mimic natural genetic recombination or conjugation albeit aimed at a specific result. That result can be a unique combination of δ-endotoxins not previously noted in a *B. thuringiensis* strain or the expression of an altered protein toxin that is more efficiently produced or has enhanced activity for a specific target pest. An important risk issue that surfaces is the effect that the genetic alterations has on the toxicity or host range to nontarget species (especially insect species) when compared to the original *B. thuringiensis* strain.

Data and information on the composition of the final product needs to be submitted to support the registration of each pesticide product. This statement of composition, or confidential statement of formula (CSF), includes the name and percentage by weight of each active and inert ingredient (e.g., surfactants, diluents, carriers, and preservatives) as it will be found in commerce and is considered confidential business information. The safety of the final end-use product, which includes the intentionally added inerts, is examined separately from the pathogenicity and toxicity considerations for the technical grade active microbial ingredient. For example, specific tests for eye and skin irritation as well as for inhalation and oral toxicity are performed or the data cited to provide toxicity information for product precautionary labeling.

For this case study, the taxonomic description of *B. thuringiensis* is relatively straightforward as this bacterium has been examined for years and has extensive biochemical and physical characteristics that aid in its identification. The taxonomic identification may highlight problems with closely related species and it allows an assessment of those characteristics that may need closer scrutiny in the species to be registered. For example, *B. thuringiensis* is biochemically related to the bacterium *B. cereus* and less so to *B. anthracis* based on traditional taxonomic classification of the genus *Bacillus*. This relationship dictates that certain health hazards related to these other bacteria be addressed during both initial registration and during routine production of the *B. thuringiensis* insecticidal products.

The major feature distinguishing *B. thuringiensis* from *B. cereus* using traditional taxonomy is the presence of visible crystals produced during bacterial sporulation. The production of these crystals is known to be encoded on bacterial plasmids. For commercial purposes, the presence of the crystal is an expression of potential insecticidal activity since it is well established that *B. thuringiensis* insecticidal activity is primarily due to these crystal proteins more commonly referred to as δ-endotoxins (Aronson et al., 1986). However, *B. cereús* is also known to be occasionally isolated as an insect pathogen so there may be some possibility for the *B. thuringiensis* bacterium itself to express some insect pathogenesis.

Classification of *B. thuringiensis* at the subspecies level includes a combination of flagellar antigen typing and a determination of the specific insecticidal activity against a range of target insect pests. The various δ-endotoxins or Cry proteins (for crystal proteins) have also been classified based on their spectrum of insecticidal activity: CryI toxins against lepidopteran species, CryII toxins against lepidopteran and dipteran, and CryIII against coleopteran species. Since these specific δ-endotoxins were encoded on the bacterial plasmids, a profile of such extrachromosomal DNA would further support the traditional taxonomic approach. However, the significance of flagellar antigen typing and plasmid profiles, which were traditionally required to identify *B. thuringiensis* strains (de Barjac and Frachon, 1990), became less important when it became evident that plasmids encoding δ-endotoxins were not uniquely identifiable and several types of δ-endotoxins could be readily found in different subspecies of *B. thuringiensis* (Höfte & Whiteley, 1989).

For registration, the determination of the exact subspecies of *B. thuringiensis* by both traditional biochemical testing and flagellar antigen reaction is essential. While the previously used taxonomical methods did not define insect specificity related to the expression of toxins, new strains or subspecies are routinely compared to other registered *B. thuringiensis* strains. This is particularly important in light of the recent reports that certain *B. thuringiensis* strains may have the ability to produce toxins similar to the *B. cereus* enterotoxins implicated as the cause of the emetic and diarrheogenic symptoms of *B. cereus* food poisoning, especially cooked rice (Jackson et al., 1995; Honda et al., 1991; Damgaard, 1995). To date none of the currently registered strains have definitively been shown to produce these enterotoxins. Therefore, any new strain that diverges significantly from registered strains, which have a history of safe use, may require a closer examination for the presence of these enterotoxins.

In addition, data should be generated to address the range of susceptible insects for the *B. thuringiensis* subspecies in question. Some species of insects required to be tested will be mentioned in the nontarget insect section under environmental safety testing. However, registrants often test in-house colonies of insect pests, which would aid in identifying the strain in question. As discussed in the *B. thuringiensis* Reregistration Eligibility Document (U.S. EPA, 1998a), registrants must now screen their isolates to determine the types and relative amounts of δ-endotoxins produced. The fact that product efficacy, as well as range of insect pests to be controlled, can be altered by adjusting the type of δ-endotoxin has led to the development of *B. thuringiensis* strains that have been genetically altered to express either different types of δ-endotoxins or altered δ-endotoxins themselves.

For strains that have been genetically altered to enhance their pesticidal activities, the registrant must submit as part of their product identification a description of the methods used to introduce the new traits. In addition, information about the origin of the new trait and what other characters were concurrently introduced needs to be presented. The method used to introduce the gene(s) as well as a discussion of the stability of the trait in the new bacterial host and the likelihood of the trait to move into other organisms from that new bacterial host must be described. This could be examined in instances where the δ-endotoxins are introduced into bacterial species outside those normally able to exchange genetic material with *Bacillus* species. When the toxins were introduced into *Pseudomonas* species, the potential for toxin escape was determined to exist, which presented a different toxin exposure with a possible adverse environmental effects. However, the introduction of similar δ-endotoxins into a *Clavibacter* species was not judged similarly, at least in part due to the limited potential for movement of the trait outside the genetically engineered bacterial host.

For the currently registered engineered products, registrants are using well-defined molecular techniques to circumvent barriers to gene transfer and mixing. For example, although the δ-endotoxin gene is found on a plasmid, the movement of that plasmid out of the *B. thuringiensis* host is a rare event. Essentially, these large plasmids encoding Cry proteins have very low horizontal mobility. Using transformation technology, a specific combination of δ-endotoxins can be inserted into a host such as *B. thuringiensis* with well-known safety and production characteristics. This is preferable to the laborious isolation and screening of environmental isolates to find a new *B. thuringiensis* strain with the right combination of toxins, then optimizing its production or using other production methods to achieve the same end.

11.8.2 Quality Assurance/Quality Control

An important aspect of any production system is the incorporation of quality control features to guarantee the efficacy of the final product and, in particular for microbial pesticides, that require pure culture fermentation to ensure that the final product is free of human pathogens. *B. thuringiensis* products often incorporate a bioassay against a target pest as part of their quality assurance/quality control process due to the historical use of standards for labeling *B. thuringiensis* products for potency. Until recently there was a standardized culture of *B. thuringiensis* subsp. *kurstaki* available from the USDA that was used by companies to determine the potency of products against a specific lepidopteran pest. This resulted in product labels with statements of bioactivity in the form of international units of activity against *Trichoplusia ni* and allowed consumers to compare the potency of available products. Currently, there are no publicly available sources of the *B. thuringiensis* standard. Some companies maintain internal standards for use in bioassays, but biopotency statements, while not false or misleading, cannot be used to compare products between companies. Some *B. thuringiensis* products may also include δ-endotoxin quantification by SDS-polyacrylamide gel electrophoresis, and colony-forming units or spore counts. Under FIFRA, every pesticide label must include a statement of active ingredient by percent weight. Other statements are required for *B. thuringiensis* products as described in the Reregistration Eligibility Document (U.S. EPA, 1998a).

For assurance that the final product is void of human pathogens registrants must perform a subcutaneous injection assay on every production batch to screen for the possibility of contamination by *B. anthracis*. Manufacturers routinely include a screening for the presence of significant microbial contamination by specific plating techniques for indicator organisms such as fecal coliforms, molds, and yeasts as well as for specific human pathogenic microorganisms such as *Salmonella*, *Shigella*, and *Vibrio*. In addition, there are reports of *B. thuringiensis* strains producing a toxin known as β-exotoxin, which has mammalian effects. Therefore, *B. thuringiensis* production lots must be tested for β-exotoxin by a housefly larvae bioassay or HPLC, and shown not to have the toxin present.

The product analysis ultimately determines what human and other nontarget toxicology studies are needed to render a satisfactory safety finding to support the registration. In part, this is based on a thorough taxonomic description of the organism, its relationship to known pathogens in the same taxonomic grouping, and its ability to produce secondary metabolites or toxins that may have adverse effects other than those intended for the target pest. Another aspect of the product characterization is historical data and/or information on wide-scale human and environmental exposure of this microbe prior to its use as a pesticide. A microbe to control foliar diseases that had been originally isolated from the depths of a pool of estuarine mud or a protozoan pathogen of a vertebrate pest would require greater examination than the introduction of a known δ-endotoxin into a previously registered *B. thuringiensis* strain. For such hypothetical cases, the history of registering similar products or the same strain allows the use of existing published safety data to aid in the risk assessment.

11.8.3 Mammalian Toxicology

All registered microbial pesticide products must be supported by mammalian toxicology data to determine the risks associated with human and domestic animal exposure. The mammalian toxicology data is generated from toxicity/infectivity studies following oral, pulmonary, and injection exposure to high doses of the microbial agent. Tests for dermal and eye irritation, as well as oral and inhalation toxicity, must be performed with the final product, which may include additional formulation ingredients with unknown toxicities. For most of the genetically engineered microbial products derived from registered pesticides, the toxicity/infectivity tests that were performed for the original unaltered microbe can be used to support the safety determination of the altered microbe. This process of using the same tests to support more than one product is known as "bridging." The rationale for bridging studies is that the toxicity/infectivity of the altered product has not been significantly altered by the introduction of the traits. Frequently this bridging decision is supported by the repetition of a subset of the toxicology studies done previously to confirm that the assumption behind bridging is valid. Bridging can also be done for the end product formulations if the additional ingredients have not changed in character or amount. To date no change in the toxicity has been found for genetically altered *B. thuringiensis* products except for toxicity to the targeted insects.

None of the mammalian toxicology studies submitted for genetically engineered *B. thuringiensis* products or their progenitor strains and reviewed to date have shown significant adverse effects that triggered examination above the Tier I level. However, some pulmonary toxicity studies have shown that the test animals were unable to clear spores from the lung by the final examination point. This unusual persistence of spores has been reported before for *B. thuringiensis* strains (McClintock et al., 1995). These findings suggest that while there is no toxicity or adverse consequences associated with this unusual persistence, a dust-filtering mask is recommended for the safe mixing and application of these and all microbial products. Overall, the findings of the mammalian toxicology studies have not indicated any specific hazards associated with genetically engineered strains of *B. thuringiensis* that would alter the safety assessment of these products.

11.8.4 Environmental Effects

A major concern with genetic modification of insect control agents like *B. thuringiensis* is that these modifications might alter the host range and presumed safety of these desirable microbial products for other organisms in the environment. For genetically engineered products, the approach to assessing the changes has been to examine the introduced characters and determine how these could alter the resulting microbial construct. The testing requirements for nontarget organisms are then decided from a consideration of these changed characters and how these may affect exposed nontarget organisms differently than the parent microbe. In some instances, the majority of the nontarget studies were repeated with the genetically altered strain to ensure the toxicity had not significantly changed. For engineered *B. thuringiensis* products reviewed to date, the following studies were provided: avian acute toxicity for bobwhite quail (*Colinus virginianus*) and mallard duck (*Anas platyrhyncos*), a 21-day *Daphnia magna* toxicity study, a rainbow trout (*Salmo gairdneri*) study, a honeybee (*Apis mellifera*) toxicity, a parasitic wasp (*Nansonia vitripennis*) toxicity study, a green lacewing (*Chrysopa* spp.) and a ladybird beetle (*Hippodamia* spp.) toxicity study. It should be noted that not all of these studies were performed for every engineered product. For example, if a previously registered product was engineered to enhance control of a coleopteran pest, then perhaps a test on a beneficial coleopteran such as the ladybird beetle would be considered to see if the toxicity had changed for nontarget insect species. In some instances the changes seen against target pests were so slight that most of the nontarget toxicity information was bridged from the parental strains. In the cases reviewed to date, there was no change in toxicity of the altered microbe when compared to the parental microbe for the surrogate species used in environmental effects testing.

As observed with the mammalian toxicity studies, the administration of the test substance may show adverse effects in the experimental animals. Consequently, it is important to consider the dosing regime used for these studies. In the Subdivision M Guidelines or OPPTS 885.4000, certain high dosing requirements are described. For nontarget insects 10 to 100 times the field rate or 100 times the LD_{50} value for the target must be tested. A 1000-fold higher than field rates expected in the top 6 inches of water or at least 10^6 CFU/ml must be used for aquatic toxicity testing to provide the maximum hazard dose and detect any possible adverse effects with use. When the field rates are already high, this 1000-fold dosing regime may occasionally give adverse effects possibly due to turbidity or physical effects of the dosing material.

Difficulty in generating such a high test dose is cited by testing laboratories. Therefore most of the aquatic studies are done at lower dose rates more feasible to administer.

As in any risk assessment, if adverse effects are observed in hazard tests, a consideration must be made of the likelihood of exposure. The field exposure levels need to predict whether the environmental use poses any significant hazard. Using information from the proposed application rate, a calculation can be done to anticipate what the expected residue levels are immediately after pesticide application based on a foliage interception value for different plants. Since the calculated residue levels is often well below the determined LC_{50} values for the nontargets species likely to be exposed during use (especially insects), the risk conclusions for these genetically engineered products are that there are minimal risks to the environment with labeled uses. It should be noted that even though there are acceptable laboratory studies that have shown a lack of toxicity to aquatic species, all pesticide labels require a warning on the label to keep out of aquatic habitats except for those approved specifically for aquatic use such as mosquito control. Similar to the findings for mammalian testing, the nontarget testing results have not indicated an unexpected toxicity for genetically engineered B. thuringiensis strains when compared to the parental strains except for enhanced activity against the target pests.

11.9   A GENERIC CASE STUDY: CORN EXPRESSING AN INSECTICIDAL PROTEIN

The use of genetic engineering in crop plants to aid in the development of new varieties has led to the introduction of pest resistance traits with exposures not previously anticipated. The majority of crops that have been examined for safety by the Agency have been altered to express a protein toxin derived from Bacillus thuringiensis. Fortunately, the Agency has experience with safety issues related to these protein toxins due to their presence in the numerous B. thuringiensis microbial pesticides and other genetically engineered microbes. Many of the same issues of mammalian and other nontarget organism safety and environmental exposure were examined for genetically engineered microbes and were used to frame the risk assessment for genetically engineered plants. The risk characterization takes the same form by addressing product characterization, mammalian toxicology, and nontarget toxicology. However, there are certain aspects that are unique to pesticidal proteins in plants that will be discussed in each section. Again the discussion will be framed in terms of a hypothetical corn cultivar expressing a toxin protein from B. thuringiensis.

11.9.1   Product Characterization

A description of the source of the introduced gene and the expressed gene product are needed to determine the nature of the new trait and any previous exposures. In this case, many of the δ-endotoxin proteins are familiar to the Agency, having been examined as part of microbial preparations. The structural similarity of the endotoxins at the primary sequence level are also well established, especially for the CryI family of proteins. Since the proposed definition of a plant-pesticide includes the genetic material encoding the active pesticidal substance, a description of the introduced genetic material and its controlling sequences should be submitted. The incorporation of the DNA for the new trait into the corn plant is usually verified by Southern analysis probing with either the whole plasmid DNA, used to introduce the trait, or a restriction endonuclease-digested DNA fragment containing the trait(s). To verify stable incorporation of the trait into the corn genome, the registrant provides Southern analyses examining subsequent generations for presence of the trait in several different genetic backgrounds.

To enhance the expression of a bacterial gene in a eukaryotic system, changes to the codons for specific amino acids are made for the actual protein encoding region. Since the genetic code is redundant for certain amino acids, changes made in the nucleic acid sequence can result in an identical amino acid sequence, especially if the altered codons still code for the same amino acid at that location in the protein. If slight changes in the amino acid sequence have occurred, the expressed protein should be shown to have activity similar to that of the source organism (i.e., bioactivity against the target pest, immunorecognition, biochemical stability). Another consideration would be if the substituted amino acid had similar side-chain chemical properties (e.g., valine compared to leucine). The similarity in biological activity between the protein in the source and its expression in the corn plant is an essential feature of the characterization, because the subsequent toxicology testing is based on information about the trait as expressed in the source organism.

Not only is the presence of the protein important but its exact levels in various plant parts is also essential to determining the expected exposure for both humans and other nontarget organisms. Expression data for corn, genetically engineered to express the δ-endotoxin, should be submitted for the foliage, pith, kernel, root, and pollen at various stages of plant maturity. This expression data is critical for assessing the environmental impact of corn expressing the δ-endotoxin as well as developing a rigorous approach for delaying the appearance of target insects resistant to the expressed protein. The subject of resistance management has been examined in depth and is an ongoing concern for the Agency. For further review and consideration, references are available on recent advances in the field (U.S. EPA, 1998b).

For toxicity testing, the ideal source of the test substance would be the substance expressed from the corn plant. However, one of the advantages of expressing the δ-endotoxin proteins in plants is that a low expression level is still efficacious. Therefore, obtaining pure δ-endotoxin protein from the corn plant would be a laborious process with limited yield. It is often easier to produce sufficient quantities of the proteinaceous material in an alternate source (e.g., the source microbe or an *E. coli* or yeast expression system). In cases where alternate production of the test substance is employed, it is imperative that the biological similarity or equivalence of this alternate source be demonstrated to be the same as that expressed in the plant. These equivalence tests are similar to the product chemistry/characterization studies and should demonstrate that the introduced DNA in the alternate source expresses a protein of the same biochemical and biological activity as the plant.

11.9.2 Mammalian Toxicology

It is essential to take into account the expected exposure to the pesticidal proteins when considering the data required to assess the safe use of plants expressing plant-pesticides. For all the δ-endotoxins expressed in corn, the primary exposure for humans and domestic animals is dietary. However, there are corn plants with δ-endotoxins where kernel expression is negligible and the highest expression is found in the foliage and pith. This low expression may lessen the dietary exposure for the expressed protein but it is not absent. Since the major exposure for these proteins is dietary, registrants have chosen to support the safety of their plant-pesticides with acute oral studies and *in vitro* digestibility studies. The acute oral study has been performed using purified protein at doses considered close to the maximum hazard (i.e., 2–5 gm/kg body weight). None of the acute toxicity studies for corn-expressed δ-endotoxin have shown adverse effects other than transient weight loss in the experimental animals during the duration of the observation period.

The *in vitro* digestibility studies are performed with purified protein in conditions that mimic natural gastric and intestinal fluids (solution preparation is available in the *U.S. Pharmacopeia*). The studies involve monitoring the degradation of the introduced protein into constituents that no longer have bioactivity by either western blot analysis or insect bioassay. Some registrants chose to demonstrate that the δ-endotoxin was rapidly degraded even in the presence of lower than expected digestive enzyme concentrations or simply in acidic pH. The results have generally shown a lack of stability in gastric solutions and high stability in intestinal solutions. This is not unexpected for the δ-endotoxin proteins that are either activated or stable in insect midguts, which tend to be alkaline and characterized by a trypsin-type digestion. Rapid degradation in the stomach adds weight to the presumption that these proteins are not different from the myriad of uncharacterized proteins that constitute the human diet. In addition, rapid digestion of these proteins is a feature that could be used to distinguish them from the common proteins known to induce food allergy. Food allergens tend to be more stable to acid or heat and expressed at higher levels in the food plant.

Knowledge of the exact amino acid sequence discussed in the product characterization also provides another mechanism to evaluate toxicity. Since the amino acid sequence for many proteins is known and published in the scientific literature, a comparison of the sequence homology of the δ-endotoxin expressed in the corn plant can be performed. It is especially important to examine for regions of potential homology between known toxins and allergens as these are the proteins that have the highest potential for causing adverse effects in nontarget organisms. The amino acid homology comparison is not a toxicity test but rather a guide for subsequent toxicity or allergenicity examinations. Homology provides information about the potential of a protein to cause an adverse effect, since structurally similar proteins tend to have similar properties. The level of homology with a toxin or allergen that would trigger a closer examination is not clear, nor is the appropriate algorithm for judging homology. One comparison used a sequential analysis of eight contiguous amino acids for the length of the δ-endotoxin, since the literature on food allergy has indicated that as few as eight amino acids provide an epitope for recognition and reaction in sensitive individuals (Rothbard and Gefter, 1991; Metcalfe et al., 1996). To date, none of the δ-endotoxins expressed in plants as plant-pesticides have shown significant homology to any proteins other than related δ-endotoxins.

Overall the assessment for toxicity has found no significant adverse effects in the acute oral study, no significant homology with any protein known to be a food allergen or mammalian toxin, and in general a rapid degradation in simulated gastric fluids. These results have led to a finding of a lack of toxicity for mammalian species by the oral route of exposure for the δ-endotoxins expressed in corn.

11.9.3 Environmental Effects

As a route of introducing a pesticide into the environment, the expression of a δ-endotoxin in corn plants provides certain advantages. There is less chance for nontarget organisms to be directly exposed to the δ-endotoxin compared to a spray application. The microbial *B. thuringiensis* spray products have a short environmental lifetime, need repeated sprays for acceptable control, and generally lack activity against later larval instars. The continual expression of the δ-endotoxin proteins by the plant and the δ-endotoxin's presence in specific tissue has enhanced the effectiveness of the δ-endotoxin. However, this higher expression level may lead to a greater exposure to δ-endotoxin in some species directly feeding on the plant material and has led to consideration of the increased possibility of selecting for resistance to the δ-endotoxins in the target pests. This increased selection for resistance has been noted as a concern by users of *B. thuringiensis*–based products to control insect pests on other plants. Although resistance management is not discussed here references are available on recent advances in the field and are cited in summaries of EPA workshops and scientific panels (U. S. EPA, 1998b).

A careful examination was also made of the possible nontarget species exposed to the δ-endotoxins expressed in corn. The general range of species to be examined follows those outlined for microbial or biochemical pesticides. However, special consideration is given to test only those species that represent reasonable routes of exposure to the pesticidal substance as it is expressed in the plant. Another important difference in the examination of hazards associated with environmental effects from that for addressing human health hazards is the choice of test substance. Plant tissue actually expressing the pesticidal substance is used in many of the studies for environmental effects. Avian species were tested with corn kernels, since residues of corn left after harvest is an important food source for some wild avian species. Since corn is used for catfish feed, the grain was tested in aquatic species for potential adverse effects. Pollen was used as a test substance for honeybee and *Daphnia* tests. Pollen was chosen for *Daphnia*, since aquatic exposure during pollen shed will occur and it was chosen for honeybee, since it can be a source of food for bee larvae.

One unique aspect of the transgenic corn assessments compared to other pesticide safety assessments has been the consideration of soil invertebrates' exposure to the pesticidal substance. For the corn products expressing δ-endotoxins the exposure of soil fauna to the pesticidal substance was viewed as very different from that presented by conventional *B. thuringiensis* microbial sprays where the δ-endotoxins apparently have a short half-life in the soil. Therefore, toxicity studies on earthworms and springtails or collembola (*Folsomia* or *Xenylla* spp.) were performed to address this new exposure. The test substance used in these studies was usually leaf tissue expressing the δ-endotoxin but often purified δ-endotoxin alone or mixed with leaf tissue was included as a treatment. These studies supplemented an extensive range of nontarget studies, which included tests on adult and larval honeybees, ladybird beetles (*Hippodamia* spp.), parasitic wasps (*Brachymeria* spp.), green lacewings (*Chrysopa* spp.), *Daphnia magna*, trout, catfish, and bobwhite quail (*Colinus virginianus*).

Other nontarget studies not specifically addressed above are examined by considering the likelihood of exposure through expression in plant tissue. This negligible plant exposure consideration was the reason no aquatic species were tested for δ-endotoxin expressed in potatoes. However, with corn there could be significant exposure in adjacent aquatic environments to corn pollen during anthesis and corn meal is employed as a fish food. Thus the *Daphnia* and fish studies were required. The fact that the pesticidal substance cannot survive the heat and pressures generated during processing for fish food was used to justify the lack of exposure for fish species in some cases.

The nontarget plant studies are not like those found in the guidelines for other pesticides. The issue is really more accurately defined as a biological escape from the transgenic plant with expression in neighboring crops or movement into wild relatives of the transgenic plant. For corn, there are no wild relatives extant within the continental United States. Therefore, the issue of gene escape can be addressed by a discussion of the reproductive biology of corn and the geographic distribution of close corn relatives. In most instances, this will be the method of addressing the issue of gene escape for transgenic agronomic plants. This may not be possible for agronomic species such as sunflower, squash, or blueberries, which were derived from native North American plant species and have wild relatives growing in close proximity. It may also not be possible for longer lived plants such as small fruits, orchard and forest trees.

Overall the assessment for environmental toxicity and effects to nontarget species has found no significant adverse effects. It is important to note that in some instances this safety finding is limited in its applicability to the expression level of the currently registered product line, since the tests did not include purified δ-endotoxin treatment to provide for extrapolation for other expression levels.

11.10 SUMMARY

The OPP at the EPA has recognized that certain categories of pesticides require less data and/or information to support a finding of no significant adverse effects to humans and the environment. Biopesticides, which include microbial and biochemical pesticides, as well as plant pesticides, are included in this category (see EPA's biopesticide webpage at http://epa.gov/oppbppd1/biopesticides). The current registration requirements for naturally occurring and genetically altered biopesticides are intended to generate necessary data and information on the identity, composition, and potential adverse effects of a specific pesticide. Case studies were presented to demonstrate the mechanisms of the regulatory process for biopesticidal products that are either naturally occurring or are derived from biotechnology.

Small-scale field tests with genetically modified or nonindigenous microorganisms require the EPA to be notified and data to be submitted with the notification. Although not mandatory, a presubmission conference with the appropriate Registration Action Leader of BPPD is recommended before developing the required data and preparing the application. The presubmission conference also provides the registrant an opportunity to develop a proposed data set with an agency scientist who will address the perceived risks associated with an active pesticidal ingredient. The data is then evaluated by EPA scientists to determine whether an EUP is necessary prior to conducting the small-scale field tests. When applying for an EUP, a registration, or amended registration, the registrant must satisfy a generic set of data requirements that are associated with the active pesticidal ingredient. Additional data are usually required to address concerns associated with the end-use formulation.

The Agency currently provides incentives for the development and commercialization of biological pesticides. Such incentives include tiered data requirements, reduced data requirements based on the nature and historical use of the product, expedited review times, and in some cases tolerance fee waivers. OPP also developed a set of guidance criteria for a subset of pesticides that are naturally occurring food components/food additives and that would limit the kinds of hazard evaluation data required for both human health effects and nontarget organisms. In fact, if the pesticidal food component met certain criteria, the required data base might be minimized or even totally waived. The rationale was based on the historical use and available safety information, which suggested that such food components did not pose any additional or unreasonable risks to humans and the environment when used as a pesticide.

Other existing mechanisms or incentives to registration and commercialization include, but are not limited to, (1) 250 acre limit before an EUP is required for many pheromones, (2) tolerance exemptions for certain classes of pesticides rather than for each compound, (3) specific exemptions based on formulations (e.g., pheromones in traps are exempt under 40 CFR Section 25(b)), and (4) 25(b) exemption for certain, historically safe naturally occurring compounds (see Table 11.4 for a historical perspective).

REFERENCES

American Cyanamid Corporation. VEGTDEL-AaIT Notification to Conduct Small Scale Field Trials. EPA No. 2A1-NMP-E, 1994.

Aronson, A. I., W. Beckman, and P. Dunn. *Bacillus thuringiensis* and Related Insect Pathogens. *Microbiological Reviews* 50, 1–24, 1986.

Ayres, M. D., S. C. Howard, J. Kuzio, M. Lopez-Ferber, and R. D. Possee. The Complete DNA Sequence of *Autographa californica* Nuclear Polyhedrosis Virus. *Virology* 202, 586–605, 1994.

Bishop, D. H. L. Genetically Engineered Viral Insecticides — A Progress Report 1986–1989. *Pesticide Science* 27, 173–189, 1989.

Blissard, G.W. Baculovirus-Insect Cell Interactions. In *Insect Cell Cultures: Fundamental and Applied Aspects*. Vlak, de Gooijer, Tramper, and Miltenburger, Eds., Kluwer Academic Publishers, Dordrecht, Boston, London, 73–93, 1996.

Carbonell, L. F., M. R. Hodge, M. D. Tomalski, and L. K. Miller. Synthesis of a Gene Coding for an Insect-Specific Scorpion Neurotoxin and Attempts to Express it Using Baculovirus Vectors. *Gene* 73, 409–418, 1988.

Clem, R. J. Regulation of Programmed Cell Death by Baculoviruses. In *The Baculoviruses*. L. K. Miller, Ed., Plenum Press, New York, London, 237–266, 1997.

Damgaard, P. H. Diarrhoeal Enterotoxin Production by Strains of *Bacillus thuringiensis* Isolated from Commercial *Bacillus thuringiensis*–Based Insecticides. *FEMS Immunology and Medical Microbiology* 12, 245–250, 1995.

deBarjac, H. and E. Frachon. Classification of *Bacillus thuringiensis* Strains. *Entomophaga* 35, 233–240, 1990.

DuPont Agricultural Products. Notification for Small Scale Field Trials of *Autographa californica* NPV LqhIT2 Gene Inserted Baculoviruses: AcNPV-Lq (IE1) and AcNPV-Lq (p10). EPA No. 352-NMP-4, 1996.

DuPont Agricultural Products. Notification to Conduct a Small Scale Field Test with Two Genetically Modified Microbial Pesticides. EPA No. 352-NMP-L. 1997.

Evans, H. F. Ecology and Epizootiology of Baculoviruses. In *The Biology of Baculoviruses*, *Vol. II*. R. R. Granados and B. A. Federici, Eds., CRC Press, Boca Raton, FL, 89–132, 1986.

Friesen, P. D. Regulation of Baculovirus Early Gene Expression. In *The Baculoviruses*. L. K. Miller, Ed., Plenum Press, New York, London, 141–170, 1997.

Fuxa, J. R., S. A. Alaniz, A. R. Richter, L. M. Reilly, and B. D. Hammock. Capability of Recombinant Viruses for Environmental Persistence/Transport. In *Biotechnology Risk Assessment: U.S.EPA/USDA/Environmental Canada/Agriculture and Agri-Food Canada*. Risk Assessment Methodologies, Proceeding of the Biotechnology Risk Assessment Symposium, June 23–25, 1996, Ottawa, Ontario, Canada. M. Levin, C. Grim, and S. Angle, Eds., Chapter 9. University of Maryland Biotechnology Institute, 315–325, 1997.

Godfray, H. C. J. Field Experiments with Genetically Manipulated Insect Viruses: Ecological Issues. *TREE* 10, 465–469, 1995.

Groener, A. Specificity and Safety of Baculoviruses. In *The Biology of Baculoviruses, Vol. I*. R.R. Granados and B.A. Federici, Eds., CRC Press, Boca Raton, FL, 1986.

Groener, A., R. R. Granados, and J. P. Burand. Interaction of *Autographa californica* Nuclear Polyhedrosis Virus with Two Non Permissive Cell Lines. *Intervirology* 21, 203–209, 1984.

Hartig, P. C., M. A. Chapman, G. G. Hatch, and C. Y. Kawanishi. Insect Virus: Assays for Toxic Effects and Transformation Potential in Mammalian Cells. *Applied Environmental Microbiology* 55, 1916–1920, 1989.

Heinz, K. M., B. F. McCutchen, R. Herrmann, M. P. Parella, and B. D. Hammock. Direct Effects of Recombinant Nuclear Polyhedrosis Viruses on Selected Nontarget Organisms. *Journal of Economic Entomology* 88, 259–264, 1995.

Herrmann, R., L. Fishman, and E. Zlotkin. The Tolerance of Lepidopterous Larvae to an Insect-Selective Toxin. *Insect Biochemistry* 20, 625–637, 1990.

Herrmann, R., H. Moskowitz, E. Zlotkin, and B. D. Hammock. Positive Cooperativity Among Insecticidal Scorpion Neurotoxins. *Toxicon* 33, 1099–1102, 1995.

Höfte, H. and H. R. Whiteley. Insecticidal Crystal Proteins of *Bacillus thuringiensis*. *Microbiological Reviews* 53, 242–255, 1989.

Honda, T., A. Shiba, S. Seo, J. Yamamoto, J. Matsuyama, and T. Miwatami. Identity of Hemolysins Produced by *Bacillus thuringiensis* and *Bacillus cereus*. *FEMS Microbiology Letters* 79, 205–210, 1991.

Ignoffo, C. The Nuclear Polyhedrosis Virus of *Heliothis zea* (Boddie) and *Heliothis virescens* (Fabricius). IV Bioassay of virus activity. *Journal Invertebrate Pathology* 7, 315–319, 1965.

Ignoffo, C. M. Evaluation of *In Vivo* Specificity of Insect Viruses. In *Baculoviruses for Insect Pest Control: Safety Considerations*. Summers et al. Eds., American Society for Microbiology, Washington, D.C. 52–66, 1975.

Jackson, S.G., R. B. Goodbrand, R. Ahmed, and S. Kasatiya. *Bacillus cereus* and *Bacillus thuringiensis* Isolated in a Gastroenteritis Outbreak Investigation. *Letters in Applied Microbiology* 21, 103–105, 1995.

Jarvis, D. L. Baculovirus Expression Vectors. In *The Baculoviruses*. L. K. Miller, Ed., Plenum Press, New York, London, 389–429, 1997.

Jarvis, D. L., J. A. Fleming, G. R. Kovacs, M. D. Summers, and L. A. Guarino. Use of Early Baculovirus Promoters for Continuous Expression and Efficient Processing of Foreign Gene Products in Stably Transformed Lepidopteran Cells. *BioTechnology* 8, 950–955, 1990.

Krapcho, K. J., R. M. Kral, Jr., B. C. Vanwagenen, K. G. Eppler, and T. K. Morgan. Characterization and Cloning of Insecticidal Peptides from the Primitive Weaving Spider *Diuguetia canities*. *Insect Biochemistry and Molecular Biology* 9, 991–1000, 1995.

Lu, A., P. J. Krell, J. M. Vlak, and G. F. Rohrmann. Baculovirus DNA Replication. In *The Baculoviruses*. L. K. Miller, Ed., Plenum Press, New York, London. 171–191, 1997.

Lu, A. and L. K. Miller. Regulation of Baculovirus Late and Very Late Gene Expression. In *The Baculoviruses*. L. K. Miller, Ed., Plenum Press, New York, London, 193–216, 1997.

Luckow, V. A. and M. D. Summers. Trends in the Development of Baculovirus Expression Vectors. *BioTechnology* 6, 47–55, 1988.

Maeda, S. Increased Insecticidal Effect by a Recombinant Baculovirus Carrying a Synthetic Diuretic Hormone Gene. *Biochemical and Biophysical Research Communication* 165, 1177–1183, 1989.

McClintock, J. T. Regulatory Requirements: Biochemicals. In *Proceedings of the USDA IR-4/EPA Minor Use Biopesticide Workshop*, Washington, D.C., National Foundation for Integrated Pest Management Education, Austin, TX, 29–39, 1995.

McClintock, J. T., J. L. Kough, and R. D. Sjoblad. Regulatory Oversight of Biochemical Pesticides by the U.S. Environmental Protection Agency: Health Effects Considerations. *Regulatory Toxicology and Pharmacology* 19, 115–124, 1994.

McClintock, J. T., C. R. Schaffer, and R. D. Sjoblad. A Comparative Review of the Mammalian Toxicity of *Bacillus thuringiensis*–Based Pesticides. *Pesticide Science* 45, 95–105, 1995.

McClintock, J. T., R. D. Sjoblad, and R. Engler. Are Genetically Engineered Pesticides Different? Potential EPA Data Requirements for Toxicological Evaluation. *Chemtech* 21, 490–494, 1991.

McCutchen, B. F., P. V. Choudary, R. Crenshaw, D. Maddox, S. G. Kamita, N. Palekar, S. Volrath, E. Fowler, B. D. Hammock, and S. Maeda. Development of a Recombinant Baculovirus Expressing an Insect-Selective Neurotoxin: Potential for Pest Control. *BioTechnology* 9, 848–852, 1991.

McCutchen, B. F., R. Herrmann, K. M. Heinz, M. R. Parella, and B. M. Hammock. Effects of Recombinant Baculoviruses on a Nontarget Endoparasitoid of *Heliothis virescens*. *Biological Control* 6, 45–50, 1996.

McIntosh, A. H. and R. Shamy. Biological Studies of a Baculovirus in a Mammalian Cell Line. *Intervirology* 13, 331–341, 1980.

McNitt, L., K. E. Espelie, and L. K. Miller. Assessing the Safety of Toxin-Producing Baculovirus Biopesticides to a Nontarget Predator, the Social Wasp *Polistes metricus* Say. *Biological Control* 5, 267–278, 1995.

Metcalfe, D. D., J. D. Astwood, R. Townsend, H. A. Sampson, S. L. Taylor, and R. L. Fuchs. Assessment of the Allergenic Potential of Foods Derived from Genetically Engineered Crop Plants. *Critical Reviews in Food Science and Nutrition.* 36(S), S165–S186, 1996.

Miller, L. K. Genetically Engineered Insect Virus Pesticides: Present and Future. *Journal of Invertebrate Pathology* 65, 211–216, 1995.

Miller, L. K. and A. Lu. The Molecular Basis of Baculovirus Host Range. In *The Baculoviruses.* L. K. Miller, Ed., Plenum Press, New York, London, 217–237, 1997.

Murphy, F. A., C. M. Fauquet, D. H. L. Bishop, S. A. Ghabrial, A. W. Jarvis, G. P. Martelli, M. A. Mayo, and M. D. Summers. *Virus Taxonomy: Sixth Report of the International Committee on Taxonomy of Viruses.* Springer Verlag, Wien, New York, 104–113, 1995.

O'Reilley, D. R. Auxiliary Genes of Baculoviruses. In *The Baculoviruses.* L. K. Miller, Ed., Plenum Press, New York, London, 267–300, 1997.

Possee, R. D. Cell-Surface Expression of Influenza Virus Haemaglutinin in Insect Cells Using a Baculovirus Vector. *Virus Research* 5, 43–59, 1986.

Possee, R. D., M. Hirst, L. D. Jones, D. H. L. Bishop, and P. J. Cayley. Field Tests of Genetically Engineered Baculoviruses. In *Opportunities for Molecular Biology in Crop Protection. BCPC Monograph* 55, 23–33, 1993.

Quistad, G. B. and W. S. Skinner. Isolation and Sequencing of Insecticidal Peptides from the Primitive Hunting Spider, *Plectreurys tristis* (Simon). *Journal of Biological Chemistry* 269, 11098–11101, 1994.

Reiman, A. and H. G. Miltenburger. Cytogenetic Studies in Mammalian Cells After Treatment with Insect Pathogenic Viruses (Baculoviridae). *Entomophaga* 28, 33–44, 1983.

Rothbard, J. B. and M. L. Gefter. Interactions Between Immunogenic Peptides and MHC Proteins. *Annual Review of Immunology* 9, 527, 1991.

Sjoblad, R. D. Potential Requirements for Immunotoxicity Testing for Pesticides. *Toxicology and Industrial Health* 4, 391–394, 1988.

Smith, G. E., M. D. Summers, and M. J. Fraser. Production of Human Beta Interferon in Insect Cells Infected with a Baculovirus Expression Vector. *Molecular and Cellular Biology* 3, 2156–2165, 1983.

Stewart, L. M. D., M. Hirst, M. Lopez Ferber, A. T. Merryweather, and P. J. Cayley. Construction of an Improved Baculovirus Insecticide Containing an Insect-Specific Toxin Gene. *Nature* 52, 85–88, 1991.

Tomalski, M. D. and L. K. Miller. Insect Paralysis by Baculovirus-Mediated Expression of a Mite Neurotoxin Gene. *Nature* 352, 82–85, 1991.

U.S. Environmental Protection Agency, Office of Pesticides and Toxic Substances. *Pesticide Assessment Guidelines Subdivision M Biorational Pesticides.* EPA 540/9-82-028. Document No. PB83153965. National Technical Information Service, U.S. Department of Commerce, Springfield, VA, 1983.

U.S. Environmental Protection Agency, Office of Pesticides and Toxic Substances. *Subdivision M of the Pesticide Testing Guidelines: Microbial and Biochemical Pest Control Agents.* Document No. PB89-211676. National Technical Information Service, U.S. Department of Commerce, Springfield, VA, 1989.

U. S. Environmental Protection Agency, Office of Prevention, Pesticides and Toxic Substances. Reregistration Eligibility Decision (RED) *Bacillus thuringiensis.* EPA 738-R-98-004, p. 157, 1998a.

U. S. Environmental Protection Agency, Office of Prevention, Pesticides and Toxic Substances. The Environmental Protection Agency's White Paper on *Bacillus thuringiensis* Plant — Pesticide Resistance Management. EPA 739-S-98-001, p. 86, 1998b.

van Beek, N. A. M., and P. R. Hughes. The Response Time of Insect Larvae Infected with Recombinant Baculoviruses. *Journal of Invertebrate Pathology* 72, 338–347, 1998.

Washburn, J. O., B. A. Kirkpatrick, and L. E. Volkman. Insect Protection Against Viruses. *Nature* 383, 767, 1996.

Wood, H. A., P. R. Hughes, and A. Shelton. Field Studies of the Co-Occlusion Strategy with a Genetically Altered Isolate of the *Autographa californica* Nuclear Polyhedrosis Virus. *Environmental Entomology* 23, 211–219, 1994.

Zlotkin, E., M. Eitan, V. P. Bindokas, M. E. Adams, M. Moyer, W. Burkhart, and E. Fowler. Functional Duality and Structural Uniqueness of Depressant Insect-Selective Neurotoxins. *Biochemistry* 30, 4814–4821, 1991.

Zlotkin. E., M. Gurevitz, E. Fowler, and M. Adams. Depressant Insect Selective Neurotoxins from Scorpion Venom: Chemistry, Action and Gene Cloning. *Archives of Insect Biochemistry and Physiology* 22, 55–73, 1993.

Index

A

ABCP, *see* African-wide Biological Control
    Project
Abiotic factors, 182–183
AcMNPV-LqhIT2, 340–341, *see also*
    Baculoviruses
Active ingredients, 319, 320–321, *see also*
    Biochemical pesticides
Acute toxicity/pathogenicity tests, 315–317
Aedes spp., 156
Aequorea victoria, 252
Aerial release, 146
Aeroisolation, 315
Aerosols, 84
Africa, 12–13
African-wide Biological Control Project (ABCP),
    12
Age, traps, 72
Agglutinins, 224
Aggregation pheromones, 69, 73, 89, *see also*
    Pheromones, and other semiochemicals
Agricultural pests, 6
Agricultural systems, 13
Agriculture, environmentally friendly, 295–298
Agroecosystems, 339–340
Agrotis ipsilon, 256
Air-sampling, 85–86
Alarm pheromones, 70, 90, *see also* Pheromones,
    and other semiochemicals
Alfalfa, 19
Allelochemicals, 175, 181
Allergenicity, 350
Amber disease, 41
Ambylseius finlandicus, 246
American bioassay, 41
Amyelois transitella, 80
a-Amylase inhibitors, 177, 223–224, 285–286,
    219
Anagasta kuehniella, 40, 41

Androctonus australis, 286
Anemotaxis, 67–68
Animals, 233, 313–315
Anisoplia austriaca, 51
Anopheles spp., 156
Antibiosis, 174–175, 184
Anticarsia gemmatalis, 46
Antifeedants, 124, 129–132, 287, *see also*
    Azadirachtin
Antioxidants, 106
Antixenosis, 174–175
Ants, 70, *see also Oecophylla smaragdina*
Aphidius spp., 5
Aphids, 5, 70, 225
Aphids, woolly apple, *see Eriosoma lanigerum*
Aphis gosspii, 19–21
Aphtis mytelaspidis, 6
Apis mellifera, 35–36
Apoanagryus lopezi, 12
Apples, 7, 74–75, 76, 172
Aquatic studies, 317, 347
Arachnid venom, 286
Armyworm, African, *see Spodoptera exempta*
Arthropods, *see also* Individual entries
    augmentation biological control, 13, 14
    biotypes and resistance in plants, 183, 184
    growth regulation and juvenile hormone
        agonists, 128–129
    transformation and genetic engineering,
        244–245
Ascogaster quadridentata, 126
Associational resistance, 172, *see also* Plant,
    resistance to insects; Resistance
Attracticides, 79–80
Attraction, 71–73
Augmentation biological control, *see also*
    Biological control
    characterization, 4

359

Milton Keynes UK
Ingram Content Group UK Ltd.
UKHW021822071024
449327UK00021B/1393